Ion Flux in Pulmonary
Vascular Control

NATO ASI Series

Advanced Science Institutes Series

A series presenting the results of activities sponsored by the NATO Science Committee, which aims at the dissemination of advanced scientific and technological knowledge, with a view to strengthening links between scientific communities.

The series is published by an international board of publishers in conjunction with the NATO Scientific Affairs Division

A	**Life Sciences**	Plenum Publishing Corporation
B	**Physics**	New York and London
C	**Mathematical and Physical Sciences**	Kluwer Academic Publishers
D	**Behavioral and Social Sciences**	Dordrecht, Boston, and London
E	**Applied Sciences**	
F	**Computer and Systems Sciences**	Springer-Verlag
G	**Ecological Sciences**	Berlin, Heidelberg, New York, London,
H	**Cell Biology**	Paris, Tokyo, Hong Kong, and Barcelona
I	**Global Environmental Change**	

Recent Volumes in this Series

Series A: Life Sciences

Ion Flux in Pulmonary Vascular Control

Edited by

E. Kenneth Weir

Veterans Administration Medical Center
Minneapolis, Minnesota

Joseph R. Hume

University of Nevada
Reno, Nevada

and

John T. Reeves

University of Colorado
Denver, Colorado

Springer Science+Business Media, LLC

Proceedings of a NATO Advanced Research Workshop:
The Sixth Grover Conference on the Pulmonary Circulation,
held October 15–18, 1992,
in Sedalia, Colorado

NATO-PCO-DATA BASE

The electronic index to the NATO ASI Series provides full bibliographical references (with keywords and/or abstracts) to more than 30,000 contributions from international scientists published in all sections of the NATO ASI Series. Access to the NATO-PCO-DATA BASE is possible in two ways:

—via online FILE 128 (NATO-PCO-DATA BASE) hosted by ESRIN, Via Galileo Galilei, I-00044 Frascati, Italy

—via CD-ROM "NATO Science and Technology Disk" with user-friendly retrieval software in English, French, and German (©WTV GmbH and DATAWARE Technologies, Inc. 1989). The CD-ROM also contains the AGARD Aerospace Database.

The CD-ROM can be ordered through any member of the Board of Publishers or through NATO-PCO, Overijse, Belgium.

Library of Congress Cataloging-in-Publication Data

Ion flux in pulmonary vascular control / edited by E. Kenneth Weir,
Joseph R. Hume, and John T. Reeves.
    p.   cm. -- (NATO ASI series. Series A, Life sciences ; v.
251)
    "Proceedings of a NATO advanced research workshop, the Sixth
Grover Conference on the Pulmonary Circulation, held October 15-18,
1992, in Sedalia, Colorado"--T.p. verso.
    "Published in cooperation with NATO Scientific Affairs Division."
    Includes bibliographical references and index.
    ISBN 978-1-4613-6016-2
    1. Pulmonary circulation--Regulation--Congresses. 2. Pulmonary
endothelium--Congresses. 3. Ion channels--Congresses. 4. Vascular
smooth muscle--Congresses. 5. Vascular endothelium--Congresses.
6. Pulmonary hypertension--Congresses.   I. Weir, E. Kenneth.
II. Hume, Joseph Randy, 1947-   . III. Reeves, John T. IV. North
Atlantic Treaty Organization. Scientific Affairs Division.
V. Grover Conference on the Pulmonary Circulation (6th : 1992 :
Sedalia, Colo.) VI. Series.
    [DNLM: 1. Ion Channels--congresses. 2. Pulmonary Circulation-
-physiology--congresses. 3. Endothelium, Vascular--physiology-
-congresses.   WF 600 I64 1992]
QP107.I56 1993
612.2--dc20
DNLM/DLC
for Library of Congress   93-5628
                                                       CIP

ISBN 978-1-4613-6016-2   ISBN 978-1-4615-2397-0 (eBook)
DOI 10.1007/978-1-4615-2397-0

©1993 Springer Science+Business Media New York
Originally published by Plenum Press,New York in 1993
Softcover reprint of the hardcover 1st edition 1993

PREFACE

Ions can pass through a single membrane channel at a rate of 10^6 ions/second. Over the last decade the ability to measure ion flux so precisely and to document the opening and closing of individual ion channels has provided a powerful tool to those working on smooth muscle physiology and vascular reactivity. The use of potassium channel blockers by Tom Lloyd in the 1960s and calcium channel blockers by Ivan McMurtry in the 1970s indicated the importance of ion flux in regulating pulmonary vascular tone. Recent advances in technology, principally the patch-clamp technique and fluorescent ion-sensitive dyes, now permit a more detailed description of physiologic mechanisms.

This volume arises from the Sixth Grover Conference on the Pulmonary Circulation, a NATO Advanced Research Workshop, held in Colorado in October 1992. A group of international scientists who are leaders in the field of ion flux focused their attention on the problems of the pulmonary vasculature. The chapters in this book describe the present state of knowledge of the movement and storage of ions in vascular endothelial and smooth muscle cells. Those who are not familiar with the techniques of patch clamping and calcium imaging will find an introduction to these methods in the chapters by Leblanc and Wan and Archer et al.

The role of potassium channels in oxygen sensing illustrates the rapid progress which the study of ion currents has made possible. Several chapters describe the concept that hypoxic inhibition of one or more potassium channels may lead to membrane depolarization and signal a fall in oxygen tension. The possibility that oxygen tension alters gating of the potassium channels through changes in redox status is discussed by Post et al.

The endothelium modulates pulmonary and systemic vascular reactivity and permeability. Five chapters examine the function of ion channels and ion flux in endothelial cells. Measurement of intracellular calcium and patch-clamp techniques have provided new insights into endothelial control mechanisms, such as the action of endothelium-dependent vasodilators. These dilators increase cytosolic calcium by inducing both calcium entry through receptor-operated channels and calcium release from intracellular stores. Calcium-dependent potassium channels are then activated, leading to hyperpolarization which generates the electrochemical gradient for further calcium entry. Differences in the handling of ions between cultured and freshly dispersed endothelial cells remain to be resolved.

This NATO workshop provided the opportunity for the interchange of an immense amount of information. The book is the distillation of that information, written immediately after the conference, with the advantage of that intense period of discussion. The organizing committee is grateful to the sponsors of the meeting, especially the Scientific Affairs Division of NATO, the National Heart, Lung and Blood Institute of the National Institutes of Health, Axon Instruments, Inc., Burroughs Wellcome Company, Marion Merrell Dow Inc., Merck Sharp & Dohme, Pfizer Laboratories, the Upjohn Company, and the Pulmonary Circulation Foundation.

E. Kenneth Weir
Joseph R. Hume
John T. Reeves

CONTENTS

POTASSIUM CHANNELS IN VASCULAR SMOOTH MUSCLE

POTASSIUM CHANNELS AND OXYGEN SENSING

ION FLUX IN VASCULAR ENDOTHELIAL CELLS

PULMONARY HYPERTENSION: FROM PHENOMENOLOGY TO A MOLECULAR UNDERSTANDING OF DISEASE MECHANISMS

Norbert F. Voelkel

Pulmonary and Critical Care Division
University of Colorado Health Sciences Center
4200 E. 9th Ave.
C-272
Denver, Colorado 80262

Science is the art of shaping questions and identifying the underlying assumptions. Nature has no secrets. The obstacles to furthering knowledge lie in the formulation of new, probing questions. Artful questions often require new technology from unrelated disciplines.*

INTRODUCTION

The pulmonary circulation is anatomically defined as the vasculature between the right ventricle and the left atrium; under normal conditions this circulation accommodates the entire cardiac output, yet it is a low pressure, - low resistance system. Leaving the right ventricle, the blood enters the central conduit arteries that have Windkessel - function, then enters rapidly branching muscular arteries, and finally the precapillary resistance vessels. Whereas flow and pressure have been studied with great intensity, cell-cell interactions in the lung capillaries are only recently gaining attention (1-3). Flow-passive opening of reserve capillaries is a major regulatory factor in the control of the lung circulation and there is the general consensus that the lung circulation does not possess a significant myogenic tone (4). The lung vessels are innervated by both sympathetic and parasympathetic fibers, and it is evident that small changes in the pulmonary venous tone can have devastating effects on capillary fluid filtration - leading to edema formation (5). Large amounts of new information regarding the control of lung vessels under normal

---

* Tribute to Robert F. Grover

conditions and conditions of lung injury have been gained from the studies of isolated lung preparations (6). The reductionist approach has also lead, in recent years to utilize isolated large and small pulmonary artery preparations and most recently to the use of single pulmonary artery smooth muscle cells (7,8). Endothelial cells in culture are being used in order to investigate cell-cell interactions relevant to the adult respiratory distress syndrome (ARDS), to sickle cell lung syndrome and questions related to neonatal pulmonary hypertension. It is recognized that ARDS (9,10) with an estimated patient number of 300,000 patients per year and a mortality of 50%, and primary pulmonary hypertension (PPH) with an estimated number of 1,000 new cases per year and virtually a mortality of 100%, are the major clinical problems involving the pulmonary circulation. Although pulmonary embolism is an important clinical problem involving the lung circulation, this disease is not a primary disorder of the lung circulation. In situ thrombosis in the lung circulation may most frequently occur in the sickle cell lung syndrome (11).

PATHOGENETIC FACTORS OF SEVERE PULMONARY HYPERTENSION

Primary (or unexplained) pulmonary hypertension is the most devastating form of pulmonary vascular disease with an estimated survival time of 10 months in the presence of a mean pulmonary arterial pressure in excess of 85mmHg or an elevated right atrial pressure in excess of 20mmHg. Vasoconstriction and vascular remodeling are the two cardinal factors regarded as important in the pathogenesis of this disorder. Shear stress and stretch may be more important in the development of pulmonary hypertensive disorders associated with high flow states as they are encountered in Eisenmenger disorders. Products of coagulation, platelet factors and products of inflammation may be contributing to forms of secondary hypertension, in particular in collagen-vascular disorders. A *genetic predisposition* leading in some unknown way to a hyperactive vasculature- perhaps in analogy to the hyperactive bronchial smooth muscle in asthma- or leading to an uncontrolled vascular proliferative process and subsequent (secondary) pressure elevation is likely the *sine qua non* of severe pulmonary hypertension. Given such a genetic predisposition, it is extremely important to identify risk factors and factors that cause the expression of "pulmonary hypertension genes". A synopsis of the clinical presentation and the pathogenetic mechanisms is provided in the form of a "vertical hierarchy" tabulation (Table 1).

TABLE 1. SEVERE PULMONARY HYPERTENSION

	Synopsis of the Clinical Presentation and the Pathogenetic Mechanisms "Vertical Hierarchy"	
Patient	dyspnea, syncope	clinical medicine
System physiology	pulmonary hypertension CO, DLCO, PaCO$_2$	physiology
Organ	vessel remodeling	pathology histology
Cell	hypertrophy, phenotypic switch	cell biology
Biopolymers	matrix proteins	biochemistry
Molecules	mediators, cytokines, growth factors, ion channels	"
Gene expression	elastin mRNA	molecular biology

Typically the symptoms presented by the patient with severe pulmonary hypertension occur late in the course and represent right ventricular failure. Therefore, at the time of diagnosis the pulmonary artery pressure is high, oftentimes the cardiac output (CO), and the pulmonary diffusion capacity (DLCO) are reduced and the pulmonary histology shows small arteries in various states of remodeling. The small arteries show hypertrophy of the vascular smooth muscle, extensive changes in the adventitia due to increase in collagen and elastin, and the intima may be hypertrophied to the point of lumen obliteration (12). Our knowledge of these processes on a molecular level is still very limited. In animal models of neonatal, severe pulmonary hypertension (13) there is evidence of reduction in pulmonary endothelial cell prostacyclin synthesis (14) and increase in elastin message (15). As mentioned, regarding the time course of the development of pulmonary hypertension, it is unclear whether "twitchy vessels", i.e. vascular hyperactivity lead to the vascular remodeling or vice versa.

REGULATION OF PULMONARY VASCULAR TONE

Species differences in hypoxic vasoconstriction (16-18) indicate strongly that tone regulation in the lung circulation is under genetic control. More recently studies with susceptible and nonsusceptible cattle, and studies using various rat strains provide compelling indirect evidence for inherited pulmonary vascular tone regulation (19,20). Whether the cellular oxidant/antioxidant balance is part of the genetic makeup determining lung vascular smooth muscle contractile behavior is still unclear. It is known, however, that lung vascular tone is altered in the injured lung due to the influence of vasodilator - and vasoconstrictor mediators. Prostacyclin, hydrogen peroxide, platelet activating factor (PAF) and nitric oxide (NO) are all likely to be produced by cell-cell interactions in the capillary microenvironment. Pulmonary vascular reactivity is altered in the hypertensive, remodeled lung circulation (21), and rats fed a fish oil diet have been shown to have a reduced vascular response to hypoxia and angiotensin II (22). One of the remaining challenging research questions is the elucidation of the molecular mechanisms that bridge increased vascular tone and lung vascular remodeling. It appears that at least in an animal model of chronic hypoxic pulmonary hypertension chronic treatment with agents that are anti-inflammatory and devoid of hemodynamic activity, inhibits pulmonary hypertension and lung vascular remodeling (23).

The other remaining challenge to modern lung physiology is the mechanism of oxygen-sensing - in the lung circulation the mechanism of hypoxic vasoconstriction (24-26). The teleology of this response may have its "origin" in the fetal circulation and during the adult life the response is required for proper. ventilation/perfusion matching - according to the principle: what is not ventilated must not be perfused. Yet we are still lacking important details regarding the molecular mechanisms of this response. The putative sequence of events is shown in the schematic below.

$$\downarrow P_{O_2} \rightarrow \downarrow E_M \rightarrow \uparrow Ca^{++} \text{ flux} \rightarrow \uparrow VSM \text{ contraction}$$

A decrease in the *alveolar* P_{O_2} leads to a change in the membrane potential (depolarization) of the vascular smooth muscle cell of precapillary arteries and to an influx of calcium that generates the contractile response. How exactly the signal of a decrease in oxygen tension is transduced remains unclear. Because pharmacological agents that block the entry of calcium into cells via the L-type Ca^{++} channel are extremely effective in blocking hypoxic vasoconstriction, it is surmised that calcium entry rather than intracellular calcium release is *the* key factor in the hypoxic pressor response. Studies performed with isolated small pulmonary arteries (27-29) and more recently even with isolated smooth muscle cells from small pulmonary arteries (7,8) suggest that the muscle cells that are strategically located in the resistance arteries contain the oxygen sensor and the entire "equipment" for signal transduction. The opening and closing of ion channels may be secondarily affected during changes of oxygen tension or may in fact be an the essential part of the oxygenation, - dependent contraction- signal transduction. The hypoxic pressor response has been successfully investigated in isolated lung preparations where it can be shown that the onset of pulmonary vasoconstriction occurs within a few seconds after reduction of alveolar oxygen concentration (30); in fact, hypoxic vasoconstriction may be regulated on a breath-by-breath basis. The hypoxic pressor response is altered by the redox state of the cells, in particular hypoxic vasoconstriction is drastically reduced by agents that alter intracellular glutathione (31-33). A recent study describes the inhibition of hypoxic pulmonary vasoconstriction by diphenyleneiodonium (34) an inhibitor of NADPH oxidase. This heme protein has been considered as a putative candidate for the O_2 sensor (35). Alternatively, a direct effect of oxygen on the performance of K^+ channels has been suggested by Lopez-Barnes and colleagues (36).

MECHANICAL FORCES

The principle mechanical forces in the circulation are shear stress (37) and stretch.

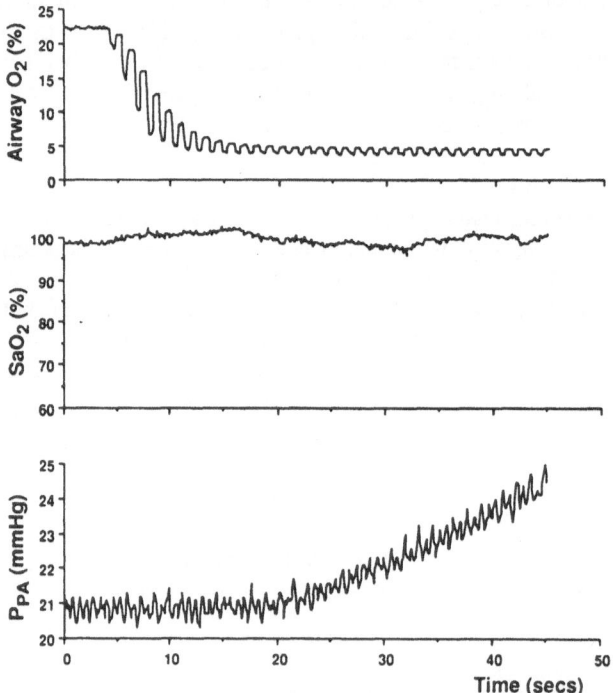

FIG. 1. Signals for airway O_2 concentration (breath-by-breath changes), pulmonary arterial inflow line O_2 saturation (Sa_{O2}), and Ppa (pulmonary artery pressure), recorded during airway challenge with hypoxic gas mixture (O_2 concentration 3%). Reprinted from Klaus et al. J. Appl. Physiol. (30).

4

Endothelial- and smooth muscle cells are the cells that have been studied and are believed to contain mechanosensors (38-41). A step increase in shear stress stimulates a transient increase in intracellular calcium in bovine aortic endothelial cell monolayers. The increase of intracellular calcium appears to be saturable and increasing logarithmically with increasing shear stress (42). The effect of shear stress and stretch on the opening probability of ion channels of endothelial and smooth muscle cells is discussed in detail in other chapters of this volume.

ION CHANNELS IN THE PULMONARY CIRCULATION

McMurtry et al (43) demonstrated 1976 that verapamil decreased the magnitude of the hypoxic pulmonary vasoconstriction in the isolated rat lung preparation and that this inhibitory effect was rather preferential for hypoxia- induced contractile responses. Subsequently it has been shown that high perfusate calcium (perhaps by stabilizing smooth muscle cell membranes), inhibits hypoxic vasoconstriction (44), and that a compound that increases the calcium transport through the L-type Ca^{++} channels (Bay K 8644), potentiates hypoxic vasoconstriction (45-46). Harder and co-workers (29,47) showed in 1985 that hypoxia causes membrane depolarization and induces calcium dependent action potentials in small pulmonary arteries. Since then research has been focused on the behavior of potassium channels (48-52) and to some extent also on the transmembrane sodium gradient (53). Table 2 provides a brief synopsis of findings relevant to the performance of the lung circulation and the involvement of the ion channels.

Figure 2 summarizes some of the pharmacological interventions and their effect on the magnitude and the shape of the hypoxic pressor response curve in isolated lung preparations.

TABLE 2. CHANNEL STUDIES IN THE PULMONARY CIRCULATION

	Ca^{++} - entry blocker inhibits hypoxic vasoconstriction
	(McMurtry et al 1976)
	High Ca^{++} inhibits hypoxic vasoconstriction
	(Voelkel et al 1980)
Ca^{++}	
	Bay K 8644 (voltage-regulated Ca^{++} channel) potentiates hypoxic vasoconstriction
	(McMurtry 1985, Tolins et al 1986)
	Hypoxia induces Ca^{++} - dependent action potentials in small pulmonary arteries (Harder et al 1985)
Na^+	Transmembrane Na^+ gradient alters responses of pulmonary arteries
	(Salvaterra et al 1989)
	K^+ channel openers inhibit pulmonary vasoconstriction
	(Several authors)
	K^+ channel opener inhibits small pulmonary artery hypoxic contraction
	(Yuan et al 1990)
	Pulmonary vasodilation after ET blocked by glibenclamide (Lippton et al 1991)
K^+	
	K^+_{ATP} modulate hypoxic response in ferret lungs (Wiener et al 1991)
	Ca^{++}/Mg^{++} K^+_{ATP} in rat pulmonary artery (Robertson 1992)
	Role of Ca^{++} - sensitive K^+ channel in hypoxic vasoconstriction. (Post et al, 1992)
	Diamide (SH reagent) increases K^+ outward current in canine pulmonary artery SMC (Weir et al 1992)

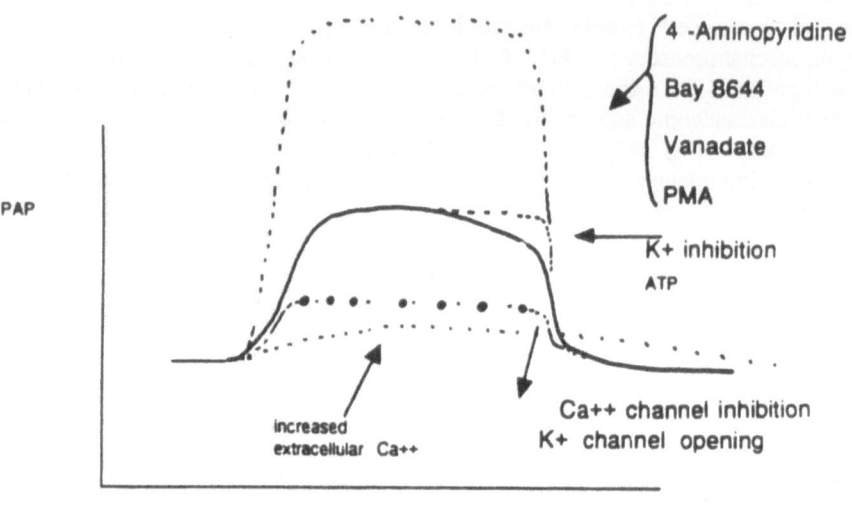

4 -Aminopyridine

Bay 8644

Vanadate

PMA

K+ inhibition

ATP

increased
extracellular Ca++

Ca++ channel inhibition
K+ channel opening

Time

Figure 2

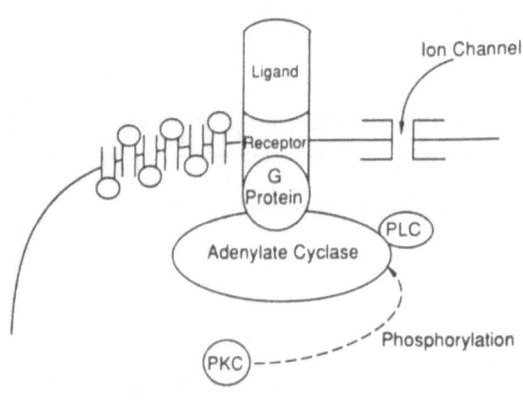

Figure 3

The hypoxic pressor response can be potentiated by 4-aminopyridine Bay K 8644, PMA and vanadate (54-56). Inhibition of calcium entry, as well as opening of potassium channels by agents like cromacalim or lemacalim inhibit the magnitude of the response (57,58), increasing extracellular calcium inhibits the magnitude of the pressor response and delays the onset as well as the vasodilation after hypoxic vasoconstriction (44). Inhibition of a K^+_{ATP} channel by glibenclamide prevents the "role off" of the pressure response under conditions of severe hypoxia"(49). The schematic below provides a concept of the signal transduction pathways involved in vascular smooth muscle contraction.
Agonists are bound to specialized receptors in the cell membrane and G-proteins,

phospholipases, and phosphorylation steps are all involved in the reactions that cause an increase in intracellular calcium. Alternatively, the sequence of events can be initiated by opening of cell membrane ion channels and calcium influx. Very little is still known about the interplay between membrane ion channel dynamics, release of calcium from the endoplasmic reticulum and the intracellular cytosolic calcium levels. In the following, two examples are provided for pharmacological alterations of pulmonary vascular tone investigated in the isolated, salt solution perfused rat lung preparation. Figure 4 shows the effect of suramin that is known to block the interaction of ATP with the P_2 - purinoceptor (59).

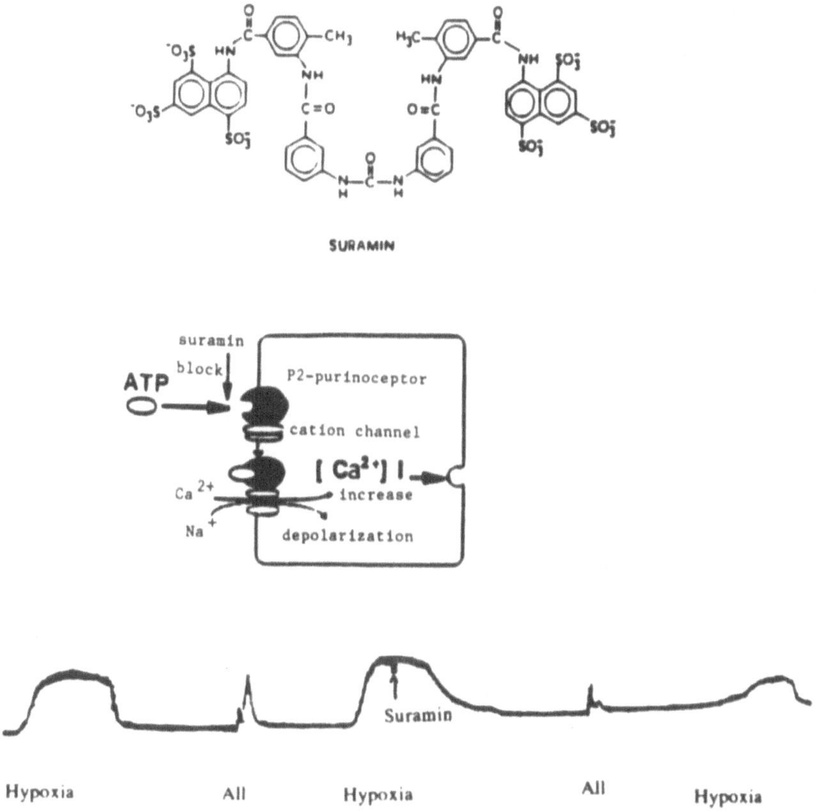

SURAMIN

Figure 4

Interaction of ATP with this receptor results in the opening of a cation channel and increase in intracellular calcium. Blockade of this receptor by suramin may therefore decrease Ca^{++} - influx. Suramin, added to the perfusate of the isolated lung preparation, causes inhibition of ongoing hypoxic vasoconstriction and subsequently impairment of lung vasoreactivity. Figure 5 shows the effect the of addition of the marine sponge product

7

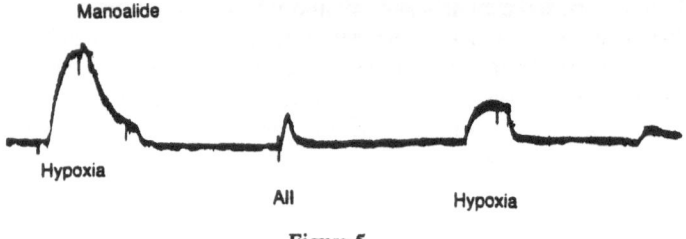

Phospholipase A$_2$ Inhibitor (Manoalide) Inhibits Pulmonary Vasoconstriction

Manoalide

Hypoxia

AII

Hypoxia

Figure 5

manoalide, an inhibitor of phospholipase A$_2$. Again, as in Figure 4, manoalide, added during ongoing hypoxic vasoconstriction, inhibits the pressor response and subsequent vascular reactivity.

Thapsigargin, a sesquiterpene lactone extracted from an umbelliferous plant, causes an increase in intracellular calcium by inhibition of the endoplasmatic calcium ATPase (50,61). Addition of this compound (10^{-6}M) to the perfusate of isolated rat lungs causes a small rise in baseline perfusate pressure and impairment of vascular reactions thereafter. Even the potentiation of hypoxic vasoconstriction by vanadate (56) is abolished in the presence of thapsigargin.

In recent years, due to the use of molecular biology techniques, (cloning, sequencing, gene mutation and expression of mutated genes in the oocyte system), our knowledge of the structure and function of ion channels has increased exponentially. Perhaps the greatest amount of information is currently available about K$^+$ channels. Pertinent information regarding channels is provided in Tables 3-5.

TABLE 3. K$^+$ CHANNELS

ubiquitous (VSMC, lymphocytes, mitochondrial inner membrane) (62-66)
several gene families (67)
boundaries between voltage - gated and K$^+_{Ca++}$ may be blurred (68)
gene expression can be altered (example: estrogen) (69)
can be mutated (70,71)
activated by G proteins (72,73)
activity regulated by phosphorylation (PKC, [Ca$_i$] (74-77)
activity regulated by redox state of SH groups, (GSH) (78)
can be expressed (xenopus oocytes) (79,80)
blockade of K$^+$ channels has been shown
to affect: cAMP formation (81)
        :   cell proliferation (lymphocytes)
        :   activity of natural killer cells (82)
        :   flow - mediated vasodilation (40,83)
                (endothelial NO release)
        :   cell volume (mitochondrial volume) (66)
            regulation

TABLE 4. K$^+_{ATP}$- ACTIVATION INVOVLED VASOREGULATION CAUSED BY

Ach
VIP
CGRP
EDRF
EDHF

TABLE 5. K$^+$ CHANNEL PHARMACOLOGICAL MODULATION (84)

Cromakalim		BK$^+_{Ca}$
	Open	
Diazoxide		K$^+_{ATP}$
		Others?
4 Aminopyridine	block	I$_A$, transient outward
		A current
		ACH$_m$, K$^+_{Na}$
TEA	block	ROC
		Delayed rectifier
		Inward rectifier
		BK$_{Ca}$
Charybdotoxin	block	BK$_{Ca}$, IK,
Dendrotoxin	block	Transient I$_A$ outward
Quinidine	block	
Quinine		BK$_{Ca}$, IK

K$^+$ channels are ubiquitous, they can alter their properties following point mutation, they can be regulated by phosphorylation, by intracellular calcium and by the redox state of the cell. Table 5 lists several of the described K$^+$ channels and currently used pharmacological tools. As can be seen, agents like tetraethylammonium (TEA) affect several known K$^+$ channels- although in different dose ranges.
It is now apparent that K$^+$ channels are tetrameres; each of the four units has six hydrophobic membrane- spanning regents; incase of voltage- gated channels the fourth unit (S4) contains the membrane- potential sensing apparatus. There is now recent evidence for cooperative interactions of these membrane potential sensing units in K$^+$ channel gating (85). There are some important interactions between receptors and channels. For example alpha1- Adrenoceptor agonists suppress the transient outward A-current (I$_A$) (86), and it has been sh own that Alpha 2- Adrenoceptor agonists inhibit voltage dependent calcium currents and increase inward rectifying K$^+$ currents (87). On the other hand, any procedure that decreases intracellular calcium also blocks calcium activated K$^+$ channels (84).

From a clinical-therapeutic point of view it is important to mention that diazoxide which activates K$^+_{ATP}$ channels has been used in a small number of patients in the chronic treatment of severe pulmonary hypertension (88).

Specific questions that the student of the lung circulation will ask about ion channel modulation and their potential contribution to pulmonary vascular pathogenesis are:

Which of the many ion channels are important in vascular disease?

Do different vessels have different sets of ion channels? Conduit- versus resistance vessels?

Which of the many ion channels are involved in oxygen- sensing? In vascular contractility?

Are ion channels important in cell proliferation, in gene expression? In matrix protein synthesis? (89). In the regulation of cell adhesion molecules?

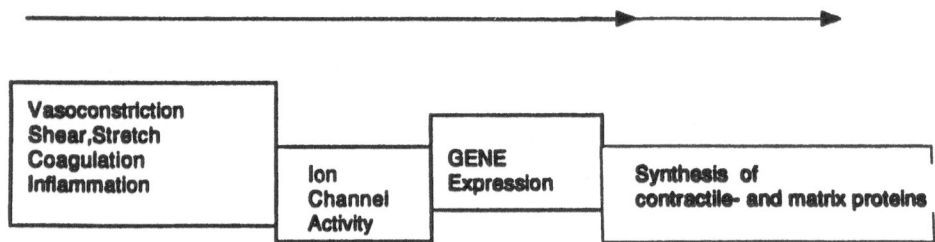

The schematic proposes a flow of signals from hemodynamically relevant events like vasoconstriction, shear and stretch towards gene expression and the evolution of a hypertensive vascular wall cell phenotype. It is likely that ion channel activities are relevant in pulmonary hypertension and therefore critical targets for new treatment modalities.

REFERENCES

1. **Tonnesen M.G., L.A. Smedly, P.M. Henson.** Neutrophil-endothelial interactions: modulation of neutrophil adhesiveness induced by complement fragments C5a and C5a des-arg and formylmethionyl-leu cyl-phenylalanine in vitro. *J. Clin. Invest.* 74:1581, 1984.
2. **Warner, A.E. and J.d. Brain.** The cell biology and pathogenic role of pulmonary intravascular macrophages. *Am. J. Physiol.* 258:L1-L12, 1990.
3. **Voelkel, N.F., j. Czartolomna, J. Simpson, and R.C. Murphy.** FMLP causes eicosanoid-dependent vasoconstriction and edema in lungs from endotoxin-primed rats. *Am. Rev. Respir. Dis.* 145:701-711, 1992.
4. **Pain, M.C.F. and J.B. West.** Effect of the volume history of the isolated lung on distribution of blood flow. *J. Appl. Physiol.* 21(5):1545-1550, 1966.
5. **Dauber, I.M., and J.V. Weil.** Lung injury edema in dogs: influence of sympathetic ablation. *J. Clin. Invest.* 72, 1977.
6. **Chang, S.W. and N.F. Voelkel.** The isolated perfused lung preparation as a research tool. "Comparative Biology of the Normal Lung," CRC Press, Baton Rouge (1992).
7. **Murray, T.R., L. Chen, B.E. Marshall, and E.J. Macarak.** Hypoxic contraction of cultured pulmonary vascular smooth muscle cells. *Am. Respir. Cell Mol. Biol.* 3: 457-465, 1990.
8. **Madden, J.A., M.S. Vadula, and V.P. Kurup.** Effects of hypoxia and other vasoactive agents on pulmonary and cerebral artery smooth muscle cells. *Am. J. Physiol.* 7(3):L384-L393, 1992.
9. **Ashbaugh, D.G., D.B. Bigelow, T.L. Petty, and V.E. Levine.** Acute respiratory distress in adults. *Lancet.* 2:319-323, 1967.
10. **Zapol, W.M., M.A. Rie, M. Frikker, M.T. Snider, and D.A. Quinn.** Pulmonary circulation during adult respiratory distress syndrome, *in*: Zapol WM, Falke KJ (eds) Acute respiratory failure. "Lung Biology in Health and Disease," Marcel Dekker, New York (1985).

11. Collins, F.S. and E.P. Orringer. Pulmonary hypertension and corpulmonale in sickle hemoglobinopathies. *Am. J. Med.* 73:814-821, 1982.

12. Wagenvoort, C.A., and N. Wagenvoort. "Pathology of Pulmonary Hypertension," John Wiley and Sons, New York, London, (1977).

13. Stenmark, K.R., J. Fasules, D.M. Hyde, N.F. Voelkel, J. Henson, A. Tucker, H. Wilson, and J.T. Reeves. Severe pulmonary hypertension and arterial adventitial changes in newborn calves at 4,300 m. *J. Apppl. Physiol.* 62(2): 821-830, 1987.

14. Badesch, D.B., E.C. Orton, L.M. Zapp, .Y. Westcott, J. Hester, N.F. Voelkel, and K.R. Stenmark. Decreased arterial wall prostaglandin production in neonatal calves with severe chronic pulmonary hypertension. *Am. J. Respir. Cell Mol. Biol.* 1:489-498, 1989.

15. Mecham, R.P., L.A. Whitehouse, D.S. Wrenn, W.C. Parks, G.L. Griffin, R.M. Senior, E.C. Crouch, K.R. Stenmark, and N.F. Voelkel. Smooth muscle-mediated connective tissue remodeling in pulmonary hypertension. *Science* 237:423-426m 1987.

16. Tucker, A., I.F. McMurtry, J.T. Reeves, A.F. Alexander, A.H. Will, and R.F. Grover. Lung vascular smooth muscle as a determinant of pulmonary hypertension at high altitude. *Am. J. Physiol.* 228:762-7, 1975.

17. Walker, B.R., N.F. Voelkel, I.F. McMurtry, and E.M. Adams. Evidence for diminished sensitivity of the hamster pulmonary vasculature to hypoxia. *J. Appl. Physiol.* 52: 1571-1574, 1982.

18. Peake, M.D., A.L. Harabin, N.J. Brennan, and J.T. Sylvester. Steady state vascular responses to graded hypoxia in isolated lungs of five species. *J. Appl. Physiol.* 51(5): 1214-1219, 1981.

19. Kazuhiko, S., S. Webb, A. Tucker, M. Rabinovitch, R.F. O'Brien, I.F. McMurtry, and T.J. Stelzner. Factors influencing the idiopathic development of pulmonary hypertension in the fawn hooded rat. *Am. Rev. Respir. Dis.* 145:793-797, 1992.

20. Lishan, H.E., S.W. Chang, and N.F. Voelkel. Pulmonary vascular reactivity in Fischer rats. *J. Appl. Physiol.* 70(4): 1861-1866, 1991.

21. Rabinovitch, M., W. Gamble, A.S. Nades, O.S. Miettinen, and L. Reid. Rat pulmonary circulation after chronic hypoxia: hemodynamics and structural features. *Am. J. Physiol.* H818-H827, 1979.

22. Morganroth, M.L., W.C. Pickett, S. Worthen, M. Mathias, J.T. Reeves, and N.F. Voelkel. Decreased pulmonary vascular responsiveness in rats raised on an essential fatty acid deficient diet. *Prostaglandins.* 33(2): 181-197, 1987.

23. Ono, S., J.Y. Westcott, and N.F. Voelkel. PAF antagonists inhibit pulmonary vascular remodeling induced by hypobaric hypoxia in rats. *J. Appl. Physiol.* 73(3): 1084-1092, 1992.

24. von Euler, U.S., and G. Liljestrand. Observations on the pulmonary arterial blood pressure in the cat. *Acta. Physiol. Scand.* 12:301-20, 1946.

25. Fishman, A.P. Hypoxia on the pulmonary circulation: how and where it acts. *Circ. Res.* 38:221-31, 1976.

26. Voelkel, N.F. Mechanisms of hypoxic pulmonary vasoconstriction. *Am. Rev. Respir. Dis.* 133:1186-1195, 1986.

27. Suzuki, H., and B.M. Twarog. Membrane properties of smooth muscle cells in pulmonary hypertensive rats. *Am. J. Physiol.* 242:H907-H915, 1982.

28. Suzuki, H., and B.M Twarog. Membrane properties of smooth muscle cells in pulmonary arteries of the rat. *Am. J. Physiol.* 242:H900-H906, 1982.

29. Harder, D. J.A. Madden, and C. Dawson. Hypoxic induction of Ca^2-dependent action potentials in small pulmonary arteries of the cat. *J. Apply Physiol.* 59:1389-93, 1985.

30. Klaus, S.J., A.J. Micco, J. Czartolomna, L. Latham, and N.F. Voelkel. Rapid onset of hypoxic vasoconstriction in isolated lungs. *J. Appl. Physiol.* 72(5): 2018-2023, 1992.

31. Weir, E.K., J.E. Eaton, and E. Chesler. Redox status and pulmonary vascular reactivity. *Chest* 88:249S-252S, 1985.

32. Archer, S.L., D. Peterson, D.P. Nelson, E.G. McMaster, B. Kelly, J.W. Eaton, and E.K. Weir. Oxygen radicals and antioxidant enzymes alter pulmonary vascular reactivity in the rat lung. *J. Appl. Physiol.* 66:102-111, 1989.

33. Peters-Golden, m.C. Shelly, and M.L. Morganroth.Inhibition of rat lung glutathione synthesis attenuates hypoxic pulmonary vasoconstriction and the associated leukotriene C_4 production. *Am. Rev. Respir. Dis.* 140: 1210-1215, 1989.

34. Thomas, III, H.M., R.C. Carson, E.D. Fried and R.S. Novitch. Rapid communication. *Biochem. Pharmacol.* 42(7):R9-R12, 1991.

35. Cross A.R., L. Henderson, O.T.G. Jones, M.A. Delpiano, J. Hentschel, and H. Acker. Involvement of an NAD(P)H oxidase as a PO_2 sensor protein in the rat carotid body. *Biochem. J.* 272: 743-747, 1990.

36. Lopez-Lopez, J., C. Gonzalez, J. Urena and J. Lopez-Barneo. Low pO_2 selectively inhibits K channel activity in chemoreceptor cells of the mammalian carotid body. *J. Gen. Physiol.* 93:1001-1015, 1989.

37. Van Grondelle, A., D. Ellis, m.M. Mathias, R.C. Murphy, R.J. Strife, j.T. Reeves, and N.F. Voelkel. Altering hydrodynamic variables influences PGI_2 production by isolated lungs and endothelial cells. *J. Appl. Physiol.* 57 (2): 388-395, 1984.

38. Lansman, J.B., T.J. Hallam, and T.J. Rink. Single stretch-activated ion channels in vascular endothelial cells as mechanotransducers. *Nature Lond.* 325:811-813, 1987.

39. Olesen, S.P., D.E. Clapham, and P.F. Davies. Hemodynamic shear stress activates a K^+ current in vascular endothelial cells. *Nature Lond.* 331:168-170, 1988.

40. Cooke, J.P., E. Rossitch, Jr., N.A. Andon, J. Loscalzo, and V.J. Dzau. Flow activates an endothelial potassium channel to release an endogenous nitrovasodilator. *J. Clin. Invest.* 88:1663-1671, 1991.

41. Singer, J.J., and J.V. Walsh, Jr. Characterization of calcium-activated potassium channels in single smooth muscle cells using the patch-clamp technique. *Pfluegers Arch.* 408:98-111, 1987.

42. Shen, J., F.W. Luscinskas, A. Connolly, C.F. Dewey, Jr., and M.A. Gimbrone, Jr. Fluid shear stress modulates cytosolic free calcium in vascular endothelial cells. *Am. J. Physiol.* 262: C384-C390, 1992.

43. McMurtry, I.F., A.B. Davidson, J.T. Reeves, and R.F. Grover. Inhibition of hypoxic pulmonary vasoconstriction by calcium antagonists in isolated rat lungs. *Cir. Res.* 38:99-104, 1976.

44. Voelkel, N.F., K.G. Morris, I.F. McMurtry, and J.T. Reeves. Calcium augments hypoxic vasoconstriction in lungs from high-altitude rats. *J. Appl. Physiol.* 49(3): 450-455, 1980.

45. McMurtry, I.F. BAY K 8644 potentiates and A23187 inhibits hypoxic vasoconstriction in rat lungs. *Am. J. Physiol.* 249: H741, 1985.

46. Tolins, M., E.K. Weir, E. Chesler, D.P. Nelson, and A.H.L. From. Pulmonary vascular tone is increased by a voltage-dependent calcium channel potentiator. *J. Appl. Physiol.* 60(3): 942-948, 1986.

47. Madden, J.A., C.A. Dawson, and D.R. Harder. Hypoxia-induced activation in small isolated pulmonary arteries of the cat. *J. Appl. Physiol.* 59: 113-118, 1985.

48. Yuan, X.J., M.L. Tod, L.J. Rubin, and M.P. Blaustein. Contrasting effects of hypoxia on tension in rat pulmonary and mesenteric arteries. *Am. J. Physiol.* H281-H289, 1990.

49. Wiener, C.M., A. Dunn, and J.T. Sylvester. ATP-dependent K^+ channels modulate vasoconstrictor responses to severe hypoxia in isolated ferret lungs. *J. Clin. Invest.* 88:500-504, 1991.

50. Lippton, J.L., G.A. Cohen, I.F. McMurtry, and A.L. Hyman. Pulmonary vasodilation to endothelin isopeptides in vivo is mediated by potassium channel activation. *J. Appl. Physiol.* 70(2): 947-952, 1991.

51. Robertson, B.E., P.R. Corry, P.C.G. Nye, and R.Z. Kozlowski. Ca^{2+} and Mg-ATP activated potassium channels from rat pulmonary artery. *Eur. J. Physiol.* ,1992.

52. Salvaterra, C.G., L.J. Rubin, J. Schaeffer, and M.P. Blaustein. The influence of the transmembrane sodium gradient on the responses of pulmonary arteries to decreases in oxygen tension. *Am. Rev. Respir. Dis.* 139: 933-939, 1989.

53. McMurtry, I.F. Angiotensin is not required for hypoxic constriction in salt solution-perfused rat lungs in salt solution-perfused rat lungs. *J. Appl. Physiol.* 56(20: 375-380 1984.

54. Orton, E.C., B. Raffestin, and I.F. McMurtry. Protein kinase C influences rat pulmonary vascular reactivity. *Am. Rev. Respir. Dis.* 141: 654-658, 1990.

55. Voelkel, N.F., and J. Czartolomna. Vanadate potentiates hypoxic pulmonary vasoconstriction. *J. Pharma. Exper. Thera.* 259(2): 666-672

56. Weir, E.K., J.M. Post, S.L. Archer and J.R. Hume. "The redox status of a potassium channel may serve as a pulmonary vascular oxygen sensor". Intern Conference on Pulmonary Vascular Remodeling. London, September 7-9, 1992. (Abstract).

57. Eltze, M. Glibenclamide is a competitive antagonist of cromakalim, pinacidil, and RP 49356 in guinea-pig pulmonary artery. *Eur. J. Pharmacol.* 165:231-239.

58. Voelkel, N.F. Regulation of pulmonary vascular tone. *Europ. Resp. J.* (In press)

59. Inoue, K., K. Nakazawa, M. Ohara-Imaizumi, T. Obama, K. Fujimori, and A. Takanaka. Selective and competitive antagonism by suramin of ATP-stimulated catecholamine-secretion from PC-12 pheochromocytoma cells. *Br. J. Pharmacol.* 102: 581-584, 1991.

60. Mikkelsen, E.O, O. Thastrup, and S.B. Christensen. Effects of thapsigargin in isolated rat thoracic aorta. *Pharmacol. & Toxicol.* 62:7-11, 1988.

61. Ely, J.A., C. Ambroz, A.J. Baukal, S.B. Christensen, T. Balla, and K.J. Catt. Relationship between agonist- and thapsigargin-sensitive calcium pools in adrenal glomerulose cells. *J. Biol. Chem.* 266:18635-18641, 1991.

62. Jan, L.Y., and Y.N. Jan. Voltage-sensitive ion channels. *Cell* 56:13-25, 1989.

63. Noack, T.H., P. Deitmer, E. Lammel. Membrane currents in single smooth muscle cells of the guinea pig aortal vein and possible effect of prostacyclin. *Clin. Pharma.* 7:24-30,1990.

64. Deutsch, C., D. Krause, and S.C. Lee. Voltage-gated potassium conductance in human T lymphocytes simulated with phorbol ester. *J. Physiol.* 372: 405-423, 1986.

65. Grissmer, S., B. Dethlefs, J.J. Wasmuth, A.L. Goldin, G.A. Gutman, M.D. Cahalan, and K.G. Chandy. Expression and chromosomal localization of a lymphocyte K^+ channel gene. *Immunology* 87:9411-9415, 1990.

66. Inoue, I., H. Nagase, K.Kishi, and T. Higuti. ATP-sensitive K^+ channel in the mitochondrial inner membrane. *Nature* 352:244-248, 1991.

67. Kamb, A., M. Weir, B. Rudy, H. Varmus, and C. Kenyon. Identification of genes from pattern formation, tyrosine kinase, and potassium channel families by DNA amplification. *Biochem.* 86: 4372-4376, 1989.

68. Quast, U., N.S. Cook. Moving together K^+ channels openers and ATP-sensitive K^+ channels. *TIPS* 10: 431-435, 1989.

69. Pragnell, M., K.J. Snay, J.S. Trimmer, N.J. MacLusky,F. Naftolin, L.K. Kaczmarek, and M.B. Boyle. Estrogen induction of a small, putative K^+ channel mRNA in rat uterus. *Neuron.* 4: 807-812, 1990.

70. MacKinnon, R., and G. Yellen. Mutations affecting TEA blockade and ion permeation in voltage-activated K^+ channels. *Science* 250:276-279, 1990.

71. Papazian, D.M., L.C. Timpe, Y.N. Jan, and L.Y. Jan. Alteration of voltage-dependence of shaker potassium channel by mutations in the S4 sequence. *Nature* 349: 305-310, 1991.

72. Kim, D., D.L. Lewis, L. Graziadei, E.J. Neer, D. Bar-Sagi, and D.E. Clapham. G-protein beta, gamma-subunits activate the cardiac muscarinic K^+-channel via phospholipase A_2. *Nature* 337: 557-559, 1989.

73. **Brown, A.M., and L. Birnbaumer.** Direct G protein gating of ion channels. *Am. J. Physiol.* 254: H401-H410, 1988.

74. **Sullivan, S.K., K. Swamy, N.R. Greenspan, and M. Field.** Epithelial k channel expressed in Xenopus oocytes is inactivated by protein kinase C. *Physiol.* 87:4553-4556, 1990.

75. **Ämmälä, C., O. Larsson, P.O. Berggren, K. Bokvist, L. Juntti-Berggren, H. Kindmark, and P. Rorsman.** Inositol triphosphate-dependent periodic activation of a Ca^{2+}- activated K^+ conductance in glucose-stimulated pancreatic beta- cells. *Nature* 353: 849-852, 1991.

76. **Numann, R., W.A. Catterall, T. Scheuer.** Functional modulation of brain sodium channels by protein kinase C phosphorylation. *Science* 254:115-118, 1991.

77. **Lückhoff, A., D.E. Clapham.** Inositol 1,3,4,5-tetrakisphosphate activates an endothelial Ca^{2+} -permeable channel. *Nature* 355: 356-358, 1992.

78. **Ruppersberg, J.P., M. Stocker, O. Pongs, S.H. Heinemann, R. Frank, and M. Koenen.** Regulation of fast inactivation of cloned mammalian $I_K(A)$ channels by cysteine oxidation. *Nature* 352: 711-714, 1991.

79. **Isacoff, E.Y., Y.N. Jan, and L.Y. Jan.** Evidence for the formation of heteromultimeric potassium channels in Xenopus oocytes. *Nature* 345: 530-534, 1990.

80. **Lu, L., C. Montrose-Rafizadeh, and W.B. Guggino.** Ca^{2+} -activated K^+ channels from rabbit kidney medullary thick ascending limb cells expressed in Xenopus oocytes. *J. Biol. Chem.* 265(27):16190-16194, 1990.

81. **Schultz, J.E., S. Klumpp, R. Benz, W.H. CH. Schürhoff-Goeters, A. Schmid.** Regulation of adenyl cyclase from Paramecium by an intrinsic potassium conductance. *Science* 255: 600-604, 1992.

82. **Schlichter, L., N. Sidell, and S. Hagiwara.** Potassium channels mediate killing by human natural killer cells. *Immunol.* 83:451-455, 1986.

83. **Nilius, B.** Regulation of transmembrane calcium fluxes in endothelium. *NIPS* 6:110-114, 1991.

84. **Cooke, N.S.** the pharmacology of potassium channels and their therapeutic potential. *TIPS* 9:21-28, 1988.

85. **Tytgat, J., and P. Hess.** Evidence for cooperative interactions in potassium channel gating. *Nature* 359:420-424, 1992.

86. **Kurachi, Y., H. Ito, T. Sugimoto, T. Shimizu, I. Miki, and M. Ui.** a-Adrenergic activation of the muscarinic K^+ channel is mediated by arachidonic acid metabolites. *Europ. J. Physiol.* 414:102-104, 1989.

87. **Surprenant, A., D.A. Horstman, H. Akbarali, L.E. Limbird.** A point mutation of the $beta_2$ -adrenoceptor that blocks coupling to potassium but not calcium currents. *Science* 257: 977-980, 1992.

88. **Honey, M., L. Cotter, N. Davies, D. Denison.** Clinical and hemodynamic effects of diazoxide in primary pulmonary hypertension. *Thorax* 35: 269, 1980.

89. **Tozzi, C.A., G.J. Poiani, A.M. Harangozo, C.J. Boyd, and D.J. Riley.** Pulmonary vascular endothelial cells modulate stretch-induced DNA and connective tissue synthesis in rat pulmonary artery segments. *Chest* 93(3):169S-170S, 1988.

90. **Post, J.M., J.R. Hume, S.L. Archer, and E.K. Weir.** Direct role for potassium channel inhibition in hypoxic pulmonary vasoconstriction. *Am J Physiol* 262(Cell Physiol. 31): C882-C890, 1992.

THE ELECTROPHYSIOLOGY OF SMOOTH MUSCLE CELLS AND TECHNIQUES FOR STUDYING ION CHANNELS

Normand Leblanc and Xiaodong Wan

Division of Cardiovascular Sciences
Department of Physiology
University of Manitoba
St. Boniface General Hospital Research Centre
Winnipeg, MAN, Canada R2H 2A6

INTRODUCTION

Considerable attention has been paid to understanding the basis of transmembrane ion flux in smooth muscle. As for all other excitable and many non excitable cells, regulation of the membrane permeability to several ions is responsible for controlling the concentration of free intracellular calcium ions ($[Ca^{2+}]_i$) which indirectly regulates the contractile properties of smooth muscle cells. In 1991, Bert Sakmann and Erwin Neher were awarded the Nobel Prize of Physiology and Medicine in part for their contribution in developing the Patch Clamp Technique. This allowed for the first time recording the activity of single transmembrane proteins whose function is to control flux of specific ions across the membrane. A little more than a decade after the main report that refined this technique was published (4), thousands of reports have now described in great detail the biophysical properties, pharmacological profiles and mechanistic implications of several classes of ion channels present in nearly all cell membranes of the body. The significance of the contribution of the patch clamp technique can be highlighted even more in the field of smooth muscle electro-physiology as the voltage clamp techniques previously used to assess ion channel function had technical difficulties and serious limitations.

Many excellent essays have already described this technique in detail and the reader who is interested in further advancing his knowledge should consult the following articles (3,20) and monographs (5,13,15,16). The general objective of this chapter is to offer a simple comprehensive description of the technique. Specifically, the reader will be familiarized with the basic steps involved in isolating healthy smooth muscle cells without which the technique could not be utilized. The reader will then be provided information on the essential components required to operate a functional patch clamp setup and be introduced to the advantages and limitations of using the various configurations of the patch clamp technique, with emphasis on their specific applications to study ion channels in smooth muscle cells. Finally, information will be provided regarding some of the basic concepts associated with the analysis and interpretation of ion channel data.

It is important to mention that other powerful techniques have been developed to investigate the properties of ion channels at the molecular level. It is now possible to take advantage of biochemical techniques to extract crude or purified ion channel proteins in which activity can be restored after incorporation of the extracted material into artificial lipid bilayers for structure-function and drug-channel interaction studies. More recently, the application of recombinant DNA techniques to clone ion channels has provided additional tools to derive and modify the amino acid sequence of these proteins and better assess their structural and functional properties. Due to space limitations, these topics will not be discussed here. However, the interested reader should consult the following recent publications (15,22).

THE PATCH CLAMP TECHNIQUE

General

The patch clamp technique can be used in a combination of configurations to record: (i) the "normal" electrical behaviour of the cell or the resting membrane potential (RMP) and may thus serve the same purposes as a standard microelectrode, (ii) the activity of all channels present in the cell membrane (often referred to as macroscopic or whole-cell currents), (iii) the activity of one or more ion channel proteins (microscopic or unitary currents); and (iv) net current produced by electrogenic carrier systems (Na^+/K^+ pump, Na^+/Ca^{2+} exchange, etc.). Before the advent of cell isolation procedures, smooth muscle electrophysiologists were forced to voltage clamp multicellular preparations to gain information on membrane ionic currents. These techniques (single- or double-sucrose gap technique, two- or three-electrode voltage clamp, single-electrode switch clamp), besides being technically difficult and inconsistent, suffered from serious limitations which were related to the specific geometric arrangement and coupling of the smooth muscle cells present in a multicellular preparation (8). Because of these limitations, the quality of the clamp was in general very poor which led to inadequate measurements of membrane current. The patch clamp technique eliminated some and minimized most of the problems inherent to voltage clamping multicellular preparations.

Cell Isolation Procedures

In parallel with the development of the patch clamp technique was the improvement of cell culturing and dispersion routines. Figure 1 outlines a general basic procedure used to isolate smooth muscle cells and is similar to that employed by many laboratories. The first step in attempting to study ion channel function in smooth muscle is to develop a Ca^{2+}-tolerant preparation. It is certainly fair to say that obtaining such a preparation on a daily basis requires knowledge about witchcraft, some luck, and a lot of trial and error. In general, the exact recipe described in published articles most likely suits best the laboratory that developed it. As for other dispersion techniques, isolation of smooth muscle cells essentially comprises four basic steps. The smooth muscle of interest is first dissected free of adherent fat, connective tissues and endothelium, and cut into smaller pieces (fig. 1-a). The latter chunks are then incubated in a nominally Ca^{2+}-free medium, either at room or at physiological temperatures for 10-30 min (fig. 1-b). This procedure has been suggested to facilitate dispersion by separation of the outer portion of the basal lamina of the surface membrane and by favouring disruption of desmosomes connecting adjacent cells (23). In our laboratory, we found the addition of 100 μM EGTA to buffer the calcium contaminants present in the solutions significantly improved the yield and Ca^{2+}-tolerance of the myocytes. The third step in the procedure is to digest the pieces of muscle with enzymes (fig. 1-c). A wide variety of enzymes have been used but collagenases and papain from various sources are still the most popular for isolating smooth muscle cells.

CELL DISPERSION TECHNIQUE

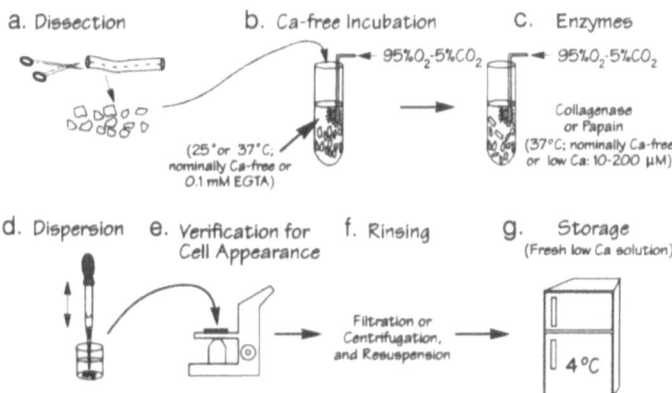

Figure 1: Basic general steps involved in the isolation of healthy single smooth muscle cells. See text for explanations.

Here the enzymatic step is usually done at 35-37°C for 20-60 min using either a nominally Ca^{2+}-free solution or solution in which Ca^{2+} has been added (10-50 μM) to improve the digestion as collagenase and other enzymes are activated by calcium ions. In smooth muscles containing a lot of connective tissue (for instance arteries), other enzymes such as protease or elastase are sometimes needed to separate the cells. A difficult task in isolating smooth muscle cells is knowing when to end the digestion. A few minutes can make a big difference in the quality of the cells. In our laboratory, a few pieces are retrieved from the enzyme medium and triturated in fresh low Ca^{2+} solution using a pasteur pipette to release the single cells (fig. 1-d). Verification under the microscope is repeatedly done every 2-5 min (fig. 1-e), approximately 15 min after the incubation with the enzymes was initiated. The incubation should be stopped when a "few" cells can be viewed following mild trituration (~1 min). At this time, all the remaining pieces (if you have any left !!!) are rinsed two or three times in fresh low Ca^{2+} medium and triturated to harvest the cells. Centrifugation at low speed (and resuspension of the cells) or filtering (through a nylon mesh) the medium containing the cells are commonly used in several laboratories to remove cellular debris that may stick to patch micropipettes during the experiments (fig. 1-f). Cells are usually kept in the cold (4°C) in low Ca^{2+} solution (10-200 μM; fig. 1-g) and can be used within 4-8 hours after isolation, and sometimes after 24 hours. Another method for keeping the cells was that of Klöckner and Isenberg (10) who kept the harvested cells in a solution containing elevated concentration of K^+ but no Ca^{2+} and low Na^+ (K-B medium); this procedure was shown to prevent cell loading of Na^+ and Ca^{2+} and enabled the use of cells for up to a week. From our experience, the technique described above is quite universal. We have succeeded in isolating very different types of smooth muscle cells (rabbit coronary artery and portal vein, guinea-pig portal vein and ileum, dog airway muscles and pulmonary artery, rat pulmonary artery). The Ca^{2+}-tolerant cells are elongated, have a clear membrane and contract upon elevation of the concentration of K^+ in the perfusate and in response to an appropriate agonist.

Typical Patch Clamp Setup

In their initial efforts, Neher and Sakmann (11) succeeded in recording for the first time the activity of unitary conductances produced by the opening and closing of the nicotinic receptor stimulated by acetylcholine. The technique consisted in simply applying a micropipette to the surface of a cell while monitoring the increase in electrical resistance between the pipette (which has a fixed resistance of a few megohms in physiological solutions) and the membrane in reference to a ground electrode placed in the bath solution. Despite considerable efforts to clean the cell surface by various enzymatic methods and screening several kinds of glass capillaries, seal resistances were usually below 100 MΩ which resulted in poor signal-to-noise ratio of the single channel recordings. It was not until the late seventies and eighties that new advancements by the same group permitted high resolution recording of not only single channel currents but also currents generated by all the channels distributed in the entire cell membrane most commonly referred to as whole-cell currents.

Figure 2 sketches a diagram of a typical modern patch clamp setup (consult ref. 15 for advanced technical details). One important element of the setup is evidently the microscope. Virtually all patch clampers use an inverted microscope, with objectives mounted underneath the mechanical stage onto which a superfusing chamber containing the single cells can be manually controlled. This simple arrangement gives enough free space above the chamber to easily bring down the micropipette and other devices (other pipettes to picospritz substances to the cell, reference electrode, suction lines, etc.) with a minimal number of obstacles. The choice of the microscope depends on the intended use. If one is planning on only studying whole-cell or single channel currents, then relatively inexpensive microscope (~$ 3,000) will do just fine. A good example is the Nikon TMS model which we have successfully used to patch clamp smooth muscle cells. A trinocular version of this model is also available for attachment of an ordinary CCD TV camera. On-line cell image can thus be viewed on a TV monitor during the experiment. However, if one plans to simultaneously conduct microfluorescence spectroscopy (to measure ion concentration using ion-sensitive fluorescent dyes such as Fura2 or Indo-1 for Ca^{2+}, or BCECF for pH, etc.) or "flash-photolysis" of caged compounds (Nitr5 or DM-Nitrophen for Ca^{2+}; caged nucleotides; others) while patch clamping cells, a stripped down version of an epifluorescence inverted microscope equipped with quartz lenses (minimum absorbency in the UV spectrum) should be purchased. In our laboratory, we have used the Nikon Diaphot to simultaneously record $[Ca^{2+}]_i$ with Indo-1 and ionic currents using the patch clamp technique in vascular smooth muscle cells. Other companies also provide excellent microscopes (Olympus, Zeiss) that are as suitable. It is therefore wise to examine in detail the specifications and price of each of the available models.

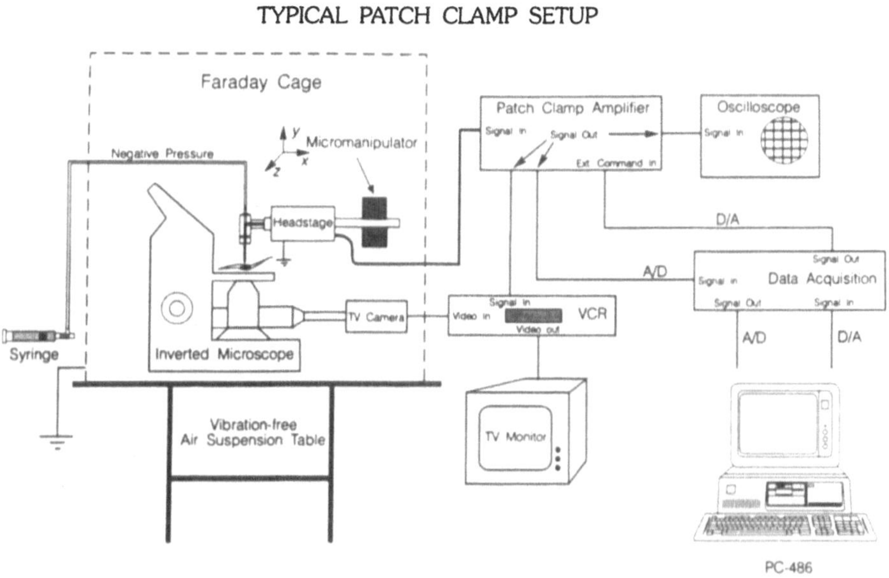

TYPICAL PATCH CLAMP SETUP

Figure 2: Diagram of the components of equipment found on a modern patch clamp setup. See text for explanations.

The microscope is itself sitting on a vibration-free table to prevent high frequency motion (horizontal and vertical) of the micropipette tip. This component is essential to ease the formation of the gigaohm seal between the pipette and the membrane, and to prevent the appearance of electrical spikes caused by mechanical oscillations which may contaminate the recording of membrane current, especially unitary currents. Home made arrangements are sometimes used (heavy stones resting on rubber pieces or tennis balls) but air suspension tables (Kinetics Systems, Micro-G, Newport) are by far the most popular types used and are probably more effective in reducing vibration in unsteady buildings.

During a patch clamp experiment, freshly isolated smooth muscle cells are deposited in the experimental chamber (built with plexi-glass) in which perfusion and

19

suction lines insure constant flow of a physiological buffer. A period of 5 to 10 min is necessary for the cells to settle and attach to the bottom of the chamber before initiating superfusion. A patch micropipette is then filled with a suitable filtered solution and tightly fixed (rubber seals) in a pipette holder. A flexible tygon tubing is attached to the back of the holder and serves to apply negative pressure to the cell membrane via a tight syringe. A chloridized silver wire or a Ag/AgCl pellet in contact with the pipette solution is connected to the headstage which is the site of preamplification of the current and voltage signals. The headstage contains the main operational amplifiers and feedback resistors (in resistive models) or capacitors (in integrating models) that will determine the input gain. Patch clamp headstages are current-to-voltage (CV) converters, that is contrary to traditional microelectrode amplifiers which are voltage followers (voltage output corresponds to voltage input), voltage output corresponds to current input; more simply, when a command voltage (called V_p) is applied through the patch pipette relative to a ground electrode placed in the bath, the headstage amplifier generates a current that is equal and opposite to the whole-cell or single channel current that keeps the voltage at its two inputs equal to V_p. This is in fact the basic concept of the voltage clamp technique elaborated more than forty years ago. The membrane is artificially forced to remain at a command voltage. The headstage ensures with great accuracy that V_p will remain steady by constantly supplying the appropriate current. The headstage, microscope and micromanipulator are completely surrounded by a grounded Faraday cage (fig. 2). This arrangement is essential to reduce the amount of electrical noise originating from various sources (60 Hz from power lines, TV monitors and computer screens, electromagnetic interference, etc.). Because of improved electronic circuitry (transistors and operational amplifiers), commercially available patch clamp amplifiers (Axon Instruments, List, Dagan and others) can now generate output signals with only a fraction of picoampere (pA) RMS noise allowing the resolution of very small unitary currents.

Once mounted on the headstage, the pipette holder and patch pipette can approach the cell with the aid of a micromanipulator; most models are equipped with coarse and fine movements. The micropipette current coming out of the headstage is then transmitted to the patch clamp amplifier for further processing (output gains, pipette and stray capacitance compensations, junction potential corrections, series resistance compensation, filtering, leak subtraction, etc.). The output signals can be viewed on-line on an oscilloscope screen, and/or permanently stored on videotape along with the cell image, through a modified VCR, and/or acquired on-line through an analog to digital (A/D) acquisition board connected to a computer in which the data can be temporarily stored on the computer hard disk or other types of storage media. In most laboratories, the latter also serves to drive the entire patch clamp experiment by controlling the voltage or current clamp protocol applied to the patch of membrane or cell via a D/A board connected to the patch clamp amplifier. Sophisticated software is now commercially available to effectively control the experiment, acquire and analyze the data, and process it in a publication format. The reader interested in building his own setup should consult a recent publication that lists the most important companies manufacturing the necessary components discussed above (21).

Configurations of the Patch Clamp Technique

The patch clamp technique exploits the simple physical property that glass can bind to biological membranes with great affinity. It is first necessary to fabricate the micropipettes using a two-stage puller which essentially consists of a heating coiled element surrounding a portion of a glass capillary (1.0-1.6 mm in outer diameter) held tight between a fixed clamp at one end and a sliding clamp at the other (fig. 3-a). Horizontal and vertical pullers are commercially available which allow the user to control the heating during the first and second pulls and in some models the amount of traction applied to the sliding clamp. Such as shown in Figure 3-c, the procedure yields two patch micropipettes, both of which can be used with relatively equal efficacy once the tip has been fire-polished using a home-made or commercially available microforge.

Figure 4 summarizes the various steps required to obtain the basic configurations of this technique. A fire-polished patch micropipette filled with an appropriate filtered solution directly in contact with a Ag/AgCl wire connected to the patch clamp amplifier headstage is applied to the surface of the smooth muscle cell viewed at high optical magnification (≥ 400 X) with the use of a high resolution micromanipulator. Many researchers routinely apply positive pressure via the syringe before touching the cell to avoid obstruction of the pipette tip by floating particles or cellular debris as the micropipette is brought down into the chamber. However from our experience,

FABRICATION OF PATCH PIPETTES

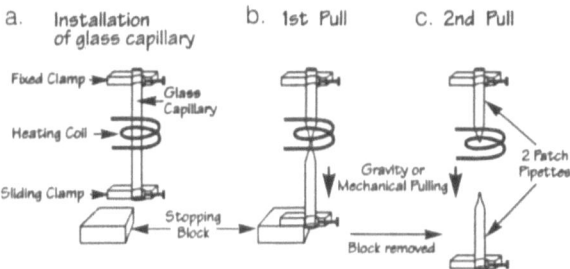

Figure 3: Diagram illustrating the principle and steps involved during two-stage vertical pulling of patch micropipettes. A glass capillary (for a list of the different types of glass capillaries that can be used, consult ref. 14) is installed as shown in (a) and held tight between a fixed (top) and a sliding (bottom) clamps. The element is then heated at a temperature above the melting point; heating allows the glass to expand in response to gravity or active traction (in some models) until the sliding clamp is arrested by a stopping block (b) placed at a desired distance (~3-7 mm). The heating coil is subsequently moved down to be aligned with the centre of the narrowed capillary. After removing the stopping block, a second pulling step (with less heat) is carried out, producing two patch micropipettes (c).

constant flush of pipette solution containing high K^+ concentrations (to measure "normal" whole-cell currents or the activity of single K^+ channels) depolarizes the smooth muscle myocytes which often contract or become thinner which renders the task of applying the pipette on the cell more difficult. Instead, once the pipette is near the cell chosen, a quick pulse of positive pressure is applied which is sufficient to clear out the pipette tip. When the pipette potential (voltage difference between the pipette and the reference or ground electrode placed in the bathing medium) has been cancelled out, the pipette can be gently applied onto the cell membrane (fig. 4A). Because the smooth muscle cells are thin (~5 μm in diameter) and the outer diameter of the pipette is not much smaller (~2 μm), it is important to aim for the center of the cell or around the nuclear region which is slightly wider. Gentle pressure applied to the cell by the micropipette usually helps minimize Ca^{2+} uptake from the bathing solution during application of suction. This latter point is important especially for the study of whole-cell currents that are Ca^{2+}-dependent such as the large conductance Ca^{2+}-activated K^+ current (K_{Ca} or B-K). After pressing the membrane, weak negative air pressure (5-10 cm H_2O) is applied via a gas tight syringe connected to pipette holder (fig. 2). Depending on how clean the surface of the membrane is, it is

PATCH CLAMP CONFIGURATIONS

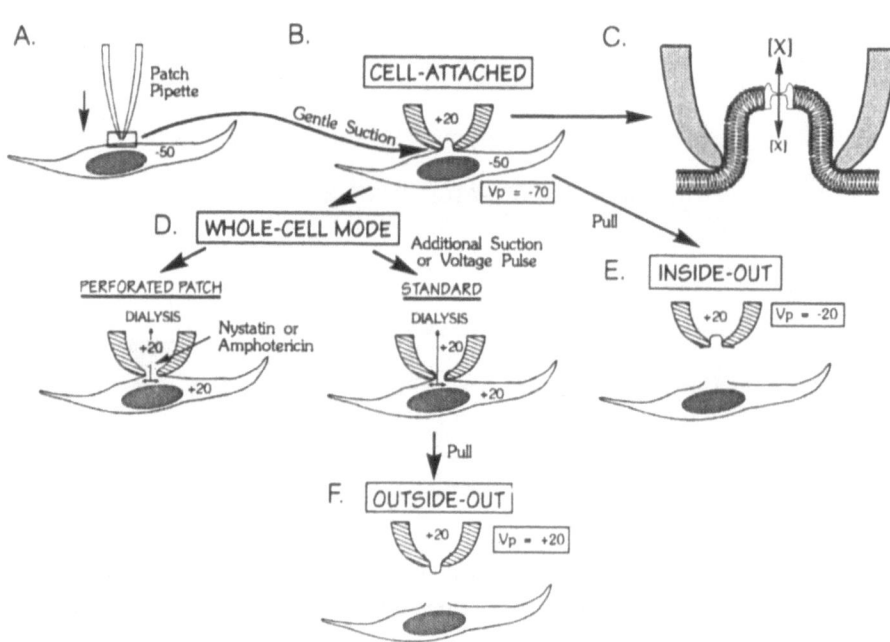

Figure 4: Diagram explaining the sequential steps necessary to obtain the various configurations of the patch clamp technique. V_p represents the net voltage across the membrane patch which results from the algebraic sum of the voltage applied ($V_{appl.}$) via the pipette and the voltage of the internal side of the membrane. In cell-attached mode (B), +20 mV is applied to the outer surface of the membrane meaning that internal voltage will be -20 mV and V_p = -20 mV + RMP (-50 mV in this example) = -70 mV. In whole-cell (D) and inside-out mode (E), $V_{appl.}$ = V_p. For outside-out patches (F), V_p = - $V_{appl.}$. [X]: concentration of a particular ion. See text for further explanations.

sometimes possible to obtain a gigaohm seal by simply touching the cell surface. Monitoring of the electrical seal is accomplished by applying repetitive voltage pulses (10-20 mV in amplitude and 10-50 msec in duration) through the pipette and recording the current. Before touching the membrane, the amplitude of the current will be a direct measure of the pipette resistance as defined by Ohm's law ($V = RI$) and is generally in the order of 1-5 MΩ. As the pipette touches the membrane, resistance to current flow increases and is viewed on the oscilloscope as a decrease in the amplitude of the current. With suction, the seal resistance increases until it reaches a very high value, that is in the order of 10^9 Ω or gigaohm which led to the term "giga-seal". With microscopy, it was found that when a giga-seal is formed, the membrane becomes invaginated within the pipette interior and forms a semi-sphere (fig. 4C; ref. 17).

When a gigaohm seal is formed, a small patch of membrane becomes electrically isolated. Following appropriate cancellation of pipette, stray and patch capacitance, individual openings of single channel proteins can be recorded with high resolution under voltage clamp conditions (fig. 4C). Indeed, when special precautions are taken to reduce pipette capacitance and 60 cycle noise (proper shielding, low level of fluid in the bath, application of Sylgard up to 100 μm from the tip, etc.), unitary currents < than 1 pA can be separated from background noise at 1-2 kHz lowpass filtering. This configuration is called "Cell-Attached" (fig. 4B). It is useful to study ion channels under "physiological" conditions since the environment of the channel(s) under investigation has been little perturbed, at least in comparison with the standard whole-cell (fig. 4D) and excised patch configurations (fig. 4E and F). It is also a valuable technique to assess whether the stimulation of an ion channel by activation of a receptor by an agonist requires a second messenger synthesized in the cytoplasm. A stimulatory effect of an agonist supplied from the surrounding medium could be an indication that a second messenger is synthesized and is involved in the response since it is generally assumed that it can not traverse the gigaohm seal. On the other hand, enhancement of channel activity by including an agonist in the pipette solution and lack of effect when superfused in the bath could suggest a direct stimulation of the channel by the agonist such as demonstrated for cardiac L-type Ca^{2+} channels which implicated "direct gating" via the G-protein G_s (1).

There are several disadvantages of using the cell-attached mode. One is that the transmembrane voltage across the patch of membrane is not solely dependent upon the voltage applied via the pipette (V_p; fig. 4B) but is instead the algebraic sum of V_p and the resting membrane potential of the cell which is an unknown variable. Based on numerous microelectrode studies in smooth muscle, it could be assumed that RMP would be about -50 mV with a potential error of ± 5-20 mV. Therefore, it would be difficult to determine the true voltage-dependence of the ion channel under the patch unless a second micropipette was used in whole-cell mode (fig. 4D) to accurately control the transmembrane potential. One method commonly used to "minimize" RMP is to superfuse the cell with an elevated K^+ concentration (150 mM) which will depolarize the myocyte near 0 mV. One major drawback of the latter method is that Ca^{2+} must be omitted (EGTA is usually added) from the bathing solution to avoid contraction of the myocyte due to activation of voltage-dependent Ca^{2+} channels by depolarization. If one wishes to study the activity and contribution of Ca^{2+}-dependent ion channels (Ca^{2+}-activated K^+ and Cl^- channels), their "physiological" role would be underestimated as intracellular Ca^{2+} depletion will inevitably occur with time in Ca^{2+}-

free medium. Another disadvantage of the cell-attached mode is that the solutions on both sides of the membrane can not be changed. This complicates the analysis of agonist-channel interactions in terms of concentration-dependence and time-dependence of activation or inhibition of an ion channel. Attempts have been made to alter the pipette solution during the course of an experiment but success rate is in general very low.

Because of the high stability of the membrane-glass interaction in a giga-seal, Hamill et al. (4) showed that it is possible to excise the membrane patch and keep the seal practically intact. As illustrated in figure 4E and F, once in the cell-attached mode and provided that the cell probed is well attached to the bottom surface of the chamber, the pipette can be lifted upward thus tearing off the patch of membrane from the cell while keeping the giga-seal virtually intact. With this manipulation, the inside or cytoplasmic side of the membrane now faces the bath solution and the excised path of membrane is said to be in the "Inside-Out" configuration (fig. 4E). This method of recording unitary currents from single channel proteins is technically relatively straightforward and allows one to easily alter the composition of the solution facing the inner mouth of the channel pore. It that respect, it has been a very useful tool to investigate the permeation process (properties of the pore: selectivity, conductance), gating properties (opening and closing behaviours), and regulation by cytoplasmic constituents and second messengers of several smooth muscle ion channels. Large conductance Ca^{2+}-activated K^+ channels have been found in abundance in all smooth muscle cells and were shown to be sensitive to voltage (depolarization) and $[Ca^{2+}]_i$. Experiments with inside-out patches of smooth muscle cell membranes have been commonly used to assess the $[Ca^{2+}]_i$-dependency of K_{Ca} by superfusing the excised patch with solutions of known Ca^{2+} concentrations as determined from the properties of Ca^{2+}-EGTA mixtures (9). Evidence for the existence of ATP-sensitive K^+ channels was possible using this technique as some K^+ channels were found to be inhibited by exposing the cytoplasmic side of the membrane with various concentrations of ATP (18). With this mode, voltage applied via the pipette is the actual transmembrane voltage seen by the ion channel(s) under study, which eliminates the problem of uncertainty about RMP in the cell-attached configuration. One disadvantage of this technique is that the activity of many ion channels (for instance dihydropyridine-sensitive L-type Ca^{2+} channels) quickly runs down and often vanishes after patch excision. This phenomenon is likely produced by the loss of essential cellular regulators or by the physical disruption of the cytoskeletal elements involved in their control. Such a drawback can in turn be used as an advantage by attempting to restore lost channel function by supplying regulators (2nd messengers such as cyclic AMP or GMP) or complete pathways (protein kinase systems) suspected to control their activity.

Another practical method of the technique allows one to record the activity of channels from the entire membrane of the cell (macroscopic ionic currents) and is referred to as the "Whole-Cell" configuration (fig. 4D). There are two variants of this configuration. The "standard" whole-cell mode is the original configuration as introduced by Hamill et al. (4) whereby the membrane invagination produced by the tight seal in cell-attached mode (fig. 4C) is suddenly ruptured while maintaining the giga-seal by the application of additional negative pressure through the syringe or by supplying a strong voltage surge (1-1.5 Volt) of variable duration (0.1-10 msec) via the micropipette (a feature called the ZAP command is available on many patch clamp amplifiers). Gain of access is viewed as the sudden appearance of capacitive spikes in response to 10-20 mV repetitive voltage clamp pulses and are produced by charging of the membrane capacitance (C_m). As shown in figure 4D (Standard), rupture of the membrane permits exchange of cytoplasmic and pipette fluid or "dialysis". Because the volume of the micropipette is several orders of magnitude larger than the volume of the cell, a close approximation is that it acts as an infinite reservoir; after a period of equilibration, pipette solution and cytoplasm have a similar ionic composition. Thus the whole-cell recording mode allows one to control the composition of the cytoplasm. On the other hand, the large pipette volume also acts as a sink in which metabolic factors and factors important for ion channel function are likely to be washed out of the cell with time. Indeed, dialysis-induced run down of whole-cell Ca^{2+} current is a well known phenomenon in smooth muscle cells, especially in arterial preparations in which the peak current after breaking the membrane may often be less than 50 pA. However, as mentioned earlier for the inside-out configuration, attempts can be made to prevent the time-dependent decline of a current due to run-down by providing compounds suspected to be responsible for its regulation. It is worth mentioning that as easy as it may be to obtain a giga-seal in smooth muscle cells, the success rate of rupturing the membrane and maintaining the seal and access is dramatically lower. Even by comparison with other cell types that we have had experience with (cultured fibroblast cells, cardiac myocytes from guinea-pig, rabbit and rat), single vascular smooth muscle cells are significantly more frustrating to whole-cell clamp (besides the additional frustration of attempting to isolate Ca^{2+}-tolerant myocytes). One should not be discouraged if only one or two successful experiments can be carried out per day.

One interesting feature of the whole-cell recording mode is that it can be used to record the resting membrane potential of the cell in current clamp mode. However, one needs to be cautious when interpreting RMP with this technique. Because of dialysis, the recorded RMP is most likely different from the true RMP of the cell before membrane rupture and rather reflects the resting electrical properties of the membrane exposed to a controlled ionic and biochemical environment. Another particular problem specifically related to smooth muscle cells is that their input resistance (R_{inp}; 300 MΩ-20 GΩ) is sometimes within the same order of magnitude as the seal resistance (5-50 GΩ). This will result in shunting of transmembrane voltage across the seal resistance leading to a measured RMP that is more depolarized than its true value. For example, if the seal resistance = 5 GΩ and R_{inp} = 1 GΩ, and if the true RMP of a cell is -50 mV, the measured RMP would be shunted by ~+ 8 mV (-42 mV). Therefore small changes in seal resistance during the course of an experiment can result in RMP changes that are unrelated to the experimental conditions. Almost all smooth muscle patch clampers will agree that whole-cell current clamp experiments are a nightmare. Current clamp recordings can be unstable and it is not rare to observe slow oscillations of 5-20 mV in amplitude.

One important necessity to adequately record whole-cell ionic currents is to meet the criterion of "space clamp" or the condition of homogenous spatial distribution of transmembrane voltage under voltage clamp. In general, this condition can be easily met for small spherical cells (10-20 μm or less). However, this could become a problem when attempting to voltage clamp long (> 100 μm) and thin (5-10 μm) smooth muscle cells. Since the length constant of these cells have been determined to be on the order of several millimeters, voltage control would be considered adequate for membrane potentials at rest or even mild depolarizations as the length constant would be several-fold longer than the length of the cell. However as the membrane is further depolarized, outwardly rectifying K^+ currents become activated which will decrease R_{inp} and consequently the length constant leading to a potential loss of voltage control and spatial discrepancy of transmembrane voltage distribution. This is likely to be significant in very long smooth muscle cells such as found in the gut which are commonly longer than 200 μm. One way to minimize this problem is to patch clamp the cell at its centre (near the nucleus). Spatial homogeneity may be tested by conducting a double whole-cell experiment in the worst case situation: one patch electrode positioned at one end to record membrane voltage, and a second patch electrode at the other end to voltage clamp the cell.

A second problem during whole-cell voltage clamp experiments is the presence of a resistance in series (R_s) with membrane resistance (R_m) which results from the sum of two resistive components: the pipette resistance (which is measured in the bath before patch clamping the cell) + the access resistance (R_{acc}); the latter being the sum of the resistance to dialysis (passage of current at the tip opening) and internal resistivity (cytoplasm) of the cell. The presence of R_s introduces an error (V_{err}) in the voltage applied to the cell which is proportional to the magnitude of membrane current (I_m): $V_{err} = R_s I_m$; when recording a Ca^{2+}-activated K^+ current of 4 nA at +40 mV with an uncompensated R_s of 5 MΩ, V_{err} would be as high as 20 mV. In addition, as for simple electrical RC circuits, the introduction of a resistance in series with the parallel arrangement of R_m and C_m will slow down the charging of C_m during a voltage clamp step (fast current transients observed following step changes in voltage). Upon application of a voltage clamp command pulse (V_c) applied via the pipette, the actual membrane potential (V_m) will reach its final value according to the following exponential distribution: $V_m = V_c[1 - \exp(-t/\tau_s)]$, where τ_s is defined by the product $R_s C_m$. For a smooth muscle cell with C_m = 50 pF and R_s = 10 MΩ, τ_s = 0.5 msec. If the cell is clamped from -50 mV to +50 mV, V_m would reach 95% of its final value in 1.5 msec. This would of course be limiting if fast time-dependent currents are to be resolved (as is the case for K_{Ca} at positive potentials). All modern patch clamp amplifiers have built-in series resistance compensation circuitry that enables compensation of 80-95% of R_s. This procedure minimizes the R_s-induced voltage error, and improves C_m charging time and the effective frequency bandwidth of the recorded ionic current. The only way of reducing R_{acc} and therefore R_s is to employ lower resistance pipettes bearing the appropriate geometry.

Another method of whole-cell recording called the "Perforated Patch Method" (fig. 4D) was recently introduced by Horn and Marty (7). This method takes advantage of the ionophore or pore forming properties of two polyene antibiotics, namely nystatin and amphotericin, which are included in the pipette solution and serve to gain intracellular access. After formation of a cell-attached patch (fig. 4B), the antibiotic slowly diffuses to the invaginated semi-spherical membrane (fig. 4C) where it begins to partition in the membrane by forming channels that are permeant to small mono-valent ions and uncharged molecules < 0.8 nm (6), and impermeant to multivalent ions such as Ca^{2+} and Mg^{2+}; the polyene-induced pores are about nine times more permeant to cations such as Na^+ and K^+ when compared to Cl^-. How can intracellular access be monitored? As for the standard whole-cell mode, repetitive voltage clamp pulses ($\pm$ 10-20 mV; 5-20 msec in duration) are applied to the cell attached patch. As partitioning of the antibiotic occurs, a slow capacitive current (I_C) produced by the charging of C_m slowly appears reflecting access to the cell's interior. As the number of pores formed increases with time, I_C becomes larger and its time course faster indicating a decrease in R_{acc}. A steady-state is usually reached within 20-30 min after seal formation. By optimizing the geometry of the pipette tip both in size (the largest pipette tip that allows formation of a gigaohm seal) and shape (one that favours the typical Ω shape of the invaginated membrane), it is possible to routinely obtain values of R_s that are comparable to the standard whole-cell technique (< 10 MΩ). The perforated patch technique presents several advantages over the standard method: (i) because enzymes and second messenger systems are not washed out, it slows down and often eliminates run-down of ion channels that are regulated by cell signalling mechanisms; (ii) once partitioning is complete, R_{acc} is more stable so that R_s and C_m compensations are more easily achieved; (iii) longer whole-cell recordings are possible (up to 3 hrs) and this is particularly true for smooth muscle cells; (iv) since divalent cations are not dialysed, homeostasis of ions such as Ca^{2+} or Mg^{2+} can be studied with fluorescent probes (Fura2/AM, Indo-1/AM, MagFura/AM) simultaneously with whole-cell currents with little perturbation of the ion transporters that regulate their activity. There are two main limitations of using this variant: (i) longer times are required to obtain whole-cell access and when a lot of measurements are quickly needed, the standard configuration would be a better choice; (ii) since large molecular weight substances can not be provided to the cytoplasm, it is then not possible to investigate the role of a key regulator suspected to influence the activity of an ion channel.

Useful information have been obtained with both whole-cell methods on the ionic channels that generate the electrical activity of smooth muscle cells. It is fair to say that they are the most popular variants of the patch clamp technique used because they allow one to rapidly acquire and analyze macroscopic (average behaviour of all the channels present in the membrane) ionic currents in single cells. The choice of one variant over the other will depend on a compromise between the need to control the intracellular milieu (standard) and preserve the cytoplasm (perforated patch).

The final configuration which also permits the recording of individual channels is the excised patch mode called "Outside-Out" (fig. 4F). It is the most difficult configuration to obtain and certainly the least stable in terms of maintaining a constant gigaohm seal. It can be obtained from the standard whole-cell recording mode (fig. 3D) by slowly withdrawing the pipette from the cell, provided that the latter is well attached to bottom of the chamber. The principle is rather simple. In whole-cell mode, the membrane is adhering to the internal side of the pipette forming a funnel-like structure. As the pipette is lifted upward, the tunnel part of the funnel narrows as the pulling force extends the continuous membrane and whole-cell access decreases. During this entire process, the outer leaflet of the membrane, which forms the outer surface of the tunnel is in direct contact with the bathing solution. At some critical point when the tunnel becomes very thin, the membrane reseals itself exposing the outside of the membrane to the bathing fluid. This technique is suitable for studying the action of transmitter substances that bind to surface receptors, and pharmacological agents or ions that influence ion channels when applied from the external side.

INFORMATION CONTAINED IN PATCH CLAMP DATA

Channels have been categorized in two general classes: (i) channels that are influenced by the transmembrane electrical field and are called "voltage-gated" or "voltage-operated" (e.g. in smooth muscle, T- and L-type Ca^{2+} channels, nearly all K^+ channels, and many chloride channels); (ii) channels that are activated by binding of an agonist to a receptor somewhat coupled to an ion channel are named "ligand-gated" or "receptor-operated" (ROC) channels (e.g. non-selective cation channels that are stimulated by purinergic, adrenergic or cholinergic receptors); this definition should perhaps be broadened to include physical factors as potential ligands as many types of stretch-activated ion channels have been recently identified in smooth muscle cells. Moreover, most voltage-dependent ion channels are now known to be modulated by protein kinases in response to activation of several types of receptors (e.g. many are coupled to L-type Ca^{2+} channels in smooth muscle) or by second messengers or ions (K_{Ca} is both activated by depolarization and $[Ca^{2+}]_i$).

There are two processes that control the flow of ions across a single channel pore. The first process is the permeation pathway determined by the three-dimensional structure of the protein's amino acid chain sequences which affect the size, geometry and net charge of its permeating pore. These features in turn determine the rate of ion flux when the channel is in the open state (see fig. 6B; amplitude of unitary current) and act as a filter to select which ion passes through it. By analogy, the permeation property could be viewed as a door separating two rooms. The door could be of variable size and hence allow more people to go through it (rate of exchange or conductance); in addition, it could delineate the shape of a human body and thus prevent other creatures from traversing it (selectivity).

The second property of the channel is the process of opening and closing of the channel. Because ion channels are proteins that obey the rules of thermal molecular motion, their conformation switches randomly between two discrete levels (all-or-none) of conduction: an open state (O) allowing passage of ions or current flow and a closed state (C) that prevents ion flux (fig. 6A). Coming back to the door example, the gating mechanism could be visualised as a revolving entrance door seen in many buildings. Because of the spinning action, people could enter the building only at discrete times and at a rate that would depend on spinning velocity. If the spinning velocity was not constant but varying with time, the number of people entering per cycle would vary.

What makes ions move across a channel? Despite the fact that an ion channel might be in the open state, if there is no driving force or electrochemical energy applied on the ions, no net current flow will ensue. The driving force is defined by the expression (V_m - E_{ion}), where V_m is the transmembrane potential and E_{ion} is the equilibrium potential of an ion as defined by the Nernst equation: E_{ion} = RT/zF log ($[X]_o/[X]_i$) where R is the gas constant, T the absolute temperature, z the valence of the ion, and F the Faraday constant, and $[X]_o/[X]_i$ the ratio of extracellular to intracellular concentrations of the ion X. At 37°C, the term RT/zF is a constant which is ~equal to 61 mV. Therefore E_{ion} depends only on the concentration gradient and represents the electrical energy in Volts that would be required to maintain the gradient constant. When V_m is forced away from E_{ion} (as is the case under voltage clamp condition), net current can be recorded which is produced by ion flux through all the channels in the membrane (whole-cell), or through a single channel in a membrane patch.

Example of a Whole-Cell Experiment in a Vascular Smooth Muscle Cell

When using the standard or perforated patch whole-cell recording mode, many parameters can be measured. As mentioned above, the resting membrane potential of the cell can be obtained in current clamp mode. This is a useful indicator of the "physiological" electrical behaviour of the cell but it gives little information on the ion channels that generate these electrical potentials. The voltage clamp technique on the other hand allows the recording of ionic currents by clamping the cell at a desired voltage (fig. 5A). During voltage clamp pulses (bottom family of traces in fig. 5A), membrane current can be recorded as a function of time at each voltage applied. From the initial capacitive current (very fast spikes; I_C) resulting from charging of the membrane capacitance, a good approximation of the cell surface can be obtained and this is because the cell membrane bears properties similar to that of a simple capacitor with the insulating lipid bilayer interfacing two conductive media (cytoplasm and bath solution). Since the strength of a capacitor is directly proportional to the charging surface, cell surface can be easily obtained by measuring C_m. Electrophysiological experiments have determined that the specific capacitance of biological membranes is relatively uniform from cell to cell and is $\approx 1 \ \mu F/cm^2$ (F or Farad is the unit of capacitance and is equal to 1 coulomb/volt). Membrane capacitance and indirectly cell surface can be estimated by integrating I_C. This is useful because it allows measurement of current density (often expressed in picoamperes/picofarad for a single cell) which enables comparison of ionic currents in cells of different sizes.

Whole-cell and single channel experiments have confirmed that K^+ channels predominate in smooth muscle cell membranes (9). Figure 5 gives an example of a standard whole-cell experiment carried out on a single rabbit coronary artery smooth muscle. The pipette solution contained "normal" concentrations of ions that approximate the cell content: 140 mM K^+, 10 mM Na^+, 40 mM Cl^-, and 5 mM Mg.ATP at pH 7.2. The cell was superfused with a standard Hepes buffer (pH 7.4) and was held under voltage clamp at a holding potential of -60 mV. A two step protocol was used as depicted in figure 5A (bottom family of traces). Two hundreds and fifty millisecond steps were imposed in sequence from -100 mV to +40 mV in 10 mV increment steps. After the initial pulse, membrane potential was stepped to -25 mV to record a tail current (see text below). Very small currents could be recorded in this cell at potentials from -100 mV to -30 mV (traces are virtually overlapping on this scale). These currents did not appear to vary with time during the pulse. However, for step potentials positive to -30 mV, time-dependent currents were activated with their onset becoming faster as the membrane was made more positive. Upon stepping down to -25 mV, steps that triggered time-dependent currents were associated with the appearance of currents decaying with time, so called tail currents, which reflected the reverse process of activation, that is deactivation or more simply channel closure. From examination of the traces shown in panel A, it can be concluded that the underlying ion channels that generate these macroscopic currents are clearly voltage-dependent. One way of showing this is to construct a current-voltage (I-V) relationship as shown in panel B. Membrane current was measured at the end of the first pulse and plotted as a function of the command voltage pulse. By convention, an outward current which reflects the movement of positive charges from inside to outside (like K^+) is always plotted positive. The resulting I-V relationship is typical of many smooth muscle cells and exhibits outwardly rectifying properties; that is current flow in the outward direction is favoured over current flowing in the inward direction. It is also evident from panel B that from -100 mV to ~-30 mV, membrane conductance is very low (and therefore the input resistance of the cell is very high) which suggests that few

ANALYSIS OF WHOLE-CELL CURRENT

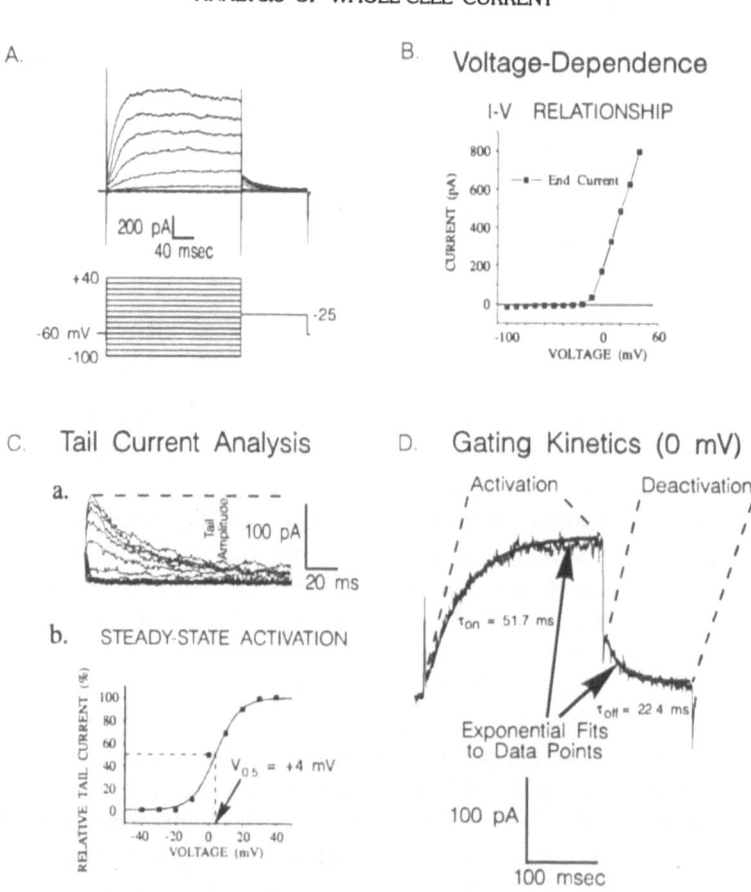

Figure 5: Example of analysis of whole-cell K^+ currents recorded from a single rabbit coronary artery smooth muscle cell. (A) Family of whole-cell currents (top traces) elicited by a series of voltage clamp pulses from a holding potential of -60 mV (bottom traces). (B) Current-voltage (I-V) relationship constructed by plotting the amplitude of the current measured at the end of the initial test pulse as a function of the command voltage. (C) a. Deactivating tail currents evoked at -25 mV (panel A) were redrawn on expanded scales and measured as indicated; b. construction of a steady-state activation curve based on normalization of the tail currents shown in a; the line passing through the data points is a least-square fit to a Boltzmann distribution; as indicated, half-maximal activation ($V_{0.5}$) occurred at +4 mV. (D) Analysis of kinetics of activation and deactivation at 0 mV. Activation (τ_{on}) at 0 mV and deactivation (τ_{off}) at -25 mV were well fitted by single exponential functions (lines passing through the current traces).

channels are open near RMP. This is very typical of many smooth muscle cells. Beyond -30 mV, ion channels become activated by membrane depolarization and current flow increases in a non-linear fashion.

Panel C shows an example how tail current analysis can provide additional information about the steady-state voltage-dependence of the underlying channel. The tail currents recorded at -25 mV in panel A were redrawn in panel C-a on an expanded scale. The amplitude of the tail currents can be measured and normalized against the highest amplitude ($\{tail/tail_{max}\}*100$)), and plotted as a function of voltage during the initial step (panel C-b). A steady-state activation curve can thus be constructed. The line passing through the data points is a least-square fit to a Boltzmann distribution with its parameters ($V_{0.5}$ and slope of the relationship) reflecting the intrinsic voltage-dependent properties of the fully activated ion channel. This relationship signifies that the channel under investigation (provided that it is composed of only on type of channel) is closed for potentials below -30 mV and is fully activated (maximum open probability) at potentials beyond +30 mV. Along with other tests (pharmacology, single channel data, kinetics), we have determined that this time-dependent current has properties similar to delayed rectifier K^+ channels identified in many smooth muscle cells.

Analysis of whole-cell current also allows assessment of the kinetics of the ionic current. Shown in figure 5D is a reproduction of the outward current elicited at 0 mV as shown in panel A. The two lines passing through the trace are least-square exponential fits with calculated time constants of activation (τ_{on}) and deactivation (τ_{off}) of 51.7 and 22.4 ms, respectively. In this example, the two processes were well fitted by a single exponential function which could indicate a simple scheme of transitions from closed to open states.

Many other voltage clamp protocols can be used to assess the ionic selectivity, conductance, and dynamic behaviour of time-dependent and time-independent whole-cell currents. The whole-cell recording mode offers an easy and quick way of investigating the mechanisms of drug action on ion channels. Most substances targeted to alter ion channel function usually inhibit or enhance their activity by modulating their gating mechanism although compounds may also influence their permeation (as is the case for tetraethylammonium chloride or TEA which inhibits K_{Ca} by reducing its single channel conductance with little effect on gating). By similar types of analyses as that exemplified in figure 4 and others, it is often possible to determine the mechanism of drug action. However, one must remember to interpret whole-cell data with caution because the analysis relies on the assumption that the particular ion channel of choice has been effectively isolated from all other channels by appropriate ion replacements, pharmacological tools and/or specific voltage clamp protocols. The final test should be the reconstruction of the macroscopic current from isolation of the suspected ion channel as can be measured using cell-attached and excised patch methods.

Example of Analysis of Single Channel Data

Figures 6 and 7 give an example of a single channel experiment and the kind of information that can be derived from membrane patch unitary current measurements. In this experiment, the pipette solution contained 150 mM K^+. Figure 6A shows continuous recording of BK channel activity (K_{Ca}; the most abundant K^+ channel in smooth muscle cells) at three different pipette potentials in a cell-attached patch from a single smooth muscle cell from the rabbit left descending coronary artery. Since in this experiment membrane potential is held constant, ion channel activity is recorded in the steady-state and the ensuing type of analysis is called "stationary analysis" (2,12).

In this patch, there appears to be a minimum of two "identical" BK channels jumping randomly from closed (lowest level) to open state (dashed lines). Of course

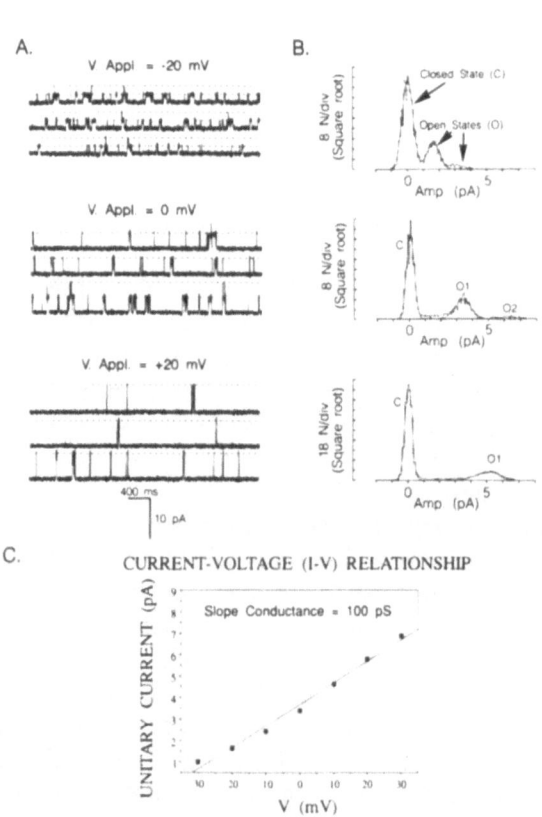

Figure 6: Example of a cell-attached patch experiment conducted on a rabbit coronary artery myocyte. As indicated, the pipette solution contained 150 mM K^+. The cell was bathed in a normal Hepes-buffered solution containing 5.4 mM K^+ and 1.8 mM Ca^{2+}. (A) Continuous recording of single Ca^{2+}-activated K^+ channels at three different potentials (V. Appl.) as illustrated. Channel activity at 0 mV reflects the contribution of K_{Ca} at the resting membrane potential. For V. Appl. = -20 mV and +20 mV, the patch was respectively depolarized (inside membrane more positive) and hyperpolarized (inside membrane more negative). (B) All points amplitude histograms generated from the corresponding single-channel recording shown in A. Notice that the number of events are plotted on a square root ordinate. For each histogram, the lines passing through the bins are least square fits to a double Gaussian distribution. (C) Construction of the current-voltage (I-V) relationship by plotting the amplitude of single-channel currents estimated from the histograms as a function of V. Appl. The data points were fitted by linear regression yielding a slope conductance of 100 pS which is an index of the permeation process of K_{Ca} under these conditions.

at one given command potential (for instance -20 mV), the probability of observing simultaneous opening of the two channels is significantly lower than a single opening from either channel. Assuming that there are only two channels in the patch, behaving independently but having equal open probabilities (P_{open}; fraction of time spent in the open state), then the probability of observing simultaneous opening of the two channels would be defined by $(P_{open})^2$.

Two differences can be noticed when the pipette potential is changed. First the amplitude of the discrete unitary levels increases with membrane hyperpolarization (since voltage is applied to the outside of the membrane, application of positive charges outside translates into accumulation of negative charges inside or hyperpolarization). As mentioned earlier for the cell-attached configuration, the patch transmembrane voltage is the sum of the applied voltage + RMP. In the example, clamping the patch at 0 mV (middle traces of fig. 6A) means that in fact it is clamped at RMP and therefore gives an indication of the contribution of K_{Ca} at RMP. Assuming that RMP $\approx$ -50 mV and $[K^+]_i \approx$ 150 mM, then the predicted equilibrium potential for K^+ would be near 0 mV. As the patch is hyperpolarized (V. appl. = +20 mV), the driving force for K^+ increases and so is the single channel amplitude. Secondly, channel openings were less frequent and briefer with membrane hyperpolarization. Indeed, at V. Appl. = +20 mV, double openings could no longer be observed indicating a decrease in P_{open}.

Because of the stochastic nature of single channel events, not all openings have the same amplitude. Measurement of single channel amplitudes can be done by generating amplitude histograms as shown in figure 6B. At the three potentials tested, there were longer periods of time in which there were no openings of either channel thus explaining the large peak at 0 pA (C). At -20 mV and 0 mV, two open states (O1 and O2) were evident but the relative magnitude of each peak was higher at -20 mV indicating a higher P_{open}. This was even more evident at +20 mV where a single open state (O1) could be detected with the histogram. Another feature of these graphs is that the distance between the peaks (or amplitude of current in pA) increased with hyperpolarization of the patch (from V. Appl. = -20 mV to +20 mV). The lines passing through the histograms are Gaussian fits of which parameters determine the mean signal channel amplitude at each potential. Mean peak current can then be plotted as function of patch potential to construct an I-V curve such as the one displayed in panel C. In this patch the slope of the I-V relationship or "slope conductance" of the channel can be derived by linear regression. As shown, the slope conductance of the channel was 100 pS. This is a direct index of the permeation properties of the channel. Slope conductance and reversal potential measurements of the unitary currents in inside-out patches with different ion species on both sides of the membrane would provide additional information on its permeation properties and selectivity. Amplitude histograms can also be used to calculate open state probability using the following equation:

$$NP_{open} = (a_1 + 2a_2 + 3a_3 + ... + na_n)/(a_0 + a_1 + a_2 + a_3 + ... + a_n) \qquad (1)$$

where NP_{open} is the total number of channels in the patch (N) times open probability (P_{open}), a_1, a_2, a_3,..., and a_n are the areas under each peak reflecting the discrete open levels (for instance O1 and O2 in fig. 6B), and a_0 the area under the closed state peak.

So far, we have only dealt with methods to gain information about the permeation process of an ion channel but have not discussed how to extract information on its gating properties or kinetics. Despite the fact that there are only two measurable conducting states (open or close), a channel protein may in fact transit between several more stable conformational states with each transition involving a jump step over an energy barrier. Despite the fact that these conformational states are not directly observable, they become evident when long periods of single channel activity are monitored over time at a constant voltage. By analogy to equilibrium chemical reactions, the stochastic behaviour of single channel proteins can be analyzed by simple kinetic modelling. In the simplest case of a channel with single closed (C) and open states (O), the state diagram would be as follows:

$$C \underset{\alpha}{\overset{\beta}{\longleftrightarrow}} O \qquad (2)$$

where α and β are the rate constants of the forward and reverse transitions. Models of channel kinetics can be constructed using several assumptions about stochastic processes (2,12): (i) channel lifetimes are much longer than the time required for channel opening or closing; (ii) once the channel is in a given state, the probability that a channel transit to the alternate state depends only on the current state, not its past history (Markov process); (iii) the rate of transition from one state to another depends only on the rate constant (α or β) which is assumed to be constant at a given temperature and voltage (in the case of a receptor-operated channel, the concentration of the ligand would be the determining factor). The consequence of these assumptions is that for any given state, the lifetimes distribution is described by a probability density function the resolution of which is a single exponential function. For the simple reaction scheme depicted above (Equation 2), the rate constant α can be derived by determining the time constant τ, or "mean open time", of the open lifetimes distribution which is represented in a frequency histogram.

Unfortunately, the simple reaction scheme (Equation 2) is more the exception than the rule for most ion channels as they may have more than one open and closed states and sometimes one or more inactivated states (fast TTX-sensitive Na^+ channels, L-type Ca^{2+} channels, several K^+ channels). Figure 7 gives an example of kinetic analysis of single channel data. The analysis was performed on the same data as that shown in figure 6. Frequency histograms of open (a) and closed (b) times were constructed. For each histogram, the lines are least-square fits for single or double exponential functions from which the time constants can be obtained (τ_f and τ_s). These time constants or mean open or close times can then be plotted as a function of voltage to examine their voltage-dependence (fig. 7B). Ideally, kinetic analysis should be done on patches containing only one channel. When simultaneous openings of ion channels occur, it is not known which of the channel closes first. In addition, many computer programs used for discriminating between openings and closings will discard multiple opening events which may yield an erroneous estimation of τ_{open} especially when the open probability is high. In the current example, the data would suggest that these BK channels have at least two open and two closed states of which the derived rate constants will be highly voltage-dependent (fig. 7B). Therefore, a more complex scheme would have to be proposed to explain the gating mechanism of these K^+ channels. However, more than one scheme would be possible:

$$C_1 \longleftrightarrow C_2 \longleftrightarrow O_1 \longleftrightarrow O_2 \qquad (3)$$

$$C_1 \longleftrightarrow C_2 \longleftrightarrow O_1$$
$$\updownarrow \qquad\qquad (4)$$
$$O_2$$

$$....... \qquad (5)$$

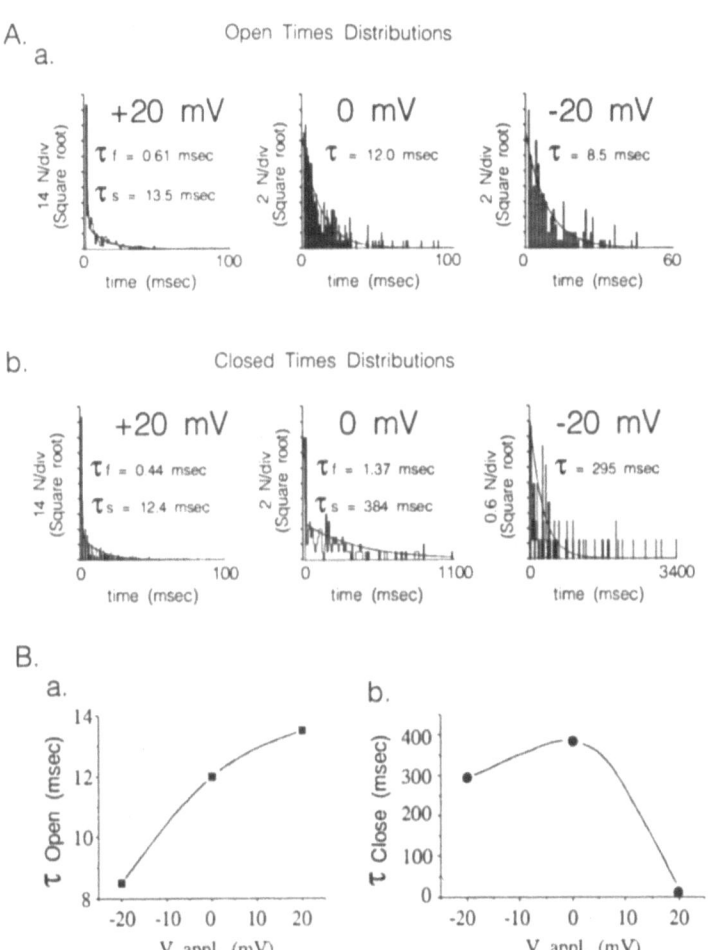

Figure 7: Example of extraction of gating kinetics of single channel currents. (A) Open (a) and close (b) times distributions at three different potentials (true membrane polarity) were compiled by analyzing the single channel data depicted in figure 6. Each histogram was fitted by a single or double exponential distribution of which time constant(s) (τ_f or fast; τ_s or slow) reflect(s) the mean open or close lifetime(s). As for figure 6, the number of events are plotted on a square root ordinate. (B) Voltage-dependence of τ_s of open (a) and close (b) states.

From rate constant determinations, it would be possible to discard many possibilities as predictions for each reaction scheme can be made and tested experimentally. This type of analysis along with ion transfer properties analysis (fig. 6), are extremely useful to identify a particular ion channel and characterize its functional molecular characteristics. Most drugs and hormones modulate ion channel activity by altering the rate constants of transitions between the different conformational states. These kinetic changes induced by various agents which are often time- and voltage-dependent can be accurately described by analysis of single channel data. Of course ion channel activity is often more complicated than that pictured in this chapter (for instance many channels open in bursts, despite the fact that voltage is maintained constant throughout the experiment) and more sophisticated analytical routines are required to extract the desired information. For more information on the analysis of patch clamp data, the reader should consult the following textbook chapters (2,12).

Correlation Between Whole-Cell and Single Channel Currents

In the whole-cell experiments shown in figure 5A, the current traces were very smooth and this is because they are generated by the opening and closing of a large population of ion channels so that it is not possible to detect the discrete levels of the underlying single channel events. Another difference was that voltage was stepped from a holding potential to various levels to examine the time and voltage-dependence of the delayed rectifier K^+ current. A similar type of analysis could be done at the single channel level by clamping a membrane patch at a holding voltage and stepping to a new level. If one or more channels are present and they are voltage- and time-dependent, they might open with a delay after the pulse, and then close and reopen several times during the pulse. By applying repetitive pulses to the same voltage, each sweep will show a different pattern of opening and closing. By averaging 100-200 sweeps to the same voltage (only one sweep is required when recording a whole-cell current), a procedure called "ensemble averaging", it is possible to reconstruct a "macroscopic" current of which the time constant of activation (and possibly inactivation) can be derived, compared and correlated to a whole-cell macroscopic current recorded under similar conditions. The mean calculated current reflects time-dependent changes in open state probability after a step change in voltage. This type of analysis may be used to test the various kinetic models derived from stationary analysis. Non-stationary analysis is essential to learn about channels that undergo inactivation (e.g. TTX-sensitive Na^+ channels. L-type Ca^{2+} channels) as channel openings would not be detected if the patch voltage was held constant as in figure 6.

Hodgkin and Huxley (for a review, consult ref. 5) recorded and analyzed macroscopic Na^+ and K^+ currents in the giant squid axon and showed that the time-dependent current changes could be fitted with the following general equation:

$$I_i\ (t) = f_{(t)} \times G_{max} \times (V - E_{ion}) \tag{6}$$

where $I_i\ (t)$ is the magnitude of the current at time t after the onset of the pulse, $f_{(t)} \times G_{max}$ is the fraction ($0 \leq f \leq 1$) of maximum conductance (G_{max}) of the ionic current at time t, and ($V - E_{ion}$) the driving force. In this case, the dynamic behaviour of the ionic current would only be determined by the value of f since at a constant test potential, the terms G_{max} and $(V - E_{ion})$ would be constant; in others words, f would be reflecting the time-dependent changes of open probability of the sum of all the channels in the membrane.

Using a similar concept, a macroscopic current derived from single channel activity would be as follows:

$$I_i(t) = N \times P_{open}(t) \times i \qquad (7)$$

and
$$i = g_{ion} / (V - E_{ion}) \qquad (8)$$

where N is the total number of functional channels present in the membrane, $P_{open}(t)$ is the channel open probability at time t, i the amplitude of the unitary current (pA), and g_{ion} the slope conductance of the channel (fig. 6C). At constant voltage, P_{open} would be the only time-dependent variable.

CONCLUSIONS

Since the advent of the patch clamp technique, tremendous progress has been made in identifying and characterizing the structure and function of ion channels in smooth muscle. The main goals of this chapter were to provide a simple description of the various configurations of this technique, and the type of information that can be extracted from patch clamp data.

The patch clamp technique is a powerful tool that allows the accurate study of the biophysical properties and/or modulation of voltage-activated or ligand-gated ion channels present in native membranes of freshly dispersed or cultured smooth muscle cells. Combined with other techniques (ion microfluorometry, "flash-photolysis", biochemical techniques, etc.), it can be used to assess the role of a particular ion channel in smooth muscle excitation-contraction coupling.

Many challenges are awaiting in the field of ion channels in smooth muscle. Future studies will identify the physiological role of the channels described so far and identify novel channels and their regulation by neurotransmitters, hormones and endothelial factors. These studies will also provide information on more specific agonists and antagonists, especially for K^+ and Cl^- channels, identification of the genes coding for the identified channels and how they are expressed during development, and delineate the properties of ion channels present in the membrane of cytoplasmic organelles. In the following chapters, expert researchers experienced with the techniques outlined in this chapter and others will provide novel information about the structure, function, and regulation of ion channels identified in smooth muscle, endothelial and other cells, and how this information can advance our understanding about the physiology and pathophysiology of pulmonary vasculature.

Acknowledgements

This work was supported by a Grant-In-Aid from the Medical Research Council of Canada. X. Wan was supported by a Faculty Fund (Medicine), University of Manitoba. N. Leblanc is a MRC Scholar of Canada.

References

1. Brown, A.M., and L. Birnbaumer. Direct G protein gating of ion channels. Am. J. Physiol. 254: H401-H410, 1988.
2. Colquhoun, D., and A.G. Hawkes. The principles of the stochastic interpretation of ion-channel mechanisms. In: "Single-Channel Recording", ed. by B. Sakmann and E. Neher, Plenum Press, New York, 135-175, 1983.
3. Franciolani, F. Patch clamp technique and biophysical study of membrane channels. Experentia 42: 589-594, 1986.
4. Hamill, O.P., Marty , A., Neher, E., Sakmann, B. and F.J. Sigworth. Improved patch-clamp techniques for high-resolution current recording from cells and cell-free membrane patches. Pflügers Arch. 391: 85-100, 1981.
5. Hille, B. "Ionic Channels of Excitable Membranes", Sinauer Associates, Sunderland, Massachusetts, 1984.
6. Horn, R. and S.J. Korn. Prevention of rundown in electrophysiological recording. In: "Ion Channels", Methods in Enzymology, vol. 207, ed. by Rudy, B., and L.E. Iverson, Academic Press, New York, 149-155, 1992.
7. Horn, R., and A. Marty. Muscarinic activation of ionic currents measured by a new whole-cell recording method. J. Gen. Physiol. 92: 145-159, 1988.
8. Johnson, E.A., and M. Lieberman. Heart: excitation and contraction. Ann. Rev. Physiol. 33: 479-532, 1971.
9. Kirber, M.T., R.W. Ordway, Clapp, L.H., Sims, S.M., Walsh Jr., J.V., and J.J. Singer. Voltage, ligand, and mechanically gated channels in freshly dissociated smooth muscle cells. In: "Potassium Channels: Basic Function and Therapeutic Aspects", ed. by T.J. Colatsky, Alan R. Liss, Inc., New York, 123-143, 1988.
10. Klöckner, U., and G. Isenberg. Action potentials and net membrane currents of isolated smooth muscle cells (urinary bladder of the guinea-pig). Pflügers Arch. 405: 329-339, 1985.
11. Neher, E., and B. Sakmann. Single channel currents recorded from membrane of denervated frog muscle fibres. Nature 260: 799-802, 1976.
12. Patlak, J. The information content of single channel data. In: "Membranes, Channels, Noise", ed. by R.S. Eisenberg, M. Frank, and C.F. Stevens, Plenum Press, New York, 197-234, 1984.
13. Plonsey, R., and R.C. Barr. Membrane biophysics. In: "Bioelectricity - A Quantitative Approach", Plenum Press, New York, 165-203, 1988.
14. Rae, J.L., and R.A. Levis. Glass technology for patch clamp electrodes. In: "Ion Channels", Methods in Enzymology, vol. 207, Plenum Press, New York, 66-92, 1992.
15. Rudy, B., and L.E. Iverson, eds. "Ion Channels", Methods in Enzymology, vol. 207, Academic Press, New York, 1992.
16. Sakmann, B., and E. Neher, Eds. "Single-Channel Recording". Plenum Press, New York, 1983.
17. Sakmann, B., and E. Neher. Geometric parameters of pipettes and membrane patches. In: "Single-Channel Recording", ed. by B. Sakmann and E. Neher, Plenum Press, New York, 37-51, 1983.
18. Standen, N.B., Quayle, J.M., Davies, N.W., Brayden, J.E., Huang, Y., and M.T. Nelson. Hyperpolarizing vasodilators activate ATP-sensitive K^+ channels in arterial smooth muscle. Science 245: 177-180, 1989.

20. Stevens, C.F. Biophysical studies of ion channels. Science 225: 1346-1350, 1984.
21. Supplement to Science. Guide to scientific products, instruments and services. Science 258, 41-159, 1992.
22. Talvenheimo, J.A. The purification of ion channels from excitable cells. J. Membr. Biol. 87: 77-91, 1985.
23. Trube, G. Enzymatic dispersion of heart and other tissues. In: "Single-Channel Recording", ed. by B. Sakmann and E. Neher, Plenum Press, New York, 69-76, 1983.

CALCIUM CHANNELS IN ARTERIAL SMOOTH MUSCLE CELLS

J.M. Quayle and M.T. Nelson

Department of Pharmacology
Medical Research Facility
University of Vermont,
55A South Park Drive
Colchester
Vermont 05446

INTRODUCTION

Maintained calcium influx into vascular smooth muscle through voltage-dependent calcium channels plays a key role in the control of the intracellular free calcium ion concentration ($[Ca^{2+}]_i$), and hence the contractile state of arterial smooth muscle cells (e.g. 12). Agents that depolarize (e.g. elevated extracellular potassium, pressurization, many vasoconstrictors) cause arteries to constrict. This tone is often reduced or abolished by removing extracellular calcium, organic calcium entry blockers or membrane hyperpolarization (11, 12, 13). Although other processes are clearly important in regulating vascular tone (for example regulation of intracellular calcium release, calcium extrusion, or the calcium-force relationship), these results strongly suggest that calcium entry through dihydropyridine-sensitive calcium channels is necessary for maintained tone development in many cases, and that the membrane potential dependence of arterial tone arises from the voltage-dependence of the calcium channel (12).

In this review we will outline some of our recent results on calcium channel permeation, voltage-dependent gating, pharmacology, and activation by vasoconstrictors. All studies were conducted on the dihydropyridine-sensitive calcium channel.

PERMEATION

The study of ion permeation through single calcium channels in arterial smooth muscle cells, as in other cells, has been limited by the small single channel currents and short lived openings of the calcium channel. Barium ions pass through calcium channels at a higher rate than calcium ions, and many single channel studies have used high concentration of barium as charge carrier to maximize the single channel current amplitude (e.g. 6, 11, 13, 15). In addition, a dihydropyridine calcium channel agonist (e.g. Bay K 8644 or Bay R 5417) is often present to prolong single channel openings. In these recording conditions, the single channel conductance lies in the range of 18 to 28 pS (e.g. 4, 11, 13, 15, 18), similar to that

Ion Flux in Pulmonary Vascular Control, Edited by
E.K. Weir *et al.,* Plenum Press, New York, 1993

reported for the channel in other muscle preparations (6, 14). Calcium passes through the channel less freely than barium, for example the single channel conductance in rabbit basilar arteries (negative to -20 mV) is 24.6 pS in 80 mM $[Ba^{2+}]$ and 15.1 pS in 80 mM $[Ca^{2+}]$ (18).

Earlier studies of unitary calcium currents through the dihydropyridine-sensitive calcium channel in heart cells indicated a single saturable binding site for calcium, with half saturation constant of 13.9 mM, predicting unitary calcium currents at physiological calcium concentrations of <0.1 pA, immeasurably small with available techniques (6). However, more recent data indicated that currents through the cardiac channel were surprisingly large at low barium concentrations (7 pS with 1 mM $[Ba^{2+}]$) (17). Recently, we have recorded single calcium channel currents in cerebral arteries at low divalent cation concentrations, and were able to resolve single channels in 2 mM calcium at physiological membrane potentials (between -60 and -30 mV) of between 0.3 and 0.1 pA (5).

Recording calcium current over the whole cell allows currents to be measured at physiological calcium concentrations in the absence of dihydropyridine agonists. For example, Figure 1A illustrates calcium current (2.6 mM $[Ca^{2+}]$) recorded from a cell from rabbit basilar artery on depolarizing from a holding potential of -70 mV to a test potential of 0 mV. Calcium entry into a single smooth muscle cell, measured as calcium current, will be the product of the single channel current (i), the number of channels (N), and the channel open probability (P_{open}). As the cell is depolarized, calcium current increases due to voltage-dependent activation of channels (see below), reaching a peak at around +10 mV in this cell (Figure 1B). Current declines at more positive potentials; channels are maximally activated, but the driving force for calcium entry, and therefore the single channel current, falls as the membrane potential approaches the electrochemical reversal potential for the calcium channel.

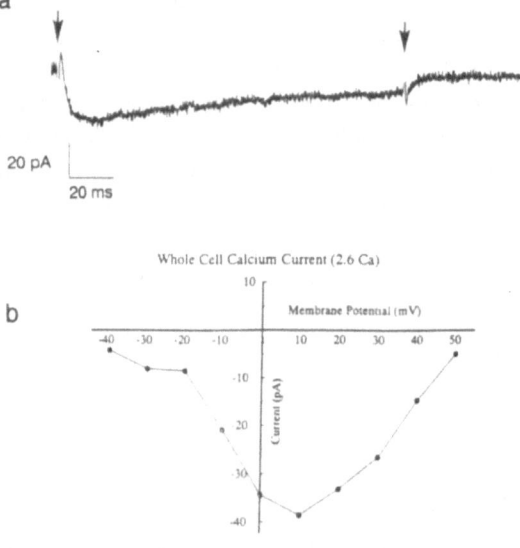

Figure 1. A: Calcium current from a rabbit basilar artery cell, recorded on depolarizing from -70 mV to 0 mV (indicated by arrows). The extracellular solution contained (in mM) :138 NaCl, 5.6 KCl, 1 $MgCl_2$, 2.6 $CaCl_2$, 10 HEPES, 1 4-Aminopyridine, 20 Tetraethylammonium Cl, pH 7.4. The pipette solution contained (in mM): 75 Cs_2SO_4, 40 CsCl, 7 $MgCl_2$, 10 HEPES, pH 7.2. Currents were recorded using the amphotericin permeabilized patch mode of whole cell recording. Series resistance was 15.5 Mohms, and cell capacitance was 15.1 pF. B: Peak calcium current-voltage relationship for the same cell.

GATING

The dihydropyridine-sensitive calcium channel is very sensitive to membrane potential, channel open probability (P_{open}) increasing exponentially with membrane depolarization over the physiological range of membrane potentials. Thus any physiological variable causing a depolarization (e.g. vasoconstrictors or pressurization) will open the channel, and any agent causing membrane hyperpolarization (e.g. vasodilators) will cause channel closure. To gain insights into the voltage-dependence of arterial smooth muscle tone it is important to understand the steady state properties of dihydropyridine-sensitive calcium channels.

We have recently analysed the voltage-dependence of calcium channels in rat and rabbit cerebral arteries (15, 18). Channel properties (80 mM barium, 0.5 μM Bay R 5417) were studied using a conventional approach, in response to depolarizing steps from a negative holding potential, and also in the steady state at different membrane potentials, which is presumably more relevant to conditions of steady state tone development in arteries. In response to voltage steps, channel open probability increases approximately e-fold for a 4.5 mV depolarization at membrane potentials negative to -20 mV. Channel activity was also recorded in the steady state, and channels showed similar voltage-dependence for pulsed and steady state data, though absolute levels of open probability were lower in the steady state (because of voltage-dependent inactivation).

The calcium channel can be considered to reside in a finite number of states, with rate constants governing transitions between states. Our single channel data in rat cerebral arteries is consistent with a simple linear scheme comprising two closed states, an open state, and an inactivated state.

$$C_r \underset{k_{-1}}{\overset{k_1}{\rightleftharpoons}} C_s \underset{k_{-2}}{\overset{k_2}{\rightleftharpoons}} O \underset{k_{-3}}{\overset{k_3}{\rightleftharpoons}} I$$

The voltage-dependence of channel activation arises mainly from transitions k_1 and k_{-1}. Inactivation is not inherently voltage-dependent, but does increase with depolarization, as a result of the voltage-dependence of channel activation (i.e. the channel must open before it can inactivate). Although such a scheme is clearly an oversimplification, and describes channel gating under unphysiological conditions, it is similar to that proposed in a recent review of calcium channels in muscle cells based on data from a wide variety of techniques (14).

Although these observations represent a first step in understanding the steady-state properties of calcium current in arterial smooth muscle cells, an issue of obvious importance for steady-state tone development, a better knowledge of the steady-state properties of calcium current in more physiological conditions is required. Whole cell calcium currents have been recorded in the steady state in single smooth muscle cells, but the exact relationship between membrane potential and calcium current in more physiological conditions remains to be determined (7).

PHARMACOLOGY

Dihydropyridine-sensitive calcium channels have a pharmacology which allows them to be distinguished from other classes of calcium channel, being inhibited by organic calcium entry blockers belonging to the dihydropyridine, phenylalkylamine and benzothiazipine classes, and by certain divalent cations (e.g. cadmium) (14).

Arterial calcium currents, both at the whole cell and single channel level, can be blocked by organic calcium entry blockers (13, 16, 18). Single channels (80 mM [Ba^{2+}], 1

μM Bay K 8644 or 500 nM Bay R 5417) in mesenteric and cerebral arteries are blocked by dihydropyridines with high affinity (13 ,18). Channel inhibition by dihydropyridines is complex, being dependent on the membrane potential. For example, in mesenteric arteries, the half inhibition constant for nisoldipine inhibition of single channels (80 mM [Ba^{2+}, 1 μM Bay K 8644) in response to a test pulse to 0 mV changes from 12.1 nM at a holding potential of -100 mV to 1.9 nM at a holding potential of -55 mV. This is because dihydropyridines bind the inactivated state of the calcium channel with highest affinity, and depolarization favours entry into the inactivated state. In mesenteric arteries, inhibition of single channels and of arterial tone induced by elevated extracellular potassium can be explained on the basis of a half inhibition constant of 0.07 nM for the inactivated state and 3 nM for the closed state(s) of the calcium channel. Single calcium channels (80 mM [Ba^{2+}], 1 μM Bay K 8644) in rabbit mesenteric artery are also inhibited by extracellular cadmium, with a dissociation constant of 36 μM at -20 mV, similar to that reported for the channel in cardiac myocytes (14, 19). Whole cell calcium currents recorded from single cells from rabbit pulmonary artery are also completely blocked by the dihydropyridine nifedipine (1 μM) and cadmium (100 μM) (2).

Although the ability of organic calcium channel antagonists such as the dihydropyridines to inhibit arterial tone implicates the voltage-dependent calcium channel in these contractions, quantitative data on inhibition of arterial calcium channels by calcium entry blockers is still lacking in many cases.

The dihydropyridine calcium channel agonist Bay R 5417 activates calcium channels in cerebral arteries (15, 18). Bay R 5417 (0.5 μM) has little effect on single channel amplitude, increases channel open times and shifts the voltage-dependence of channel activation by about 15 mV in the hyperpolarizing direction (15, 18).

MODULATION

The activity of dihydropyridine-sensitive calcium channels can be modulated by vasoactive agents. For example, noradrenaline activates channels in cells isolated from rabbit mesenteric and ear arteries and from rat portal vein (1, 10, 11). In mesenteric arteries, single channel recordings show that noradrenaline shifts the voltage-dependence of channel activation to more negative potentials (11). However, whole cell recordings from ear artery and portal vein suggest that noradrenaline may also increase whole cell current by increasing the number of functional channels, with no shift in the voltage-dependence of individual channels (1, 10). A shift in the voltage-dependence of activation of calcium channels in the hyperpolarizing direction also occurs for serotonin in rabbit basilar arteries (Figure 2; 18). In the example illustrated, channel open probability increased 18 fold at 0 mV on addition of 10 nM serotonin. Such a shift may be a general feature of calcium channel modulation in smooth muscle, for example, in rat bronchus the cholinergic agonist carbachol (10 μM) shifts the activation curve 9 mV in the hyperpolarizing direction without changing the steepness or maximum open probability of the channel (8).

Many vasoconstrictors activate phospholipase C, which will generate diacylglycerol which activates protein kinase C. Membrane permeable activators of protein kinase C (e.g. phorbol esters) increase calcium currents in vascular smooth muscle cells, and protein kinase C may therefore underlie activation of calcium channels by vasoconstrictors (e.g. 3, 10).

44

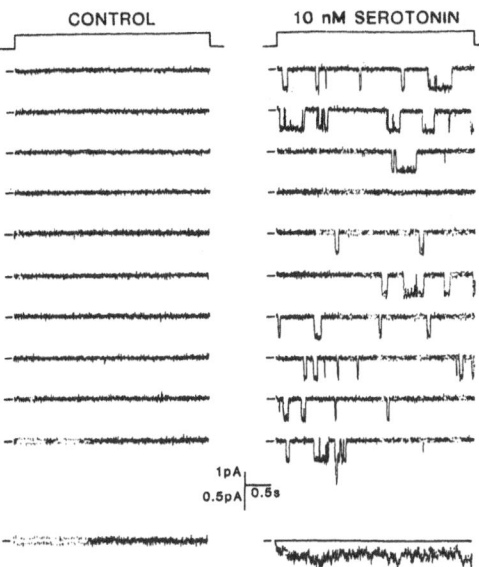

Figure 2: Serotonin activation of single channels (80 mM [Ba^{2+}], 0.5 μM Bay K 8644). Channels recorded on depolarizing from a holding potential of -70 mV to a test potential of 0 mV. Average currents are shown below original records. Reproduced, with permission, from Worley et al. (18).

CHANNEL SUBTYPES

Calcium channels in smooth muscle cells, as in other cells, have been sub-divided into a number of classes by voltage-dependence of channel gating, single channel properties (e.g. conductance), and sensitivity to calcium channel antagonists. Dihydropyridine-sensitive calcium channels have been detected in all vascular smooth muscle cells examined, and can be sub-divided into more than one component in some cells. For example, in rabbit basilar and mesenteric arteries in addition to the most frequently observed channel (conductance about 25 pS negative to -10 mV with 80 mM [Ba^{2+}] as charge carrier), there is a second component with a single channel conductance of about 12 pS (13, 18). This channel shows similar voltage-dependence and dihydropyridine sensitivity to the dominant form (13, 18). Cloning of the dihydropyridine-sensitive calcium channel from rat aorta also indicates the presence of multiple isoforms of the channel in arterial smooth muscle (9).

A component of calcium current insensitive to the organic calcium channel blockers has been described in some cells (e.g. 4, reviewed in ref. 14). This component is activated and inactivated at more negative potentials than the dihydropyridine-sensitive channel, and has a smaller single channel conductance. The channel is likely to be fully inactivated at physiological membrane potentials in arteries, and it's physiological role is uncertain (4).

CONCLUSIONS

Because of technical difficulties in resolution of single calcium channels, most studies to date have been conducted in unphysiological conditions (high concentrations of barium, in the presence of dihydropyridine calcium channel agonists). Despite these limitations, direct observation of single channels provide information about the basic processes of channel permeation, gating kinetics and modulation by physiological and pharmacological agents with higher resolution than other techniques. Measurements of whole cell currents can provide confirmation of these single channel observations at physiological concentrations of charge carrier.

The dihydropyridine-sensitive calcium channel appears to play a central role in regulating vascular tone. We propose that the voltage-dependence of the calcium channel underlies the membrane potential dependence of vascular tone (12).

ACKNOWLEDGMENTS

This work was supported by grants to MTN from the NSF (DCB-8702476 and DCB-9019563) and the NIH (HL44455).

REFERENCES

1 BENHAM, C.D. & R.W. TSIEN. Noradrenaline modulation of calcium channels in single smooth muscel cells from rabbit ear artery. *J. Physiol.* 404: 767-784, 1988.

2 CLAPP, L.H. & A.M. GURNEY. Modulation of calcium movements by nitroprusside in isolated vascular smooth muscle cells. *Pflugers.Arch.* 418: 462-470, 1991.

3 FISH, D., G. SPERTI, W. COLUCCI, & D. CLAPHAM. Phorbol ester increases the dihydropyridine-sensitive calcium conductance in a vascular smooth muscle cell line. *Circ. Res.* 62: 1049-1054, 1988.

4 GANITKEVICH, V.Y. & G. ISENBERG. Contribution of two types of calcium channels to membrane conductance of single myocytes from guinea-pig coronary artery. *J. Physiol.* 426: 19-42, 1990.

5 GOLLASCH, M., J. HESCHELER, J.M. QUAYLE, J.B. PATLAK, & M.T. NELSON. Single Ca-channel currents of arterial smooth muscle at physiological calcium concentrations. *Am. J. Physiol. (Cell Physiol.).* In Press.

6 HESS, P., J.B. LANSMAN, & R.W. TSIEN. Calcium channel selectivity for divalent and monovalent cations. voltage and concentration dependence of single channel current in ventricular heart cells. *J. Gen. Physiol.* 88: 293-320, 1986.

7 IMAIZUMI, Y., K. MURAKI, M. TAKEDA, & M. WATANABE. Measurement and simulation of noninactivating Ca current in smooth muscle cells. *Am. J. Physiol.* 256: C880-C885, 1989.

8 KAMISHIMA, T., M.T. NELSON, & J.B. PATLAK. Carbachol modulates voltage sensitivity of calcium channels in bronchial smooth muscle of rats. *Am. J. Physiol.* 263 (*Cell Physiol.* 32): C69-C77, 1992.

9 KOCH, W.J., P.T. ELLINORE, & A. SCHWARTZ. cDNA cloning of a dihydropyridine-sensitive calcium channel from rat aorta. *J. Biol. Chem.* 265: 17786-17791, 1990.

10 LOIRAND, G., P. PACAUD, C. MIRRONEAU, & J. MIRRONEAU. GTP-binding proteins mediate noradrenaline effects on calcium and chloride currents in rat portal vein myocytes. *J. Physiol.* 428: 517-529, 1990.

11 NELSON, M.T., N.B. STANDEN, J.E. BRAYDEN, & J.F. WORLEY. Noradrenaline contracts arteries by activating voltage-dependent calcium channels. *Nature.* 336: 382-385, 1988.

12 NELSON, M.T., J.B. PATLAK, J.F. WORLEY, & N.B. STANDEN. Calcium channels, potassium channels and voltage-dependence of arterial smooth muscle tone. *Am. J. Physiol.* 259: C3-C18, 1990.

13 NELSON, M.T. & J.F. WORLEY. Dihydropyridine inhibition of single calcium channels and contraction in rabbit mesenteric arteries depends on voltage. *J. Physiol.* 412: 65-91, 1989.

14 PELZER, D., S. PELZER, & T.F. McDONALD. Properties and regulation of calcium channels in muscle cells. *Rev. Physiol. Biochem. Pharmacol.* 114: 107-207, 1990.

15 QUAYLE, J.M., J.G. McCARRON, J.R. ASBURY, & M.T. NELSON. Single calcium channels in resistance-sized cerebral arteries from rats. *Am. J. Physiol. (Heart Circ. Physiol.).* In Press.

16 SIMARD, J.M. Calcium channel currents in isolated smooth muscle cells from the basilar artery of the guinea pig. *Pfluegers Arch.* 417: 528-536, 1991.

17 YUE, D.T. & E. MARBAN. Permeation in the dihydropyridine-sensitive calcium channel. Multi-ion occupancy but no anomalous mole-fraction effect between Ba^{2+} and Ca^{2+}. *J. Gen. Physiol.* 95: 911-939, 1990.

18 WORLEY, J.F., J.M. QUAYLE, N.B. STANDEN, & M.T. NELSON. Regulation of single calcium channels in cerebral arteries by voltage, serotonin, and dihydropyridines. *Am. J. Physiol.* 261 (*Heart Circ. Physiol.* 30): H1951-H1960, 1991.

19 HUANG, Y., J.M. QUAYLE, J.F. WORLEY, N.B. STANDEN & M.T. NELSON. External cadmium and internal calcium block of single calcium channels in smooth muscle cells from rabbit mesenteric artery. *Biophys. J.* 56: 1023-1028, 1990.

INTERACTIONS OF INTRACELLULAR CA²⁺ POOLS
IN VASCULAR SMOOTH MUSCLE

Cornelis van Breemen and Qian Chen

Department of Molecular and Cellular Pharmacology
University of Miami School of Medicine
Miami, Florida 33101

TECHNIQUES USED IN THE MEASUREMENT OF INTRACELLULAR CALCIUM POOLS

Contraction

On the basis of Heilbrunn and Wiercinsky's discovery[1] that muscle contraction was initiated by an increase in $[Ca^{2+}]_i$, transients in tension development have been used as indicators of Ca^{2+} release from intracellular pools. The report by Filo et al.[2] of detergent skinned smooth muscle contraction induced by increments of $[Ca^{2+}]$ in the bathing solution validated this method for this tissue. However the more recent discovery of modulation of myofilament Ca^{2+} sensitivity by agonists does complicate the quantitative aspects of this indirect method[3]. This is especially true if different agents are used to mobilize the intracellular pools.

⁴⁵Ca²⁺ flux

In vascular smooth muscle the release of intracellular Ca^{2+} is accompanied by a stimulation of Ca^{2+} efflux[4] and it seems likely that this process reflects the changes in $[Ca^{2+}]_i$ produced by Ca^{2+} mobilizing agonists more faithfully than contraction[5]. Because all of the Ca^{2+} in the intracellular physiologically important pools can be labelled, this method is capable of providing estimates of the actual pool sizes[5]. Especially in skinned fibers $^{45}Ca^{2+}$ fluxes have been employed to provide reliable quantitative measurements of the magnitude of Ca^{2+} release and uptake by the sarcoplasmic reticulum (SR)[6,7].

Ca²⁺ sensitive K⁺ currents

The third technique for recording properties of intracellular Ca^{2+} pools, which is successfully pursued at the present, is the whole cell patchclamp recording of fluctuations in Ca^{2+} regulated K^+ currents[8]. Because the Ca^{2+} activated K^+ channels are located in the plasmalemma and may be clustered, this technique records only the local $[Ca^{2+}]_i$ in the periphery of the cells. It has the advantage over the other methods that the effects of non membrane permeable molecules on intracellular regulatory sites can be tested by including them in the whole cell patch pipette and dialyzing them into the intact cells[9].

Direct measurements of intracellular calcium

At present x-ray analytical electron microscopy is still the only method capable of direct measurement of organellar calcium[10]. Electron microprobe analysis of smooth muscle cryosections has the great advantage of high spatial resolution for the localization of intracellular calcium. On the other hand it has the disadvantages of not differentiating between ionic and bound Ca^{2+} and being so labor intensive as to limit kinetic measurements.

Intracellular ionic calcium is now routinely measured using Ca^{2+} sensitive dyes such as fura-2 and Indo-1[11]. When introduced as the lipophilic acetoxymethyl ester hydrolyzable by cytoplasmic esterases, some uncertainty exists related to their possible accumulation into the organelles[11], but when they are injected or dialyzed into the cells as free acids a reliable measure of free cytoplasmic Ca^{2+} is obtained. However the activity and content of the intracellular stores are still gaged indirectly by agonist-induced transient fluorescent signals observed when Ca^{2+} entry from the extracellular space is prevented by Ca^{2+} entry blockers or removal of extracellular Ca^{2+}.

INTRACELLULAR ORGANELLES INVOLVED IN THE PHYSIOLOGICAL REGULATION OF $[Ca^{2+}]_i$

There are two major systems of intracellular membranes capable of Ca^{2+} transport in smooth muscle. Clearly the major physiological role is played by a network of interlacing tubes of smooth membranes termed the sarcoplasmic reticulum[12,13], which is continuous both with the rough endoplasmic reticulum and the lumen of the double membrane nuclear envelope[13]. The latter has recently been shown to play a role in the differential regulation of the nuclear Ca^{2+} concentration[14]. The SR volume varies from 0.5-7.5 % of the total cell volume and has been defined in terms of peripheral SR which is closely associated with the plasmalemmal invaginations (caveoli), and central SR[15].

The mitochondria are clearly separate organelles although they are often closely associated with the SR and caveoli. The mitochondrial Ca^{2+} pump has a much lower Ca^{2+} affinity than the SR Ca^{2+} pump and only removes Ca^{2+} from the cytoplasm if the concentration exceeds 3-5 μM^6, which is above the physiological range of values. However it is possible that local $[Ca^{2+}]_i$ during agonist induced Ca^{2+} release from the SR, or stimulated Ca^{2+} entry may transiently exceed these values, in which case mitochondria would function as local buffers.

The SR membranes contain an active Ca^{2+} pump for accumulation of Ca^{2+} and two types of excitable Ca^{2+} channels for rapid Ca^{2+} release into the cytoplasm. In addition it has a slower leak , which does not appear to be regulated. The Ca^{2+} storage capacity of smooth muscle SR is enhanced by the presence of a cardiac-like form of calsequetrin a low affinity high capacity Ca^{2+} binding protein[16]. Casteels and collaborators[17] identified the Ca^{2+} pump in smooth muscle SR as a 100 kDa Ca^{2+}-Mg^{2+}-ATPase, which is homologous to the SR pump in cardiac and slow skeletal muscle, but distinct from the plasmalemmal Ca^{2+}-Mg^{2+}-ATPase and the SR Ca^{2+} pump in fast twitch skeletal muscle[18]. As in cardiac muscle the SR Ca^{2+} pump in at least some smooth muscle is regulated by the protein kinase substrate phospholamban. Its inhibitory effect on the pump is relieved by phosphorylations mediated by cAMP-, cGMP-, and Ca^{2+}-calmodulin-dependent protein kinases[19,20,21].

Rapid Ca^{2+} release from the SR into the cytoplasm is effected by one set of Ca^{2+} channels activated by Ca^{2+} and ATP[22,23,24] and another set which is activated by Ins(1,4,5)P$_3$[25,26,27]. The first set of release channels, responsible for Ca^{2+} induced Ca^{2+}

release (CICR) are blocked by procaine and high concentrations of Mg^{2+} [28,24]. They are stimulated to open to a sub-conductance state by ryanodine[29,30], and their Ca^{2+} sensitivity is greatly increased by caffeine. The effectiveness of caffeine to rapidly discharge the SR Ca^{2+} in smooth muscle has made it a useful tool to assess SR content under various experimental conditions.

The $Ins(1,4,5)P_3$ receptor-channel molecules have polypeptide chains which are rather homologous to those of the ryanodine receptors (CICR channels). Although they represent different channel proteins, with different activating mechanisms, they appear to have certain regulatory sites in common. Iino[31,32] has reported that both adenine nucleotides and cytoplasmic Ca^{2+} facilitate $Ins(1,4,5)P_3$ induced release, and Missiaen et al. recently demonstrated that SR luminal Ca^{2+} has the same effect[33]. The $Ins(1,4,5)P_3$ receptors are specifically blocked by heparin[34].

COMPARTMENTALIZATION OF FUNCTIONAL CALCIUM POOLS IN SMOOTH MUSCLE

The first interesting properties revealed about agonist releasable intracellular Ca^{2+} pools are that they are of limited size and are to a large, but probably variable extend, refilled from the extracellular space[35,36]. One application of a maximally effective dose of a full agonist (e.g. norepinephrine in the rabbit aorta) in the absence of an extracellular Ca^{2+} supply will prevent Ca^{2+} release by subsequently applied physiological agonists [37]. Under Ca^{2+} free conditions partial agonists will partially deplete the stores, as indicated by a reduction of a subsequent full agonist induced Ca^{2+} release. Interestingly one partial agonist may not diminish the response resulting from a subsequent exposure to another partial agonist. These findings suggest functional Ca^{2+} pools whose sizes are based on the density of the specific types of receptors. If each activated receptor is able to effect SR Ca^{2+} release only within a certain radius of the receptor, probably due to limited diffusion of $Ins(1,4,5)P_3$ in a cytoplasm containing $Ins(1,4,5)P_3$-phosphatase[38], then it is conceivable that the releasable Ca^{2+} domains of densely distributed receptors responding to a full agonist would overlap the domains of all other types of receptors. However little overlap would exist between the domains belonging to two sets of sparsely distributed receptors sensitive to partial agonists[37]. Thus it may be possible to have distinct functional pools based on the location of the SR with respect to the receptors, rather than on the difference of basic properties of various SR fractions. Alternatively the identity of different Ca^{2+} pools may be based on the type of release mechanism present in separate ER fractions as will be discussed below. As is common for other smooth muscle characteristics, variations in functional Ca^{2+} pools exist when different types of smooth muscle are compared. In the rabbit aorta there is a complete overlap between caffeine-sensitive and norepinephrine-sensitive Ca^{2+} pools[5], while in small mesenteric arteries of the rabbit, norepinephrine releases only a fraction of the caffeine-sensitive Ca^{2+} pool[28]. Since the Ca^{2+} releasable by norepinephrine was lost more rapidly into a Ca^{2+} free extracellular solution than was the case for the average caffeine releasable Ca^{2+}, and also was more rapidly restored it was postulated that in the small mesenteric artery the stimulated production of $Ins(1,4,5)P_3$ was sufficient to reach the peripheral SR but was hydrolysed before it could reach the deep SR[39]. An alternative explanation would be that norepinephrine induced Ca^{2+} release only through activation of $Ins(1,4,5)P_3$ receptors and that only part of the SR was endowed with such receptors, while all of the SR contained CICR receptors. In the intact and patch clamped smooth muscle cells isolated from guinea-pig taenia caecum, carbachol and caffeine depleted the same Ca^{2+} pool, but IP3 administered through the patch pipette was able to further increase the cytoplasmic Ca^{2+} signal in the absence of external Ca^{2+} [40]. These results were interpreted as evidence for the presence of two SR Ca^{2+} compartments: one containing CICR and $Ins(1,4,5)P_3$-sensitive channels, and another containing only $Ins(1,4,5)P_3$-sensitive channels. However caution is warranted in the interpretation of net changes in Ca^{2+} store contents, under conditions were the Ca^{2+} pump is active, since the latter may prevent complete emptying of the SR even though some Ca^{2+} channels are activated in all of the SR membranes. A more direct experimental approach to the elucidation of smooth muscle Ca^{2+} compartmentalization is to study Ca^{2+} movements in skinned or permeabilized smooth muscle preparations[24,41,6]. Unidirectional $^{45}Ca^{2+}$ efflux from saponin skinned cultured aortic smooth muscle cells indicated that although in these cells caffeine induced Ca^{2+} release was much slower than that induced by $Ins(1,4,5)P_3$, both types of release channels were present in 80%-90% of the total Ca^{2+} pools[42].

Further studies on freshly isolated permeabilized smooth muscle preparations are required for the definition of functionally distinct Ca^{2+} pools in smooth muscle.

INTERACTIONS BETWEEN CALCIUM FLUXES ACROSS THE SR MEMBRANES AND CALCIUM FLUXES ACROSS THE PLASMALEMMA

The well established functions of smooth muscle SR are to release Ca^{2+} into the cytoplasm during activation and to accumulate Ca^{2+} during relaxation. Primary events in the plasmalemma signal SR Ca^{2+} release in two ways. The receptor initiated cascade of G-protein and phospholipase C (PLC) activation generates IP3, which activates its receptors in the SR as described above[43], and Ca^{2+} influx may activate the CICR channels. CICR initiated by plasmalemmal Ca^{2+} currents underlying smooth muscle action potentials had been postulated on the basis that insufficient Ca^{2+} would enter during an action potential to explain the amplitude of the mechanical twitch[44]. Ganitkevich and Isenberg[45] were recently able to demonstrate CICR from the SR of guinea-pig urinary bladder smooth muscle cells by simultaneous measurements of indo1 fluorescence and Ca^{2+} currents in whole cell patchclamped cells. They showed that the phasic component of the Ca^{2+} transient resulting from a depolarization pulse depended on the state of SR filling, and could be abolished by continuous application of caffeine, ryanodine or thapsigargin (an inhibitor of the SR Ca^{2+} pump).

The converse of CICR; namely, a buffering of plasmalemmal Ca^{2+} influx by the SR has also been demonstrated. This may be best demonstrated by first depleting the SR by a transient caffeine application in Ca^{2+} free medium followed by stimulation of Ca^{2+} influx[46]. Under these conditions contraction of tonic smooth muscle was delayed due to SR accumulation of Ca^{2+} which entered the smooth muscle cells. Only after the SR had taken up its physiological complement of Ca^{2+}, did the stimulated Ca^{2+} influx contribute to aortic contraction[46]. When $[Ca^{2+}]_i$ was measured during a similar protocol applied to the inferior vena cava of the rabbit, a biphasic $[Ca^{2+}]_i$ signal was observed during the delayed contraction[47]. Thus as had been shown before with the use of aquorin[48] Ca^{2+} uptake by the depleted SR caused a delay between the initial appearance of the Ca^{2+} signal and the development of tension. Continuous application of caffeine abolished this delay and greatly increased the initial elevation in cytoplasmic Ca^{2+}[47]. These results indicate that, at least in some types of smooth muscle, Ca^{2+} enters from the extracellular space into a restricted cytoplasmic space located immediately below the plasmalemma, from where it is partially taken up into the superficial SR, and in part diffuses into the bulk of the cytoplasm, where it activates contraction. This hypothetical model also referred to as the superficial buffer barrier has been described in detail elsewhere[47,49]. The SR buffer barrier function appears to contribute to regulation of the steady state $[Ca^{2+}]_i$, since prevention of SR Ca^{2+} accumulation by either caffeine, ryanodine or thapsigargin leads to maintained elevation of $[Ca^{2+}]_i$ in smooth muscle of the rabbit vena cava. In this tissue these agents lead to a decrease in Ca^{2+} extrusion (Chen and van Breemen, unpublished results) rather then an increase in Ca^{2+} entry. The mechanism whereby SR Ca^{2+} transport contributes to Ca^{2+} extrusion is not clear at this time, but may involve vectorial Ca^{2+} release into very narrow junctional regions between SR and plasmalemmal membranes, with subsequent extrusion via the Na^+/Ca^{2+} exchanger and 130 kDa $Ca^{2+}-Mg^{2+}$-ATPase. In support of this idea, Stehno-Bittel and Sturek[50] have recently shown an increase in the activity of Ca^{2+} activated K channels during "unloading" of SR, while the $[Ca^{2+}]_i$ measured with fura2 was at resting level.

An alternative mechanism whereby SR depletion may raise $[Ca^{2+}]_i$ in other smooth muscle is termed "capacitative Ca^{2+} entry"[51], which links ER discharge to enhanced plasmalemmal Ca^{2+} permeability. Pacaud and Bolton[52] presented some support for the capacitative Ca^{2+} entry hypothesis in the guinea-pig jejunal smooth muscle cells. They found that caffeine markedly amplified the fluctuations in $[Ca^{2+}]_i$ obtained when the membrane potential was varied between 50 and -50 mV. These fluctuations were similar to those seen in the presence of carbachol during Em steps between 50 and -50 mV. However the effects seen in the presence of carbachol were accompanied by an increase in membrane conductance, whereas caffeine did not significantly increase the membrane conductance. Since the fluctuations in $[Ca^{2+}]_i$ with the above changes in Em were dependent on the presence of $[Ca^{2+}]_o$ it was concluded that both agents opened the same Ca^{2+} permeant cation channels. However the results may not rule out the possibility that caffeine amplified the $[Ca^{2+}]_i$ fluctuations by eliminating the buffering effect of the SR. In this study the identification of a purely capacitative Ca^{2+} entry mechanism was further complicated by the presence of cation channels which were activated or potentiated by increases in $[Ca^{2+}]_i$. Such Ca^{2+} sensitive cation channels are activated upon Ca^{2+} release from the SR and represent yet another interaction between the two membrane systems.

It may be concluded that smooth muscle SR constitutes a dynamic intracellular calcium pool(s) with highly regulated Ca^{2+} release and Ca^{2+} uptake mechanisms. Whether it removes Ca^{2+} from the cytoplasm or releases it, either in graded quantities or regeneratively, depends on the state of activation of the plasmalemma and the cytoplasmic and luminal Ca^{2+} concentrations.

REFERENCES

1. L.V. Heilbrunn, and F.J. Wiercinsky, The action of various cations on muscle protoplasm, *J. Cell. Comp. Physiol.* 29:15-32 (1947).
2. R.S. Filo, D.F. Bohr, and J.C. Ruegg, Glycerinated skeletal and smooth muscle: calcium and magnesium dependence, *Science* 147:1581-1283(1965).
3. J.P. Morgan, and K.G. Morgan, Vascular smooth muscle:the first recorded Ca^{2+} transients. *Pflugers Arch* 395:75-77(1984).
4. C. van Breemen, Transmembrane calcium transport in vascular smooth muscle, *in*:"Vascular Neuroeffector Mechanism", J.A. Bevan, ed., S. Karger, Basel (1976).
5. P.A.A. Leijten, and C. van Breemen, The effects of caffeine on the noradrenaline-sensitive calcium store in rabbit aorta, *J. Physiol.* 357:327-339(1984).
6. H. Yamamoto, and C. van Breemen, Ca^{2+} compartments in saponin skinned cultured vascular smooth muscle cells, *J. Gen. Physiol.* 87:369-389 (1986).
7. L. Missiaen, I. Delerck, G. Droogmans, L. Plessers, H. De Smedt, L. Raeymaekers and R. Casteels, Agonist dependent Ca2+ and Mn2+ entry dependent on the state of refilling of Ca2+ stores in aortic smooth muscle cells of rat, *J. physiol.* 427:171-186 (1990).
8. C.D. Benham, and T.B. Bolton, Spontaneous transient outward currents in single visceral and vascular smooth muscle cells of the rabbit, *J. Physiol.* 381:385-406 (1986).

9. S. Komori, and T.B. Bolton, Inositol trisphosphate released stored calcium to block voltage dependent calcium channels in single smooth muscle cells, *Pflugers Arch* 418:437-441 (1991).

10. M. Bond, H. Shuman, A.P. Somlyo, and A.V. Somlyo, Total cytoplasmic caalcium in relaxed and maximally contracted rabbit portal vein smooth muscle, *J. Physiol.* 357:185-201 (1984).

11. G. Grynkiewicz, M. Poeni, and R.Y. Tsien, A new generation of Ca2+ indicators with greatly improved fluorescence properties, *J. Biol. Chem.* 260:3440-3450 (1985).

12. C.E. Devine, A.V. Somlyo, and A.P. Somlyo, Sarcoplasmic reticulum and excitation-contraction coupling in mammalian smooth muscles, *J. Cell. Biol.* 52:690-718(1972).

13. A.V. Somlyo, Ultrastructure of vascular smooth muscle, *in*:"The Handbook of Physiology, the Cardiovascular System, vol II: Vascular Smooth Muscle", D.F. Bohr, A.P. Somlyo, and H.V. Sparks, eds. American Physiological Society, Washington, DC, (1980).

14. B. Himpens, H. De Smedt, G. Droogmans, and R. Casteels, Differences in the regulation between nuclear and cytoplasmic Ca2+ in cultured smooth muscle cells, *Am. J. Physiol.* 263:C95-C105 (1992).

15. A.P. Somlyo, and B. Himpens, Cell Calcium and its Regulation in Smooth Muscle, *FASEB J.* 3:2266-2276 (1989).

16. F. Wuytack, L. Raeymaekers, J. Verbist., L.R. Jones, and R. Casteels, Smooth-muscle endoplasmic reticulum contains a cardiac-like form of calsequestrin, *Biochim. Biophys. Acta.* 899:151-158 (1987).

17. F. Wuytack, L. Raeymaekers, J. Verbist, H. De Smedt, and R. Casteels, Evidence for the presence in smooth muscle of two types of Ca^{2+}-transport ATPase, *Biochem. J.* 224:445-451 (1984).

18. F. Wuytack, Y. Kanmura, J.A. Eggermont, *et al.*, Smooth muscle expresses a cardiac/slow muscle isoform of the Ca^{2+}-transport ATPase in its endoplasmic reticulum, *Biochem. J.* 257:117-123 (1989).

19. L. Raeymaekers, J.A. Eggermont, F. Wuytack, and R. Casteels. Effects of cyclic nucleotide dependent protein kinases on the endoplasmic reticulum Ca^{2+} pump of bovine pulmonary artery, *Cell Calcium* 11:261-268 (1990).

20. L. Raeymaekers, and L.R. Jones, Evidence for the presence of phospholamban in the endoplasmic reticulum of smooth muscle. *Biochim. Biophys. Acta* 882:258-265 (1986).

21. L. Raeymaekers, F. Hofmann, and R. Casteels, Cyclic GMP-dependent protein kinase phosphorylates phospholamban in isolated sarcoplasmic reticulum from cardiac and smooth muscle. *Biochem. J.* 252:269-273 (1988).

22. M. Endo, M. Tanaka, and Y. Ogawa, Calcium induced release of calcium from the sarccoplasmic reticulum of skinned skeletal muscle fibres, *Nature* 228:34-36 (1970).

23. M. Endo, S. Yagi, and M. Iino, Tension-pCa relation and sarcoplasmic reticulum responses in chemically skinned smooth muscle fibers, *Fed. Proc.* 41:2245-2250 (1982).

24. K. Saida, Intracellular Ca release in skinned smooth muscle, *J. Gen. Physiol.* 80:191-202 (1982).

25. M.J. Berridge, and R.F. Irvine, Inositol trisphosphate, a novel second messenger in cellular signal transduction, *Nature* 312:315-321 (1984).

26. E. Suematsu, M. Hirata, T. Hashimoto, and H. Kuriyama, Inositol 1,4,5-trisphhosphate releases Ca^{2+} from intracellular store sites in skinned single cells of porcine coronary artery, *Biochem. Biophys. Res. Commun.* 120(2):481-485 (1984).

27. B.E. Erlich, and J. Watras, Inositol 1,4,5-trisphosphate activates a channel from smooth muscle sarcoplasmic reticulum, *Nature* 336:583-586 (1988).

28. K. Saida, and C. van Breemen, Characteristics of the norepinephrine-sensitive Ca^{2+} store in vascular smooth muscle, *Blood Vessels* 21:43-52 (1984).

29. K. Hwang, and C. van Breemen, Ryanodine modulation of [45]Ca efflux and tension in rabbit aortic smooth muscle, *Pflugers Arch* 408:343-350 (1987).

30. E. Rousseau, and G. Meisner, Single cardiac sarcoplasmic reticulum Ca2+-release channels: activated by caffeine, *Am. J. Physiol.* 256:H328-H333, (1989).

31. M. Iino, Effects of adenine nucleotides on ionositol 1,4,5-trisphosphate-induced calcium release in vascular smooth muscle cells, *J. Gen. Physiol.* 98:681-698 (1991).

32. M. Iino, Biphasic Ca2+ dependence of inositol 1,4,5-trisphosphate-induced Ca release in smooth muscle cells of guinea-pig taenia Caeci, *J. Gen. Physiol.* 95:1103-1122 (1990).

33. L. Missiaen, H. De Smedt, G. Droogmans, and R. Casteels Ca^{2+} release induced ba inositol 1,4,5-trisphosphate is a steady state phenomenon controlled by luminal Ca^{2+} in permeabilized cells, *Nature* 357:June 18 (1992).

34. S. Kobayashi, A.V. Somlyo, A.P. Somlyo, Heparin inhibits the inositol 1,4,5-trisphosphate-dependent, but not the independent, calcium release induced yb guaanine nucleotide in vascular smooth muscle, *Biochem. Biophys. Res. Commun.* 153:625-631 (1988).

35. C. van Breemen, B.R. Farinas, P. Gerba, E.D. McNaughton, Excitation-contraction coupling in rabbit aorta studied by the lanthanum method for measuring cellular calcium influx, *Circ. Res.* 30:44-54 (1972).

36. P.A. Leijten, and C. van Breemen, The relationship between noradrenaline-induced teension development and stimulated [45]Ca efflux in rabbit mesenteric small artery, *British J. Pharmacol.* 87:739-747 (1986).

37. R. Loutzenhiser, and C. van Breemen, Mechanism of activation of isolated rabbit aorta by PGH_2 analogue U-44069, *Am. J. Physiol.* 241:C243-C249 (1981).

38. A.P. Somlyo, and A.V. Somlyo, Flash photolysis studies of excitation-contraction coupling, regulation, and contraction in smooth muscle, *Annu. Rev. Physiol.* 52:857-874 (1990).

39. C. van Breemen, and K. Saida, Cellular mechanisms regulating $[Ca^{2+}]_i$ smooth muscle, *Annu. Rev. Physiol.* 51:315-3229 (1989).

40. T. Yamazawa, M. Iino, and M. Endo, Presence of functionally different compartments of the Ca^{2+} store in single intestinal smooth muscle, *FEBS letters* 301:181-184 (1992).

41. M. Endo, Mechanism of action of caffeine on the sarcoplasmic reticulum of skeletal muscle, *Proc. Jpn. Acad.* 511:479-484 (1975).

42. C. van Breemen, K. Saida, H.Yamamoto, K. Hwang, and C. Twort, Vascular smooth muscle sarcoplasmic reticulum:Function and Mechanisms of Ca^{2+} release, *Ann. N. Y. Acad. Sci.* 522:60-73 (1988).

43. S. Komori, and T.B. Bolton, The role of G-proteins inmuscarinic receptor inward and outward currentsin rabbit jejunal smooth muscle, *J. Physiol.* 427:359-419 (1990).

44. T.B. Bolton, Mechanisms of action of transmitters and other substances on smooth muscle, *Physiol. Rev.* 220:607-718, (1979).

45. V. Ya. Ganitkevich, and G. Isenberg, Contribution of Ca^{2+} induced Ca^{2+} release to the $[Ca^{2+}]_i$ transients in myocytes from the guinea-pig urinary bladder, *J. Physiol.* 548:119-137 (1992).

46. C. van Breemen, S. Lukeman, P. leijten, H. Yamamoto, and R. Loutzenhiser, The role of superficial SR in modulating force development induced by Ca entry into arterial smooth muscle, *J. Cardiovasc. Pharmacol.* 8(SUPPL 8):s111-s116 (1986).

47. Q. Chen, M. Connell, and C. van Breemen, The superficial buffer barrier in smooth muscle. *Can. J. Physiol. Pharmacol.* 70:509-514 (1992).

48. C. Rembold, Desensitization of swine arterial smooth muscle to transplasmalemmal Ca^{2+} influx, *J. Physiol.* 416:273-290 (1989).

49. Q. Chen, and C. van Breemen, Function of Smooth Muscle Sarcoplasmic Reticulum, *in:* "Advances in Second Messenger and Phosphoprotein Research", J.W. Putney Jr., ed., Raven Press, Ltd., New York (1992).

50. L. Stehno-Bittel, and M. Sturek, Spontaneous sarcoplasmic reticulum calcium rrelease and extrusion from bovine, not porcine, coronary artery smooth muscle, *J. Physiol.* 451:49-78 (1992).

51. J.W. Putney Jr., A model for receptor-regulated calcium entry, *Cell Calcium* 7:1-112 (1986).

52. P. Pacaud, and T.B. Bolton, Relation between muscarinic receptor cationic current and internal calcium in guinea-pig jejunal smooth muscle cells, *J. Physiol.* 441:477-499 (1991).

Na-Ca EXCHANGE IN ISOLATED CORONARY MYOCYTES:
A STUDY COMBINING VOLTAGE CLAMP AND Ca^{2+} FLUOROMETRY

Gerrit Isenberg and Vladimir Ganitkevich

Department of Physiology
University of Cologne
5000 Köln 41, Germany

INTRODUCTION

Although the importance of Na-Ca exchange for vascular tone has been suggested as early as 1973, there is still much controversy about. Originally, it was thought[16] that the reduction of the transmembraneous [Na$^+$] gradient stimulates Ca^{2+} influx through the Na-Ca exchanger with the result that the increased amount of activator Ca^{2+} increases the degree of contractile activation. However, when the [Na$^+$] gradient was reduced by removal of extracellular sodium, contraction was not always observed. Hence, the hypothesis was reformulated[2], it was suggested that the Ca^{2+} inflowing through the Na-Ca exchanger is sequestered into the SR, the greater extent of SR Ca^{2+} load providing a more intense release of activator Ca^{2+} during application of an agonist or caffeine.

In this study, the possible contribution of Na-Ca exchange to the calcium balance was studied in myocytes isolated from the circumflex artery of the guinea-pig. The experimental protocol is basically a reduction or reversal of the sodium gradient by rapid removal of extracellular sodium. Some problems of this method, such as the possible release of neurotransmitters from presynaptic nerve endings[15], can be avoided in the isolated cell. Other secondary effects, as membrane depolarization leading to activation of Ca^{2+} influx through dihydropyridine-sensitive channels, were avoided by performing the experiments under the conditions of the voltage-clamp, the membrane potential being held at -50 mV. The putative effects of sodium removal on the cellular Ca^{2+} balance were monitored by the changes in

Ion Flux in Pulmonary Vascular Control, Edited by
E.K. Weir *et al.*, Plenum Press, New York, 1993

the concentration of calcium ionized in the cytosol ($[Ca^{2+}]_c$) as it can be evaluated from Indo-1 fluorescence. In order to test the postulated influence of the Na-Ca exchanger on the Ca^{2+} load of the SR, this parameter was screened by fast applications of 10 mM caffeine. That is, the peak amplitudes and the decay of caffeine-induced Ca^{2+} transients were compared under conditions where the sodium gradient was varied. The functional importance of Na-Ca exchange increases with the elevation of $[Na^+]_i$[2,19], therefore, part of the experiments were performed with myocytes dialyzed with a pipette solution containing Na^+ at a concentration $[Na^+]_p = 150$ mM that should saturate the intracellular Na^+ binding sites of the Na-Ca exchanger ($K_D = 28$ mM[19]).

Our results suggest for the isolated coronary myocyte that the Na-Ca exchange is of minor importance for cellular calcium balance when $[Na^+]_p$ is 10 mM. For elevated $[Na^+]_p$ the results do reveal effects of Ca^{2+} influx through Na-Ca exchange. The results suggest that only a minor part of the measured changes in $[Ca^{2+}]_c$ is a direct consequence of Ca^{2+} influx. A large part is suggested to result from Ca^{2+} release which can be triggered by the Ca^{2+} influx through Na-Ca exchange if this mechanism is stimulated.

The full length paper of this data will appear somewhere else[9].

METHODS

Details of the methods have been described previously, for cell isolation[6], recording of whole-cell currents[6], microfluospectroscopy of $[Ca^{2+}]_c$[7] fast solution change[8]. Briefly, the experiments were performed as follows. At 36 °C, the cells were continuously superfused with a physiological salt solution composed of (mM) 150 NaCl, 2.5 CaCl$_2$, 1.2 MgCl$_2$, 5.4 KCl, 20 glucose, 5 HEPES, pH 7.4. For sodium-removal experiments, 150 mM Na^+ was replaced by 150 mM N-methyl-glucosamine$^+$. The patch pipette were filled with an intracellular solution containing (mM) 140 KCl, 2 Na$_2$ATP, 3 MgCl$_2$, 10 HEPES, 6 NaOH, pH 7.2 plus 100 μM Indo-1 (potassium salt); in a series of experiments, 140 mM NaCl was used instead of 140 mM KCl. Indo-1 loading through the patch pipette preceded the measurements for at least 2 min which turned out to be sufficient because of the small volume of the cells. A problem in the use of small cells, however, was the run-down of the $[Ca^{2+}]_c$ transients which limited the experimental time to 5 min.

RESULTS AND DISCUSSION

Resting $[Ca^{2+}]_c$ of coronary myocytes

The resting $[Ca^{2+}]_c$ in coronary myocytes was 166 ±62 nM on average (mean ± S.D., n=25). The resting value is somewhat higher than the 150 nM published for myocytes from rabbit ear artery[3] or 120 nM for myocytes from the guinea-pig urinary bladder[7]. The resting $[Ca^{2+}]_c$ was insensitive to changes of the holding potentials between -50 and -100 mV but it increased when the holding potential was -40 mV or more positive (compare[7]). Since $[Ca^{2+}]_c$ is the balance of Ca^{2+} release and Ca^{2+} reuptake by the SR and Ca^{2+} efflux and Ca^{2+} influx through the sarcolemma, the effect of membrane potential suggests that the contribution of Ca^{2+} influx can be neglected at -50 mV and more negative holding potentials.

Testing SR Ca^{2+} load with caffeine

Caffeine is known to activate reconstituted Ca^{2+} release channels isolated as ryanodine-receptors from cardiac[17,18] or vascular SR[10]. It is likely that the caffeine activation is the result of a sensitization of the release channel to Ca^{2+}, inducing Ca^{2+} induced Ca^{2+} release (CICR)[18]. Caffeine-induced Ca^{2+} release has been widely used to define the "caffeine-sensitive Ca^{2+} store" which, in coronary myocytes, may largely overlap with the IP_3-sensitive Ca^{2+} store[6]. The caffeine-induced Ca^{2+} release may be used for testing the filling of the SR with releasable Ca^{2+}, as described in the literature[14]. Here, the amount of release is measured with the changes in $[Ca^{2+}]_c$ ("caffeine-induced Ca^{2+} transients"). The method may be preferable compared to caffeine-induced force transients because Ca^{2+} sensitivity of the myofilaments may not stay constant.

Upon a fast application of 10 mM caffeine for 2 s, $[Ca^{2+}]_c$ increased within 1-2 s to 1620 ±490 nM (n=25, compare Fig. 1). When $[Ca^{2+}]_c$ was washed out, $[Ca^{2+}]_c$ always fell to concentrations below the resting value, i.e. an undershoot of $[Ca^{2+}]_c$ occurred[5,8] (100 nM verse 160 nM in Fig. 1). It was important to apply caffeine with a system for the fast solution change ("picospritzer"); when caffeine was applied more slowly through the bath, the Ca^{2+} transient rose much more slowly and reached lower amplitudes[8,14].

When caffeine was applied 30 s after the first exposure, the second $[Ca^{2+}]_c$ transient peaked to only 30% of the first response. The result suggests that 70% of the Ca^{2+} available for the first SR Ca^{2+} release, was extruded into the extracellular space and not available for the next Ca^{2+} release. Depletion of releasable Ca^{2+} from the SR was even more prominent when caffeine was applied repeatedly at short intervals or after application of caffeine for 30 s. The results are compatible with the hypothesis that Ca^{2+} reuptake and Ca^{2+} extrusion can compete for the released Ca^{2+}. In the continuous presence of caffeine the SR leaks Ca^{2+}, and Ca^{2+} extrusion is favored.

After a short application of caffeine, $[Ca^{2+}]_c$ decayed during caffeine wash out because Ca^{2+} release was ended by closure of the Ca^{2+} release channels. During the decay of the Ca^{2+} transient, part of the released Ca^{2+} is reloaded back into the SR by the Ca^{2+}-ATPase of the

SR (SERCa), and part is extruded into the extracellular space by the Ca^{2+}-ATPase of the plasmalemma (PMCa)[12]. Ca^{2+} extrusion by PMCa causes loss of cellular Ca^{2+}, as indicated by the attenuation of the following Ca^{2+} transients. The result that, after the wash out of caffeine, $[Ca^{2+}]_c$ fell to a sub-resting $[Ca^{2+}]_c$ ("undershoot") led us to postulate[8] that the low intraluminal SR $[Ca^{2+}]$ stimulates the reuptake of Ca^{2+} by the SR Ca^{2+}-ATPase[11]. Stimulation of SERCa by caffeine-mediated inhibition of phosphodiesterase has been excluded recently[8].

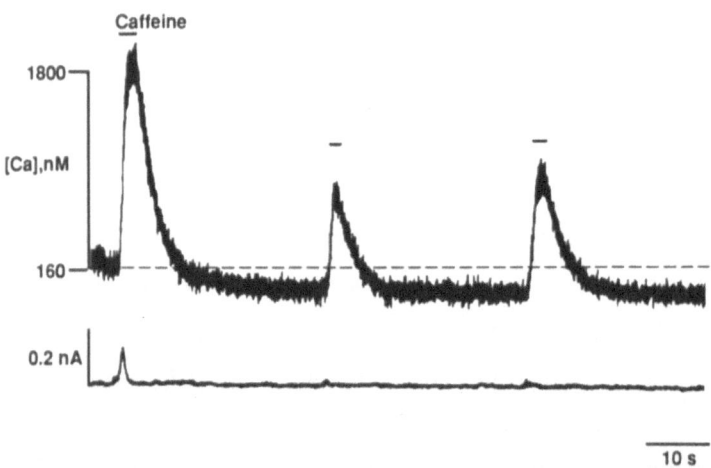

Figure 1. Caffeine-induced increase in $[Ca^{2+}]_c$ (upper trace) and Ca^{2+} activated K^+ current (I_{KCa}, lower trace). On-line pen-recording of the ratio of fluorescence at 410 nm to 470 nm, the $[Ca^{2+}]_c$ scale is non-linear. The 2 s exposure time to 10 mM caffeine is marked by bar. Note: Upon wash-out of caffeine, $[Ca^{2+}]_c$ fell to an undershoot of 100 nM.

During the caffeine-induced $[Ca^{2+}]_c$ transient, an outward current I_{KCa} appeared at the holding potential of -50 mV (Fig. 1). We attribute I_{KCa} to the Ca^{2+} activation of BK channels[6]. At -50 mV, the Ca^{2+} sensitivity of this channel is low, therefore the local $[Ca^{2+}]$ activating the BK channel should be much higher than the concentration $[Ca^{2+}]_c$ measured by Indo-1 globally in the cytosol. This discrepancy fits with the result that the time course of I_{KCa} was always faster than the time course of the $[Ca^{2+}]_c$ transient. Although a quantitative comparison of the signals is not possible because the Ca^{2+} dependence of I_{KCa} is non-linear and essentially unknown for the present experiments, the difference strongly suggests that $[Ca^{2+}]_c$ in smooth muscle cells is inhomogeneous during the high rates of Ca^{2+} release, as it was postulated for skeletal muscle[4] and cardiac muscle[20].

Effects of sodium removal at 10 mM [Na$^+$]$_p$

Assuming an intracellular [Na$^+$]$_i$ of 10 mM, a reversal potential of -20 mV is estimated for the Na-Ca exchanger. That is, at -50 mV holding potential and resting [Ca^{2+}]$_c$, the Na-Ca exchanger should operate in the Ca^{2+} efflux mode. Substitution of extracellular sodium by the less permeable cation N-methyl-glucosamine$^+$ (NMG) is thought to shift the reversal potential of the Na-Ca exchanger to strongly negative potentials, thereby reversing the Na-Ca exchanger from the Ca^{2+} efflux to the Ca^{2+} influx mode. When the effects of sodium removal on [Ca^{2+}]$_c$ were tested, 7 of 10 cells did not respond at all, i.e. [Ca^{2+}]$_c$ remained constant. In 3 cells, sodium removal induced a small increment in [Ca^{2+}]$_c$ (71 ±11 nM) within 15-20 s. Sodium removal did nod induce I$_{KCa}$.

The failure of sodium-removal in inducing a significant increase in [Ca^{2+}]$_c$ could be due to a "Ca^{2+} buffering effect" of the superficial SR, i.e. the Ca^{2+} inflowing through Na-Ca exchange may have been sequestered into the SR[2] without significant binding to the Indo-1. In such a case, the SR should have been loaded with releasable Ca^{2+} to a higher extent, and caffeine should be able to induce larger Ca^{2+} transients. This possibility was tested by 2 s caffeine applications in the absence of [Na$^+$]$_o$. The caffeine-induced Ca^{2+} transients peaked to concentrations not different from the control (p≥0.05), i.e. the hypothesis of enhanced Ca^{2+} loading of the SR could not be verified.

In the absence of extracellular sodium, the caffeine-induced Ca^{2+} transients fell with a similar half time as during control, also, they fell to an undershoot that was not significantly different from the one recorded in the presence of 150 mM [Na$^+$]$_o$. When caffeine was tested 30 s later a second time, the continuous absence of sodium had no influence on the peak [Ca^{2+}]$_c$ transient. These results suggest that the Ca^{2+} efflux that contributes to the decay of [Ca^{2+}]$_c$ is not through the Na-Ca exchange but through the plasmalemmal Ca^{2+}-ATPase.

Effects of sodium removal at 150 mM [Na$^+$]$_p$

The absence of significant effects of sodium removal on resting [Ca^{2+}]$_c$ or on the caffeine-induced Ca^{2+} transients suggests that the Ca^{2+} fluxes mediated by the Na-Ca exchanger are relatively unimportant in comparison to those due to channels or Ca^{2+}-ATPases. The relative unimportance may be due to a low density of exchanger molecules in the myocytes from coronary artery. However, it could also be a consequence of the cell dialysis with 10 mM Na$^+$; it has been suggested[2] that sodium removal does not induce appreciable Ca^{2+} influx unless [Na$^+$]$_i$ is raised to concentrations above the K$_D$ of the internal

Na$^+$ binding sites (28 mM[19]). Hence, all following experiments were performed with a pipette solution containing a $[Na^+]_p$ of 150 instead of 10 mM.

The small volume of the coronary myocytes made it possible that their intracellular ionic composition could be rapidly modified. The I_{KCa} recorded at -50 mV reversed within 3 min polarity, i.e. turned from an outward into an inward current indicating the reversal of the potassium driving force, i.e. a fall of intracellular $[K^+]$ to a value lower than the 5.4 mM $[K^+]_o$. The dialysis of 150 mM $[Na^+]_p$ did not significantly increase resting $[Ca^{2+}]_c$. However, at 150 mM $[Na^+]_p$ the $[Ca^{2+}]_c$ response to sodium removal was stimulated.

With 150 mM $[Na^+]_p$, the $[Ca^{2+}]_c$ response to sodium removal was variable in amplitude and rate (see below). Fig. 2, top, shows that $[Ca^{2+}]_c$ increased from 160 to 320 nM within 12 s, then $[Ca^{2+}]_c$ remained constant. In comparison to the following caffeine-induced Ca^{2+} transient, the sodium-removal response is slow, "tonic" and of low amplitude. When $[Ca^{2+}]_c$ had stabilized at 320 nM, the 2 s short caffeine application induced Ca^{2+} transients that reached similar peak $[Ca^{2+}]_c$ as those in the control cells dialyzed with 10 mM $[Na^+]_p$ (Fig. 2 top). This result suggests that the elevated $[Ca^{2+}]_c$ did not inactivate the caffeine-induced SR Ca^{2+} release. It further suggests that the preceding sodium-removal did not significantly enhance the amount of caffeine-releasable Ca^{2+}.

The decay of the caffeine-induced Ca^{2+} transient occurred with half-times that were not different to the half-times for controls analyzed for 10 mM $[Na^+]_p$ above. The decay of $[Ca^{2+}]_c$ did not reach the elevated pre-caffeine level (320 nM), instead it fell to an undershoot of 80 nM (Fig. 2 top). In 10 cells, the absence of extracellular sodium could not prevent the undershoot (107 ±40 nM). The result suggests that Ca^{2+} handling by the Ca^{2+}-ATPases of the SR and the plasmalemma is powerful enough to even overcome the Ca^{2+} influx through the stimulated Na-Ca exchanger. The elimination of the tonic increase in $[Ca^{2+}]_c$ by the caffeine-application could be attributed to a stimulation of the SERCa that sequestered the extra Ca^{2+} (having entered via Na-Ca exchange) into the emptied SR. An other possible explanation would be that the increase in $[Ca^{2+}]_c$ due to sodium removal is not only due to Ca^{2+} influx but has a component of Ca^{2+} release from the caffeine-sensitive store.

The effect of sodium removal was also tested during the period of the undershoot. In 3 of 10 cells, sodium removal slowly increased $[Ca^{2+}]_c$ to an elevated tonic level. Under those circumstances, a second caffeine application induced a Ca^{2+} transient whose peak was similar to that during the first caffeine application. The result supports the view that the decay of the Ca^{2+} transient induced by a short caffeine application is rate limited by Ca^{2+} reuptake by SERCa but that sodium removal can interfere with this process.

Possible contribution of SR Ca^{2+} release to the sodium-removal Ca^{2+} transients

With 150 mM [Na$^+$]$_p$, the rise of [Ca^{2+}]$_c$ during sodium removal was subject to variability; increments as small as 200 nM (Fig. 2, top) contrasted with transient

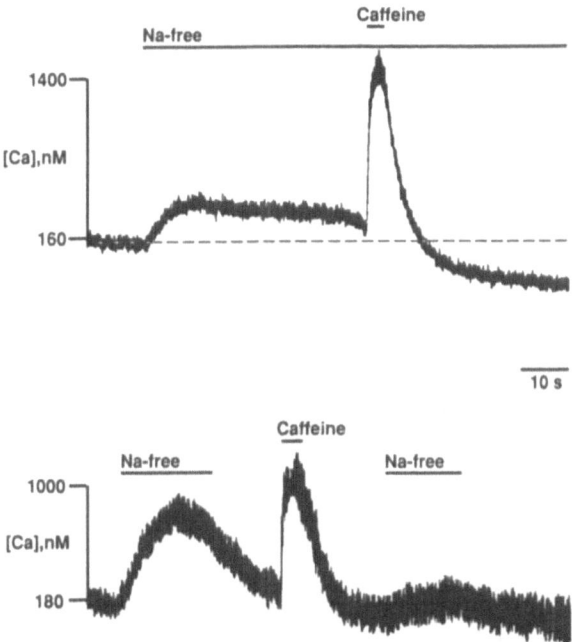

Figure 2. Effects of extracellular sodium removal in [Ca^{2+}]$_c$ recorded from myocytes dialyzed with 150 mM Na$^+$. Top: Example for an increase to a low sustained [Ca^{2+}]$_c$ value. Note: the response to caffeine is not modified by the absence of extracellular sodium. Bottom: Example for a transient increase to high [Ca^{2+}]$_c$ peak. Note: The second sodium removal, following the caffeine-transient, is strongly attenuated.

"regenerative" responses that peaked up to 2400 nM. Fig. 2, bottom,, shows an example where [Ca^{2+}]$_c$ rises transiently, i.e. in the constant absence of [Na$^+$]$_o$ a peak of almost 1000 nM is reached from which [Ca^{2+}]$_c$ decays. The high variability may suggest that in addition to the Ca^{2+} influx through the Na-Ca exchanger other mechanisms are involved in the effects of sodium removal.

We suggest that the SR can modify the Ca^{2+} influx through Na-Ca exchange as it is induced by sodium removal. The effect of Ca^{2+} influx on the [Ca^{2+}]$_c$ signal could be attenuated by sequestration of Ca^{2+} into the SR, and it could be amplified if it would trigger Ca^{2+} induced Ca^{2+} release from the SR. The latter possibility is suggested by results from cells whose SR Ca^{2+} load was reduced by pretreatment with caffeine (Fig. 2). In all cells the pretreatment reduced the amplitude of the Ca^{2+} transients due to sodium removal; the reduction was by 70% on average (n=4). The possibility that Ca^{2+} influx through Na-Ca exchange can trigger SR Ca^{2+} release is further supported by the finding that sodium-removal Ca^{2+} responses of large and transient amplitudes (Fig. 2 bottom, first response) attenuated the peak of a following caffeine-induced Ca^{2+} transient, as if the sodium-removal response had deprived the caffeine sensitive store of releasable Ca^{2+}.

In the continuous presence of caffeine, the Ca^{2+} release function of the SR is strongly suppressed. Under those circumstances, sodium-removal incremented $[Ca^{2+}]_c$ by only 100 nM which is low compared to the 620 nM increment recorded in caffeine-free control experiments. The comparison does not favor the view that the Ca^{2+} transients due to sodium removal are a direct consequence of Ca^{2+} influx. Instead, they support the hypothesis that the response comprises a large component due to SR Ca^{2+} release which is triggered by the Ca^{2+} influx through the Na-Ca exchanger. Probably, the contribution of Ca^{2+} release to the sodium-removal response varied in dependence on the degree of SR Ca^{2+} loading and the synchronization of the Ca^{2+} release channels.

Ryanodine attenuates Ca^{2+} transients due to sodium removal

Ryanodine is known to interact with the SR Ca^{2+} release channels more specifically than caffeine does. For reconstituted vascular preparations, the release channels were shown to be put in a permanently open low conductance state[17]. As a consequence, the SR Ca^{2+} is deprived of releasable Ca^{2+}, i.e. the Ca^{2+} induced Ca^{2+} release from SR is abolished. In coronary myocytes, 5 µM ryanodine, dialyzed out of the pipette solution, blocked the caffeine-induced Ca^{2+} transients within 2-3 min (Fig. 3). The inability of caffeine to raise $[Ca^{2+}]_c$ suggests that the caffeine-sensitive Ca^{2+} store was functionally removed. Under those conditions, sodium removal still increased $[Ca^{2+}]_c$ (Fig. 3). The increase was slow, and the amplitude of the sustained $[Ca^{2+}]_c$ values were 150 nM on average (n=5). This residual Ca^{2+} transient should be attributed to the direct effects of Ca^{2+} influx on $[Ca^{2+}]_c$ through the Na-Ca exchanger stimulated by the high $[Na^+]_p$ of 150 mM.

Our results suggest for the isolated coronary myocyte that the Na-Ca exchange is of minor importance for cellular calcium balance when $[Na^+]_p$ is 10 mM. For cells dialyzed

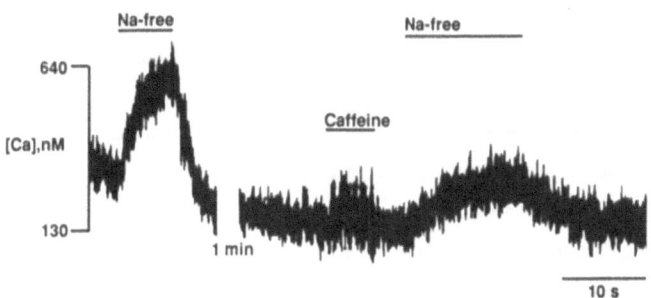

Figure 3. Ryanodine suppresses the $[Ca^{2+}]_c$ transient due to sodium-removal. Cell dialysis with 150 mM Na^+ plus 5 µM ryanodine. First response to sodium-removal was recorded before onset of the ryanodine-effect. Note: after 3 min of ryanodine-loading the $[Ca^{2+}]_c$ response to caffeine is blocked whereas the response to sodium-removal is largely reduced.

with solutions of high sodium content, we could demonstrate the effects of sodium removal on Ca^{2+} influx through stimulated Na-Ca exchange. However, sodium removal or reversal of the sodium gradient is a condition that will not occur in vivo and also 150 mM intracellular sodium is unlikely. Since the dialysis of high $[Na^+]_p$ did not significantly enhance resting $[Ca^{2+}]_c$, did not stimulate the peak of the caffeine-induced Ca^{2+} transients, or retard the decay of the caffeine-induced Ca^{2+} transients, a rather minor role of Na-Ca exchange is suggested for the Ca^{2+} balance of these coronary myocytes. This conclusion is in line with the low density of the molecules of Na-Ca exchanger in vascular tissue. It does not rule out that Na-Ca exchange is important for Ca^{2+} fluxes through the membranes of smooth muscle cells from other tissues such as the ureter[1]. If Ca^{2+} fluxes through Na-Ca exchange were the causal event for coronary vasoconstriction in man, these cells should have properties different to those described here.

REFERENCES

1. P.I. Aaronson and C.D. Benham, Alterations in $[Ca^{2+}]_i$ mediated by sodium-calcium exchange in smooth muscle cells isolated from the guinea-pig ureter, J. Physiol. (Lond.) 416:1 (1989).

2. T. Ashida and M.P. Blaustein, Regulation of cell calcium and contractility in mammalian arterial smooth muscle: the role of sodium-calcium exchange, J. Physiol. (Lond.) 392: 617 (1987).

3. C.D. Benham, ATP-activated channels gate calcium entry in single smooth muscle cells dissociated from rabbit ear artery, J. Physiol. (Lond.) 419: 689 (1989).

4. M.B. Cannell and D.G. Allen (1984) Model of calcium movements during activation in the sarcomere of frog skeletal muscle. Biophys. J. 45: 913 (1984).

5. D.D. Friel and R.W. Tsien, A caffeine- and ryanodine-sensitive Ca^{2+} store in bullfrog sympathetic neurones modulates effects of Ca^{2+} entry on $[Ca^{2+}]_i$, J. Physiol. (Lond.) 450: 217 (1992).

6. V.Ya. Ganitkevich and G. Isenberg, Isolated guinea pig coronary smooth muscle cells. Acetylcholine induces hyperpolarization due to sarcoplasmic reticulum calcium release activating potassium channels. Circ. Res. 67: 525 (1990).

7. V.Ya. Ganitkevich and G. Isenberg, Depolarization-mediated intracellular calcium transients in isolated smooth muscle cells of guinea-pig urinary bladder. J. Physiol. (Lond.) 435: 187 (1991).

8. V.Ya. Ganitkevich and G. Isenberg, Caffeine-induced release and reuptake of Ca^{2+} by Ca^{2+} stores in myocytes from guinea-pig urinary bladder, J. Physiol. (Lond.) 458: 99 (1992).

9. V.Ya. Ganitkevich and G. Isenberg, Contribution of Ca^{2+} -induced Ca^{2+} release to the $[Ca^{2+}]_i$ transients in myocytes from guinea-pig urinary bladder, J. Physiol. (Lond.) 458: 119 (1992).

10. V.Ya. Ganitkevich and G. Isenberg, Ca^{2+} entry through augmented Na-Ca exchange releases Ca^{2+} from caffeine-sensitive stores in guinea-pig coronary myocytes, J. Physiol. (Lond.), (1993).

11. A. Herrmann-Frank, E. Darling and G. Meissner, Functional characterization of the Ca^{2+} -gated Ca^{2+} release channel of vascular smooth muscle sarcoplasmic reticulum, Pflügers Arch. 418: 353 (1991).

12. G. Inesi and L. DeMeis, Regulation of steady state filling in sarcoplasmic reticulum. Roles of back-inhibition, leakage, and slippage of the calcium pump. J. Biol. Chem. 246: 5929 (1989).

13. L. Missiaen, F. Wuytack, L. Raeymaekers, H. DeSmedt, G. Droogmans, I. DeClerk and R. Casteels, Ca^{2+} extrusion across plasma membrane and Ca^{2+} uptake by intracellular stores. Pharm. Therap. 50: 191 (1992)

14. M.J. Mulvany, C. Aalkjaer and P.E. Jensen, Sodium-calcium exchange in vascular smooth muscle, Ann. N.Y. Acad. Sci. 639: 498 (1991).

15. S.C. O'Neill, P. Donoso and D.A. Eisner, The role of $[Ca^{2+}]_i$ and $[Ca^{2+}]$ sensitization in the contracture of rat myocytes; measurements of $[Ca^{2+}]_i$ and $[caffeine]_i$. J. Physiol. (Lond.) 425: 55 (1990)

16. C.M. Rembold, H. Richard and X.L. Chen, Na^+-Ca^{2+} exchange, myoplasmic Ca^{2+} concentration, and contraction of arterial smooth muscle. Hypertension 19: 308 (1992).

17. H. Reuter, M.P. Blaustein and G. Haeusler (1973), Na-Ca exchange and tension development in arterial smooth muscle. Phil. Trans. Roy. Soc. Lond B 265: 87 (1973).

18. E. Russeau and G. Meissner, Single cardiac sarcoplasmic reticulum Ca^{2+}-release channels: activation by caffeine, Am. J. Physiol. 256: H328 (1989).

19. R. Sitsapesan and A.J. Williams (1990) Mechanisms of caffeine activation of single calcium-release channels of sheep cardiac sarcoplasmic reticulum. J. Physiol. (Lond.) 423: 425 (1990).

20. B.J. Smith, R.M. Lyu and L. Smith, Sodium-calcium exchange in aortic myocytes and renal epithelial cells: Dependence on metabolic energy and intracellular sodium. Ann. N.Y. Acad. Sci. 639: 505 (1992).

21. W.G. Wier and D.T. Yue, Intracellular calcium transients underlying the short-term force-interval relationship in ferret ventricular myocardium, J. Physiol. (Lond.) 376: 507 (1986).

ROLE OF CALCIUM ATPASES IN PULMONARY VASCULAR REACTIVITY

Imad S. Farrukh and John R. Michael

Departments of Medicine, the Veterans Affairs Medical Center, and the
University of Utah Health Science Center,
Suite 4R240, 50 N. Medical Drive, Salt Lake City, Utah. 84312

INTRODUCTION

Changes in cytosolic Ca^{2+} $[Ca^{2+}]_i$ primarily determine the contraction/relaxation cycle of vascular smooth muscle. Agonists that increase $[Ca^{2+}]_i$ and induce vasoconstriction are counteracted by cellular Ca^{2+} transport mechanisms that decrease $[Ca^{2+}]_i$ and modulate vasoconstriction. In vascular tissue, the Ca^{2+}-ATPases play a crucial role in regulating $[Ca^{2+}]_i$ and tone. Pulmonary vascular smooth muscle and endothelial cells possess two forms of Ca^{2+}-ATPases: plasma membrane (PM) Ca^{2+} pumps (130-150 kd) and sarcoplasmic or endoplasmic reticulum (SER) Ca^{2+} pumps (100-115 kd) (12, 16, 17). The PM Ca^{2+} pumps along with Na^+/Ca^{2+} exchange are the major mechanisms for Ca^{2+} extrusion from the cell. The SER Ca^{2+}-ATPase pumps buffer an increase in $[Ca^{2+}]_i$ by reloading intracellular stores with Ca^{2+} (16, 36).

Both the PM and SER Ca^{2+}-ATPases are classified as members of the P-type ATPases (28), meaning that these enzymes form a covalently phosphorylated obligatory intermediate that arises from the transfer of the γ-phosphate of ATP to a specific aspartate residue at the catalytic site. Both pumps have a high affinity for Ca^{2+}, implying that they are active at low levels of $[Ca^{2+}]_i$ (6). The exact stoichiometry between Ca^{2+} ion transport and ATP hydrolysis is uncertain. In reconstituted purified enzyme preparations, a 1:1 Ca^{2+}:ATP stoichiometry has been found, i.e. one mole of Ca^{2+} is transported by the hydrolysis of one mole of ATP (27, 40). A stoichiometry approaching two has also been measured (34). Thus, regulation of $[Ca^{2+}]_i$ via Ca^{2+}-ATPases requires a significant expenditure of energy by the cell. Reconstituted Ca^{2+}-ATPases are active only in the presence of phospholipids, and negatively charged phospholipids can stimulate the Ca^{2+}-ATPases, illustrating the importance of the lipid environment to the function of these enzymes (40).

The PM and SER Ca^{2+}-ATPases have significant differences in their characteristics. The PM pumps are larger than the SER pumps, because they contain a regulatory domain that the SER pumps lack (38). Calmodulin increases the activity of the PM Ca^{2+} pumps, although an isoform in hepatocytes may be insensitive to calmodulin (1, 22). Calmodulin does not directly affect the activity of the SER Ca^{2+} pumps. It, however, may indirectly influence these pumps by activating phosphodiesterases which reduce intracellular levels of cAMP or cGMP, thereby possibly decreasing SER Ca^{2+}-ATPase activity (17). Phospholamban does not influence the PM Ca^{2+} pumps, but it regulates the activity of many, but not all, isoforms of the SER pump (16, 20, 32). Unphosphorylated phospholamban inhibits the SER Ca^{2+}-ATPases. Phosphorylation of phospholamban by the cAMP- or cGMP-dependent protein kinases relieves this inhibition, thus increasing the activity of the SER Ca^{2+} pumps.

At least four and possibly five different genes encode different isoforms of the PM Ca^{2+}-ATPase (PMCA 1 through PMCA 4) and at least three different genes (SERCA 1 through SERCA 3) encode SER Ca^{2+}-ATPase isoforms (17, 38). Expression of these genes combined with alternative splicing of their mRNAs lead to a variety of Ca^{2+}-ATPase isoforms that may be tissue-specific and differentially regulated (3). Human fetal lung tissue, for example, appears to contain primarily PMCA 1b, 4a, and 4b isoforms (3). As an example of differential regulation, phospholamban regulates the activity of the enzymes produced by two of the genes for the SER pump (SERCA 1 and SERCA 2), but the enzymes produced from the third gene (SERCA 3) appear to lack the binding site for phospholamban and may be regulated directly by cAMP-dependent protein kinase (4, 17, 20). Potential sites for alternative splicing of the mRNAs for the PM Ca^{2+}-ATPase pumps have been identified near the calmodulin binding domain and near the serine that is phosphorylated by the cAMP-dependent protein kinase. Insertions or deletions at these sites may significantly affect the regulation of the PM pump by these two signal transduction mechanisms (3, 39). In fact, isoforms of PM Ca^{2+}-ATPase that are insensitive to regulation by calmodulin (1, 6) or cAMP-dependent protein kinase have been identified (3, 23, 39).

EFFECT OF VANADATE ON PULMONARY VASCULAR TONE

To investigate the physiological effects of inhibiting Ca^{2+}-ATPases in the pulmonary circulation, we have used sodium orthovanadate (13). Vanadate is the classic inhibitor of the P-type ATPases; it can inhibit both the PM and SER pumps, blocking the PM Ca^{2+}-ATPases at lower concentrations than are required to inhibit the SER Ca^{2+} pumps (19). Vanadate is thought to inhibit the Ca^{2+}-ATPases by acting as a transition state analog of phosphate. Vanadate binds to the Ca^{2+}-ATPases in their E_2 configuration, thereby blocking the last step of the reaction cycle that returns the enzyme to the starting E_1 conformation. Vanadate, however, can also block other P-type ATPases, such as the Na^+/K^+-ATPase. Additionally, vanadate in mM concentrations can produce a variety of effects in cell experiments, including possible activation of GTP-binding proteins (21) and inhibition of phosphotyrosine phosphatases (15). Thus, vanadate clearly inhibits Ca^{2+}-ATPases, but it may also have additional effects that could influence vascular tone.

We studied the effect of vanadate on pulmonary vascular tone in the isolated ferret lung perfused at constant flow with equal amounts of autologous blood and Krebs-Henseleit buffer. We first investigated the possibility that vanadate might increase tone by inhibiting the Na^+/K^+-ATPase. Three pieces of evidence indicate that vanadate does not inhibit Na^+/K^+-ATPase in the isolated ferret lung. First, ouabain, which inhibits Na^+/K^+-ATPase (37), and vanadate produce additive increases in pulmonary arterial pressure (Fig. 1). Second, ouabain likely increases vascular tone by increasing Ca^{2+} entry via Na^+/Ca^{2+} exchange. This occurs because blocking Na^+/K^+-ATPase raises intracellular Na^+ (Na^+_i), thereby enhancing Na^+_i-dependent Ca^{2+} entry. If ouabain and vanadate act via different mechanisms, then an inhibitor of Na^+/Ca^{2+} exchange such as amiloride should prevent the effect of ouabain but not vanadate (35). Amiloride prevents and reverses the increase in pulmonary arterial pressure caused by ouabain, but does not affect the vasoconstriction produced by vanadate. Third, 8 bromo cGMP prevents the increase in pressure caused by vanadate but not the vasoconstriction produced by ouabain. Thus, vanadate does not increase pulmonary arterial pressure by inhibiting Na^+/K^+-ATPase. Other investigators have also found that vanadate can inhibit Ca^{2+}-ATPases without blocking Na^+/K^+-ATPase (18, 19). This is thought to occur because of the intracellular conversion of the pentavalent vanadate ion which inhibits Na^+/K^+-ATPase to the vanadyl ion which does not block this enzyme (5, 42).

We have also found that vanadate increases $[Ca^{2+}]_i$ in ferret and rabbit pulmonary vascular smooth muscle cells loaded with Indo-1. The increase in $[Ca^{2+}]_i$ is slow and persistent. In blood perfused isolated ferret lungs ventilated with a normoxic gas mixture (21% O_2, 5% CO_2, balance N_2), vanadate causes a dose-dependent increase in pulmonary arterial pressure compared to a time control (Fig. 2) (13). Ventilation of the lungs with a hypoxic gas (4% O_2, 5% CO_2, balance N_2) causes hypoxic pulmonary vasoconstriction (Fig. 2). Vanadate also produces a dose-dependent increase in pulmonary arterial pressure during hypoxia (Fig. 2) (13). Vanadate and hypoxia produce additive effects, implying that hypoxia does not act primarily by inhibiting Ca^{2+}-ATPase activity (Fig. 2).

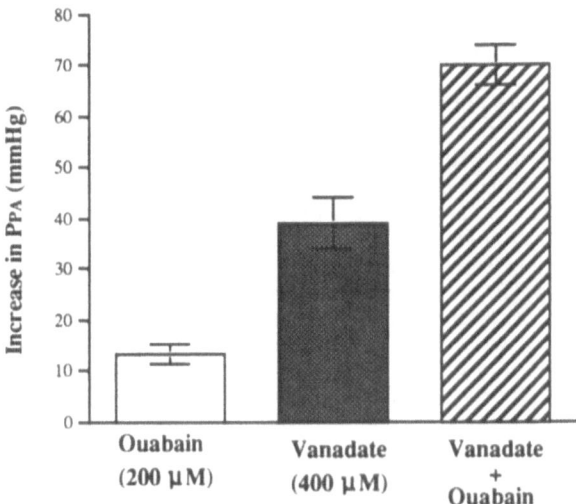

Figure 1. Increase in pulmonary arterial pressure (P_{PA}) caused by the maximally effective doses of ouabain and vanadate given separately or together. Experiments performed in isolated ferret lungs perfused at constant flow. Ouabain (200 µM) and vanadate (400 µM) produce an additive or greater increase in pressure. Values are mean ± SEM. Baseline pulmonary arterial pressure was 10 ± 1 mmHg in all groups.

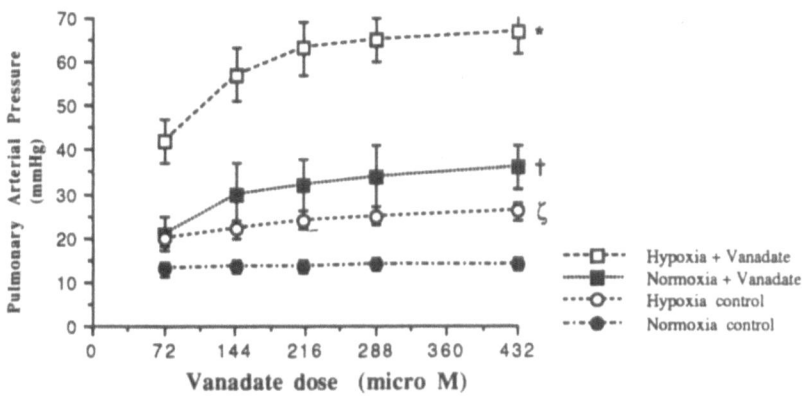

Figure 2. During normoxia (closed squares), vanadate causes a dose-dependent increase in pulmonary arterial pressure († $P<0.025$ compared with the normoxic time control group (closed circles). Ventilating the lungs with 4% O_2 increases pressure in the hypoxic time control group (open circles) (ζ $P<0.02$ compared to normoxic time control). Vanadate causes a further increase in pressure in hypoxic lungs (open squares) (* $P<0.02$ compared with the hypoxic time control). Reprinted with permission from Am. Rev. Respir. Dis. 145: 1389-1397, 1992.

If one changes the ventilating gas from a hypoxic to a normoxic mixture, pulmonary arterial pressure decreases to the normoxic level in less than five minutes in untreated lungs. In vanadate treated lungs, removing hypoxia also decreases pulmonary arterial pressure but to the higher level present in the normoxic vanadate treated lungs. Recovery from hypoxic pulmonary vasoconstriction in lungs treated with vanadate requires two to three times longer than in untreated lungs (13). Thus, vanadate delays recovery from hypoxic pulmonary vasoconstriction, consistent with its ability to inhibit Ca^{2+}-ATPases. This implies that the presence of functional Ca^{2+}-ATPases is crucial to the regulation of pulmonary vascular tone.

Vanadate also increases pulmonary arterial pressure in lungs perfused with a Ca^{2+}-free buffer and EDTA. Vanadate significantly increases pulmonary arterial pressure in both normoxic and hypoxic lungs perfused without extracellular Ca^{2+} (Fig. 3). This increase in pressure suggests that vanadate in addition to blocking the PM Ca^{2+}-ATPases also inhibits the SER Ca^{2+}-ATPases. Otherwise, we would anticipate that active SER Ca^{2+} pumps would overcome the physiological effect of inhibiting the PM Ca^{2+} pumps. Interestingly, the vanadate-induced increase in pressure is similar in normoxic and hypoxic lungs. The addition of extracellular Ca^{2+} to the perfusate further increases pulmonary arterial pressure in normoxic lungs (Fig. 3). Adding extracellular Ca^{2+} dramatically increases pressure in the hypoxic lungs treated with vanadate. These results imply that hypoxic lungs treated with vanadate, in which the Ca^{2+}-ATPases have been inhibited, have a difficult time handling the increase in $[Ca^{2+}]_i$ arising from the enhanced Ca^{2+} entry induced by hypoxia.

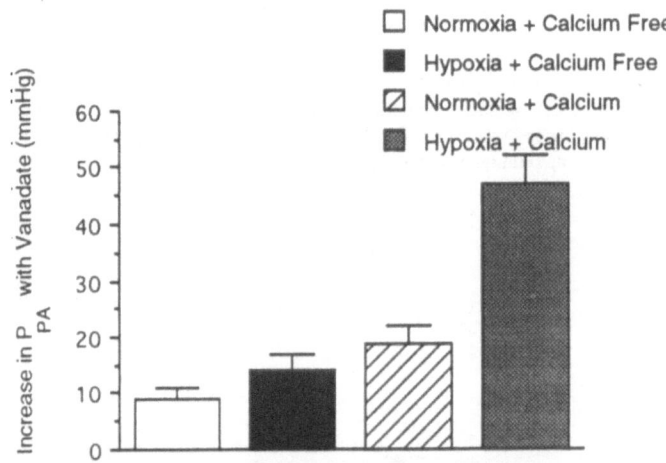

Figure 3. This figure illustrates the increase in pulmonary arterial pressure (P_{PA}) caused by vanadate (432 µM) in isolated ferret lungs perfused with or without extracellular Ca^{2+} (Ca^{2+}--free buffer plus 1 mM EDTA). Baseline pressure was 10 ± 1 mmHg in all groups.

We also investigated the ability of cyclic nucleotides to counteract the effect of vanadate (Fig. 4) (13). 8 bromo cGMP almost completely reverses the increase in pressure caused by vanadate without preventing the increase in pressure caused by ouabain. Dibutyryl cAMP posttreatment causes a small decrease in the response to vanadate (Fig. 4). The cGMP analog, however, produces a significantly greater relaxation than does the cAMP analog (decrease in pulmonary arterial pressure with 8 bromo cGMP is 30 ± 2 vs 6 ± 1 mm Hg with the cAMP analogue, $P<0.001$). In addition, pretreatment with 8 bromo cGMP significantly reduces the vasoconstriction caused by vanadate (Fig. 5). In contrast, pretreatment with dibutyryl cAMP does not significantly reduce the increase in pressure produced by vanadate (Fig. 5).

We have also found similar results with vanadate in the isolated rabbit lung. In addition, we have examined the effect of vanadate on $[Ca^{2+}]_i$ and tone in human pulmonary arterial cells and vascular rings. Vanadate produces similar physiological effects in human tissue as in the rabbit and ferret lung.

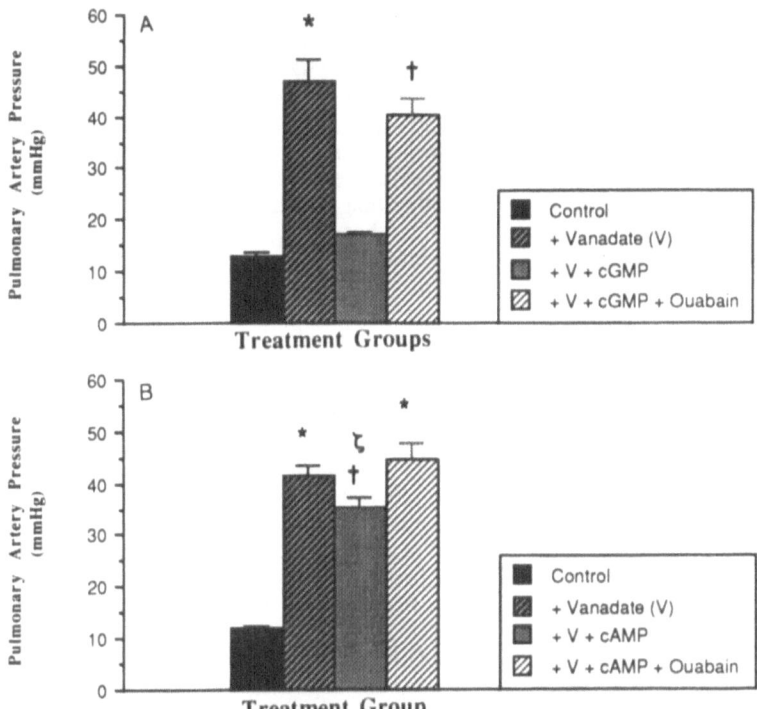

Figure 4. In panel A, vanadate (200 μM) increases pulmonary arterial pressure. Post-treating with 8 bromo cGMP (1 mM) significantly reduces the increase in pressure caused by vanadate. Infusing an additional dose of vanadate (200 μM) does not increase pressure (not shown), but infusion of ouabain (200 μM) further increases pressure. * P <0.008 compared with control or vanadate + cGMP. † P<0.0045 compared with vanadate + cGMP. In separate experiments shown in panel B, post-treating with dibutyryl cAMP (1 mM) only slightly reverses the vasoconstriction produced by vanadate (200 μM). * P< 0.001 compared with control. † P<0.01 compared to vanadate. ζ P< 0.01 compared to control. Reprinted with permission from Am. Rev. Respir. Dis. 145: 1389-1397, 1992.

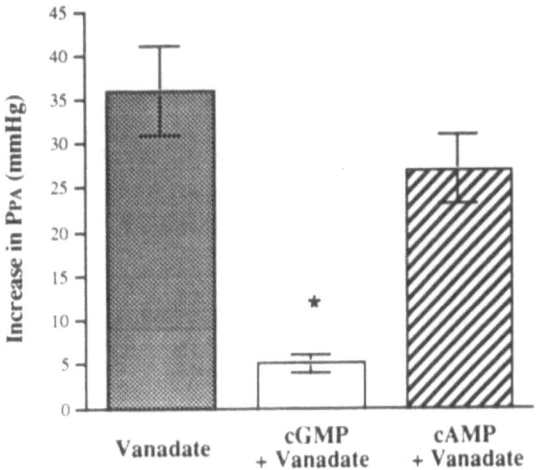

Figure 5. Pretreatment with 8 bromo cGMP (1 mM), but not dibutyryl cAMP (1 mM), prevents the increase in pulmonary arterial pressure (P_{PA}) caused by vanadate (400 μM). Baseline pressure was 10 ± 1 mm Hg in all groups.

DISCUSSION

Although vanadate is the classic inhibitor of the Ca^{2+}-ATPases, as noted above, it may have additional effects that could influence vascular tone. Our results indicate that vanadate does not increase vascular tone by inhibiting Na^+/K^+-ATPase. The inability of amiloride to affect the increase in tone also suggests that vanadate does not act by altering bidirectional Na^+/Ca^{2+} exchange. Vanadate doubles pulmonary arterial pressure in lungs perfused without extracellular Ca^{2+}, indicating that the effect of vanadate does not require Ca^{2+} entry. In the isolated rat lung, vanadate has been proposed to increase pulmonary vascular reactivity by activating protein kinase C (41). In the isolated rabbit and ferret lung, we have found at least three major differences between the response to activators of protein kinase C and vanadate (13, 26): first, activators of protein kinase C do not increase pulmonary arterial pressure in isolated lungs perfused with Ca^{2+}-free buffer and EDTA while vanadate does; second, pretreatment with dibutyryl cAMP prevents the increase in pressure caused by protein kinase C activators but does not prevent the response to vanadate; and third, 8 bromo cGMP does not reverse the vasoconstriction caused by protein kinase C activators but almost completely reverses the effect of vanadate. Thus, in our experiments, vanadate does not appear to act by stimulating protein kinase C.

A striking feature of our results is the ability of a cGMP analog to prevent and reverse the vasoconstriction caused by vanadate, while a cAMP analog is ineffective. An increase in intracellular cGMP has been proposed to reduce $[Ca^{2+}]_i$ and vascular tone by a variety of mechanisms, including diminished Ca^{2+} influx (9), decreased release of stored intracellular Ca^{2+} (9), or increased SER (10, 12, 31) or PM Ca^{2+}-ATPase activity (12, 14, 19, 29, 31, 33).

Previous studies have found that an increase in intracellular cGMP with subsequent activation of the cGMP-dependent protein kinase augments PM Ca^{2+}-ATPase activity in vascular smooth muscle (12, 14, 19, 33). Furukawa and coworkers, for example, report that compounds that increase intracellular cGMP enhanced the activity of the PM Ca^{2+}-ATPase in rat aortic smooth muscle cells (14). In contrast, treatment with forskolin or dibutyryl cAMP does not increase PM Ca^{2+}-ATPase activity. A similar difference between the effect of compounds that increase cGMP and cAMP on the PM Ca^{2+} pump has also been noted in pig coronary arterial smooth muscle (29). As discussed above, the inability of compounds capable of increasing cAMP to augment PM Ca^{2+} pump activity may reflect the presence of cAMP-insensitive isoforms (3, 17, 23). Interestingly, cGMP-dependent protein kinase does not appear to activate directly the PM Ca^{2+} pumps (2, 19). Although its mechanism of action is unknown, cGMP-dependent protein kinase may act by increasing the availability of certain membrane phospholipids. Phosphatidylinositol, phosphatidylserine, and phosphatidylinositol 4,5-bisphosphate (PIP_2), for example, augment PM Ca^{2+}-ATPase activity; in contrast, the metabolic products of PIP_2 do not stimulate the PM Ca^{2+} pump (7, 8).

In a variety of cells, both cAMP- and cGMP-dependent protein kinases can activate SER Ca^{2+} pumps that are regulated by phospholamban. Phosphorylation of phospholamban by these kinases leads to an increase in the activity of the SER Ca^{2+} pumps. A recent report, however, suggests that compounds that raise intracellular cGMP, but not compounds that raise cAMP, increase SER Ca^{2+}-ATPase activity in rat aortic vascular smooth muscle cells (10). The authors speculate that this differential effect may occur because only the cGMP-dependent protein kinase is present in membrane compartments containing phospholamban and the SER Ca^{2+} pumps (10). A number of investigators believe that increasing Ca^{2+}-ATPase activity is an important mechanism by which an increase in cGMP leads to vascular relaxation (10, 14, 29, 33). The ability of a cGMP analog to prevent and reverse preferentially the increase in vascular tone caused by vanadate further emphasizes the potential physiological significance of this mechanism of action.

Our results imply that small changes in the activity of either the SER or PM Ca^{2+} pumps may significantly affect vascular tone. The effect of agonists or physiological processes on Ca^{2+}-ATPase activity has been a relatively neglected area for investigation. In hepatocytes, classic agonists, such as vasopressin, angiotensin II, and epinephrine, may increase $[Ca^{2+}]_i$ in part by blocking PM Ca^{2+}-ATPase activity (30). The importance of this mechanism to the ability of these vasoconstrictors to increase $[Ca^{2+}]_i$ in vascular smooth muscle is unknown. The multiplicity of genes that encode for the PM and SER pumps and the potential formation of different isoforms from each gene due to alternate mRNA splicing gives rise to the possibility of cell- and tissue-specific Ca^{2+}-ATPase pumps. Physiological processes can also change the expression of these genes or alter the predominant type of isoform produced. Chronic low frequency electrical stimulation, for example, can convert a fast twitch skeletal muscle to a slow twitch muscle. This change coincides with switching off SERCA 1 gene expression and increasing SERCA 2 gene transcription (24). In the heart, pressure overload hypertrophy or end-stage heart failure reduces the amount of the Ca^{2+} pump protein SERCA 2a (11, 25). The effect of important physiological conditions, such as chronic alveolar hypoxia, on the expression of Ca^{2+}-ATPase isoforms in pulmonary vascular cells has not been explored.

In summary, our results indicate that vanadate, the classic Ca^{2+}-ATPase inhibitor, raises $[Ca^{2+}]_i$ in pulmonary vascular smooth muscle cells, increases basal normoxic pulmonary vascular tone, and augments and delays recovery from hypoxic pulmonary vasoconstriction. Interestingly, a cGMP analog can prevent or reverse the dramatic increase in vascular tone caused by vanadate; in contrast, a cAMP analog is significantly less effective. The ability of cGMP, but not cAMP, to reverse the vasoconstriction caused by vanadate suggests that cGMP-dependent protein kinase, rather than cAMP-dependent protein kinase, regulates Ca^{2+}-ATPase activity in pulmonary vascular smooth muscle.

Acknowledgments

Supported by Medical Research Funds from the DVA. Dr. Farrukh is a recipient of a Career Development Award from the Veterans Administration and an Edward Livingston Trudeau Scholar Award from the American Lung Association. Dr. Michael is a recipient of a Research Career Development Award K04 HL-02297 from the National Institutes of Health.

REFERENCES

1. Bachs, O., K. S. Famulski, F. Mirabelli, and E. Carafoli. ATP-dependent Ca^{2+}-transport in vesicles isolated from the bile canalicular region of the hepatocyte plasma membrane. *Eur. J. Biochem.* 147: 1-7, 1985.

2. Baltensberger, E., E. Carafoli, and M. Chiesi. The Ca^{2+}-pumping ATPase and the major substrate of the cGMP-dependent protein kinase in smooth muscle sarcolemma are distinct entities. *Eur. J. Biochem.* 172: 7-16, 1988.

3. Brandt, P., R. L. Neve, A. Kammesheidt, R. E. Rhoads, and T. C. Vanaman. Analysis of the tissue-specific distribution of mRNAs encoding the plasma membrane calcium-pumping ATPases and characterization of an alternately spliced form of PMCA4 at the cDNA and genomic levels. *J. Biol. Chem.* 267: 4376-4385, 1992.

4. Burk, S. E., J. Lytton, D. H. MacLennan, and G. E. Shull. cDNA cloning, functional expression, and mRNA tissue distribution of a third organellar Ca^{2+} pump. *J. Biol. Chem.* 264: 18561-18568, 1989.

5. Cantley, L. C., and P. Aisen. The fate of cytoplasmic vanadate: implications on (Na$^+$,K$^+$)-ATPase inhibition. *J. Biol. Chem.* 254: 1781-1784, 1979.

6. Carafoli, E. The calcium pumping ATPase of the plasma membrane. *Annu. Rev. Physiol.* 53: 531-547, 1991.

7. Carafoli, E., and M. Zurini. The Ca^{2+}- pumping ATPase of plasma membranes: purification, reconstitution, and properties. *Biochim. Biophys. Acta* 683: 279-301, 1982.

8. Choquette, D., G. Hakim, A. G. Filoteo, G. A. Plishken, J. R. Bostwick, and J. T. Penniston. Regulation of plasma membrane Ca^{2+} ATPases by lipids of the phosphatidylinositol cycle. *Biochem. Biophys. Res. Commun.* 125: 908-915, 1984.

9. Collins, P., T. M. Griffith, A. H. Henderson, and M. J. Lewis. Endothelium-derived relaxing factor alters calcium fluxes in rabbit aorta: a cyclic guanosine monophosphate-mediated effect. *J. Physiol.* 381: 427-437, 1986.

10. Cornwell, T. L., K. B. Pryzwansky, T. A. Wyatt, and T. M. Lincoln. Regulation of sarcoplasmic reticulum protein phosphorylation by localized cyclic GMP-dependent protein kinase in vascular smooth muscle. *Mol. Pharmacol.* 40: 923-931, 1991.

11. de la Bastie, D., D. Levitsky, L. Rappaport, J. J. Mercadier, F. Marotte, C. Wisnewsky, V. Brovkovich, K. Schwartz, and A. M. Lompre. Function of the sarcoplasmic reticulum and expression of its Ca^{2+}-ATPase gene in pressure overload-induced cardiac hypertrophy in the rat. *Circ. Res.* 66: 554-564, 1990.

12. Eggermont, J. A., M. Vroliz, F. Wuytack, L. Raeymaekers, and R. Casteels. The $(Ca^{2+}-Mg^{2+})$-ATPases of the plasma membrane and of the endoplasmic reticulum in smooth muscle cells and their regulation. *J. Cardiovasc. Pharmacol.* 12 Suppl. 5: S51-S55, 1988.

13. Farrukh, I. S., and J. R. Michael. Cellular mechanisms that control pulmonary vascular tone during hypoxia and normoxia: possible role of Ca^{2+}ATPases. *Am. Rev. Respir. Dis.* 145: 1389-1397, 1992.

14. Furukawa, K.-I., Y. Tawada, and M. Shigekawa. Regulation of the plasma membrane Ca^{2+} pump by cyclic nucleotides in cultured vascular smooth muscle cells. *J. Biol. Chem.* 263: 8058-8065, 1988.

15. Gordon, J. A. Use of vanadate as protein-phosphotyrosine phosphatase inhibitor. *Methods Enzymol.* 201: 477-482, 1991.

16. Grover, A. K. Ca-pumps in smooth muscle: one in plasma membrane and another in endoplasmic reticulum. *Cell Calcium* 6: 227-236, 1985.

17. Grover, A. K., and I. Khan. Calcium pump isoforms: diversity, selectivity and plasticity. *Cell Calcium* 13: 9-17, 1992.

18. Hudgins, M., and G. H. Bond. Alteration by vanadate of contractility in vascular and intestinal smooth muscle preparations. *Pharmacology* 23: 156-164, 1981.

19. Imai, S., Y. Yoshida, and H.-T. Sun. Sarcolemmal $(Ca^{2+} + Mg^{2+})$-ATPase of vascular smooth muscle and effects of protein kinases thereupon. *J. Biochem.* 107: 755-761, 1990.

20. James, P., M. Inui, M. Tada, M. Chiesi, and E. Carafoli. Nature and site of phospholamban regulation of the Ca^{2+} pump of SR. *Nature* 342: 90-92, 1989.

21. Kanaho, Y., P. Chang, J. Moss, and M. Vaughan. Mechanism of inhibition of transducin guanosine triphosphatase activity by vanadate. *J. Biol. Chem.* 263: 17584-17589, 1988.

22. Kessler, F., F. Bennardini, O. Bachs, J. Serratora, P. James, A. J. Caride, P. Gazzotti, J. T. Penniston, and E. Carafoli. Partial purification and characterization of the Ca^{2+}-pumping ATPase of the liver plasma membrane. *J. Biol. Chem.* 265: 16012-16019, 1990.

23. Khan, I., and A. K. Grover. Expression of cyclic nucleotide sensitive and insensitive isoforms of plasma membrane Ca pump in smooth muscle and other tissues. *Biochem. J.* 277: 345-349, 1990.

24. Kirschbaum, B., J. Simoneau, A. Bar, P. Barton, M. Buckingham, and D. Pette. Chronic stimulation-induced changes of myosin light chains at the mRNA and protein levels in rat fast-twitch muscle. *Eur. J. Biochem.* 179: 23-29, 1989.

25. Mercadier, J. J., A. M. Lompre, P. Duc, K. R. Boheler, J. B. Fraysse, C. Wisnewsky, P. D. Allen, M. Komajda, and K. Schwartz. Altered sarcoplasmic reticulum Ca^{2+}-ATPase gene expression in the human ventricle during end-stage heart failure. *J. Clin. Invest.* 85: 305-309, 1990.

26. Michael, J. R., J. Yang, I. S. Farrukh, and G. H. Gurtner. Protein kinase C-mediated pulmonary vasoconstriction in the rabbit: role of extracellular calcium, arachidonic acid metabolites, and effect of vasodilators. *J. Appl. Physiol.* (In Press).: 1992.

27. Niggli, V., E. S. Adunyah, J. T. Penniston, and E. Carafoli. Purified $(Ca^{2+} - Mg^{2+})$ ATPase of the erythrocyte membrane: reconstitution and effect of calmodulin and phospholipids. *J. Biol. Chem.* 256: 395-401, 1981.

28. Pedersen, P. L., and E. Carafoli. Ion motive ATPases. I. Ubiquity, properties, and significance to cell function. *Trends Biochem. Sci.* 12: 146-150, 1987.

29. Popescu, L. M., C. Panoiu, M. Hinescu, and O. Nutu. The mechanism of cGMP-induced relaxation in vascular smooth muscle. *Eur. J. Pharmacol.* 107: 393-394, 1985.

30. Prpic, V., K. C. Green, P. F. Blackmore, and J. H. Exton. Vasopressin-, angiotensin II-, and α1-adrenergic- induced inhibition of Ca^{2+} transport by rat liver plasma membrane vesicles. *J. Biol. Chem.* 259: 1382-1385, 1984.

31. Raeymaekers, L., J. A. Eggermont, F. Wuytack, and R. Casteels. Effects of cyclic nucleotide dependent protein kinases on the endoplasmic reticulum Ca^{2+} pump of bovine pulmonary artery. *Cell Calcium* 11: 261-268, 1990.

32. Raeymaekers, L., F. Hofmann, and R. Casteels. Cyclic GMP-dependent protein kinase phosphorylates phospholamban in isolated SR from cardiac and smooth muscle. *Biochem. J.* 252: 269-273, 1988.

33. Rashatwar, S. S., T. L. Cornwell, and T. M. Lincoln. Effects of 8-bromo-cGMP on Ca^{2+} levels in vascular smooth muscle cells: possible regulation of Ca^{2+}-ATPase by cGMP-dependent protein kinase. *Proc. Natl. Acad. Sci.* 84: 5685-5689, 1987.

34. Sarkadi, B., I. Szasz, A. Garloczi, and G. Gardos. Transport parameters and stoichiometry of active calcium ion extrusion in intact human red cells. *Biochim. Biophys. Acta* 464: 93-107, 1977.

35. Schellenberg, G. D., L. Anderson, E. J. Crague Jr., and P. D. Swanson. Inhibition of synaptosomal membrane Na^+/Ca^{++} exchange transport by amiloride and amiloride analogues. *Mol. Pharmacol.* 27: 537-543, 1985.

36. Shima, H., and M. P. Blaustein. Modulation of evoked contractions in rat arteries by ryanodine, thapsigargin, and cyclopiazonic acid. *Circ. Res.* 70: 968-977, 1992.

37. Slaughter, R. S., J. L. Shevell, J. P. Felix, M. L. Garcia, and G. J. Kaczorowki. High levels of sodium-calcium exchange in vascular smooth muscle sacrolemmal membrane vesicles. *Biochemistry* 28: 3995-4002, 1989.

38. Strehler, E. E. Recent advances in the molecular characterization of plasma membrane Ca^{2+} pumps. *J. Membrane Biol.* 120: 1-15, 1991.

39. Strehler, E. E., M.-A. Strehler-Page, G. Vogel, and E. Carafoli. mRNAs for plasma membrane calcium pump isoforms differing in their regulatory domain are generated by alternative splicing that involves two internal donor sites in a single exon. *Proc. Natl. Acad. Sci. USA* 86: 6908-6912, 1989.

40. Verbis, J., F. Wuytack, G. DeSchutter, L. Raeymaekers, and R. Casteels. Reconstitution of the purified calmodulin-dependent $(Ca^{2+}-Mg^{2+})$-ATPase from smooth muscle. *Cell Calcium* 5: 253-263, 1984.

41. Voelkel, N. F., and J. Czartolomna. Vanadate potentiates hypoxic pulmonary vasoconstriction. *J. Pharmacol. Exp. Ther.* 259: 666-672, 1991.

42. Vyskocil, G., J. Teisinger, and H. Dlouha. The disparity between the effects of vanadate (V) and vanadyl (IV) ions on (Na^+K^+)-ATPase and K^+-phosphatase in skeletal muscle. *Biochem. Biophys. Res. Commun.* 100: 982-987, 1981.

FAST NA$^+$ CURRENT AND CA^{2+} CURRENTS IN SMOOTH MUSCLES

Nicholas Sperelakis, Zhiling Xiong, Yoshihito Inoue, Yusuke Ohya, Keiichi
Shimamura, David Bielefeld, and John Lorenz

Department of Physiology & Biophysics
University of Cincinnati
College of Medicine
Cincinnati, OH 45267-0576

INTRODUCTION

Most smooth muscle (SM) cells normally do not possess fast Na$^+$ channels, and inward current for the action potential (AP) is carried primarily through slow (L-type) Ca^{2+} channels. The slow Ca^{2+} channel activity is regulated by several mechanisms, as is well known for myocardial cells. In myocardial cells, cyclic AMP and cAMP-dependent protein kinase (cA-PK) stimulate slow Ca^{2+} channel activity, whereas cyclic GMP and cGMP-dependent protein kinase (cG-PK) inhibit channel activity (reviewed in Sperelakis et al., 1992). Phosphorylation by PK-C also stimulates activity of the myocardial slow Ca^{2+} channels (Domenici & Rogers, 1988). Intracellular ATP also modulates slow Ca^{2+} channel activity in myocardial cells, ATP being obligatory for channel activity (Sperelakis & Schneider, 1976; O'Rourke et al., 1992). Acidosis rapidly, reversibly, and rather selectively inhibits the slow Ca^{2+} channels in cardiac muscle (Vogel & Sperelakis, 1977; Belardinelli et al., 1979; Irisawa and Sato, 1987). Gating of myocardial Ca^{2+} slow channels by G$_s$-protein (GTP-activated alpha subunit) has also been demonstrated (Yatani et al., 1988). Intracellular ATP also modulates activity of slow Ca^{2+} channels in SM cells (Ohya and Sperelakis, 1989a, b), and phosphorylation by either cAMP or cGMP-mediated pathways inhibit slow Ca^{2+} channel activity (Bkaily et al., 1988a).

Fast Na$^+$ channels in cardiac muscle seem to be less regulated by metabolic status (e.g., [ATP]$_i$, pH$_i$), second messengers, and phosphorylation by the various PKs. However, it has been demonstrated that fast Na$^+$ channels appear in myometrial SM cells during the latter half of pregnancy in the rat, presumably under steroid hormone control (Ohya & Sperelakis, 1989c; Inoue & Sperelakis, 1991). In addition, we have recently found functioning fast Na$^+$ channels in rat and human intestinal (colonic) SM cells (Xiong et al., 1992) and in rat portal vein cells (Xiong & Sperelakis, unpublished) under normal conditions.

This review-type article, to be published as part of the proceedings of this conference on ion channel aspects of the pulmonary circulation, provides a brief summary of (a) the changes in fast Na$^+$ channels and slow Ca^{2+} channels in myometrial SM cells (from longitudinal layer) during pregnancy in the rat (Ohya & Sperelakis, 1989; Inoue & Sperelakis, 1991); (b) the existence of functional fast Na^+ channels in colonic SM cells (from circular layer) of proximal (ascending) colon in both adult rat and human (Xiong et al., 1992); (c) the presence of functional fast Na^+ channels in vascular SM cells from rat portal vein (Xiong & Sperelakis, unpublished); and (d) evidence for the regulation of the slow Ca^{2+} channels of vascular SM cells by cAMP, cGMP, and diacyl glycerol (DAG) via phosphorylation by cA-PK, cG-PK, and PK-C (Bielefeld & Sperelakis, unpublished observations; Xiong, Sperelakis and Fenoglio-Preiser, unpublished observations).

UTERINE SMOOTH MUSCLE CELLS

Fast Na^+ Channels in Uterine Smooth Muscle Cells from Pregnant Rat

The existence of fast Na^+ channels has been reported in pregnant rat myometrium (Ohya & Sperelakis, 1989; Martin et al., 1990; Inoue and Sperelakis, 1991) and in cultured human pregnant myometrial SM cells (Young & Herndon-Smith, 1991).

Single SM cells were obtained from the longitudinal layer of pregnant and postpartum rats (5, 9, 14, 18 and 21 days of gestation, and 1-day postpartum) using enzymatic digestion. Whole-cell voltage clamp was performed at room temperature (20-22°C). Membrane capacitance was determined from the current amplitude in response to a hyperpolarizing voltage ramp pulse of 0.2V/s, applied from a holding potential (HP) of 0 mV (to avoid contamination by any time-dependent ionic currents). The fitting of data to each equation was performed using the nonlinear least-squares method. The data are given as means ± SE.

To isolate the inward currents, the pipette was filled with high Cs^+ solution of the following composition (in mM): 130 Cs^+, 10 Na^+, 5 ATP, 5.2 Mg^{2+}, 10 EGTA, 30 Cl^-, 112 glutamate, 10 HEPES, pH 7.2. The bath solution contained 3 mM 4-aminopyridine. For recording the Na^+ current, the bath solution also contained: 0 Ca^{2+}, 0 K^+, 0 EGTA, 150 Na^+, 2 Mn^{2+}, 10 HEPES, 154 Cl^-, 10 glucose, pH 7.3. For the recording of both Na^+ and Ca^{2+} currents, 2 mM $CaCl_2$ was used instead of $MnCl_2$. For isolating Ca^{2+} channel current, the bath solution contained: 0 Na^+, 0 K^+, 100 Tris, 50 TEA^+, 2 Ca^{2+} (or Ba^{2+}), 10 HEPES, 154 Cl^-, 10 glucose, pH 7.3.

Two types of inward currents (fast and slow) were recorded from freshly-isolated single SM cells of pregnant (18-day) rat uterus. Figure 1 shows typical current traces of the fast and slow inward currents, evoked by command potentials to -50 mV and above from a HP of -90 mV. Note the initial fast current followed by a slower current (e.g., at command pulses to -10 and 0 mV). When external Na^+ was replaced with $Tris^+$ and TEA^+, only slow current ($I_{Ca(L)}$) remained (Fig. 2A). In contrast, when Ca^{2+} was replaced with Mn^{2+}, only the fast inward current ($I_{Na(f)}$) was observed (Fig. 2B). This fast Na^+ current is very sensitive to tetrodotoxin (TTX): 0.1 μM nearly completely suppressed the Na^+ current (Fig. 2C). The dose-response relationship for the inhibitory action of TTX on the Na^+ current gave a K_i value of 27 nM. I / V relationships for both inward $I_{Na(f)}$ and $I_{Ca(L)}$ currents are shown in Fig. 2D. The current density of $I_{Na(f)}$ is about one-half of that of $I_{Ca(L)}$.

Steady-state inactivation curves were obtained using a double-pulse protocol (Fig. 3). The inactivation curve for fast Na^+ current was more negative, compared with the slow Ca^{2+} current. The voltage for half-inhibition of the Na^+ current was -64 mV, and that for the Ca^{2+} current was -39 mV.

The steady-state window current for I_{Na} ($I_{Na(win)}$) was determined by plotting the activation and inactivation curves (Fig. 4). The region of overlap represents $I_{Na(win)}$, and gives the voltage range in which a small sustained inward current occurs: -55 to -25 mV. Since this voltage range over which $I_{Na(win)}$ exists is close to the resting potential of -50 to -60 mV, this window current may contribute to automaticity as a background inward (depolarizing) current.

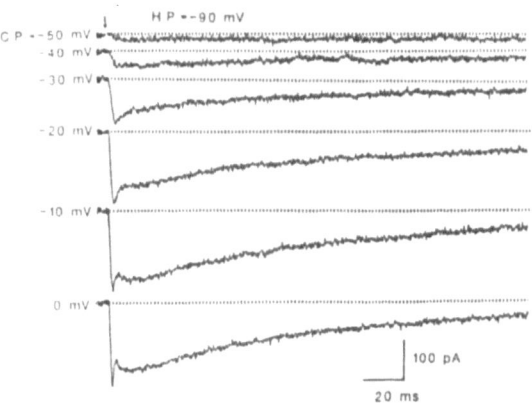

Fig. 1. Two types of inward current recorded from isolated single myometrial cells from pregnant rat using patch pipette technique. Depolarizing potential steps (-50 to 0 mV) were applied from a holding potential (HP) of -90 mV. Arrow indicates beginning of voltage step that continues to end of trace. Bath contained K^+-free solution with 150 mM Na^+ and 2 mM Ca^{2+}. Pipette solution contained high Cs^+. Cell capacitance was 80 pF. CP, command potential. (Taken with permission from Ohya & Sperelakis, 1989c.)

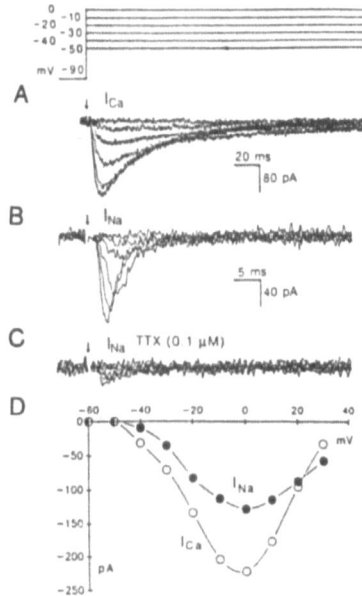

Fig. 2. Isolated fast Na^+ and slow Ca^{2+} inward currents. **A:** Slow Ca^{2+} inward current (I_{Ca}) recorded in a Na^+-free, K^+-free bath solution containing 2 mM Ca^{2+}. **B:** Fast Na^+ current (I_{Na}) recorded in a Ca^{2+}-free, K^+-free bath solution containing 150 mM Na^+. **C:** Suppression of Na^+ current by 0.1 μM TTX. Depolarizing voltage steps were applied in 10 mV increments from -50 to 0 mV from a HP of -90 mV. **D:** Current/voltage relationships for Ca^{2+} and Na^+ currents. Peak current amplitudes are plotted against command potentials. Pipette solution was the same as used in Fig. 2. All data were obtained from one cell (capacitance of 76 pF). (Taken with permission, from Ohya & Sperelakis, 1989c.)

79

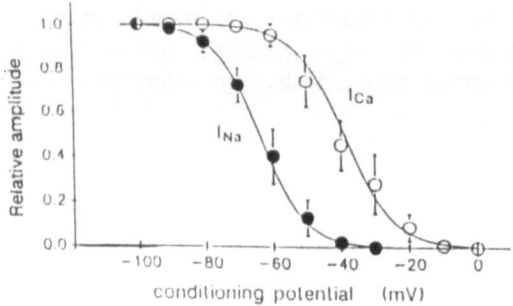

Fig. 3. Steady-state inactivation curves for Ca^{2+} (unfilled circles) and Na^+ (filled circles) currents. Conditioning pulses of various amplitudes were applied for 5 s, followed by a test pulse to -10 mV (for Ca^{2+} current) or to -20 mV (for Na^+ current). The two curves plotted were obtained by fitting data to the Boltzmann distribution. Each point represents the mean ± SE of 2-4 values from 4 cells. (Taken with permission, from Ohya & Sperelakis, 1989c.)

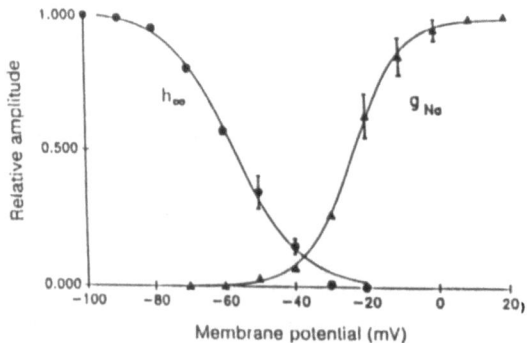

Fig. 4. Steady-state activation and inactivation curves for $I_{Na(f)}$. For the inactivation curve, a conditioning pulse to various amplitudes was applied for 2 sec before a test pulse to 0 mV. For the activation curve, data were obtained by measuring peak amplitude of I_{Na} elicited by various command potentials from a HP of -90 mV. Two curves were obtained by fitting data to the Boltzmann distribution. Each point represents mean ± SE (n = 7 (for inactivation) or 5 (for activation)). (Sperelakis et al., 1992.)

The steady-state window current for I_{Na} ($I_{Na(win)}$) was determined by plotting the activation and inactivation curves (Fig. 4). The region of overlap represents $I_{Na(win)}$, and gives the voltage range in which a small sustained inward current occurs: -55 to -25 mV. Since this voltage range over which $I_{Na(win)}$ exists is close to the resting potential of -50 to -60 mV, this window current may contribute to automaticity as a background inward (depolarizing) current.

Gain of Fast Na^+ Channels During Pregnancy

The change in fast Na^+ current during pregnancy was examined in rat myometrial SM cells during much of gestation and 1-day postpartum. A fraction of the cells did not possess $I_{Na(f)}$. For example, at day 14 of gestation, two types of cells were found: (a) cells with fast I_{Na} [$I_{Na}(+)$ cell] and (b) cells without fast I_{Na} [$I_{Na}(-)$ cell] (Fig. 5). The $I_{Na}(+)$ cells displayed both fast and slow inward currents (Fig. 5A). In contrast, the $I_{Na}(-)$ cells showed only a slow inward Ca^{2+} current (Fig. 5B).

The probability of existence of I_{Na} increased almost linearly during gestation (Fig. 6A). At day 5, none of the cells examined possessed I_{Na}, and at day 9 only 17% of the cells possessed I_{Na}. At day 14, I_{Na} could be detected in almost 50% of cells, and at days 18-21, the probability increased to ~85%. At 1 day post-partum, I_{Na} disappeared completely.

The averaged current densities, normalized by cell capacitance, were calculated from all cells examined (including those cells in which no I_{Na} was recorded) (Fig. 6B). The averaged density of I_{Na} increased almost linearly as term approached, and quickly decreased after parturition.

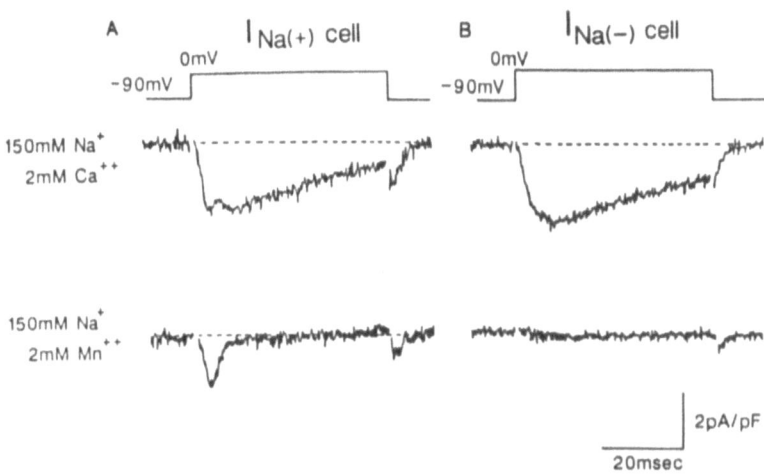

Fig. 5. Typical current traces recorded from two types of myometrial smooth muscle cells at day 14 of gestation: **A:** Cell with fast Na$^+$ current [I_{Na}(+)]. **B:** Cell without fast Na$^+$ current [I_{Na}(-)]. Depolarizing pulses to 0 mV were applied from a HP of -90 mV. Cell capacitances were 59.6 (A) and 68.0 pF (B). Bath solution was K$^+$ free and contained 150 mM Na$^+$ and 2 mM Ca^{2+}; isolated fast Na$^+$ current was obtained using Ca^{2+}-free bath solution containing 2 mM Mn^{2+} (to block any Ca^{2+} channels). **A:** Two types of inward currents (fast Na$^+$ and slow Ca^{2+} current) are present. **B:** Only slow inward Ca^{2+} current was present in control solution (top trace), and no inward current was present in Ca^{2+}-free solution (bottom trace). (Taken with permission, from Inoue & Sperelakis, 1991.)

Figure 6C shows the current densities of I_{Na} obtained from only those cells that possessed visible I_{Na}. The current densities obtained from such cells did not change during gestation.

It is proposed that Na$^+$ channel protein is synthesized and incorporated into the myometrial cell membrane during gestation. Once the gene for Na$^+$ channels is turned on, the synthesis proceeds very quickly, because the averaged current densities obtained from I_{Na}(+) cells did not change during gestation (Fig. 6C). That is, a larger and larger fraction of the cells gain fast Na$^+$ channels during pregnancy, in an apparent all-or-none manner, reaching nearly 100% of the cells near term.

A preliminary saxitoxin (STX) binding study (Tomsig et al., unpublished observations) supports the gain of fast Na$^+$ channels during pregnancy. The specific binding of STX to uterine smooth muscle increases during gestation, from a very low amount on day 5, to maximal values on days 18-21, and decreases again in postpartum. The general trend is consistent with the changes in $I_{Na(f)}$ observed in the voltage-clamp experiments.

A functional role for these newly-gained fast Na^+ channels may be proposed. (a) The insertion of fast Na^+ channels into the cell membrane during gestation should allow for faster and more complete propagation over the entire uterus by increasing max dV/dt of the AP. This factor, combined with the increase in number of gap junctions (Miller et al., 1989), may result in faster propagation of excitation, and hence more forceful contractions because of the series elastic element. (b) The increased $[Na]_i$ produced by Na^+ influx through fast Na^+ channels during excitation may cause elevation of $[Ca]_i$ by the Na^+ / Ca^{2+} exchange system (reverse mode) to potentiate myometrial contraction (Savineau et al., 1987). (c) It is possible that Na^+ current may contribute to the pacemaker potential because a substantial amount of $I_{Na(win)}$ (or slowly-inactivating inward Na^+ current ($I_{Na(si)}$)) exists in myometrial cells (Fig. 4).

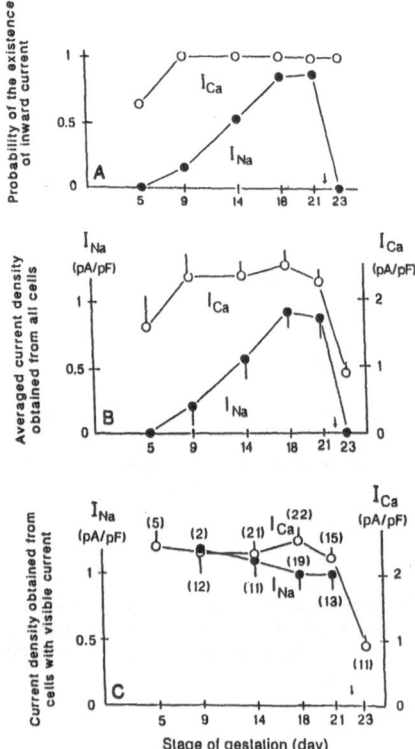

Fig. 6. Changes in the averaged densities for Na^+ and Ca^{2+} channels in myometrial smooth muscle cells at different stages of gestation. **A:** Change in probability of the existence of the inward Na^+ and Ca^{2+} currents during gestation. Probability (the fraction of cells with visible inward current) was plotted against days of gestation. Note that the fraction of cells exhibiting I_{Ca} was very high (1.0) beginning on day 9 of gestation; in contrast, the fraction of cells possessing an I_{Na} was 0 at day 5 and increased dramatically to > 0.8 at day 18-21; the fraction became zero at 1-day postpartum. **B:** Averaged current densities (normalized by cell capacitance) for Na^+ and Ca^{2+} currents obtained from all cells examined were plotted against days of gestation. These values include those cells in which there was zero I_{Na} or I_{Ca}. Each point represents mean ± SE. **C:** Current densities for Na^+ and Ca^{2+} currents obtained from only those cells that displayed visible inward current were plotted against days of gestation. Each point represents means ± SE. Note that the densities of I_{Na} and of I_{Ca} were similar in the cells from all stages of gestation examined; the Ca^{2+} current density was decreased 1 day postpartum. Modified from (Inoue & Sperelakis, 1991), by addition of unpublished data for 1-day postpartum.

Summary of Ca^{2+} Channels in Uterine Smooth Muscle

Various types of ion channels are known to exist in myometrial SM cells. Two types of Ca^{2+} channels (L and T) exist in human pregnant myometrium (Inoue et al., 1990). On the other hand, only one type of Ca^{2+} channel (the L-type) exists in pregnant rat myometrium (Ohya and Sperelakis, 1989; Inoue and Sperelakis, 1991).

Gain of Slow Ca^{2+} Channels During Pregnancy

All myometrial cells of day 9 and later possessed I_{Ca}, but some cells at day 5 did not display visible I_{Ca} (Fig. 6A). Thus, the probability of existence of Ca^{2+} channels changed less markedly during gestation than that of the fast Na^+ channels.

The averaged density of I_{Ca} of the cells at day 5 of gestation was smaller than those of cells at later stages of gestation (Fig. 6B). At 1 day post-partum, the average current density decreased dramatically, although all cells examined possessed I_{Ca} (Fig. 6B). The current density obtained from cells with visible I_{Ca} also did not change during gestation; however, it decreased dramatically after parturition (Fig. 6C).

Effects of Tocolytic and Uterotonic Agents on Ion Channels

(a) Isoproterenol (ISO). Beta-agonists are clinically used for prevention and treatment of preterm labor. High dose of ISO (10 μM) did not produce significant change in the Ca^{2+} current, in I_{Na}, or in I_K. Thus, beta-agonists must relax the uterine muscle by mechanisms other than inhibition of the Ca^{2+} channels or Na^+ channels. Sakai et al. (1992) has suggested that phosphorylation by cA-PK closes gap junction channels.

(b) Mg^{2+}. Mg^{2+} applied in the bath dose-dependently inhibited the Ca^{2+} current, with half-inhibition at 12 mM Mg^{2+}. This concentration of Mg^{2+} is not much different than the plasma levels attained during Mg^{2+} tocolysis, and so the effect of Mg^{2+} can be largely explained by a direct action on the Ca^{2+} slow channels.

(c) Nifedipine. Nifedipine inhibited the Ca^{2+} channel currents dose-dependently, with half-inhibition at 3.3 nM. Inhibition of nifedipine was greater at higher (more positive) HPs and higher command potentials. This direct action of nifedipine on the Ca^{2+} slow channels can explain its tocolytic action.

(d) Adenosine. Adenosine acts on a purinergic (P_1) receptor, and inhibits vascular SM contraction by elevating intracellular cAMP levels. However, adenosine had no effect either on $I_{Ca(L)}$, or on $I_{Na(f)}$ of uterine SM cells.

(e) Oxytocin. Oxytocin is often used in order to induce uterine contractions in cases of prolonged labor. Substantial oxytocin receptors are present at 18 days pregnancy in rat. In pregnant (18-day) rat myometrial SM cells, cultured with 1 μM estradiol, oxytocin enhanced I_{Ca} (Mironneau, 1990). However, in freshly-dispersed cells from the same species, oxytocin unexpectedly decreased I_{Ca} slightly (Sperelakis et al., 1992). Inhibition was not observed when Ba^{2+} was used as the charge carrier, suggesting the slight decrease in I_{Ca} was due to inhibition of the Ca^{2+} channel by Ca^{2+} release and influx. It has been reported that oxytocin increases IP_3 and DAG production, and IP_3-induced Ca^{2+} release from SR could account for its stimulation of contraction. Therefore, stimulation of uterine contraction by oxytocin cannot be explained by a direct effect on the Ca^{2+} channel.

I_{Na} was not affected by oxytocin in 18-day pregnant rat myometrium. Therefore, the stimulatory effect of oxytocin on uterine contraction cannot be explained by an effect on the fast Na^+ channels.

I_K was not affected by oxytocin in 18-day pregnant rat myometrium. Therefore, I_K may not have any important role in oxytocin-induced uterine contraction.

(f) ATP. ATP, an agent known to produce contraction of uterine SM cells, was added to the bath (1 - 1,000 µM) to activate the purinergic P_2 receptors. At 1,000 µM, ATP rapidly produced marked inhibition (about 40%) of $I_{Ca(L)}$ (Fig. 7B, C). ATP concentrations as low as 10 µM and 1 µM produced a similar, but smaller, inhibition of $I_{Ca(L)}$. Since the cell was simultaneously contracted, the inhibition of $I_{Ca(L)}$ may be mediated in part by IP_3-induced Ca^{2+} release from the SR and consequent Ca^{2+} inhibition of the Ca^{2+} channels. Inhibition of $I_{Ca(L)}$ is probably also caused by a decrease in the net driving force for Ca^{2+} influx (due to elevation of $[Ca]_i$). In addition, the holding current was shifted by ATP (Fig. 7A), representing a net inward sustained current, reflecting ATP activation of a receptor-operated non-selective cation channel, which allows Na^+ and Ca^{2+} entry (Honoré et al., 1989).

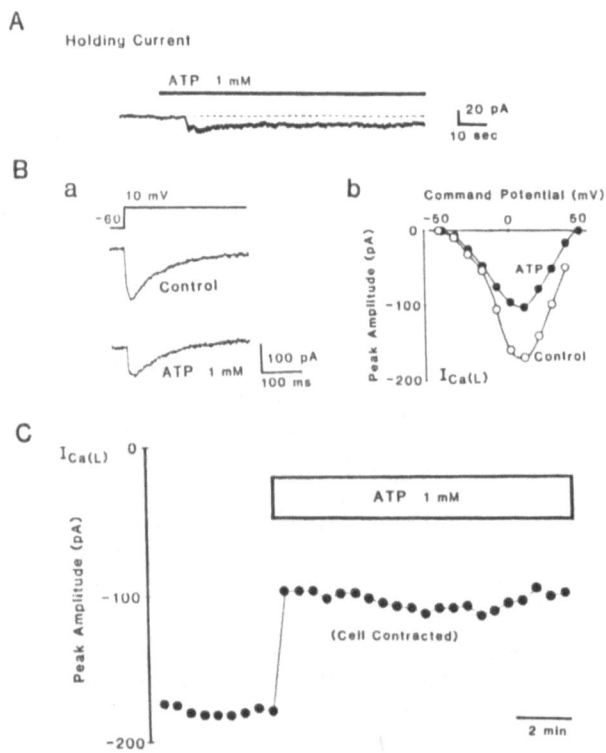

Fig. 7. Effect of ATP on I_{Ca} in longitudinal, myometrial smooth muscle cell from 18-day pregnant rat. Patch pipette contained 130 mM Cs^+ and nystatin (200 µg/ml) and bath solution 2 mM Ca.
A: Change in holding current by application of ATP (1 mM). **Ba:** $I_{Ca(L)}$ induced by voltage step from -60 to +10 mV was decreased by ATP. **Bb:** I-V relationship for $I_{Ca(L)}$ in the absence (unfilled circles) and presence (filled circles) of ATP. **C:** Typical time course of peak I_{Ca} change after ATP application. (Shimamura and Sperelakis, unpublished data.)

(g) Phorbol Ester and Protein Kinase-C. A phorbol ester (phorbol 12,13-dibutyrate; PDB), a good activator of PK-C, was added to the bath (0.03 - 0.3 µM) and its effect on $I_{Ca(L)}$ was determined. PDB substantially increased $I_{Ca(L)}$ within 5 min. (Fig. 8). This stimulation was reversed by addition of H-7, a non-specific protein kinase inhibitor (Fig. 8B). Staurosporine (0.1 µM), another and more specific inhibitor of PK-C, also reversed or prevented the stimulation of $I_{Ca(L)}$ produced by PDB.

K+ Currents

Voltage-dependent K^+ currents were recorded in Ca^{2+}-free, Na^+-free bath solution containing 6 mM K^+, and the pipette was filled with high-K^+ solution (130 mM). The presence of $I_{K(ATP)}$ and $I_{K(Ca)}$ channels could not be expressed because the pipette contained EGTA (10 mM) and ATP (5 mM). About two-thirds of all cells examined exhibited only a single component of I_K, namely a slow sustained component, and the remaining one-third exhibited a second component, namely an initial fast transient component. High dose of forskolin (30 μM) inhibited the slow component, but had little or no effect on the fast component. 4-aminopyridine inhibited primarily the fast transient component, whereas TEA inhibited primarily the slow sustained component.

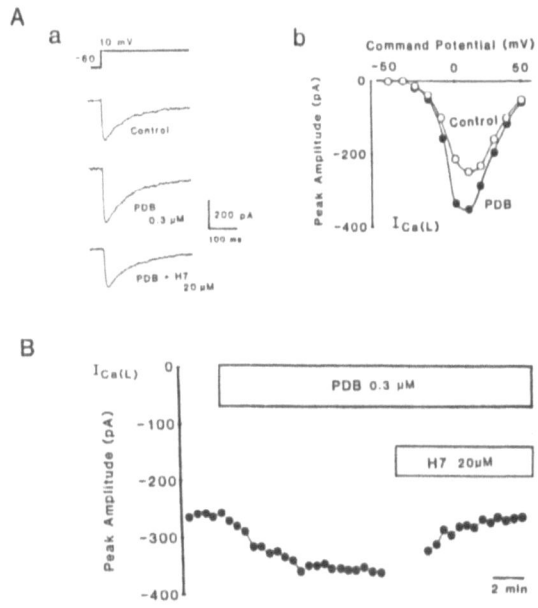

Fig. 8. Effect of phorbol 12,13-dibutyrate (PDB) on I_{Ca} in longitudinal, myometrial smooth muscle cell from 18-day pregnant rat. Patch pipette contained 130 mM Cs^+ and nystatin (200 μg/ml) and bath solution 2 mM Ca. **Aa:** $I_{Ca(L)}$ induced by voltage step from -60 to +10 mV was increased by PDB (0.3 μM). This effect was reversed by the presence of H7 (20 μM). **Ab:** I-V relationship for $I_{Ca(L)}$ in the absence (unfilled circles) and presence (filled circles) of PDB. **C:** Typical time course of peak $I_{Ca(L)}$ change after application of PDB alone or in combination with H7. (Shimamura and Sperelakis, unpublished data.)

Summary of Possible Mechanisms

Figure 9 summarizes the possible mechanisms by which uterine contractility can be modulated. In contrast to vascular SM, neither ISO nor adenosine, which produce elevation of cyclic AMP, affected I_{Ca} and I_{Na}. Therefore, no arrow can be drawn between cA-PK/cG-PK and the Ca^{2+} slow channel. Although oxytocin, which produces DAG and should activate PK-C, did not stimulate $I_{Ca(L)}$ (actually produced a small inhibition), an arrow can be drawn between PK-C and the Ca^{2+} slow channel based on the stimulation produced by phorbol ester and its prevention/reversal produced by PK-C inhibition.

Oxytocin may fail to exhibit a stimulation of $I_{Ca(L)}$ because of the predominance of the inhibition produced by the $[Ca]_i$ increase.

Since the fast Ca^{2+} channel (T-type) current could not be found in rat uterine muscle, as it was in human myometrium (Inoue et al., 1990), no arrows could be drawn between the protein kinases and the fast Ca^{2+} channel. Since isoproterenol and oxytocin had no effect on $I_{Na(f)}$, arrows could not be drawn between the protein kinases and the fast Na^+ channel. Very little is known about the effects of phosphorylation by the various PKs on regulation of (a) the K^+ channels, (b) the Ca/Na exchange system, and (c) the sarcolemmal Ca pump. Therefore, it is unclear through what mechanism(s) beta-agonists relax uterine smooth muscle. In contrast, oxytocin may stimulate uterine contraction through increasing IP_3 production and thereby Ca^{2+} release from the SR. Some utero-active agents may also act to alter the Ca^{2+} sensitivity of the contractile proteins and myosin light chain kinase (or phosphatase).

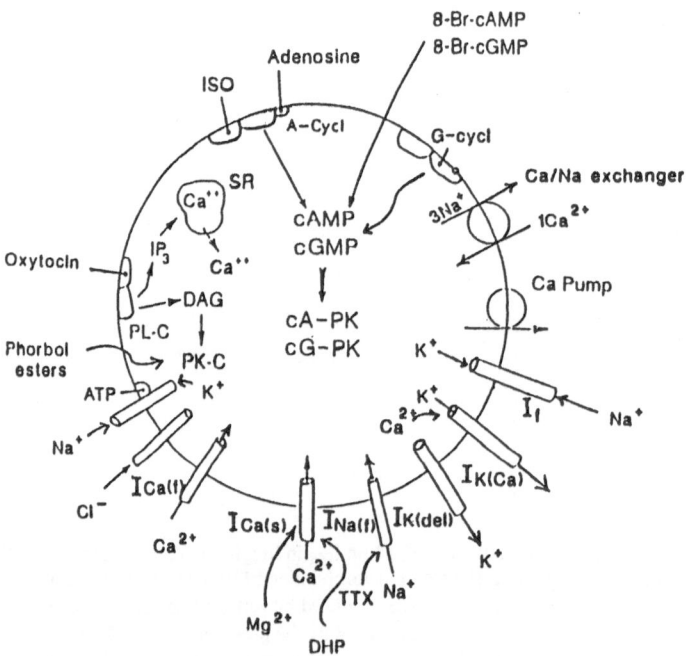

Fig. 9. Diagrammatic representation of a uterine smooth muscle cell from a pregnant rat myometrium. As shown, cAMP and cGMP do not produce relaxation via inhibition of the Ca^{2+} slow channels; PK-C also has no effect on $I_{Ca(L)}$. As depicted, Mg^{2+} ion and dihydropyridine (DHP) Ca antagonist drug do inhibit $I_{Ca(L)}$. The other ion channels, exchangers, and pumps known to be present are also depicted. (Modified from Sperelakis et al., 1992.)

COLONIC SMOOTH MUSCLE CELLS

Whole-cell voltage clamp was carried out on freshly-dispersed single smooth muscle cells from adult rat and human colons to investigate the possible presence of fast Na^+ channels and the regulation of the Ca^{2+} channels. We did find the existence of a fast Na^+ channel current (Fig. 10). With normal physiological salt solution (PSS) plus 4-aminopyridine (3 mM) in the bath and high Cs^+ solution in the pipette to inhibit outward K^+

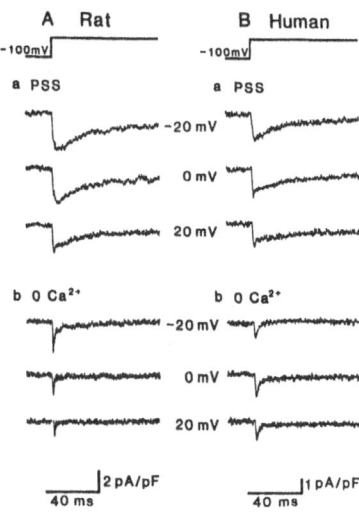

Fig. 10. Two components of inward current recorded from one colonic smooth muscle cell from rat (A) and one from human (B). Three different command potential steps (-20, 0, and +20 mV) were applied from a HP of -100 mV. Pipettes were filled with high Cs^+ solution. Bath contained PSS (Aa & Ba) and 0 Ca^{2+} solutions (Ab & Bb). The cell capacitance was 64.0 pF for the rat cell (240 day male), and 90 pF for the human cell (50 year man). Note that in PSS there were two components of the inward current: an initial fast component and a delayed slow component, and that in 0 Ca^{2+} solution, the second component disappeared, leaving only the initial transient component. (Xiong, Sperelakis, Noffsinger, and Fenoglio-Preiser, unpublished data.)

currents, an inward current possessing fast and slow components was observed when the cell membrane was depolarized more positive to -20 mV from a HP of -100 mV. When Ca^{2+} ion was removed from the PSS or when nifedipine (10 µM) and Ni^{2+} (30 µM) were simultaneously applied, the slow component disappeared and the fast component remained. The fast current component became almost completely inactivated within 10 ms. This fast component was dependent on $[Na]_o$ and was inhibited by TTX dose-dependently (IC_{50} of 130 nM in rat and 14 nM in human) (Fig. 11).

These results suggest that the slow component of inward current was a Ca^{2+} channel current, whereas the fast component was a TTX-sensitive fast Na^+ channel current. The threshold voltage, the voltage for peak current, and the reversal potential for the fast Na^+ current were, respectively, about -50, -20, and +50 mV in rat, and -40, 0, and +60 mV in human. The incidence of cells possessing fast Na^+ currents depended on the region of the colon: In rat proximal colon, the incidence was 64%; in distal colon, it was 10%. In humans, the incidence in the ascending colon was 73%, and descending colon was 22%. The densities of fast Na^+ and Ca^{2+} currents were 3.2 and 4.5 pA/pF in rat; 1.0 and 1.4 pA/pF in human, respectively. The ratio of both current densities (Na^+ vs Ca^{2+}) was 0.71, in both rat and human. We conclude that the major ion channels associated with the generation of inward currents in the circular SM cells of rat and human colon are voltage-dependent Ca^{2+} channels and fast Na^+ channels. The fast Na^+ current may facilitate propagation of excitation. Evidence for presence of fast Na^+ current in intestinal (rat ileum) SM cells was also given by Smirnov et al. (1992).

The developmental changes of Ca^{2+} channel currents in single SM cells were investigated in cells freshly isolated from the circular layer of distal colon from rat using the whole-cell voltage clamp technique. Under physiological conditions, the averaged total Ca^{2+} current density increased markedly from 1.25 pA/pF in newborn rat to 6.46 pA/pF in

60-day old rat, then gradually declined thereafter with age. There were two types of Ca^{2+} channel currents: one type, the T-type, possessed more negative threshold potentials (-65 mV) and inactivated quickly. The voltage for peak current was about -15 mV, and the reversal potential was about +65 mV. This current was highly sensitive to low concentration of Ni^{2+} (30 μM), but was resistant to nifedipine, diltiazem, Cd^{2+}, and TTX.

In contrast, the other type, the L-type, possessed more positive threshold potentials (-40 mV) and inactivated more slowly. The voltage for peak current was 0 mV, and the reversal potential was about +65 mV. This current was insensitive to low concentration of Ni^{2+}, but highly sensitive to nifedipine, diltiazem, and Cd^{2+}.

Figure 12 shows that the current densities of both $I_{Ca(L)}$ and $I_{Ca(T)}$ increased with development. $I_{Ca(T)}$ remained high during aging, whereas $I_{Ca(L)}$ declined.

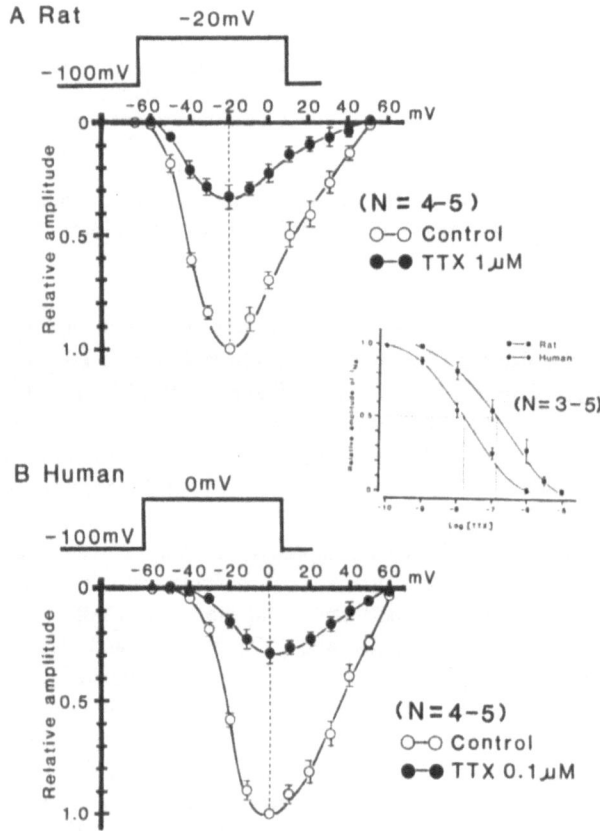

Fig. 11. Effects of tetrodotoxin (TTX) on fast I_{Na}. (A) and (B) are the averaged I/V relationships in the absence (open circles) and presence (filled circles) of TTX (1 μM in rat and 0.1 μM in human). Peak amplitude of control current (without TTX) is normalized as 1.0. Each point represents 4 to 5 values from 5 different cells. Note that the voltage for peak $I_{Na(f)}$ was -20 mV in rat, and 0 mV in human. Dose/response relationship for TTX action on peak $I_{Na(f)}$ was also presented in the middle of figure. As shown, the IC_{50} was 130 nM for rat (filled squares), and 14 nM for human (filled circles). (Xiong, Sperelakis, Noffsinger, and Fenoglio-Preiser, unpublished data.)

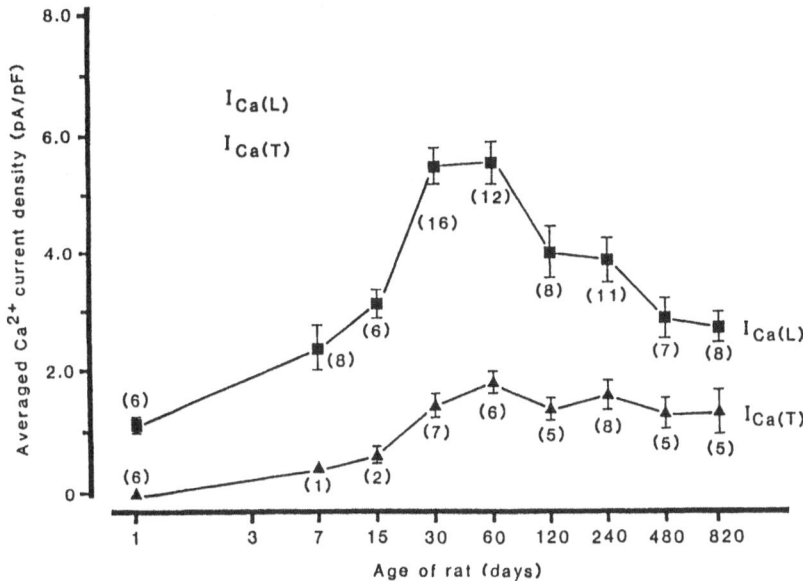

Fig. 12. Changes in averaged T-type and L-type Ca^{2+} current densities in colonic smooth muscle cells at various ages of rat. The HP was -100 mV. Current densities (normalized by cell capacitance) were plotted aganist age (days). Filled squares: $I_{Ca(L)}$ density; filled triangles: $I_{Ca(T)}$ density. Each point represents mean ± SE for the number of cells indicated in parentheses. Note that both $I_{Ca(L)}$ and $I_{Ca(T)}$ densities increased during development; with aging, $I_{Ca(L)}$ density declined ; whereas $I_{Ca(T)}$ density was constant. (Xiong, Sperelakis, Noffsinger, and Fenoglio-Preiser, unpublished data.)

VASCULAR SMOOTH MUSCLE CELLS

Presence of Fast Na^+ Channels

In rat portal vein cells, evidence for existence of fast Na^+ channels was found in about 35% of cells examined. In cells containing $I_{Na(f)}$, there were two inward currents: an initial fast transient current and a second slow more-sustained current (Fig. 13A). The first component was dependent on $[Na]_o$ and was blocked by TTX (Fig. 13A). The corresponding I / V curves for the two currents are given in Figure 13B. Similar findings which demonstrated the presence of $I_{Na(f)}$ in rat portal vein cells were made by Okabe et al. (1990).

Regulation of [Ca] by Cyclic Nucleotides

There are two types of Ca^{2+} channels in vascular SM cells: slow (L-type) and fast (T-type). The properties of these two types of Ca^{2+} channels are rather similar to those in myocardial cells, with the exception of the regulatory mechanisms which are different. The two types of Ca^{2+} currents can be distinguished by differences in their kinetics, voltage ranges of activation and inactivation, and sensitivities to pharmacological agents. The slow (sustained) Ca^{2+} channel is sensitive to dihydropyridine Ca^{2+} agonists and antagonists, prefers Ba^{2+} over Ca^{2+} as charge carrier, and activates at more positive membrane potentials (high threshold) than the fast (transient) Ca^{2+} channel. The fast Ca^{2+} channel activates at more negative membrane potentials (low threshold), has an equal preference for

Ca^{2+} and Ba^{2+}, and is insensitive to the dihydropyridines. Both cAMP and cGMP elevation have been implicated in the relaxation of vascular SM in response to some vasodilators, such as PGI$_2$, EDRF, ANF, and nitroprusside.

We examined the effects of cyclic nucleotide analogs and related agents on the electrical properties of cultured rat aortic reaggregates (Ousterhout & Sperelakis, 1987). Agents that activate adenylate cyclase, such as isoproterenol and forskolin, depressed or abolished the Ca^{2+}-dependent APs and hyperpolarized the membrane. These effects are mimicked by membrane permeable analogs of cAMP (dibutyryl or 8-bromo cAMP). 8-Bromo-cGMP also depressed or abolished the APs. Atrial natriuretic factor (ANF), which activates the membrane-bound guanylate cyclase, had inhibitory effects. These results suggest that cAMP and cGMP decrease the inward Ca^{2+} current (and/or increase an outward K$^+$ current) in the vascular SM cells.

In voltage-clamp experiments on cultured single cells from rabbit aorta, the dibutyryl analogs of cAMP and cGMP suppressed the inward Ba^{2+} current carried through the Ca^{2+} slow channels (Bkaily et al., 1988a). The 8-bromo analogs of cAMP and cGMP and nitroprusside also enhanced the delayed-rectifier outward K$^+$ current. Thus, these effects of

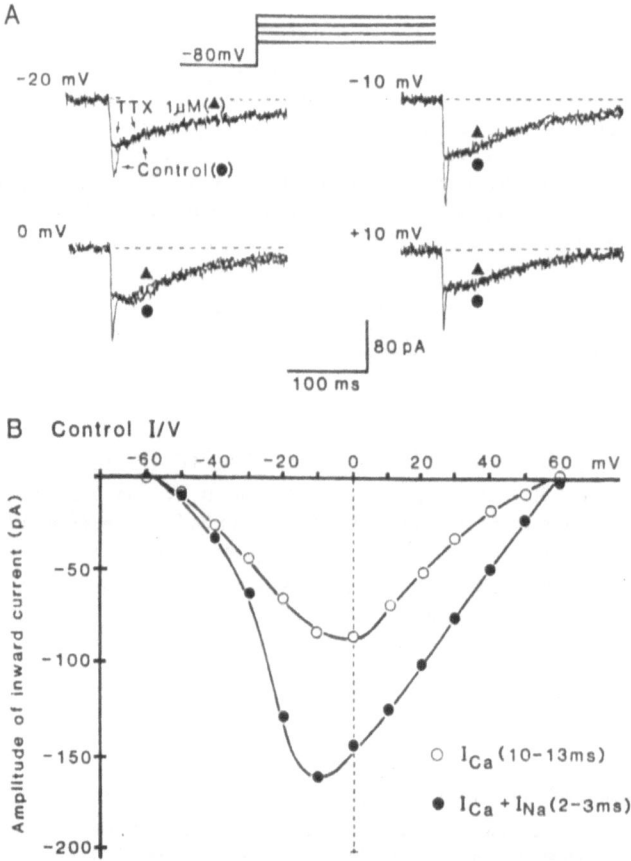

Fig. 13. Two types of Ca^{2+} channel currents recorded from freshly-dispersed single smooth muscle cells from rat portal vein. Pipette contained high Cs$^+$ and bath contained PSS; HP was -80 mV. **A:** Membrane currents evoked by 4 different CPs in the absence (filled circles) and presence (filled triangles) of TTX (1 μM). **B:** Amplitudes of inward currents (without TTX) at 2-3 ms (filled circles) and 10-13 ms (open circles) plotted at various membrane potentials. Values of (B) were from control part of (A). (Xiong, Sperelakis, and Fenoglio-Preiser, unpublished data).

cyclic nucleotides on the Ca^{2+} and K^+ channels can explain their inhibitory effects on the APs.

We propose a phosphorylation model for regulation of Ca^{2+} slow channels in vascular SM cells, similar to that for myocardial cells. In the case of vascular SM cells, cAMP and cGMP have the same effect, namely both inhibit the Ca^{2+} slow channels (Fig. 14). A second regulatory protein, when phosphorylated by protein kinase C (PK-C), stimulates the Ca^{2+} slow channels, as depicted (Fig. 14).

Figures 15 summarizes the roles played by cyclic nucleotides and PK-C in regulating the intracellular Ca^{2+} concentration ($[Ca]_i$) in VSM cells. Activation of A-cyclase or G-cyclase by appropriate membrane receptors (e.g., beta-adrenergic, prostacyclin, ANP) and G coupling proteins or directly (by agents like nitroprusside, nitric oxide free radical, or forskolin) produces elevation of cAMP and cGMP and activation of cA-PK and cG-PK. These kinases can phosphorylate the Ca^{2+} slow channel (to inhibit channel opening) and the K^+ channels (to stimulate channel opening). Both mechanisms would inhibit the Ca^{2+}-

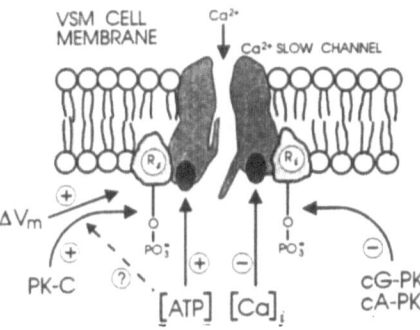

Fig. 14. Cartoon model of a Ca^{2+} slow channel protein embedded in the lipid bilayer of the cell membrane of a vascular smooth muscle cell. Depicted are two regulatory proteins, inhibitory and stimulatory, associated with the channel protein. As shown, both regulatory proteins have phosphorylation sites which affect the function of the regulatory proteins. Phosphorylation of the inhibitory protein by cAMP-protein kinase (PK-A) or by cGMP-PK (PK-G) inhibits the function of the ion channel, e.g., decreases probability of channel opening at depolarized potentials. Phosphorylation of the stimulatory protein by PK-C stimulates the Ca^{2+} slow channel, e.g., increases probability of opening. Thus, channel function is regulated by cyclic nucleotides and by diacylglycerol.

dependent APs, and thereby diminish Ca^{2+} influx and lower $[Ca]_i$. cG-PK and cA-PK also phosphorylate the sarcolemmal Ca^{2+}-ATPase and stimulate the Ca^{2+} pump in the surface membrane, thus lowering $[Ca]_i$. Lowering $[Ca]_i$ inhibits contraction and produces vasodilation.

Activation of PL-C by appropriate membrane receptors (e.g., Ang-II receptor) and G coupling proteins stimulates PI turnover with IP_3 and DAG production (Fig. 15). DAG activates PK-C to phosphorylate the Ca^{2+} slow channels and stimulate Ca^{2+} influx and raise $[Ca]_i$. Phorbol esters act, like DAG, to directly activate PK-C. IP_3 acts on the SR to release Ca^{2+} (via the Ca-release channels) and elevate $[Ca]_i$. Raising $[Ca]_i$ potentiates contraction and produces vasoconstriction.

The cyclic nucleotides also cause phosphorylation of the myosin light-chain kinase (MLCK), which diminishes its Ca^{2+} sensitivity. In addition, the cyclic nucleotides have

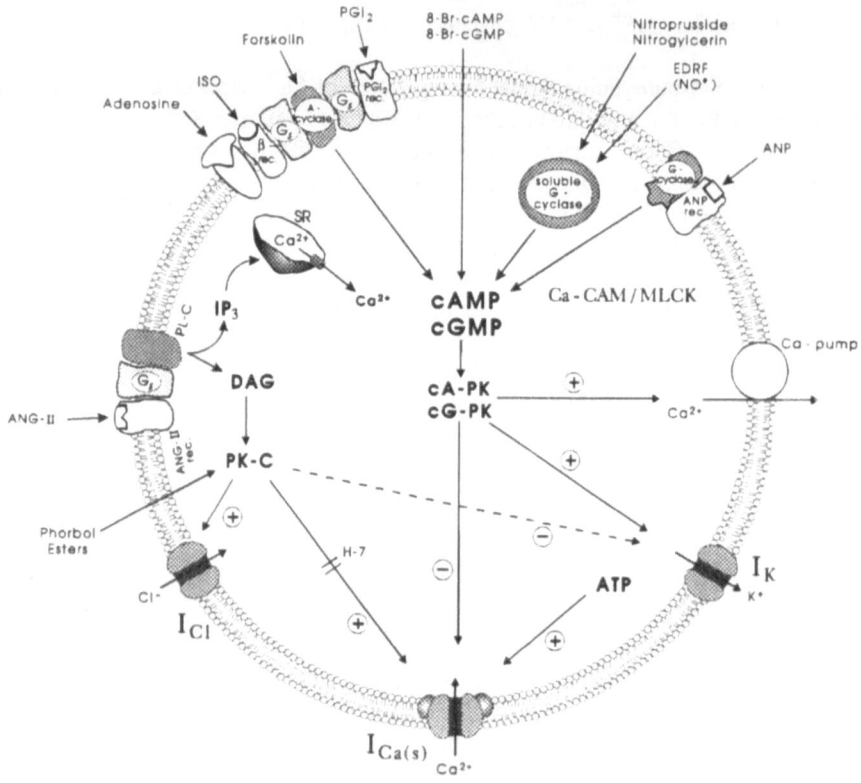

Fig. 15. Diagram of a vascular smooth muscle cell to illustrate one mechanism whereby elevation of cycli nucleotides can lead to vasodilation. Depicted are membrane receptors for stimulation of adenylate cyclase and guanylate cyclase, resulting in elevation of either cAMP or cGMP. As shown, cAMP activates PK-A and cGMP activates PK-G, which phosphorylate at least two types of ion channels and a Ca pump in the sarcolemma. Phosphorylation of the Ca^{2+} slow channel (or an associated inhibitory-type regulatory protein) produces inhibition of the channel, e.g., decreases probability of opening at depolarized potentials. Also depicted is the fact that phosphorylation by PK-C stimulates the Ca^{2+} slow channel. The delayed rectifier K^+ channel is stimulated by phosphorylation with cA-PK and cG-PK, as illustrated. Also depicted is stiumlation of the sarcolemmal Ca^{2+} pump by cA-PK and cG-PK.

been reported to inhibit phospholipase C (PL-C), and therefore production of IP_3 and DAG is inhibited. These factors would contribute to the vasodilation.

(a) **Cell lines.** 8-Br-cGMP added to the bath depressed I_{Ca} in the A10 cells (Fig. 16). As can be seen, 3 mM 8-Br-cGMP reduced I_{Ca} to less than half of control.

A similar effect was produced by 8-Br-cAMP. For example, in A7r5 cells, 3 mM 8-Br-cAMP depressed I_{Ca} by about 30% (Fig. 17). Forskolin, an agent that directly stimulates adenylate cyclase to elevate cAMP levels, also depressed Ca^{2+} channel current (carried by Ba^{2+}) (Fig. 18). In addition, forskolin shifted the reversal potential to the left, presumably due to stimulation of an outward current.

(b) Portal vein (rabbit, rat). In freshly-isolated cells from rabbit portal vein, 8-Bromo derivatives of both cAMP and cGMP substantially depressed $I_{Ca(L)}$ (Fig. 19). Similar results were obtained from rat portal vein cells. Forskolin also was able to inhibit $I_{Ca(L)}$ (Fig. 20A), and this action was prevented by pretreatment with H-7, an inhibitor of protein kinases (Fig. 20B).

Introduction of PK-A (cat. subunit) and PK-G into the cell via the patch pipette (perfused) also produced marked inhibition of $I_{Ca(L)}$. As shown, 1.76 μM PK-A inhibited

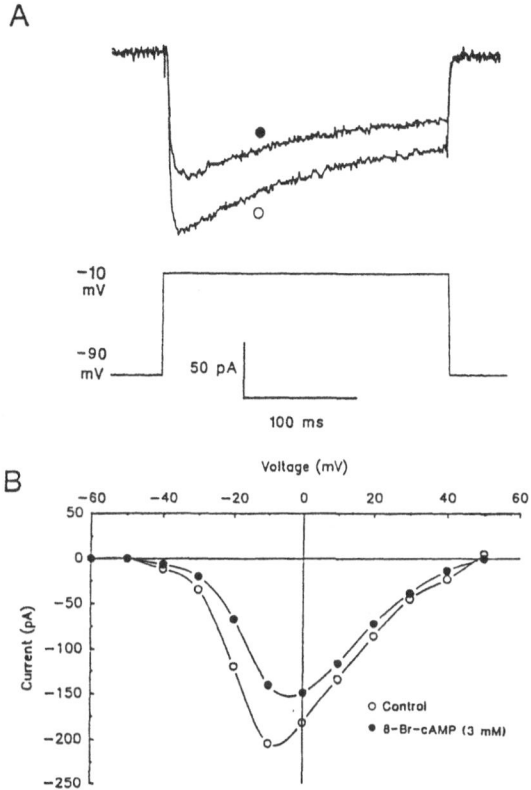

Fig. 16. Effect of 8-Br-cGMP on I_{Ba} of A10 VSM cell. Patch pipette contained high Cs^+ and nystatin (200 μg/ml) and bath contained 10 mM Ba^{2+}. **A:** Inward current (I_{Ba}) induced by a voltage step from a HP of -90 mV to 0 mV in the absence (open circle) and presence (filled circle) of 8-Br-cGMP (3 mM). **B:** I-V relationship of I_{Ba} in the absence (open circles) and presence (filled circles) of 8-Br-cGMP (3 mM). Note that peak amplitude was decreased by 8-Br-cGMP but no remarkable difference was observed in threshold, peak or reversal potential. (Bielefeld and Sperelakis, unpublished data.)

$I_{Ca(L)}$ by about 50% within 10 min. (Fig. 21A). Similar inhibition (about 50%) was produced by 50 nM PK-G (Fig. 21B).

These results provide evidence that the inhibitory effects of both cyclic nucleotides on $I_{Ca(L)}$ are exerted via the respective protein kinases and phosphorylation.

A

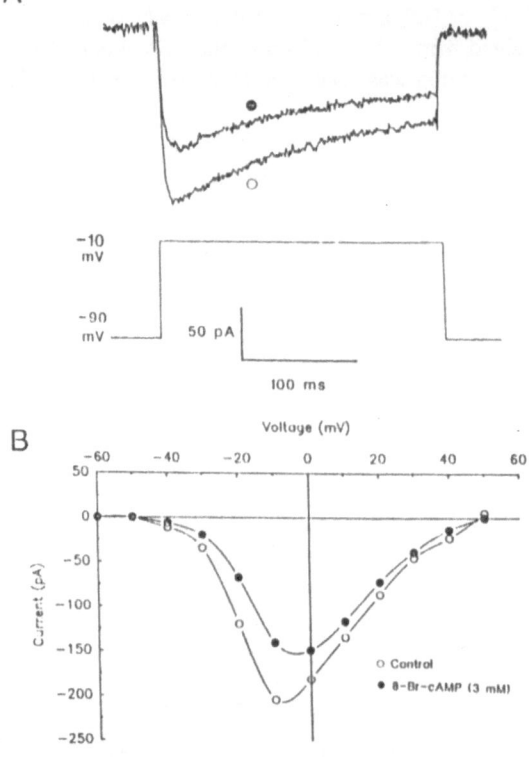

−10
mV

−90
mV

50 pA

100 ms

B

Voltage (mV)

Fig. 17. Effect of 8-Br-cAMP on I_{Ba} of A7r5 VSM cell. Patch pipette contained high Cs^+ and nystatin (200 μg/ml) and bath contained 10 mM Ba^{2+}. **A:** Inward current (I_{Ba}) induced by a voltage step from a HP of -90 mV to -10 mV in the absence (open circle) and presence (filled circle) of 8-Br-cAMP (3 mM). **B:** I-V relationship of I_{Ba} in the absence (open circles), presence of 8-Br-cAMP (filled circles). (Bielefeld and Sperelakis, unpublished data.)

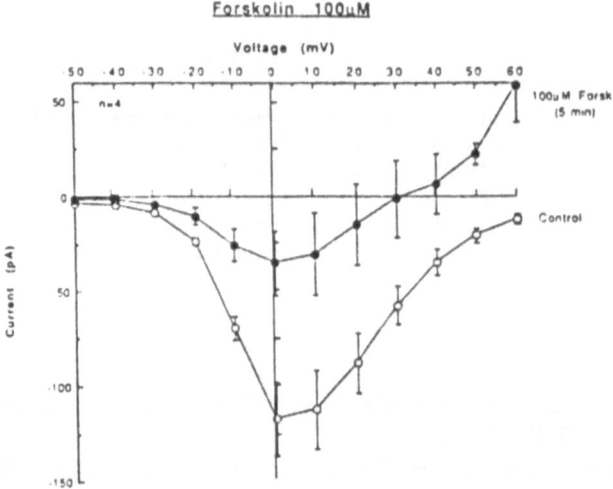

Forskolin 100μM

Voltage (mV)

Fig. 18. Effect of forskolin on I_{Ba} of A7r5 VSM cell. Patch pipette contained high Cs^+ and nystatin (200 μg/ml) and bath contained 5 mM Ba^{2+} and TEA (130 mM); the HP was -80 mV. I-V relationship in the absence (open circles) and presence (filled circles) of 100 μM forskolin. Values plotted are mean ± SEM; n = 4. (Lorenz and Sperelakis, unpublished data.)

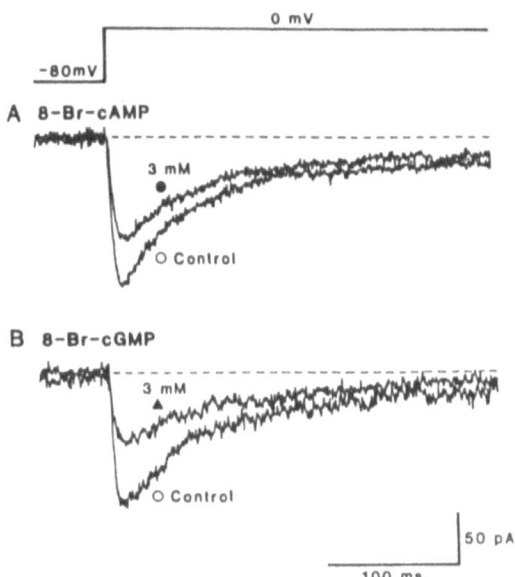

Fig. 19. Effect of 8-Br-cAMP and 8-Br-cGMP on I_{Ca} in single smooth muscle cell from rabbit portal vein. Pipette contained high Cs^+ and bath contained PSS. I_{Ca} was evoked by voltage step from a HP of -80 mV to 0 mV. **A:** Current traces recorded in absence (control) and presence of 8-Br-cAMP (3 mM). **B:** Current traces recorded in the absence (control) and presence of 8-Br-cGMP (3 mM). (A) and (B) were from different cells. (Xiong, Sperelakis and Fenoglio-Preiser, unpublished data.)

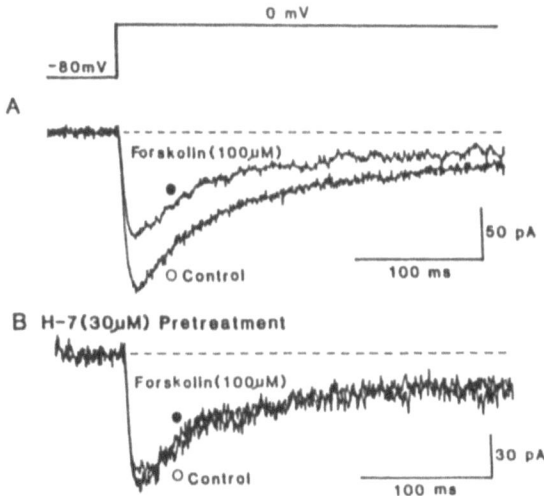

Fig. 20. Effect of forskolin on I_{Ca} in single smooth muscle cell from rabbit portal vein. Experimental conditions were same as Fig.19. **A:** Current traces recorded in absence (open circle) and presence (filled circle) of forskolin (100 μM). **B:** I_{Ca} recorded from another cell after pre-treatment with H-7 (30 μM). Note that inhibitory effect of forskolin disappeared after pre-treatment with H-7. (Xiong, Sperelakis and Fenoglia-Preiser, unpublished data.)

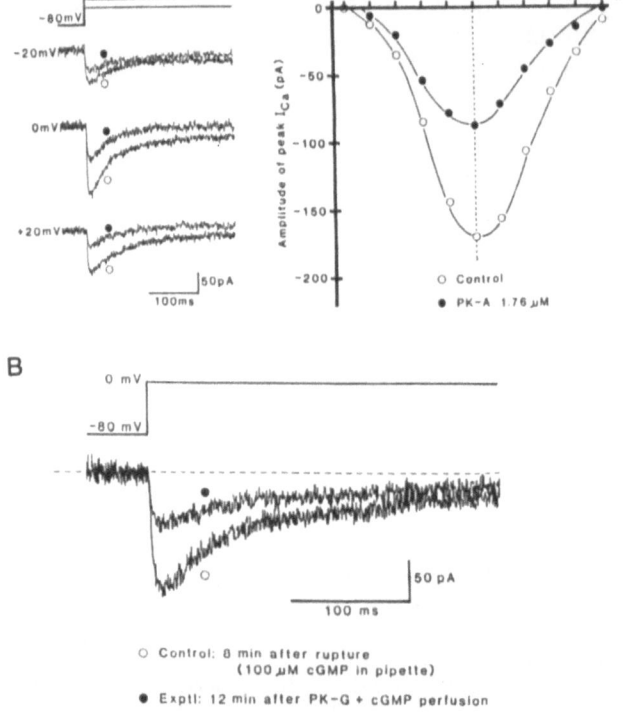

Fig. 21. Effect of cAMP-dependent protein kinase (PK-A, catalytic subunit) **(A)** and cGMP-dependent protein kinase (PK-G) **(B)** on I_{Ca} in single smooth muscle cell from rabbit portal vein. Pipette contained high Cs^+ solution and bath contained PSS. **Aa:** Typical cell showing Ca^{2+} current traces in the absence (open circles) and presence (filled circles) of 1.76 μM PK-A at 3 CPs (-20, 0 and +20 mV) from a HP of -80 mV. PK-A was applied intracellularly via perfusion technique. **Ab:** I-V curve of I_{Ca} before (open circles) and after (filled circles) application of 1.76 μ M PK-A. Note that PK-A inhibited I_{Ca} at all potentials. (Aa) and (Ab) were obtained from the same cell. **B:** I_{Ca} traces evoked by depolarizing membrane to 0 mV from a HP of -80 mV in the absence (open circles) and presence (filled circles) of PK-G (50 nM) plus cGMP (100μM). PK-G was applied by intracellular perfusion. (Xiong, Sperelakis and Fenoglio-Preiser, unpublished data.)

Regulation of Ca^{2+} Channels by ATP

We examined the properties of the two types of Ca^{2+} channels in freshly isolated cells from guinea-pig small mesenteric arteries (resistance vessels), using the whole-cell voltage clamp technique to record the fast and slow Ca^{2+} channel currents (Ohya & Sperelakis, 1989a). Injection of ATP (0.3-5 mM), using an intracellular perfusion method, modified only the slow current, but did not affect the fast current (Fig. 22). When the production of ATP was inhibited by adding cyanide and 2-deoxy-D-glucose to the bath (in absence of glucose), the Ca^{2+} slow current was abolished within 10 min, whereas the fast Ca^{2+} current was only slightly affected. These results indicate that the slow Ca^{2+} channel is metabolically dependent. The ATP dependence of the activity of the Ca^{2+} slow channels was also shown at the single-channel level (Ohya & Sperelakis, 1989b). Therefore, there may be an obligatory binding site for ATP on the inner surface of the channel which must be occupied for channel activity.

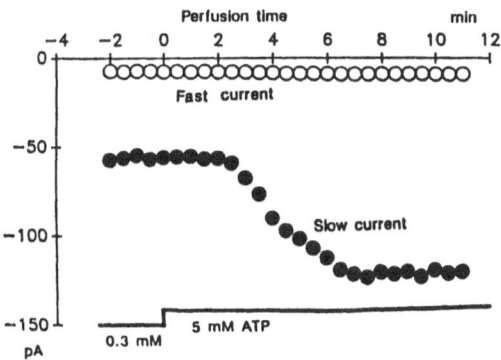

Fig. 22. Effects of intracellularly-perfused ATP on the separated fast and slow Ca^{2+} channel currents. The concentration of ATP was increased from 0.3 mM to 5 mM using the intracellular perfusion technique. The time after beginning of the perfusion is indicated on the abscissa. The slow current was isolated by using a HP of -40 mV to voltage-inactive the fast channels; command potentials to +30 mV were applied every 30 sec. The fast current was recorded using a HP of -80 mV and a command potential to -20 mV. The peak magnitudes of the fast (open circles) and slow (filled circles) Ca^{2+} channel currents were plotted as a function of time following perfusion with 5 mM ATP. The fast and slow currents were obtained from two different cells. (Data taken from Ohya and Sperelakis, 1989a.)

Modulation of Ca^{2+} Channels by Agonists

Agonists, such as norepinephrine (NE) or angiotensin-II (A-II), may induce contraction of vascular SM via several mechanisms, including (a) stimulation of Ca^{2+} influx through receptor-operated channels (ROCs); (b) stimulation of Ca^{2+} entry through voltage-dependent Ca^{2+} channels, opened indirectly by the depolarization resulting from an increase in membrane conductance for other ions; and (c) release of intracellular Ca^{2+} from store sites (for review: Bolton, 1979; Johansson & Somlyo, 1980; Kuriyama et al., 1982; Sperelakis & Ohya, 1989).

Some agonists (NE, A-II, etc.) may modify the voltage-dependent Ca^{2+} channels of vascular SM cells, as occurs in cardiac cells by beta-adrenergic agonists. NE may affect ion channel activity in vascular SM cells via direct and indirect mechanisms. One group (Benham and Tsien, 1988) reported that NE *enhanced* the Ca^{2+} channel current in rabbit ear artery, whereas another group (Droogsman et al., 1987) reported that NE *inhibited* the Ca^{2+} channel. A third group (Pacaud et al., 1987) reported that NE *inhibited* the **slow-type**, but *enhanced* the **fast-type** of Ca^{2+} channel current, in cultured rat portal vein cells. Others could not observe a significant change in mesenteric artery and dog saphenous vein (Bean et al., 1986; Yatani et al., 1987). Some intracellular factors which regulate the response to agonists might be lost during whole-cell voltage clamp.

We previously examined the effects of A-II on the electrical activity of cultured vascular SM cells from rat aorta (Zelcer & Sperelakis, 1981). A-II, applied by bolus perfusion, elicited a transient depolarization of up to 30 mV, which sometimes triggered an AP. The A-II-induced depolarization disappeared in Na^+-free solution, suggesting that it was primarily due to an increase in Na^+ conductance, with resulting Na^+ influx. Continuous superfusion of cultured aortic reaggregates with A-II (10^{-6} M) also produced depolarization, which triggered a long-lasting AP (Johns & Sperelakis, 1990). The membrane resistance was decreased by A-II, consistent with an increase in conductance for an inward depolarizing current. Low concentrations of A-II (10^{-9} M) caused a marked prolongation of ongoing Ca^{2+}-dependent APs without depolarizing the membrane.

We also demonstrated that A-II greatly stimulated the Ca^{2+} slow channel current in isolated single cells from guinea pig portal vein (Ohya & Sperelakis, 1991) (Fig. 23). This effect of A-II was mediated via a G-protein, based on inhibition of the A-II effect by GDP-β S and stimulation of the basal I_{Ca} by GTP-γS. However, this G-protein was not sensitive to pertussis toxin or to cholera toxin, hence is not G_i-like or G_s-like.

In whole-cell voltage clamp of cultured vascular SM cells from rabbit aorta, addition of A-II caused a substantial increase in the inward Ca^{2+} channel current (carried by Ba^{2+}) (Bkaily et al., 1988b). A-II ($1-5 \times 10^{-8}$ M) also suppressed the delayed-rectifier outward K^+ current. These effects of A-II on the Ca^{2+} and K^+ channel currents can explain the depolarization and AP prolongation found in cultured rat aortic cells. These observations suggest that agonists control vascular tone, at least in part, by modifying the Ca^{2+} channel and K^+ channel activities, and thereby changing $[Ca]_i$.

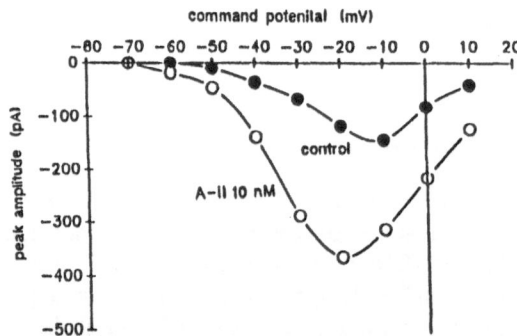

Fig. 23. Effects of angiotensin-II (A-II) on the Ca^{2+} channel current recorded from guinea-pig portal vein. Step command pulses were applied over the voltage range of -60 mV to 10 mV from a HP of -90 mV. Peak current amplitude measured before (filled symbols) and after (unfilled symbols) 10 nM A-II application, were plotted against the command potentials. Bath solution contained 2 mM Ba^{2+} as the charge carrier. Pipette solution contained high Cs with 5 mM ATP and 0.1 mM GTP. (Data taken from Ohya and Sperelakis, 1991.)

Possible Mechanisms for Agonist Modulation

(a) Regulation by C-kinase. Many neurotransmitters and hormones, whose action involve Ca^{2+} mobilization, promote the breakdown of membrane phospholipids and the formation of two putative second messengers, inositol-1,4,5-trisphosphate (IP_3) and 1,2-diacylglycerol (DAG). One of the major roles of IP_3 is to release Ca^{2+} from intracellular stores, including the SR of vascular SM cells (Somlyo et al., 1985; Suematsu et al., 1985). It has been suggested (Rasmussen et al., 1987) that activation of PK-C may be involved in agonist-induced Ca^{2+} influx, which is responsible for the tonic phase of SM contraction. The phorbol esters, which activate PK-C, produce a slowly-developing sustained contracture of vascular SM that is partially dependent on $[Ca]_o$ (Danthuluri & Deth, 1986; Chiu et al., 1987; Itoh & Lederis, 1987).

To examine a possible role of PK-C in the modulation of electrical activity in vascular SM cells, the effects of a phorbol ester, phorbol 12,13-diacetate (PDA), were determined on the resting potential and APs in cultured aortic reaggregates (Ousterhout & Sperelakis, unpublished observations). Superfusion with PDA (0.1 - 5 µM) produced a gradual depolarization and allowed the appearance of an active (plateau-type) AP response to electrical stimulation. PDA also prolonged the duration of ongoing APs. Our observations suggest that PK-C may be involved in the regulation of ion channels in vascular SM cells.

However, after prolonged exposure to higher concentrations of PDA, the APs became depressed, perhaps due to depolarization-induced inactivation of the Ca^{2+} channels. Thus, the effects of phorbol esters on the ionic currents in vascular SM cells may be complex. The depolarization produced by the phorbol esters could be due to inhibition of K^+ channels, and the induction of APs could result from this plus the activation of a Ca^{2+} conductance, possibly mediated by phosphorylation. For example, in cultured aortic (A7r5 cell line) cells, phorbol esters increased the slow (sustained) Ca^{2+} channel current (Fish et al., 1988). Werz and MacDonald (Werz & MacDonald, 1987) observed dual effects of phorbol esters on the Ca^{2+}-dependent APs in cultured neurons, and suggested that either Ca^{2+} or K^+ conductances could be depressed depending on the membrane potential.

(b) Direct Regulation by G-Protein. Direct regulation of ionic channels by GTP-binding protein (G-protein) has been identified in several tissues (for review: Brown & Birnbaumer, 1988). In some neurons, G-protein (G_o) is thought to mediate the modulation (inhibition) of Ca^{2+} channels produced by receptor activation (Hescheler et al., 1987). In cardiac cells, G-protein (G_s) was reported to directly enhance the activity of Ca^{2+} channels (Yatani et al., 1987).

We examined the effects of GTP-γS on Ca^{2+} channels in freshly-isolated single cells from guinea-pig portal vein (Ohya & Sperelakis, 1988). Single Ca^{2+} channels (conductance of 18 - 22 pS) were recorded in cell-attached patch configuration with 100 mM Ba^{2+} and Bay-K-8644 in the pipette and high K^+ solution in the bath. After opening one end of the cell, GTP-γS (0.1 mM) applied to the bath (access into the cell interior) enhanced the channel activity. This observation suggests that a G-protein may be one of the factors regulating Ca^{2+} channels in vascular SM cells also. However, we cannot exclude the possibility that G-protein acts indirectly on the Ca^{2+} channel.

When ISO (10 µM) was added to whole-cell voltage clamped rabbit portal vein cells, there was a dual effect produced (Fig. 24A, B). First, there was a transient *increase* in $I_{Ca(L)}$, followed by a sustained *decrease* in $I_{Ca(L)}$. When H-7 was applied as pretreatment (to inhibit PK-A), the second phase was prevented, namely only a sustained increase was evident (Fig. 24C). These results indicate that the second inhibitory effect was mediated by cAMP and PK-A and phosphorylation, whereas the first stimulatory effect was not. This is consistent with the first stimulatory effect being mediated by direct G-protein gating.

Conclusions

Vascular tone is regulated by a variety of neurotransmitters, vasoactive hormones and autacoids, and vasoactive drugs. These actions are mediated, at least in part, by actions on the membrane ion channels, exerted either directly or indirectly. In this section, we described evidence that three different protein kinase systems (protein kinases A, G, and C) act on and modulate the Ca^{2+} slow channels in vascular SM cells, and that ATP is required for activity of these channels (Figs. 13 and 14). Some vascular SM cells possess fast Na^+ channels.

SUMMARY AND CONCLUSIONS

Most SM cells normally do not possess fast Na^+ channels, but inward current is carried through two types of Ca^{2+} channels: slow (L-type) and fast (T-type). Whole-cell voltage clamp was done on single SM cells isolated from three types of tissue: (a) the longitudinal layer of pregnant rat uterus, (b) the circular layer of rat and human colon, and (c) vascular SM cells from rat and rabbit portal vein and cultured cell lines (A10 and A7r5). Depolarizing pulses, applied from a HP of -60 to -100 mV, evoked two types of inward

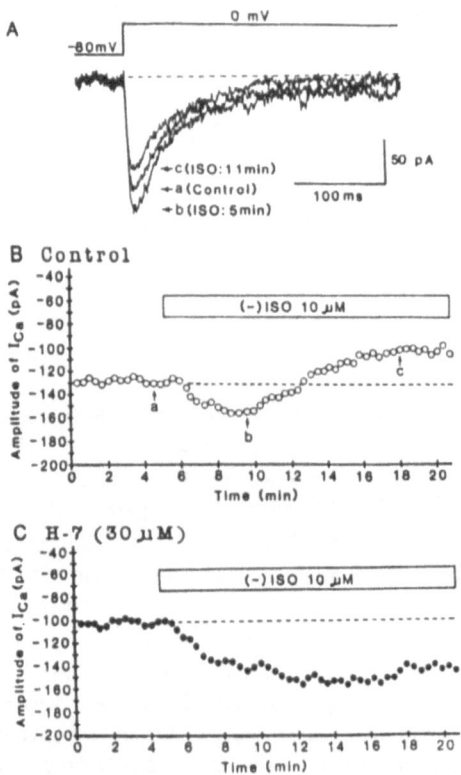

Fig. 24. Effect of isoproterenol (10 μM) on I_{Ca} in single smooth muscle cell from rabbit portal vein. **A:** Current traces recorded in absence (a, control) and presence of isoproterenol (b, 5 min; c, 11 min). **B:** Time-course of I_{Ca} during application of isoproterenol. (A) and (B) were from same cell. Note that isoproterenol initially increased I_{Ca} followed by a decrease. **C:** Time-course of I_{Ca} recorded from another cell during application of isoproterenol after pretreatment with H-7 (30 μM). Note that isoproterenol only produced a sustained increase in I_{Ca}. (Xiong, Sperelakis and Fenoglio-Preiser, unpublished data.)

current, fast and slow. The fast inward current decayed within 5 - 10 msec, depended on $[Na]_o$, and was inhibited by TTX, with $K_{0.5}$ values of 27 nM (uterine), 14 nM (human colonic), and 130 nM (rat colonic). The slow inward current decayed slowly, was dependent on $[Ca]_o$ (or Ba^{2+}), and was inhibited by nifedipine. These results suggest that the fast inward current is a fast Na^+ channel current, and that the slow inward current is a Ca^{2+} slow channel current. A fast-inactivating Ca^{2+} channel current was not evident in pregnant uterine SM, but was in colonic SM and vascular SM (guinea pig mesenteric artery). Thus, the ion channels which generate inward currents in these SM cells are primarily TTX-sensitive fast Na^+ channels and dihydropyridine-sensitive slow Ca^{2+} channels.

In pregnant rat myometrial SM cells, the number of fast Na^+ channels increased during gestation. The averaged current density increased from 0 on day 5, to 0.19 on day 9, 0.56 on day 14, 0.90 on day 18, and 0.86 pA/pF on day 21. This almost linear increase occurs because of an increase in the fraction of cells which possess fast Na^+ channels. The Ca^{2+} channel current density also was higher during the latter half of gestation. These results indicate that the fast Na^+ channels and Ca^{2+} slow channels in myometrium become more numerous as term approaches, and we suggest that the fast Na^+ current may be involved in spread of excitation.

In uterine SM cells, isoproterenol (ISO, beta-agonist) did not affect either $I_{Ca(L)}$ or $I_{Na(f)}$, whereas Mg^{2+} ($K_{0.5}$ of 12 mM) and nifedipine ($K_{0.5}$ of 3.3 nM) depressed $I_{Ca(L)}$. Oxytocin had no effect on $I_{Na(f)}$ and actually depressed $I_{Ca(L)}$ (but not I_{Ba}) to a small extent. Therefore, the tocolytic action of beta-agonists cannot be explained by an inhibition of $I_{Ca(s)}$, whereas that of Mg^{2+} and Ca antagonist drugs can be so explained. The stimulating action of oxytocin on uterine contractions cannot be explained by a stimulation of $I_{Ca(L)}$. Phorbol ester (PDB, 0.1 μM) increased $I_{Ca(L)}$ about 30%. ATP (1 μM - 1 mM) applied externally decreased $I_{Ca(L)}$ by up to 50%, concomitant with induction of contraction, suggesting Ca^{2+} inhibition of the Ca^{2+} channels.

In colonic SM cells, the threshold voltage, the voltage for peak current, and the reversal potential for the fast Na^+ current were, respectively, about -50, -20, and +50 mV in rat, and -40, 0, and +60 mV in human. The incidence of cells possessing fast Na^+ currents depended on region of the colon: In rat, the incidence was 64% in proximal colon, and 10% in distal colon. In humans, the incidence was 73% in ascending colon and 22% in descending colon. The densities of fast Na^+ and Ca^{2+} currents were, respectively, 3.2 and 4.5 pA/pF in rat, and 1.0 and 1.4 pA/pF in human. The ratio of current densities (I_{Na} / I_{Ca}) was 0.71 in both rat and human cells. The fast Na^+ current may facilitate propagation of excitation.

In vascular SM cells, about 40% of the cells tested from rat possessed a $I_{Na(f)}$, whereas in rabbit very few cells had a $I_{Na(f)}$. The current densities were approximately 2 pA/pF for $I_{Na(f)}$ and 3 pA/pF for $I_{Ca(L)}$ in rat. In rabbit, $I_{Ca(L)}$ was about 4 pA/pF. ISO (10 μM) produced a dual effect on $I_{Ca(L)}$: initially there was an increase, followed by a sustained decrease. Pretreatment with the protein kinase inhibitor, H-7, prevented the second effect of ISO (inhibition of $I_{Ca(L)}$), suggesting that the initial effect of ISO (stimulation of $I_{Ca(L)}$) was not due to phosphorylation (but perhaps reflected G-protein gating). In contrast, forskolin (FOR, 10-100 μM) only had an inhibitory action on $I_{Ca(L)}$ (ca. 30% inhibition), and this action was prevented by H-7. 8-Br-cAMP (1 - 3 mM) and 8-Br-cGMP (3 mM) added externally both depressed $I_{Ca(L)}$ by approximately 30% or 50%, respectively. Similar results were obtained with FOR, 8-Br-cGMP, and 8-Br-cAMP in the cultured cell lines.

In summary, some SM cells (e.g., colonic and some vascular) normally possess functional fast Na^+ channels, whereas in others (e.g., uterine), such current appears during pregnancy (presumably under hormonal control). Even though the relatively low resting potential in most SM cells would cause most of the fast Na^+ channels to remain in a continual state of inactivation, even a small amount of Na^+ current can contribute to propagation of excitation. In addition, the Na^+ window current would allow a sustained influx of Na^+ over the window voltage range through continuously active fast Na^+ channels, which would contribute to automaticity. With respect to Ca^{2+} currents, some SM cells (e.g., pregnant rat uterine) do not possess T-type Ca^{2+} channels, whereas others do (intestinal, some vascular). In rat colon, there is an increase in both $I_{Ca(T)}$ and $I_{Ca(L)}$ during development. In uterine SM cells, L-type Ca^{2+} current was greater during the latter half of pregnancy. In vascular SM cells, FOR, cAMP, and cGMP depressed $I_{Ca(L)}$, and H-7 prevented the effect of FOR and the second (inhibitory) effect of ISO., The first effect of ISO may reflect G-protein gating of the slow Ca^{2+} channel. The inhibitory effects of these agents presumably are mediated by phosphorylation.

REFERENCES

Bean, B.P., Sturek, M., Puga, A., and Hermsmeyer, K.: Calcium channels in muscle cells isolated from rat mesenteric arteries: Modulation by dihydropyridine drugs. Circ. Res. **59**, 229-235 (1986).

Belardinelli, L., Vogel, S.M., Sperelakis, N., Rubio, R., and Berne, R.M.: Restoration of slow responses in hypoxic heart muscle by alkaline pH. J. Mol. Cell. Cardiol. **11**, 877-892 (1979).

Benham, C.D. and Tsien, R.W.: Noradrenaline modulation of calcium channels in single smooth muscle cells from rabbit ear artery. J. Physiol. **404**, 767-784 (1988).

Bkaily, G., Peyrow, M., Yamamoto, T., Sculptoreanu, A., Jacques, D., and Sperelakis, N.: Macroscopic Ca^{2+}-Na^+ and K^+ currents in single heart and aortic cells. Mol. Cell. Biochem. **80**, 59-72 (1988a).

Bkaily, G., Peyrow, M., Sculptoreanu, A., Jacques, D., Chahine, M., Regoli, D., and Sperelakis, N.: Angiotensin II increases I_{si} and blocks I_K in single aortic cell of rabbit. Pflügers Arch. **412**, 448-450 (1988b).

Bolton, T.B.: Mechanisms of action of transmitters and other substances on smooth muscle. Physiol. Rev. **59**, 606-718 (1979).

Brown, A.M. and Birnbaumer, L.: Direct G protein gating of ion channels. Am. J. Physiol. **254**, H401-H410 (1988).

Chiu, A.T., Bozarth, J.M., Forsyth, M.S., and Timmerman, P.B.M.W.M.: Ca^{++} utilization in the constriction of rat aorta to stimulation of protein kinase C by phorbol dibutyrate. J. Pharmacol. Exp. Therap. **242**, 934-939 (1987).

Danthuluri, N.R. and Deth, R.D.: Acute desensitization to angiotensin II: Evidence for a requirement of agonist-induced diacyglycerol production during tonic contraction of rat aorta. Eur. J. Pharmacol. **126**, 135-139 (1986).

Dosemeci, A., Dhallan, R.S., Cohen, N.M., Lederer, W.J., and Rogers, T.B.: Phorbol ester increases calcium current and stimulates the effects of angiotensin II on cultured neonatal rat heart myocytes. Circ. Res. **62**, 347-357, 1988.

Droogmans, G., Declerck, I., and Casteels, R.: Effect of adrenergic agonists on Ca^{2+}-channel currents in single vascular smooth muscle cells. Pflügers Arch. **409**, 7-12 (1987).

Fish, R.D., Sperti, G., Colucci, W.S., and Clapham, D.E.: Phorbol ester increases the dihydropyridine-sensitive calcium conductance in a vascular smooth muscle cell line. Circ. Res. **62**, 1049-1054 (1988).

Hescheler, J., Kameyama, M., Trautwein, W., Mieskes, G., and Soling, H.D.: Regulation of the cardiac calcium channel by protein phosphatases. Eur. J. Biochem. **365**, 261-266 (1987).

Honoré, E., Martin, C., Mironneau, C., and Mironneau, J.: An ATP-sensitive conductance in cultured smooth muscle cells from pregnant rat myometrium. Am. J. Physiol. **257**, C297-C305 (1989).

Inoue, Y, Nakao, K., Okabe, K., Isumi, H., Kanda, S., Kitamura, K., and Kuriyama, M: Some electrical properties of human pregnant myometrium. Am. J. Obstet. Gynecol. **162**, 1090-1098 (1990).

Inoue, Y. and Sperelakis, N.: Gestational change in Na^+ and Ca^{2+} channel current densities in rat myometrial smooth muscle cells. Am. J. Physiol. **260**, C658-C663 (1991).

Irisawa, H. and Sato, R.: Intra- and extracellular actions of protons on the calcium current of isolated guinea-pig ventricular cells. Circ. Res. **59**, 348-355 (1987).

Itoh, H. and Lederis, K.: Contraction of rat thoracic aorta strips induced by phorbol 12-myristate 13-acetate. Am. J. Physiol. **252**, C244-C247 (1987).

Johansson, B. and Somlyo, A.P.: Electrophysiology and excitation-contraction coupling. In: Handbook of Physiology, Sect. 2, The Cardiovascular System, Vol. 2. Am. Physiol. Soc. pp. 301-323 (1980).

Johns, D.W. and Sperelakis, N.: Angiotensin-II stimulation of Ca^{2+}-dependent action potentials in cultured rat aortic smooth muscle cells. Eur. J. Pharmacol. **187**, 183-191 (1990).

Kuriyama, H., Ito, Y., Suzuki, H., Kitamura, T., and Itoh, T.: Factors modifying contraction-relaxation cycle in vascular smooth muscles. Am. J. Physiol. **243**, H641-H662 (1982).

Martin, C., Arnaudeau, S., Jmari, K., Rakotoarisoa, L., Sayet, I., Dacquet, C., Mironneau, C., and Mironneau, J.: Identification and properties of voltage-sensitive sodium channels in smooth muscle cells from pregnant rat myometrium. Mol. Pharmacol. **38**, 667-673 (1990).

Miller, S.M., Garfield, R.E., and Daniel, E.E.: Improved propagation in myometrium associated with gap-junctions during parturition. Am. J. Physiol. **256**, C130-C141 (1989).

Mironneau, J.: Ion channels and excitation-contraction coupling in myometrium. In: Uterine contractility, Garfield, R.E., eds., Serono Symposia, St. Louis, MO pp. 9-19 (1990).

Ohya, Y. and Sperelakis, N.: Guanosine triphosphate-dependent stimulation of L-type calcium channels of vascular smooth muscle cells. The Physiol. **31**, A88 (1988).

Ohya, Y. and Sperelakis, N.: ATP regulation of the slow calcium channels in vascular smooth muscle cells of guinea pig mesenteric artery. Circ. Res. **64**, 145-154 (1989a).

Ohya, Y. and Sperelakis, N.: Modulation of single slow (L-type) calcium channels by intracellular ATP in vascular smooth muscle cells. Pflügers Arch. **414**, 257-264 (1989b).

Ohya, Y. and Sperelakis, N.: Fast Na^+ and slow Ca^{2+} channels in single uterine muscle cells from pregnant rats. Am. J. Physiol. **257**, C408-C412 (1989c).

Ohya, Y. and Sperelakis, N.: Involvement of a GTP-binding protein in stimulating action of angiotensin II on calcium channels in vascular smooth muscle cells. Circ. Res. **68**, 763-771 (1991).

Okabe, K., Kajioka, S., Nakao, K., Kitamura, K., and Kuriyama, H.: Action of cromakalin on ionic currents recorded from single smooth muscle cells of the rat portal vein. J. Pharmacol. Exp. Therap. **252**, 832-839 (1990).

O'Rourke, B., Blackx, P.H., and Marban, E.: Phosphorylation-independent modulation of L-type calcium channels by magnesium-nucleotide complexes. Science **257**:245-248 (1992).

Ousterhout, J.M. and Sperelakis, N.: Cyclic nucleotides depress action potentials in cultured aortic smooth muscle cells. Eur. J. Pharmacol. **144**, 7-14 (1987).

Pacaud, P., Loirand, G., Mironneau, C., and Mironneau, J: Opposing effects of noradrenaline on the two classes of voltage-dependent calcium channels of single vascular smooth muscle cells in short-term primary culture. Pflügers Arch. **410**, 557-559 (1987).

Rasmussen, H., Takuwa, Y., and Park, S.: Protein kinase C in the regulation of smooth muscle contraction. FASEB J. **1**, 177-185 (1987).

Sakai, N., Tabb, T., and Garfield, R.E.: Modulation of cell-to-cell coupling between myometrial cells of the human uterus during pregnancy. Am. J. Obstet. Gynecol. **167**, 472-480 (1992).

Savineau, J., Mironneau, J., and Mironneau, C.: Influence of the sodium gradient on contractile activity in pregnant rat myometrium. Gen. Physiol. Biophysics. **6**, 535-560 (1987).

Smirnov, S.V., Zholos, A.V., and Shuba, M.F.: Potential-dependent inward currents in single isolated smooth muscle cells of the rat ileum. J. Physiol, London **454**, 549-571 (1992).

Somlyo, A.V., Bond, M., Somlyo, A.P., Scarps, A.: Inositol trisphosphate-induced calcium release and contraction in vascular smooth muscle. Proc. Natl. Acad. Sci. **82**, 5231-5235 (1985).

Sperelakis, N., Inoue, Y., and Ohya, Y.: Fast Na^+ channels in smooth muscle from pregnant rat uterus. Can. J. Physiol. Pharmacol. **70**, 491-500 (1992).

Sperelakis, N. and Ohya, Y: Electrophysiology of vascular smooth muscle. In: Physiology and Pathophysiology of the Heart, 2nd edition, Kluwer Academic Press, Boston, pp. 773-811 (1989).

Sperelakis, N. and Schneider, J.A.: A metabolic control mechanism for calcium ion influx that may protect the ventricular myocardial cell. Am. J. Cardiol. **37**, 1079-1085 (1976).

Sperelakis, N., Tohse, N., and Ohya, Y.: Regulation of calcium slow channels in cardiac muscle and vascular smooth muscle cells. In: Excitation-Contraction Coupling in Skeletal, Cardiac, and Smooth Muscle, Frank, G.B., ed., Plenum Press, New York pp. 163-187 (1992).

Suematsu, E., Hirata, M., Sasaguri, T., Hashimoto, T., and Kuriyama, H.: Roles of Ca^{2+} on the inositol 1,4,5-triphosphate-induced release of Ca^{2+} from saponin-permeabilized single cells of the porcine coronary artery. Comp. Biochem. Physiol. **82A**, 645-649 (1985).

Vogel, S. and Sperelakis, N.: Blockade of myocardial slow inward current at low pH. Am. J. Physiol. **233**, C99-C103 (1977).

Werz, M.A. and MacDonald, R.L.: Phorbol esters: Voltage-dependent effects on calcium-dependent action potentials of mouse central and peripheral neurons in cell culture. Neurosci. **7**, 1639-1647 (1987).

Xiong, Z.L., Sperelakis, N., Noffsinger, A., and Fenoglio-Preiser, C.: Fast Na^+ current in circular smooth muscle cells of the large intestine. Pflugers Arch. (submitted) (1992).

Yatani, A., Imoto, Y., Codina, J., Hamilton, S.L., Brown, A.M., and Birnbaumer, L.: The stimulatory G protein of adenylyl cyclase, G_s, also stimulates dihydroyridine-sensitive Ca^{2+} channels. J. Biol. Chem. **263**, 9887-9895 (1988).

Yatani, A., Codina, J., Imoto, Y., Reeves, J.P., Birnbaumer, L., and Brown, A.M.: A G protein directly regulates mammalian cardiac calcium channels. Science **238**, 1288-1292 (1987).

103

Yatani, A., Seidel, C.L., Allen, J., and Brown, A.M.: Whole-cell and single-channel calcium currents of isolated smooth muscle cells from saphenous vein. Circ. Res. **60**, 523-533 (1987).

Young, R.C. and Herndon-Smith, L.: Characterization of sodium channels in cultured human uterine smooth muscle cells. Am. J. Obstet. Gynecol. **164**, 175-181 (1991).

Zelcer, E. and Sperelakis, N.: Angiotensin induction of active responses in cultured reaggregates of rat aortic smooth muscle cells. Blood Vessels **18**, 263-279 (1981).

INFLUENCE OF PHYSICAL AND ENVIRONMENTAL FACTORS
ON ION CHANNELS IN ARTERIAL MUSCLE

David R. Harder, Debebe Gebremedhin and Richard J. Roman

Department of Physiology and
Cardiovascular Research Center
Medical College of Wisconsin
Milwaukee, Wisconsin

INTRODUCTION

Despite the existence of pharmacomechanical coupling in arterial muscle, it is now clear that electrophysiological events occurring at the plasma membrane, which regulate the resting membrane potential, largely influence the reactivity of intact arterial muscle cells. Factors which hyperpolarize arterial muscle inhibit voltage-operated Ca^{2+} channels reducing their ability to respond to stimuli. Similarly, factors which depolarize arterial muscle activate voltage-sensitive Ca^{2+} channels leading to activation. During the last ten years it has become apparent that one of the mechanisms responsible for the transduction of physical and environmental influences with respect to vascular reactivity is activation or inhibition of ion conductance systems setting a level of membrane potential around which vascular tone is enhanced or reduced.

This brief chapter will focus on the effect of increasing pressure on cerebral and renal arterial muscle, and the direct effect of reducing PO_2 on cerebral arterial muscle and the effect of hypoxia on modulating the response to changes in intravascular pressure. More specifically, the ability of pressure-mediated muscle cell deformation to activate membrane phospholipases and in turn to affect Ca^{2+} and K^+ channels will be discussed. The role of reduced oxygen tension (hypoxia) in affecting K^+ channel activity will be discussed in the context of modulating pressure-induced depolarization, and activation of cerebral arteries.

Ion Flux in Pulmonary Vascular Control, Edited by
E.K. Weir *et al.*, Plenum Press, New York, 1993

ELEVATION OF TRANSMURAL PRESSURE AND DEPOLARIZATION OF RENAL AND CEREBRAL ARTERIES

Elevation of transmural pressure in isolated and cannulated cerebral, renal, skeletal muscle, mesenteric and coronary arteries from a variety of animal species results in muscle membrane depolarization.[1,2,3] In the case of isolated cat cerebral arteries, the level of depolarization is substantial; from a resting level of around -68 mV at 20 mmHg to around -36 mV at 120 mmHg.[1] In cat cerebral, dog renal and guinea pig mesenteric arteries, this pressure-induced depolarization is accompanied by undershooting "spike-like" regenerative electrical activity.[1,2,3]

At least two important physiological consequences with respect to pressure-induced depolarization of arterial muscle have become apparent: 1) Under physiological levels of transmural pressure (100- 120 mmHg) the membrane potential is much closer to the activation threshold potentials for Ca^{2+} and K^+ channels than previously recognized; 2) The vascular sensitivity to a number of agonists is a markedly potentiated shift to the left of the dose response curve.[4] Thus, it is important to determine the signaling mechanisms by which changes in transmural pressure lead to activation of arterial vascular smooth muscle tone.

Recent findings that certain ion channels are activated upon physical stressing of biological membranes would appear to be a prime candidate for a transduction of physical stimuli into biological responses. Mechano-sensitive ion channels have been described in a number of cell types, and are observed under patch-clamp conditions.[5] These "stretch-activated" channels have a relatively high conductance and, therefore, could have a large effect on membrane potential if activated. Thus, mechano-sensitive ion channels may indeed play a role in pressure-induced depolarization of arterial muscle.

Several reports have demonstrated that inhibition of protein kinase C (PKC) can inhibit pressure-induced activation of arteries, and that phorbol esters augment myogenic tone.[6,7] PKC is activated by 1,2-diacylglycerol (1,2 DAG) formed via metabolism of phosphoinositide lipids.[8] Activation of phospholipase C (PLC) through a membrane receptor G protein complex, or possibly mechanical deformation, cleaves phosphatidyl inositol 4,5 biphosphate (Ptd-Ins 4,5 P_2) into the hydrophobic 1,2 DAG, and water-soluble inositol 1,4,5-triphosphate (IP_3). IP_3 elevates intracellular Ca^{2+} by its action on intracellular stores. 1,2 DAG serves as a second messenger to activate PKC, which has many functions including phosphorylation of membrane ion channels.[9,10] At present, it is not known how pressure activates PKC; however, we have preliminary data demonstrating that IP_3 and 1,2 DAG are indeed elevated in dog renal arteries as a function of transmural pressure, suggesting that pressure may stimulate PLC activity in the arterial wall.[16]

Recently, our laboratory has found that inhibition of cytochrome P-450 enzymes by a variety of mechanistically different inhibitors markedly attenuates pressure-induced myogenic tone. We have identified an endogenous P-450 metabolite of arachidonic acid from dog renal arteries, which co-elutes with 20-hydroxyeicosatetranoic acid (20-HETE) upon reverse-phase HPLC.[11,12] When 20-HETE is applied to isolated dog renal arteries, it elicits a dose-dependent activation beginning at 10^{-8} M. When applied to isolated renal arterial muscle cells, 20-HETE causes a substantial rise in $[Ca]_i$ (Fig. 1). 20-HETE (1μm) depolarizes muscle cells within pressurized intact renal arteries from -43 to -34 mV at a fixed transmural pressure of 90 mmHg. The membrane depolarization produced by 20-HETE may be partly due to the ability of 20-HETE to inhibit K^+ channel activity as seen in cell attached patches of isolated renal arterial muscle cells (Figure 2).

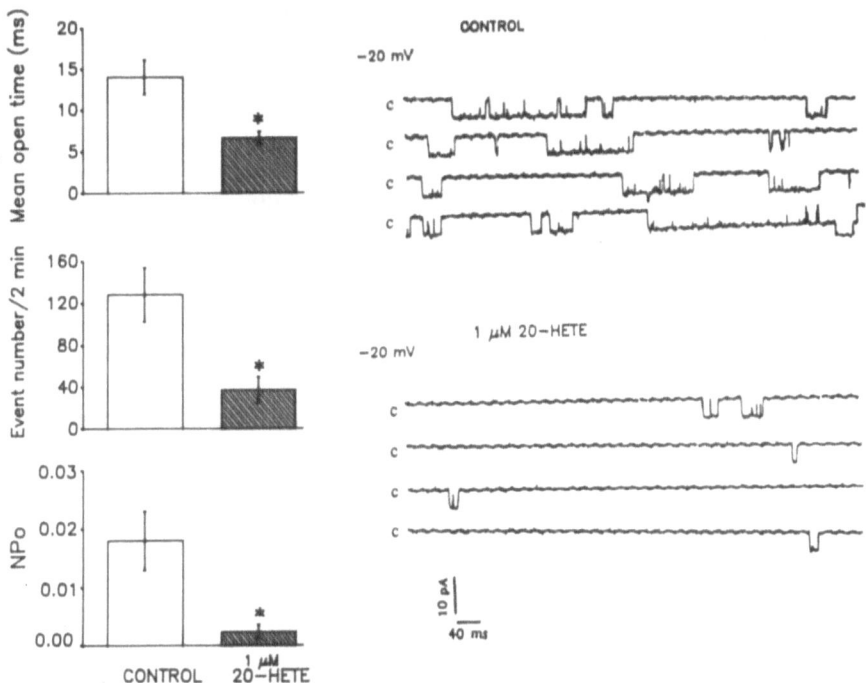

Fig. 1) Effects of 20-hydroxyeicosatetraenoic acid (20-HETE, 1 μM) on intracellular calcium concentration in vascular smooth muscle cells isolated from canine renal arcuate arteries. The inset (panel A) presents the time course of changes in intracellular calcium in a representative experiment. The bar graph (panel B) represents the mean response obtained from eight experiments with cell suspensions isolated from eight different dogs. *$p < 0.05$ compared with control value.

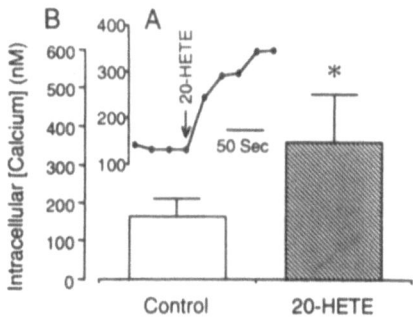

Fig. 2) Effects of 20-hydroxyeicosatetraenoic acid (20-HETE) on single potassium channel currents in smooth muscle cells isolated from canine renal arcuate arteries. Representative tracings of single potassium channel current before (top tracings) and after (bottom tracings) 20-HETE (1 μM) are shown in the right panel; c denotes closed state of the channel. Mean open time, event frequency, and open-state probability (NPo) of the channels are summarized in the left panel and are compared before (open columns) and after (hatched columns) the addition of 20-HETE (n=5 cells). *$p < 0.05$ compared with control values.

These actions of 20-HETE are probably not due to non-specific actions of fatty acids in that effects are dose-dependent in the range of 0.01 - 1 μm concentration, and the direct effect of high doses of arachidonic acid (AA) (50 μm) is to enhance, not inhibit, K^+ channel activity in excised membrane patches. Thus, 20-HETE fulfills some requirements as an endogenous vasoconstrictor formed by renal arteries and, potentially, could save as a mediator of pressure-induced myogenic activation. This last hypothesis is further strengthened by the observation that arachidonic acid potentiates the myogenic response in dog renal arteries, and this effect is blocked by P-450 inhibitors.

An important link also exists between PLC activity and the formation of 20-HETE. One of the principle sources of arachidonic acid is from DAG via the action of DAG lipase, which is regulated by $[Ca]_i$. Given that pressure enhances IP_3 and DAG levels, and that P-450 inhibition partially blocks myogenic activation, release of DAG following activation of PLC may serve as a source of arachidonic acid for the formation of 20-HETE, which could, in turn, act as a 2nd messenger to modulate K^+ channel activity, regulate membrane potential, and the influx of extracellular calcium through voltage gated channels (Figure 3).

ROLE OF PO_2 IN MODIFYING ELECTROPHYSIOLOGICAL EVENTS IN CEREBRAL ARTERIAL MUSCLE

The effects of O_2 on cerebral arterial muscle are difficult to determine. Siegel and Grote[13] have reported that reducing PO_2 from 160 to 40 torr hyperpolarizes and relaxes muscle cells within canine carotid arteries. However, there are several reports that hypoxia enhances EDRF release, and EDRF hyperpolarizes cerebral arterial muscle.[14] Thus, it is difficult to ascertain the direct effect of reducing O_2 on cerebral arterial muscle from the indirect effects of EDRF in intact vessels.

For these reasons, we have recently examined the effects of oxygen on vascular smooth muscle (VSM) cells freshly isolated from cerebral arteries. In these

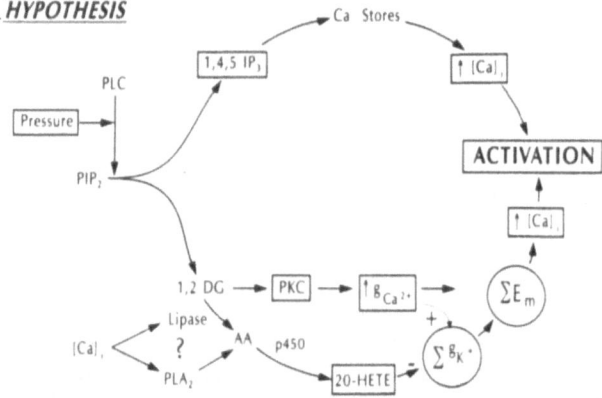

Fig. 3) An early hypothesis to explain the mechanisms of pressure-induced activation of arterial muscle.

experiments, reducing oxygen tension from 21 to $< 2\%$ torr resulted in a significant increase in the activity of a 215 pS K^+ channel in both cell attached and excised membrane patches of cat cerebral arterial muscle cells.[15] This channel is sensitive to $[Ca]_i$ between 0.088 and 0.48 μm, insensitive to $[ATP]_i$ and blocked by μm doses of TEA. Thus, this PO_2 sensitive channel displays all of the characteristics of a Ca^{2+}-sensitive K^+ channel (maxi K^+ channel). Given the large conductance of this channel, anything which modulates its activity, especially in the absence of an increase in Ca^{2+} conductance, would be expected to hyperpolarize and/or inhibit vascular tone. Given that pressure-induced activation of arteries is accompanied by depolarization and a reduction of K^+ conductance, it is possible that changes in PO_2 could modify myogenic tone through its effect on the Ca^{2+}-activated K^+ channel. This hypothesis is consistent with our previous observations in rat cerebral arteries that hypoxia can inhibit the electrical spike activity seen during pressurization and reduce the degree of pressure-induced depolarization.[17]

CONCLUSION

Our laboratory, as well as others, have found that the myogenic response to increases in transmural pressure is mediated by membrane depolarization. Biochemical and pharmacological characterization suggest that PLC is activated during step increases in pressure and this may initiate myogenic contraction via production of 2nd messengers, including IP_3, DAG and vasoactive P-450 metabolites of AA. One P-450 metabolite of AA, namely 20-HETE, is produced by renal arteries, activates and depolarizes renal arterial muscle by a mechanism involving inhibition of K^+ channel activity. Given the fact that P-450 inhibition partially blocks pressure-induced myogenic tone in dog renal arteries, it is possible that a P-450 metabolite of AA may be involved in the generation of myogenic tone. Hypoxia, through its actions on K^+ channels, also appears to modify pressure-induced myogenic tone, at least in cerebral arteries.

REFERENCES

1. Harder, D. R. Pressure-dependent membrane depolarization in cat middle cerebral artery. *Circ. Res.* 55: 197, 1984.
2. Smeda, J. S. and E. E. Daniel. Elevations in arterial pressure induce the formation of spontaneous action potentials and alter neurotransmission in canine ileum arteries. *Circ. Res.* 62: 1104, 1988.
3. Meininger, G. A. and M. J. Davis. Cellular mechanisms involved in the vascular myogenic response. *Am. J. Physiol.* 263 (Heart Circ. Physiol, 32); H647-H659, 1992.
4. Harder, D. R. Increased sensitivity of cat cerebral arteries to serotonin upon elevation of transmural pressure. *Pflugers Arch.* 411: 698, 1988.
5. Morris, C. E. Mechanosensitive ion channels. *J. Membrane Biol.* 113: 93, 1990.
6. Hill, M. A., J. C. Falcone and G. A. Meininger. Evidence for protein kinase C involvement in arteriolar myogenic reactivity. *Am. J. Physiol.* 259: H1586, 1990.

7. Laher, I. and J. A. Bevan. Staurosporine, a protein kinase C inhibitor attenuates Ca^{2+} - dependant stretch-induced vascular tone. *Biochem. Biophys. Res. Commun.* 158: 58, 1989.

8. Berridge, M. J. Inositol triphosphate and diacylglycerol: Two interacting second messengers. *Ann. Rev. Biochem.* 56: 159, 1987.

9. DeRiemer, S. A., J. A. Strong, K. A. Alpert, D. Greengard and L. K. Kaczmarek. Enhancement of calcium current in "Aplysia" neurones by phorbol ester and protein kinase C. *Nature* 313: 313, 1985.

10. Heagerty, A. M. and J. D. Ollerenshaw. The phosphoinositide signalling system and hypertension. *J. Hypertension* 5: 515, 1987.

11. Ma, Y - H, D. Gebremedhin, M. L. Schwartzman, J. R. Falck, J. E. Clark, B. S. Masters, D. R. Harder and R. J. Roman. 20-HETE is an endogenous vasoconstrictor of dog renal arcuate arteries. *Circ. Res.* 000:000 In Press, 1992.

12. Escalante, B. W., C. Sessa, J. R. Falck, P. Yadagiri and M. L. Schwartzman. Vasoactivity of 20-hydroxyeicosatetraenoic acid is dependent on metabolism by cyclooxygenase, *J. Pharmacol. Exp. Ther.* 248: 229-232, 1989.

13. Siegle, G. and J. Grote. PO_2 induced changes of membrane potential and tension in vascular smooth musculative. In: *"Oxygen Sensing in Tissues"*, H. Acker, ed., Springer-Verlag, Berlin, 1988.

14. Broyden, J. E. and G.C. Wellman. Endothelium-dependent dilation of feline cerebral arteries: Role of membrane potential and cyclic nucleotides. *J. Cerebral Blood Flow Metabolism* 9: 256, 1989.

15. Gebremedin, D., P. Bonnet, S. K. England, N. J. Rusch, J. H. Lombard and D. R. Harder. Hypoxia increases the activity of Ca^{2+}-sensitive K^+ channel in cat cerebral arterial muscle cell membranes. *Pflugers Archiv.* Submitted, 1992.

16. Harder, D. R., J. Narayanan, Y - H Ma and R. J. Roman. Activation of phospholipase C and induction of 2nd messengers in pressurized renal arteries. *FASEB Journal* 6: A1733, 1992.

17. Lombard, J. H., J. Smeda, J. A. Madden and D. R. Harder. Effect of reduced oxygen availability upon myogenic depolarization and contraction of cat middle cerebral artery. *Circ. Res.* 58: 565, 1986.

POTASSIUM CHANNELS ACTIVATED BY CALCIUM RELEASED FROM THE SARCOPLASMIC RETICULUM IN VASCULAR SMOOTH MUSCLES

Kenji Kitamura[1], Hirosi Kuriyama[1] and Hiroyuki Nabata[2]

[1.] Department of Pharmacology, Faculty of Medicine, Kyushu University, Fukuoka 812, and

[2.] Gotemba Institute, Chugai Pharmaceutical Company, Gotemba, Shizuoka 412, Japan

INTRODUCTION

It is generally thought that a contraction-relaxation cycle in visceral smooth muscles including vascular smooth muscles is regulated by the cytosolic Ca, i.e. Four Ca ions bind with calmodulin and this complex activates myosin light chain kinase (MLCK), thus causing phosphorylation of 20 kDa protein of myosin light chain (MLC). This phosphorylation of 20 kDa protein triggers actin-myosin cross bridge cycling in the presence of ATP (1-5).

Myosin phosphorylation is modified by cyclic nucleotides (cAMP and cGMP); through activation of cAMP-dependent protein kinase; A-kinase and cGMP-dependent protein kinase; G-kinase, and by protein kinase C (C-kinase) activators such as diacyl-glycerol and phorbol esters (6-17). Actin is also regulated by many elements, such as caldesmon (18), calponin (19), gelsolin (20) or leiotonin (MLCK-like substance; 21). Although activation of the contraction is induced by Ca, the amount of Ca does not proportionally enhance the amplitude of contraction due to modification of 20kDa protein phosphorylation by many factors. Such modification is termed "Ca-sensitization or -de-sensitization" (see Rev. of 22). In vascular smooth muscles, phenylephrine, histamine, endothelin and other spasmogenic agonists produce much higher phosphorylation and contraction than that induced by high K under conditions in which the amount of Ca increases to the same level. This increase in the phosphorylation and contraction is called

"Ca-sensitization", whereas application of high K or caffeine produces a high concentration of Ca in the cytosol but a smaller contraction than that expected from observed phosphorylation "Ca-desensitization".

Futhermore, in smooth muscles, Ca-independent contraction occurs during sustained contraction, i.e. during sustained contraction, Ca concentration that is measured due to agonist stimulation is rapidly reduced to a value close to the resting level. Phosphorylation of MLC and cycling rate of actin-myosin cross bridges are also lowered yet the contraction retained a level as high as that observed in the presence of high concentrations of Ca (latch-contraction; 1, 22-24). This phenomenon is explained by the presence of low crossbridge cycling of actin-myosin complex in the presence of low Ca (or low phosphorylation) following actin-myosin crossbridge cycling triggered by myosin phosphorylation.

Although the Ca requiring process for contraction is modified by various factors in physiological conditions, Ca is still an essential factor to induce contraction. In vascular smooth muscles, intracellular concentrations of Ca in the resting condition (in the presence of 2.5 mM Ca in the bathing solution) is 120-150 nM, and following removal of Ca from the bath (in the presence of 0.1 mM EGTA), it is reduced to 80-100 nM. The amount of Ca required for generation of phasic contraction is over 300 nM. The increase of Ca in the cytosol is thought to be due to influx of Ca through activation of the voltage-dependent Ca channel and receptor-operated cation channel. In addition, release of Ca occurs from the sarcoplasmic reticulum (SR) through ryanodine-sensitive and inositol 1,4,5-tris-phosphate (InsP3)-sensitive Ca channels (25-28).

In the sarcolemma, channels related to the Ca-influx into vascular smooth muscle cells are the voltage-dependent Ca channel and receptor-operated cation channel. However, increased Ca in the cytosol also regulates ion channels. Increased Ca in the cytosol activates a Ca-dependent Cl channel (depolarization; 29, 30), Ca-dependent Na channel (depolarization; 31-33, 43-45), Ca-dependent K channel (hyperpolarization; 34-42) and inactivates the voltage-dependent Ca channel (I_{Ca}; 34). It is also reported that extracellular (bath) Ca modifies the K channel (see Table 2).

To maintain Ca homeostasis, activation of the K channel plays an important role at the sarcolemma, such as determination of the resting membrane potential, repolarization of the membrane after activation of I_{Ca} inhibitory hyperpolarization induced by a neurotransmitter through activation of the K channel or excitatory depolarization induced by neurotransmitter (acetylcholine) through closing of a specific K channel (M-channel; see Table 2.). In this article, we will mainly discuss the Ca-dependent K channels regulated by Ca released from the SR in vascular smooth muscle cells.

CLASSIFICATION OF THE POTASSIUM CHANNEL

K channels in excitable cells are activated by various physical and chemical stimuli. From responses induced by these stimuli, K channels can be classified into several subtypes measured by the voltage- and patch-clamp procedures in the presence of various chemical agent (Table 1).

Table 1. Classification of the potassium channels in various excitable cells
(+: acceleration, -: inhibition)

Classification	Regulating factors	Physiological role
physical stimulation dependent type		
outward rectifying channel (about 10 pS)	depolarization (+)	shortening of action potential
inward rectifying K channel (about 15 pS)	depolarization (-)	inhibition of inward Na current
A channel (about 20 pS)	de-inactivation by hyper. and transient activation by depo..	regulation of spike frequency
SR-channel (about 150 pS)	depolarization of SR (+)	regulation of Ca release
SA-channel (40-300 pS)	stretch of the membrane	osmotic regulation
chemical stimulation dependent type		
C channel (maxi-, intermediate- and s-K; 30-300 pS)	[Ca]i (+), depolarization (+)	regulation of Ca influx and formation of rythmic activity
K_{Na} channel (about 200 pS)	[Na]i >30 mM (+)	acceleration of repolarization
K_{ACh} channel (about 40 pS)	ACh (+)	inhibition of cardiac activity
K_{ATP}	[ATP]i (-), glibenclamide (-), K channel opener (+)	inhibition of insulin secretion prevention of cardiac activity vascular relaxation
S channel (about 55 pS)	serotonin (-), FMRF amide (+),	sensitization, memory, learning
M channel (about 5 pS)	ACh (-), LHRF (-), bradykinin (-), somatostatin (+),	slow EPSP.

In vascular smooth muscles, K channels are classified into various subclasses using the whole cell and unitary current recording procedures, as in other excitable cells (physical and chemical stimulation). Phenomenologically, K currents recorded by depolarization pulses (voltage- dependent K channel; e.g. depolarization to 0 mV from a holding potential of -80 mV in isolated single smooth muscle cells of the guinea-pig portal vein using the whole cell voltage-clamp configuration) are classified into transient outward current (Ito) which occurs just after generation of the voltage-dependent inward Ca current, and sustained outward current (Iso) when a long depolarization pulse is applied (34-37). In many tissues spontaneous transient outward currents (STOCs; 38-44 or Ioo; 34) are superimposed on the Iso. Ito occurs 20-50 msec after application of a depolarization pulse and the Ioo occurrs 0.5 sec or more later following onset of the depolarization. The Ito and Ioo are due to activation of the Ca-dependent K channel and the Iso is due to activation of Ca-dependent and Ca-insensitive K channels, because the amplitude of Iso is reduced in Ca-free solution, or by tetraethylammonium (TEA) or 4-aminopyridine 4-AP) but never abolished (35). Fig. 1 shows the effects of long depolarization pulse on ionic current recorded from the guinea-pig portal vein using the voltage-clamp procedure. The membrane potential was displaced to o mV from a holding potential of -60 mV (35).

Using the current-voltage relationship, the voltage-dependent K current is classified into delayed rectifying K (brief delay following the onset of a membrane depolarization and persists while the depolarization is maintained) , inward rectifying K (anomalous rectifying; if, it mirrors the characteristics of delayed rectifier, and can pass a larger K current in the inward than in the outward direction), transient outward K (A-current; activated by

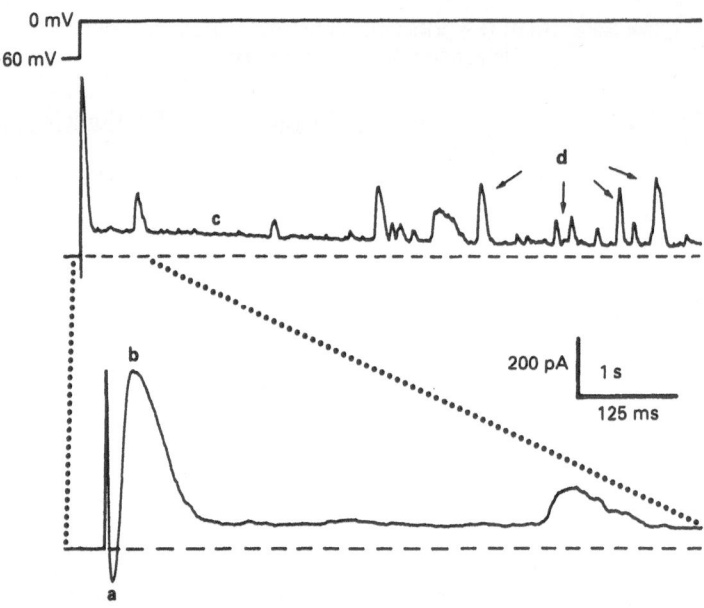

Figure 1. Membrane currents recorded in response to a depolarization pulse to 0mV from the holding
potential of -60 mV. The bottom shows the early part of the middle trace on a faster time scale. (a),
(b), (c) and (d) indicate the inward (a), transient outward (b), sustained outward (c) and oscillatory
outward (d) currents respectively. Broken lines on both traces indicate the basal membrane current
level observed at -60 mV. The pipette was filled with a high K solution and the bath solution was
physiological salt solution (PSS). Capacitative and leak currents were not subtracted. Note that the
initial upward deflection in the lower trace was a capacitative current (35).

Table 2. Characterization of potassium channels in various ssmooth muscle cells.
(macroscopic current: CTX=charybdotoxin, SP=substance P, 4-AP=4-aminopyridine)

Classification	Characteristics	Tissues	References
A. Votage-dependent K current			
(1) Transient outward current (I_{to})			
(a) Ca-dependent	CTX-sensitive, TEA sensitive	rabbit ear artery	38
	4-AP-sensitive, apamin-insensitive	rabbit portal vein	34, 40,46
(b) Ca-independent	4-AP-sensitive	rabbit pulmonary artery	46
(2) Oscillatory outward current	the same as Ca-dependent I_{to}	guinea-pig ureter	45
(STOC or Ioo)		rabbit portal vein	34
(3) Delayed outward current (I_V)			
(a) TEA-insensitive	Ca-insensitive, 4-AP-sensitive	rabbit pulmonary artery	46
	apamin-insensitive		
(b) TEA-sensitive	Ca-insensitive	rabbit portal vein	40
(4) Background K current	Ba-sensitive, Ca-sensitive	rabbit portal vein	40
B. Voltage-independent K current			
(1) ATP-sensitive K current (K_{ATP})			
(a) Ca-dependent	CTX-insensitive, TEA-sensitive	rabbit portal vein	47
		rat portal vein	
(b) Ca-independent	4-AP-sensitive, apamin-insensitive	porcine coronary artery	46-49
		rabbit portal vein	
C. Receptor-operated K current ("M")			
(1) Muscarinic M current	blocked by ACh and SP	toad stomach muscle	55
	activated by cAMP and β-agonist		
D. Non-selective cation channel			
(1) Hyperpolarization-activated current		rabbit jejunal artery	52
(I_f), anomalous rectifying K		rabbit portal vein (Okabe, K.,	
		unpublished observations)	

depolarization but decays spontaneously and rapidly while the depolarization is maintained) and background K currents, as classified on the cardiac K currents (34,35,37, 44, 54, 55 and Okabe, K. personal communication). In general, the K channel is also classified into voltage-dependent and -independent K channels. Table 2 shows some characteristics of the voltage-dependent and -independent K channel in vascular smooth muscles.

It should be mentioned here that classification of the K channel is only tentative, and no systematic analysis on relationship between the macroscopic and unitary currents has yet been made. Therefore the classification described below is only phenomenological, e.g. it is difficult to identify precisely the relationship between macroscopic delayed rectifier K current measured using the whole-cell voltage-clamp procedure and unitary K current measured with the patch-clamp procedure. With the patch-clamp procedure, several different K channel currents could be recorded with differences in the conductance of unitary current, such as large (about 150-250 pS; maxiK, BK), intermediate (50-100 pS) and small (10-30 pS; SK) conductances of K channels.These channels are also classified into Ca-dependent (intracellular Ca-senstive; maxiK and SK, and extracellular Ca-sensitive; intermediate K) and Ca-insensitive K channels (maxiK, SK and intermediate K). The K channel also classified into TEA-sensitive (mainly maxiK) and -insensitive (some SK) channels or 4-AP-sensitive (mainly SK) and -insensitive channels (maxiK and SK). Furthermore, these channels have different sensitivity to intracellular ATP, i.e. ATP-sensitive and -insensitive K channels. There are some confusion, since some invesigators have reported that the Ca-dependent (56-58) and Ca-insensitive maxi-K are sensiitve to ATP (47), and others reported that a Ca-dependent SK (48, 59, 60) or a Ca-insensitive SK is sensitive to ATP (49). For example, the ATP-sensitive K channel recorded in the guinea-pig portal vein is a Ca-dependent, TEA-insensitive, 4-AP-sensitive, charybdotoxin (CTX; a scorpion toxin)-insensitive and apamin (a bee toxin)-insensitive SK channel (48), the ATP-sensitive K channel in the rabbit portal vein is a Ca-insensitive, TEA-insensitive, 4-AP-sensitive, CTX-insensitive and apamin-insensitive SK channel (48), whereas the ATP-sensitive K channel in the rabbit mesenteric artery is Ca-insensitive, TEA-sensitive, 4-AP-insensitive, CTX-insensitive, apamin-insensitive intermediate K channel (47).

Concerning the action of 4-AP, vascular smooth muscles, especially elastic arteries such as aorta and pulmonary artery, are more sensitive to 4-AP than myogenic (resistance) vessels and visceral smooth muscles (61-64). Hara et al.(62) reported that in the guinea-pig coronary artery, 0.5 mM 4-AP is enough to depolarize the membrane but much higher concentrations of 4-AP (ten times) are required to depolarize the membrane in tracheal and gastric smooth muscles (65-67). In elastic arteries, generation of action potentials is difficult in physiological conditions but after application of 4-AP or TEA, a graded response or action potential can be evoked. In these tissues, distribution of I_{Ca} may be sparse and the 4-AP or TEA-sensitive K channel may be much more dense.

In addition, in vascular smooth muscle, M-current (muscarinic-receptor sensitive K channel) and S-current (serotonin-sensitive K channel) are postulated. These channels are shut when individual receptors are activated, hence the membrane is depolarized and produces excitatory responses. Of course these agonists also activate the receptor-operated cation channel (see Tabs. 1 and 2).

POTASSIUM CHANNELS ACTIVATED BY RELEASED CALCIUM FROM THE SARCOPLASMIC RETICULUM.

As briefly described previously, Ca is released from the SR mainly through two different routes: the InsP3-induced and ryanodine-induced Ca releases. The primary amino acid sequences and structures of the binding protein (receptor) for InsP3 are elegantly elucidated in skeletal and neural cells (68-70), and it is known that binding of InsP3 receptors (binding site) activates the Ca channel in the SR, and the amino acid sequence and primary structure of the ryanodine-sensitive Ca channel have also been elucidated (71-75), and the ryanodine-sensitive Ca channel protein is also activated by Ca itself. This phenomenon has been called " Ca-induced Ca release" (in skeletal muscle; 76 and in smooth muscles; 77-80). This Ca release mechanism is accelerated by caffeine and inhibited by procaine or other local anesthetics (77). Both the ryanodine- and InsP3-sensitive Ca channel receptors have many similarities in the primary amino acid sequences. The InsP3- and caffeine-induced contraction can be recorded from saponin- treated skinned muscle tissues in the presence of 0.1 mM Ca in the bath, repetitively applied caffeine and InsP3 with some intervals consistently releaseCa from the SR and produce the same amplitude of contractions in saponin-treated skinned muscle tissues (25). Such results were also observed using a-toxin and b-escin treated skinned muscle tissues (receptor-preserved skinned muscle tissues; 81-87).

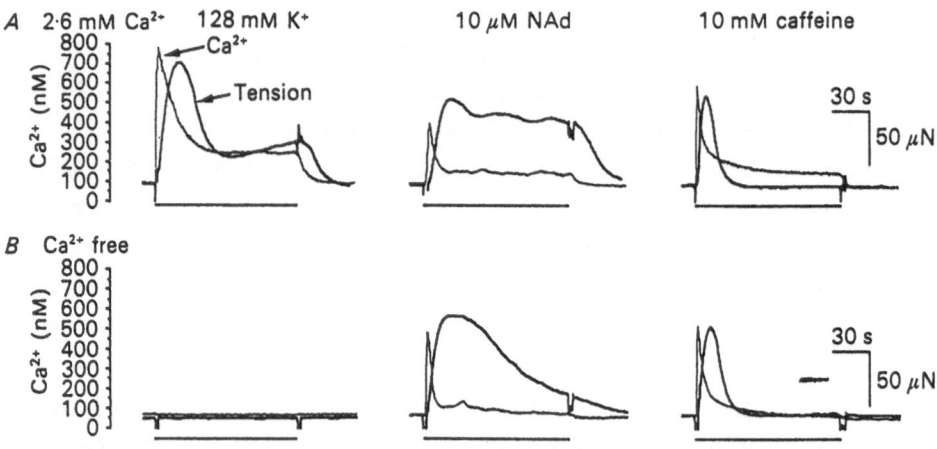

Figure 2. Effects of 128 mM K^+, 10 μM NAd or 10 mM caffeine on $[Ca^{2+}]i$ and tension in a smooth muscle strip of the rabbit mesenteric artery in the presence or absence of extracellular Ca^{2+}. A, in the presence of 2.6 mM Ca^{2+}. B, in Ca^{2+}-free solution containing 2 mM EGTA. Individual stimuli were applied for 2 min at 20 min intervals. Thinner and thicker traces indicate $[Ca^{2+}]i$ and tension, respectively. The stimulus was applied where indicated by the bar. In B, the stimuli were applied 2 min after application of Ca^{2+}-free solution containing 2 mM EGTA (91).

The amount of released Ca can also be observed using fura-2 and other fluorescence dyes. As shown in Fig. 2, on treatment with fura-2, high K (activation of the voltage-dependent Ca channel), norepinephrine (NE; an activator of InsP3-induced Ca release) and caffeine (activator of the ryanodine-sensitive Ca channel) can evoke contraction and Ca transient (91). The relationship between the amplitude of contraction and the Ca-transient differed according to the procedures described previously (sources of Ca and changes in the Ca-sensitivity of the contractile proteins; 83-86; 89-90).

In addition, the amount of Ca released from the SR could also be estimated from Ito and Ioo, i.e. depolarization of the membrane to more than -40 mV in various smooth muscle cells generates Ito and Ioo. We estimated that Ito is closely related to activation of the Ca-induced Ca release mechanism (through activations of the ryanodine-sensitive Ca channel) and Ioo is due to release of Ca through activation of both the ryanodine-sensitive and InsP3-sensitive Ca channel for the following reasons: with application of a

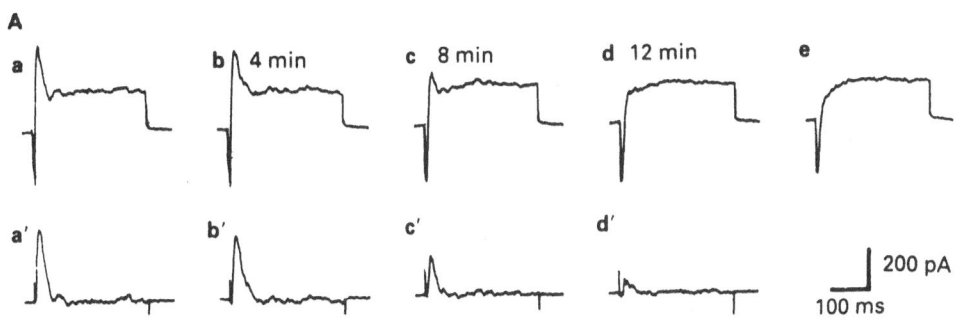

Figure 3. Effects of ryanodine on isolated transient outward current (I_{to}), and the concentration-inhibition curves on isolated I_{to}, oscillatory outward current (I_{oo}) and sustained outward current (I_{so}). (A) The membrane currents were evoked by a depolarization to 0 mV before (a) and during (b-d) application of 10 μM ryanodine. Time indicates the period after drug application. Immediately after obtaining trace (d), caffeine (3 mM) was added, and after 3 min, trace (e) was obtained. Traces (a'), (b'), (c') and (d') were obtained by subtracting the membrane current in the presence of caffeine (trace e) from membrane current before application of caffeine (traces a,b,c and d, respectively). A depolarizing pulse (300 ms in duration) was applied every 30s (35).

depolarization pulse to 0 mV from a holding potential of -60 mV in extracellular Ca-free solution (in the presence of EGTA), subsequent production of the Ito following generation of the voltage-dependent Ca current (ICa) ceased; following application of A23187 (a Ca ionophore) at low concentrations (<100 nM of this agent selectively makes the SR non-functional), there
is generation of the Ito and Ioo block; generation of I_{Ca} is a prerequisite for generation of the Ito; when Ca stored in the cell is depleted by pretreatment with caffeine or ryanodine, the I_{Ca} can be generated but not the Ito; after application of CTX, the Ito ceases. These observations indicate that influx of Ca by activation of the voltage-dependent Ca channel stimulates release of Ca from the SR through activation of the ryanodine-sensitive Ca

channel and the released Ca activates many CTX-sensitive maxiK channels simultaneously to generate the Ito. Fig. 3 shows the effects of caffeine on the depolarization induced ionic currents, especially the Ito (35).

Concerning the Ioo, the frequency of Ioo increases in proportion to the depolarization of the membrane and concentration of Ca in the bath. When Ca concentration is reduced, the frequency of Ioo is markedly reduced, especially Ioo with the small amplitude. In Ca-free solution containing EGTA, caffeine transiently enhances the frequency of Ioo generated by depolarization of the membrane but they then cease due to depletion of the stored Ca. Ryanodine shows much the same effect as those observed on application of caffeine. Fig. 4 shows the effects of Ca on the Ioo evoked by depolarization of the membrane (37).

As described previously the Ioo can be generated by depolarization of the membrane to more than -40 mV from the holding potential of -60 mV (resting membrane potential level in the rabbit and guinea-pig portal veins). However, when InsP3 is applied the Ioo can be generated at the holding potential of -60 mV without any depolarization of

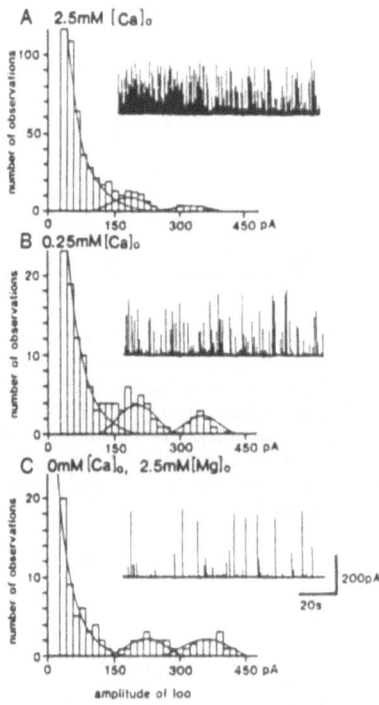

Figure 4. Amplitude histograms of Ioo (bin width, 15 pA) obtained at a membrane potential of 0 mV in solution containing three different Ca concentrations. (A) 2.5 mM $[Ca]_o$, (B) 0.25 mM $[Ca]_o$, (Ca) 0 mM $[Ca]_o$. Current trace obtained at various $[Ca]_o$ are also shown, however, fluctuation less than 30 pA have been omitted. Regression curves were fitted with following parameters; the 1st component , maximum number of observations=390 (A), 95 (B), 57 (C); mean amplitude=42 pA^2 (A), 37.5 pA^2 (B), 36 pA^2 (C): The 2nd component , maximum nunber of observations=9.4 (A), 3.8 (B), 2.0 (C): mean value =180 pA(A), 202.5 pA(B), 225 pA(C); variance=94.5 pA^2 (A), 91.5 pA2(B), 81.0 pA^2(C): The 3rd component, maximum number of observation= 2.6 (A), 2.3(B), 2.3(C); mean value =322 pA(A), 346.5 pA (B), 357 pA(C); variance = 139.5 pA^2 (A), 84.0 pA^2 (B), 261 pA^2 (C). Samling times for A and B were 1 min and that of C was 2 min. A, B and C were obtained from the same cell (37).

the membrane. InsP3 markedly enhances the frequency of Ioo with no change in the amplitude for any given depolarization pulses. On treatment with heparin, these effects of InsP3 on the Ioo are almost blocked. As shown in Fig. 5, at the holding potential of -60 mV, InsP3 generated the Ioo (37). Much the same effects on the Ioo were observed on application of NE to smooth muscle cells of the guinea-pig portal vein. When the tissue was pretreated with NE, the caffeine-induced Ioo was markedly inhibited and vice versa. This means that the Ca required for generation of Ioo comes from the Ca storage site which is sensitive to ryanodine or InsP3.

Using the cell-free patch-clamp procedures, maxiK and SK unitary currents can be recorded by depolarization of the membrane to more than -40 mV from the holding potential of -60 mV (for generation of SK but not maxiK in the rabbit portal vein application of GDP was required to prevent a run-down phenomenon but not in the guinea-pig portal vein; 49). The open probability of a maxiK channel is increased in a voltage-dependent manner, and during applications of depolarization pulse, bursts of more than 10 channel openings occur in an irregular manner.

The unitary current ceases on application of CTX, thus this large unitary current is due to activation of the Ca-dependent maxi K. These burst discharges may be composed of irregular generations of Ioo. When solely unitary current occur on depolarization, using the cell-free inside-out patch-clamp procedure, caffeine and heparin neither modify the open probablity of the maxiK channel nor the conductance of this current, whereas, when burst unitary discharges occur on depolarization, bath application of heparin or caffeine (inside-out patch) blocks the generation of burst channel openings (36). Figs. 6 and 7 show the effects of caffeine and heparin, respectively, on the unitary current and burst generations of the unitary current (36).   Presumably membrane fractions may contain a SR vesicle or

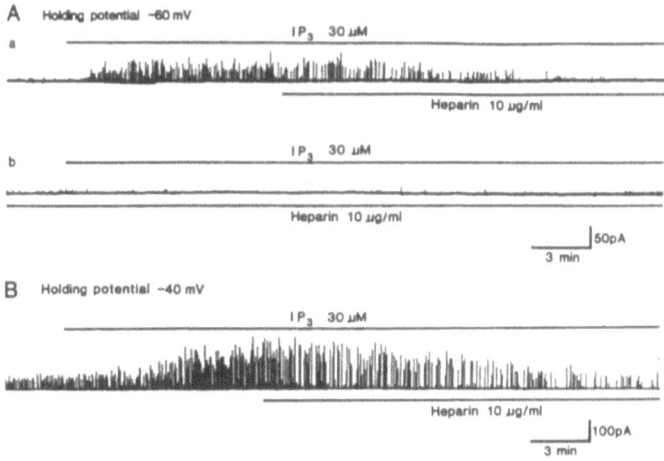

Figure 5. Effects of heparin (10 μg/ml) on InsP3 (30 μM)-induced generation of I_{oo} at two different holding potentials (-60 mV and -40 mV). InsP3 and heparin were applied intracellulary where indicated by bars. A. The membrane was held at -60 mV. Heparin was applied 11 min after (a), or 15 before (b) application of InsP3. B. The membrane was held at 40 mV. Heparin was applied 10 min after application of InsP3. Each trace was obtained from a different cell (36).

vesicles, and when Ca is eruptively released by activation of the ryanodine-sensitive Ca channel or InsP3-sensitve Ca channel, released Ca may activate many Ca-dependent K channels simultaneously. These observations indicate that in the cell-free patch-clamp procedure, some the maxiK could be activated. Both releasers themselves have no direct effect on the maxiK. In addition, some burst discharges of the unitary current do not vanish on application of heparin but are completely blocked by caffeine. This means that the SR vesicles may contain both ryanodine-sensitive and InsP3-sensitive Ca channels, but in some vesicles the Ca channels are not distributed homogeneously.

As shown in Figs. 6 and 7, caffeine consistently blocks burst unitary currents generated by depolarization but heparin does not. It is plausible to postulate that depolarization of the membrane may more potently activate the ryanodine-sensitive Ca channel, rather than the InsP3-sensitive Ca channel in the SR. In addition, after cessation of burst channel currents by caffeine or heparin, re-application of Ca induced the re-appearance of burst unitary currents. Presumably, the SR vesicle are distributed very close to the sarcolemma with a "foot" structure as observed in the triad region of skeletal muscles (86) or with skeleton proteins distributed just beneath the sarcolemma (36). Thus, some cell-free fragments prepared for the patch-clamp procedure may contain the SR vesicle or vesicles, and these attached fragments may regulate Ca mobilization (36).

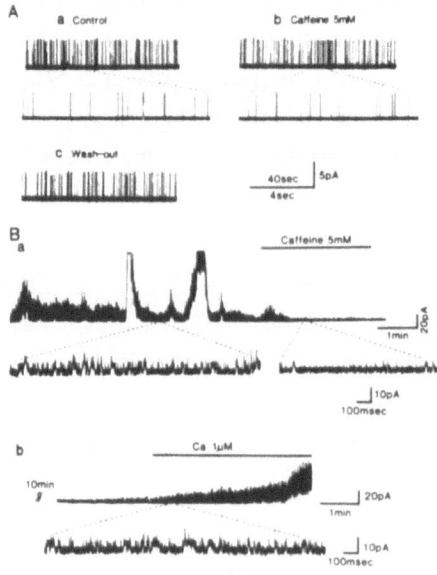

Figure 6. Effects of caffeine (5 mM) on the large conductance Ca^{2+}-dependent K+ channel recorded from the inside-out membrane patch of the rabbit portal vein, at a holding potential of -10 mV. The pipette was filled with PSS and the bath was superfused with high K+ solution containing 4 mM (A) and 0.05 mM EGTA (B). Patch electrodes with a resistance of 6 MΩ in A and 2.5 MΩ in B. (A): Traces obtained just before (a; control), with caffeine in the bath (B; 5 min after) and 5 min after caffeine removal (C). Traces with an expanded scale are also shown in A. (B): Caffeine (5 mM; a) and 1 μM free-Ca^{2+} (b) were added to the bath (intracellular side) at the time indicated by bars. (a) and (b) were obtained from the same membrane patch at a10 min interval. Traces with an expanded scale before and during application of caffeine or Ca^{2+} are shown (36).

It is thought that depolarization of the sarcolemma may not directly release Ca from the SR, because, in high K solution increase in the Ca-transient contraction occurs, but in Ca-free solution or on application of Ca antagonists, such as dihydropyridine derivatives, diltiazem or verapamil, both phenomena cease. Therefore, depolarization of the membrane induced by electrical stimulation may mainly activate the ryanodine-sensitive Ca channel in the SR following increased influx of Ca. However, heparin modifies the Ioo and burst unitary current generation induced by application of depolarization pulses to the membrane. This means that depolarization of the membrane may modify the metabolic path directly or indirectly. It was reported that K-channel openers such as nicorandil, lemakalim or pinacidil hyperpolarize the membrane through activation of the ATP-sensitive K channel, and this hyperpolarization inhibits the synthesis of InsP3 (91). In addition it has also been reported that in guinea-pig iliac muscles, depolarization of the membrane induced by high K accelerates synthesis of InsP3 (88). Furthermore, neomycin, an inhibitor of phospholipase C, inhibits Ioo generated by depolarization (37). presumably increased cytosolic Ca may accelerate the phospholipase C activity and cause the synthesis of InsP3.Conversely it is possible that heparin is not a selective inhibitor of the InsP3-sensitive Ca channel in the SR.

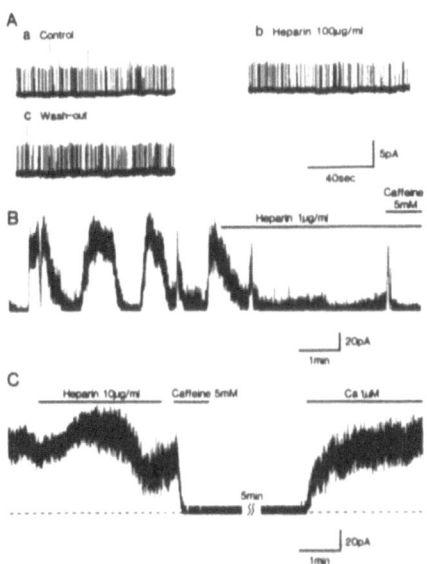

Figure 7. Effects of heparin on the large conductance Ca^{2+}-dependent K^+ channel recorded from the inside-out membrane patch of the rabbit portal vein at a holding potential of -10 mV. The pipette was filled with PSS and the bath was superfused with high K^+ solution containing 4 mM (A) and 0.05 mM EGTA (B and C). Patch electrodes with a resistance of 5 MΩ were used in A and those with 3 MΩ were used in B and C. (A): Traces obtained just before (a; control), during bath application of heparin (b; 5 min after) and 5 min after removal (c) of heparin (100 μg/ml). (B): A low concentration of heparin (1 μg/ml) was added to the bath at the time indicated by the bar and then caffeine (5 mM) was applied. (C: A high concentration of heparin (10 μg/ml) was added to the bath at the time indicated by the bar and then the same membrane was exposed to caffeine (5 mM). After removal of caffeine, 1 μM free-Ca^{2+} was added to the bath. Note that caffeine but not heparin effectively inhibited K^+ channel currents (36).

We also examined the effects of heparin, guanosine nucleotides (GTP, GTPγS and GDPβS) and protein kinase C (C-kinase) regulators (H-7and phorbol ester) on the Ca-dependent K channels using the voltage-and patch-clamp procedures, to explore the effects of C-kinase on the Ioo. Fig. 8 shows the effects of H-7 and phorbol ester on the Ioo evoked in the cell-free membrane by depolarization pulses in the rabbit portal vein. The results indicate that generation of the Ioo has a causal relation to activity of GTP-binding proteins (G-protein) and C-kinase activity (37). In the rabbit portal vein, GTP and GTPgS accelerate generation of the Ioo evoked by the depolarization pulse and these actions are prevented by GDPbS. These conclusions differ from those on Ioo by Komori & Bolton (42) and Ganitkevich & Isenberg (41). They reported that heparin and GDPgS have no effect on the Ioo generated by depolarization pulses. At present we have no idea how to solve these differences. In addition, a phorbol ester, phorbol 12, 13-dibutylate, reduced the generation of Ioo but failed to produce an outward current following application of caffeine. This action of phorbol ester is inhibited by pretreatment with H-7, an inhibitor of C-kinase. In the presence of H-7, GTPgS still enhances the generation of Ioo. Presumably, sites of action of GTP and phorbol ester may differ; phorbol ester may stimulate the Ca-depleting process from the SR and GTP may act on the process of InsP3 synthesis (37).

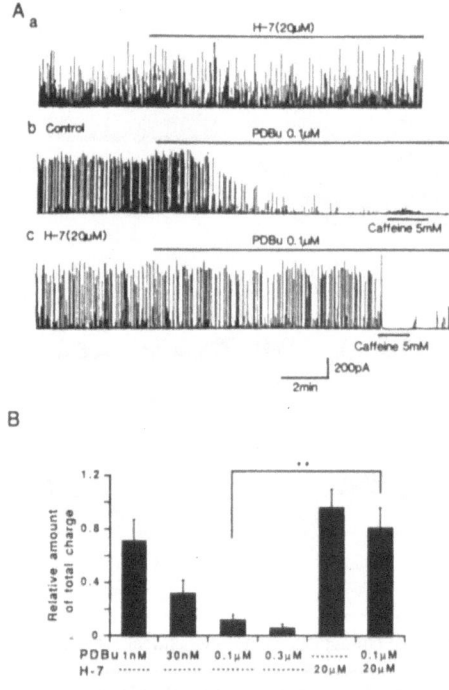

Figure 8. Effects of phorbol 12,13-dibutylate (PDBu) and H-7 on Ioo produced at a holding potential of 0 mV in the rabbit portal vein. A. Effects of 20 μM H-7 (a) and 0.1 μM PDBu or 5 mM caffeine in the absence (b) or presence of H-7 (c). In c, H-7 was applied 25 min before application of PDBu. Drugs were applied at the time indicated by bars. Each trace was obtained from a different cell. B. Relative amount of total charge carried by Ioo in various concentration of PDBu (1 nM-0.3 μM) in the presence or absence of H-7 (20 μM). Amount of total charge (2 min sampling time 0.1 msec sampling rate) in the absence of drug was normalized as 1.0 and that in the presence of PDBu or H-7 was expressed in a relative manner. Each column shows mean value of three to five observations with standard deviation. **P<0.01.

CONCLUSION

We have briefly introduced the Ca-sensitive K channel activated by release of Ca through activation of ryanodine-sensitive and InsP3-sensitive Ca channels in the SR. Using the voltage-clamp procedure, macroscopic currents related to the SR function are mainly Ito and Ioo. The former, Ito, is mainly generated by Ca released through activation of the ryanodine-sensitive Ca channel through the Ca-induced Ca release mechanism. Thus, influx of Ca induced by activation of the I_{Ca} is an essential factor, and Ca-sensitive maxiK contributes mainly to generation of the Ito. Presumably, the Ito may contribute for generation of the falling phase of the action potential as well as the after-hyperpolarization of the action potential. On the other hand, when the latter, Ioo, is generated by depolarization of the membrane, it is mainly due to activation of the ryanodine-induced Ca channel, and partly to the InsP3-induced Ca release, because the Ioo generated by depolarization is modified by agents which inhibit the release of Ca from the InsP3-sensitive Ca channel assuming that heparin is a selective blocker of this channel. When the Ioo is generated by agonists, this generation is mainly due to activations of the InsP3-induced Ca channel, and partly due to activation of the ryanodine-sensitive Ca channel following release of Ca by InsP3 through activations of the Ca-induced Ca release mechanism. The main K channel contributing to the generation of Ioo seems to be the maxiK.

Ca released from the SR shows multiple actions in vascular smooth muscle cells, and one of the actions is a negative feedback control of cell activity through hyperpolarization of the membrane by activation of the Ca-sensitive K channel activation. We have not enough knowledge to explain the physiological significance of the Ito and Ioo. Further detailed experiments are waiting, which must be carried out to clarify the features of K channels activated by Ca released from the SR.

ACKNOWLEDGEMENTS

This work was partly supported by the grants of the Ministry of Education and Welfare in Japan. We also thank to Dr. K. Creed, Murdoch Univ. for her English editing.

REFERENCES

1. Murphy, R.A. Contraction in smooth muscle cells. Annu. Rev. Physiol. 51: 275-283, 1989.
2. Sperelakis, N. and Wood, J.D. In: Frontiers in Smooth Muscle Research, New York, Wiley-Liss, 1990.
3. Sperelakis, N. and Kuriyama, H. Ion channels of Vascular Smooth Muscle Cells and Endothelial Cells, New York, Elsevier, 1991.
4. Moreland, R.S. In Regulation of smooth muscle contraction, Advances in Experimental Medicine and Biology, New York & London, Plenum Press, 304,1991.
5. Kuriyama, H. International Symposium "Smooth Muscle" -assessments of current knowledges- In Jpn. J. Pharmacol. , 58, Suppl.II, 1992.
6. Adelstein, R.S., Pato, M.D., Conti, M.A. Regulation of smooth muscle myosin kinase activity by phosphorylation and dephosphorylation. In: Muscle contraction: Its regulatory Mechanism. edited by S. Ebashi, K. Maruyama & M. Endo. Tokyo, Jpn. Sci. Soc., 303-313, 1980.

7. Ito, M., Guerriero, V. Jr., Hartshorne, D.J. Structure-function relationship in smooth muscle myosin light chain kinase. In: Regulation of Smooth Muscle Contraction. edited. by Moreland, R.S., New York and London, Plenum Press, p.3-10, 1991.
8. Nishizuka, Y. Studies and prospectives of protein kinase C. Science, 233: 305-311, 1986.
9. Tansey, M.G., Hori, M., Karaki, H., Kamm, K.E. and Stull, J.T. Okadaic acid uncouples myosin light chain phosphorylation and tension in smooth muscle. FEBS Lett., 270: 219-221, 1990.
10. Miller, J.R., Silver, P.J. and Stull, J.T. The role of myosin light chain kinase phosphorylation in Bata-adrenergic relaxation of tracheal smooth muscle. Mol. Pharmacol., 24: 235-242, 1983.
11. Fransis, S.H., Noblett, R.D., Todd, B.W., Wells, J.N. and Corbin, J.D. Relaxation of vascular and tracheal smooth muscle by cyclic nucleotide analogs that preferentially activate purified cGMP-dependent protein kinase. Mol. Pharmacol., 34: 506-517, 1988.
12. Lincoln, T.M., Cornwell, T.L. and Tayler, A.E. cGMP-dependent protein kinase mediates the reduction of Ca^{2+} by cAMP in vascular smooth muscle cells. Am. J. Physiol., 258: C399-C407, 1990.
13. Ikebe, M., Inagaki, K., Kanamura, K. and Hidaka, H. Phosphorylation of smooth muscle myosin light chain kinase by Ca^{2+}-activated, phospholipid-dependent protein kinase. J. Biol. Chem., 260: 4547-4550, 1985.
14. Nishikawa, M., Shirakawa, S. and Adelstein, R.S. Phosphorylation of smooth muscle myosin light chain kinase by protein kinase C. J. Biol. Chem., 260: 8987-8983, 1985.
15. Shearman, A.S., Sekiguchi, K. Nishizuka, Y. Modulation of ion channel activity: a key function of the protein kinase C enzyme family. Pharmacol. Rev., 41: 211-237, 1989.
16. Itoh, T., Kuriyama, H. and Suzuki, H. Differences and similarities in noradrenaline- and caffeine-induced mechanical responses in the rabbit mesenteric artery. J. Physiol., 337: 609-629, 1983.
17. Kamm, K.E., Hsu, J.-C., Kubota, Y. and Stull, J.T. Phosphorylation of smooth muscle myosin heavy and light chains. J. Biol. Chem., 264: 21223-21222, 1989.
18. Sobue, K., Muramoto, Y., Fujita, M. and Kakiuchi, S. Purification of a calmodulin-binding protein from chicken gizzard that interacts with F-actin. Proc. Natl. Acad. Sci., USA, 78: 5652-5655, 1981.
19. Takahashi, K., Hiwada, K. and Kokubu, T. Isolation and characterization of a 34000-dalton calmodulin- and F-actin-binding protein from chicken gizzard smooth muscle. Biochem. Biophys. Res. Commun., 141: 20-26, 1986.
20. Ebisawa, K. and Nonomura, Y. Enhancement of actin-activated myosin ATPase by an 84K Mr actin-binding protein in vertebrate smooth muscle. J. Biochem. 98: 1127-1130, 1985.
21. Ebashi, S. In Calcium in Human Biology. edited by Nordin BEC, London, Springer-Verlag, 317-338, 1989.
22. Rembold, C.M. Regulation of contraction and relaxation in arterial smooth muscle. Hypertension, 20: 129-137, 1992.
23. Dillon P.F., Aksoy, M.O. and Murphy, R.A. Myosin light chain phosphorylation and the cross-bridge cycle in arterial smooth muscle. Science, 211: 495-497, 1981.
24. Rembold, C.M. and Murphy, R.A. $[Ca^{2+}]$-dependent myosin phosphorylation in phorbol diester stimulated smooth muscle contraction. Am. J. Physiol., 255: C719-C723, 1988.
25. Hashimoto, T., Hirata, M., Itoh, T., Kanmura, Y. and Kuriyama, H. Inositol 1,4,5-trisphosphate activates pharmacomechanical coupling in smooth muscle of the rabbit mesenteric artery. J. Physiol., 370: 605-618, 1986.
26. Karaki, H. Ca^{2+} localization and sensitivity in vascular smooth muscle. Trends Pharmacol. Sci., 10: 320-325, 1989.
27. Iino, M. Biphasic Ca^{2+} dependence of inositol 1,4,5-trisphosphate-induced Ca release in smooth muscle cells of the guinea pig taenia caeci. J. Gen. Physiol., 95: 1103-1122, 1990.
28. Iino, M. Effects of adenine nucleotides on inositol 1,4,5-trisphosphate-induced calcium release in vascular smooth muscle cells. J. Gen. Physiol., 98: 681-698, 1991.
29. Byrne, N.G. and Large, W.A. membrane ionic mechanisms activated by noradrenaline in cells isolated from the rabbit portal vein. J. Physiol., 404: 557-573, 1988.

30. Amedee, T., Large, W.A., and Wang, Q. Characteristics of chloride currents activated by noradrenaline in rabbit ear artery cells. J. Physiol., 428: 501-516, 1990.

31. Inoue, R. and Isenberg, G. Intracellular calcium ions dodulate acetylcholine-induced inward current in guinea-pig ileum, J. Physiol., 424: 73-92, 1990.

32. Inoue, R. and Isenberg, G. Effects of membrane potential in acetylcholine-induced inward current in guinea-pig ileum. J. Physiol., 424: 57-71, 1990.

32. Inoue, R. and Isenberg, G. Acetylcholine activates nonselective cation channels in guinea pig ileum through a G protein. Amer. j. physiol., 258: C1173-1178, 1990

34. Ohya, Y., Terada, K., Yamaguchi, K., Inoue, R., Okabe, K., Kitamura, K., Hirata, M. and Kuriyama, H. Effects of inositol phosphates of the membrane activity of smooth muscle cells of the rabbit portal vein. Pflügers Arch., 412: 382-389, 1988.

35. Sakai, T., Terada, K., Kitamura, K. and Kuriyama, H. Ryanodine inhibits the Ca-dependent K current after depletion of Ca stored in smooth muscle cells of the rabbit ileal longitudinal muscle. Br. J. Pharmacol., 95: 1089-1100, 1988.

36. Xiong, Z., Kitamura, K. and Kuriyama, H. Evidence for contribution of Ca^{2+} storage sites on unitary K^+ channel currents in inside-out membrane of rabbit portal vein. Pflügers Arch., 420: 112-114, 1992.

37. Kitamura, K., Xiong, Z., Teramoto, N. and Kuriyama, H. Roles of inositol trisphosphate and protein kinase C in the spontaneous outward current modulated by calcium release in rabbit portal vein. Pflügers Arch., in press, 1992.

38. Benham, C.D. and Bolton, T.B. Spontaneous transient outward currents in single visceral and vascular smooth muscle cells of the rabbit. J. Physiol., 381: 385-406, 1986.

39. Bolton, T.B. and Lim, S.P. Properties of calcium stores and transient outward currents in single smooth muscle cells of rabbit intestine. J. Physiol., 409: 385-401, 1989.

40. Hume, J.R. and Leblanc, N. Macroscopic K+ currents in single smooth muscle cells of the rabbit portal vein. J. Physiol., 413: 49-73, 1989.

41. Ganitkevich, V. and Isenberg, G. Isolated guinea pig coronary smooth muscle cells; acetylcholine induces hyperpolarization due to sarcoplasmic reticulum calcium release activating potassium channels. Circ. Res., 67: 525-528, 1990.

42. Komori, S. and Bolton, T.B. Actions of guanine nucleotides and cyclic nucleotides on calcium stores in single patch-clamped smooth muscle cells from rabbit portal vein. Br. J. Pharmacol., 97: 973-982, 1989.

43. Komori, S. and Bolton, T.B. Role of G-proteins in muscarinic receptor inward and outward currents in rabbit jejunal smooth muscle. J. Physiol., 427: 395-415, 1990.

44. Komori, S. and Bolton, T.B. Calcium release induced by inositol 1,4,5-trisphosphate in single rabbit intestinal smooth muscle cells. J. Physiol., 433: 495-517, 1991.

45. Lang, R.J. Identification of the major membrane currents in freshly dispersed single smooth muscle cells of the guinea-pig ureter. J. Physiol., 412: 375-395, 1989.

46. Okabe, K., Kitamura, K. and Kuriyama, H. Features of 4-aminopyridine sensitive outward current observed in single smooth muscle cells from the rabbit artery. Pflügers Arch., 409: 561-568, 1987.

47. Standen, N.B., Quayle, J.M., Davis, N.W., Brayden, J.E., Huang, Y. and Nelson, M.T. Hyperpolarizing vasodilators activate ATP-sensitive K^+-channels in arterial smooth muscle. Science, 245: 177-180, 1989.

48. Kajioka, S., Oike, M. and Kitamura, K. Nicorandil opens a calcium-dependent potassium-channel in smooth muscle cells of the rat portal vein. J. Pharmacol. Exp. Ther., 254: 905-913, 1990.

49. Kajioka, S., Kitamura, K. and Kuriyama, H. Guanosine phosphate activates and adenosine 5'-triphosphate-sensitive K^+ channel in the rabbit portal vein. J. Physiol., 444: 397-418, 1992.

50. Inoue, R., Okabe, K., Kitamura, K. and Kuriyama, H. A newly-identified Ca^{2+}-dependent K^+ channel in the smooth muscle membrane of single cells dispersed from the rabbit portal vein. Pflügers Arch., 406: 138-143, 1986.

51. Inoue, I., Nakaya, Y., Nakaya, S. and Mori, H. Extracellular Ca^{2+}-activated K channelin coronary artery smooth muscle cells ànd its role in vasodilation. FEBS Lett., 255: 281-284, 1989.

52. Edwards, F.R. and Hirst, G.D.S. Inward rectification in submucosal arterioles of the guinea-pig ileum. J. Physiol., 404: 437-455, 1988.

53. Sims, S.M., Singer, J.J. and Walsh, J.V., Jr. Cholinergic agonists suppress a potassium current in freshly dissociated smooth muscle cells of the toad. J. Physiol., 367: 503-529, 1985.

54. Benham, C.D., Bolton, T.B., Lang, R.J. and Takewaki, K. The mechanism of action of Bs^{2+} and TEA on single Ca^{2+}-activated K^+ channels in arterial and intestinal smooth muscle cell membranes. Pflügers Arch., 403: 120-127, 1985.

55. Adams, D.J. and Nonner, W. Voltage-dependent potassium channels: gating ion permeation and block. In Potassium Channels. edited by N.S. Cook , New York, Ellis Horwood Ltd., Chichester, John Wiley & Sons, p.40-69, 1990.

56. Kusano, K., Barros, F., Katz, G., Garcia, M., Kaczorowski, G. and Reuben, J. P. Modulation of K channel activity in aortic smooth muscle by BRL 34915 and a scorpion toxin. Biophys. J., 51: 55a, 1987.

57. Hermesmyer, R.K. Pinacidil actions on ion channels in vascular smooth muscle. J. cardiovasc. Pharmacol., 12 (suppl. 2), 517-522, 1988.

58. Gelbend, C.H., Lodge, N.J., Talvenheime, J.A. and van breemens, C. BRL 34915 increases Popen of the large conductance Ca^{2+} activated K^+ channel isolated fom rabbit aorta in planar lipid bilayers. Biophys. J. 53: 149a, 1988.

59. Okabe, K., Kajioka, S., Nakao, K., Kitamura, K., Kuriyama, H. and Weston, A.H. Action of cromakalim on ionic currents recorded from single smooth muscle cells of the rat portal vein. J. Pharmacol. Exp. Ther., 252: 832-839, 1990.

60. Nakao, K., Okabe, K., Kitamura, K., Kuriyama, H. and Weston, A.H. Characteristics of cromakalim-induced relaxations in smooth muscle cells of guinea pig mesenteric artery and vein. Br. J. Pharmacol., 95: 795-804, 1988.

61. Ito, Y., Kuriyama, H. and Sakamoto, Y. Effect of tetraethylammonium chloride on the membrane activity of guinea-pig stomach smooth muscle. J. Physiol., 211: 445-478, 1970.

62. Hara, Y., Kitamura, K. and Kuriyama, H. Actions of 4-aminopyridine on vascular smooth muscle tissues of the guinea-pig. Br. J. Pharmacol., 68: 99-106, 1980.

63. Kajiwara, M. General feature of electrical and mechanical properties of smooth muscle cells on the guinea-pig abdominal aorta. Pflügers Arch., 393: 109-117, 1982.

64. Mirroneau, J. and Savineau, J.-P. Effects of calcium ions on outward membrane currents in rat uterine smooth muscle. J. Physiol., 302: 411-425, 1980.

65. Imaizumi, Y. and Watanabe, M. Effect of 4-aminopyridine on potassium permeability of canine tracheal smooth muscle cell membrane. Jpn. J. Pharmacol., 33: 201-208, 1983.

66. Inoue, R., Kitamura, K. and Kuriyama, K. Two Ca-dependent K-channels classified by the application of tetraethylammonium distribute on smooth muscle membranes of the rabbit portal vein. Pflügers Arch., 405: 173-179, 1985.

67. Boer, K., Bonev, A. and Papasova, M. 4-aminopyridine-induced changes in the electrical and contractile activities of the gastric smooth muscle. Gen. Physiol. Biophys., 4: 589-595, 1985.

68. Maeda, N., Minobe, M. and Mikoshiba, K. A corebellar purkinje cell marker P_{400} protein is an inositol 1,4,5-trisphosphate ($InsP_3$) receptor protein. Purification and characterization of $InsP_3$ receptor complex. EMBO J., 9: 61-67, 1990.

69. Worley, P.F., Baraban, J.M. and Snyder, S.H. Inositol 1,4,5-trisphosphate receptor binding: autoradiographie localization in rat brain. J. Neurosci., 9: 339-346, 1989.

70. Miyawaki, A., Furuichi, T., Ryou, Y., Yoshikawa, S., Nakagawa, T., Saitoh, T. and Mikoshiba, K. Structure-function relationship of the mouse inositol 1,4,5-trisphosphate receptor. Proc. Natl. Acad. Sci. USA, 88: 4911-4915, 1991.

71. Takeshima, H., Nishimura, S., Matsumoto, T., Ishida, H., Kangawa, K., Minamino, M., Matsuo, H., Ueda, M., Hanaoka, M., Hirose, T. and Numa, S. Primary structure and expression from complementary DNA of skeletal muscle ryanodine receptor. Nature, 339: 439-445, 1989.

72. Tanaka, T., Adams, B.A., Numa, S. and Beam, K.G. Repeat 1 of the dihydropyridine receptor is critical in determing calcium channel activation kinetics. Nature, 352: 800-803, 1991.

73. Tanabe, T., Mikami, A., Numa, S. and Beam, K.G. Cordiac-type excitation-contraction coupling in dysgenic skeletal muscle injected with cardiac dihydropyridine receptor cDNA. Nature, 344: 451-453, 1990.

74. Nakai, J., Imagawa, T., Hakamata, Y., Sigekawa, M., Takeshima, H. and Numa, S. Primary structure and functional expression from cDNA of the cardiac ryanodine receptor/calcium release channel. FEBS Lett 271: 169-177, 1990.

75. Fleischer, S. and Inui, M. Biochemistry and biophysics of excitation-contraction coupling. Annu. Rev. Biophys. Chem., 18: 333-364, 1989.

76. Endo, M. Ca release from the sarcoplasmic reticulum. Physiol. Rev., 57: 71-108, 1977.

77. Itoh, T., Kajiwara, M., Kitamura, K. and Kuriyama, H. Roles of stored calcium on the mechanical response evoked in smooth muscle cells of the porcine-coronary artery. J. Physiol., 332: 107-125, 1982.

78. Itoh, T., Kanmura, Y. and Kuriyama, H. A23187 increases calcium permeability of store sites more than of surface membranes in the rabbit mesenteric artery. J. Physiol., 359: 467-484, 1985.

79. Itoh, T., Kanmura, Y. and Kuriyama, H. Inorganic phosphate regulates the contraction-relaxation cyclic in skinned muscles of the rabbit mesenteric artery. J. Physiol., 376: 231-252, 1986.

80. Itoh, T., Kanmura, Y. and Kuriyama, H. Effects of a phorbol ester on acetylcholine-induced Ca^{2+} mobilization and contraction in the porcine coronary artery. J. Physiol., 397: 401-419, 1988.

81. Kubota, Y. Nomura, M., Kamm, K. E., Mumby, M.C. and Stull, T.T. GTP-gamma-S-dependent regulation of smooth muscle contractile elements. Amer. J. Physiol., 262: C405-C-410, 1992.

82. Fujiwara, T., Itoh, T., Kubota, Y. and Kuriyama, H. Effects of guanosine nucleotides on skinned smooth muscle tissue of the rabbit mesenteric artery. J. Physiol., 408: 535-547, 1989.

83. Nishimura, J., Kolber, M. and van Breemen, C. Norepinephrine and GTPgS increase myofilament Ca^{2+} sensitivity in a-toxin permeabilized arterial smooth muscle. Biochem. Biophys. Res. Commun., 157: 677-683, 1988.

84. Kobayashi, S., Kitazawa, T., Somlyo, A.V. and Somlyo, A.P. Cytosolic heparin inhibits muscarinic and a-adrenergic Ca^{2+} release in smooth muscle. J. Biol. Chem., 264: 17997-18004, 1989.

85. Kitazawa, T., Kobayashi, S., Horiuti, K., Somlyo, A.V. and Somlyo, A.P. receptor-coupled, permeabilized smooth muscle. J. Biol. Chem., 264: 5339-5342, 1989.

86. Somlyo, A.P. and Somlyo, A.V. Flash photolysis studies of excitation-contraction coupling, regulation, and contraction in smooth muscle. Ann. Rev. Physiol., 52: 857-874, 1990.

87. Somlyo, A.P., Walker, J.W., Goldman, Y.E., Trentham, D.R. and Kobayashi, S. Inositol trisphosphate, calcium and muscle contraction. Phil. Trans. Royal Soc., B320: 399-414, 1988.

88. Best, L. and Bolton, T.B. Depolarization of guinea-pig visceral smooth muscle causes hydrolysis of inositol phosphlipids. Naunyn-Schmiedeberg's Arch Pharmacol., 333: 78-82, 1986.

89. Himpens, B., Matthijis G. and Somlyo, A.P. Desensitization to cytosolic Ca 2+ and Ca2+ sensitivities of guinea-pig ileum and rabbit pulmonary artery smooth muscle. J. physiol., 413: 489-503, 1989.

90. Ozaki, H., Gerthoffer, W.T., Publicover, N.G., Fusetani, N.. and Sanders, K.M. Time-dependent changes in Ca2+ sensitivity during phasic contraction of canine antral smooth muscle. J. Physiol., 440: 207-224, 1991.

91. Itoh, T., Seki, N., Suzuki, S., Ito, S., Kajikuri, K. and Kuriyama, H. Membrane hyperpolarization inhibits agonist-induced synthesis of inositol 1,4,5-trisphosphate in rabbit mesenteric artery. J. Physiol., 451: 307-328.

ATP-SENSITIVE K$^+$ CHANNELS IN THE PULMONARY VASCULATURE

Lucie H Clapp[1] and Alison M Gurney[2]

[1]Cardiovascular Research,
 The Rayne Institute, St. Thomas's Hospital
[2]Department of Pharmacology,
 United Medical and Dental Schools,
 St. Thomas's Hospital, London SE1 7EH

INTRODUCTION

Since the discovery by Noma (1983) of a cardiac potassium-selective channel inhibited by intracellular ATP (K_{ATP}), this channel has now been described in a wide variety of cell types including smooth muscle (for review see Ashcroft and Ashcroft, 1990). At present it is not clear how the K_{ATP} channel is regulated under normal conditions, since these channels have a very low probability of opening at physiological resting ATP concentrations. What seems likely is that additional factors such as other nucleotides, G-proteins, metabolites and endogenously released hormones modulate the activity of this channel. One common feature of K_{ATP} channels is their sensitivity to block by sulphonylurea drugs such as glibenclamide and tolbutamide. In cardiac and smooth muscle cells, these channels can also be activated by a class of antihypertensive compounds, known as K$^+$-channel openers (KCOs), which include agents like pinacidil, diazoxide, nicorandil and cromakalim (Escande et al., 1989; Standen et al., 1989; Fan et al., 1990; Kajioka et al., 1990). These agents are also known to hyperpolarize vascular smooth muscle and increase K$^+$ permeability, effects which can be inhibited by glibenclamide (Weir and Weston, 1986; Quast and Cook, 1989; Standen et al., 1989; Cook and Quast, 1990). These observations have led to the view that the K_{ATP} channel is the mediator of the vasorelaxant effects of KCOs in smooth muscle. However, unlike in other tissues, it has proved difficult to identify K_{ATP} channels in smooth muscle because they appear to be present in low density. In this paper, we shall present the evidence for the existence of these channels in pulmonary artery, discussing their role in mediating the response to the hyperpolarizing vasodilator, levcromakalim (formerly known as lemakalim - the active enantiomer of cromakalim) and in regulating the resting membrane potential of pulmonary artery.

PHARMACOLOGICAL STUDIES

One feature that distinguishes KCOs from other directly acting vasodilator compounds, like for example Ca-channel blockers, is their ability to relax arteries precontracted with low but not high concentrations of extracellular K$^+$ ($[K^+]_O$). This is illustrated in figure 1A which shows the effect of levcromakalim (BRL 38227) on rabbit pulmonary artery strips precontracted with two different concentrations of $[K^+]_O$. Levcromakalim produces a substantial relaxation of the artery when it is precontracted with 20 mM $[K^+]_O$ but not with 50 mM $[K^+]_O$. This can be explained on the basis of there being virtually no transmembrane driving force on K^+ when $[K^+]_O$ is raised to 50 mM. In 50 mM $[K^+]_O$, the membrane potential depolarizes to around -26 mV (Casteels et al., 1977a) which is close to the reversal potential of -27 mV for K^+ under these experimental conditions. Thus levcromakalim is unable to produce hyperpolarization in 50 mM $[K^+]_O$, and as a result, is not able to close voltage-sensitive Ca^{2+} channels which would lead to muscle relaxation. This type of observation, together with data obtained from pharmacological and electrophysiological experiments (e.g. Weir and Weston, 1986; Cook and Quast, 1990; Standen et al., 1989) have led to the conclusion that K^+-channel activation is the major mechanism underlying the vasodilatory action of KCOs.

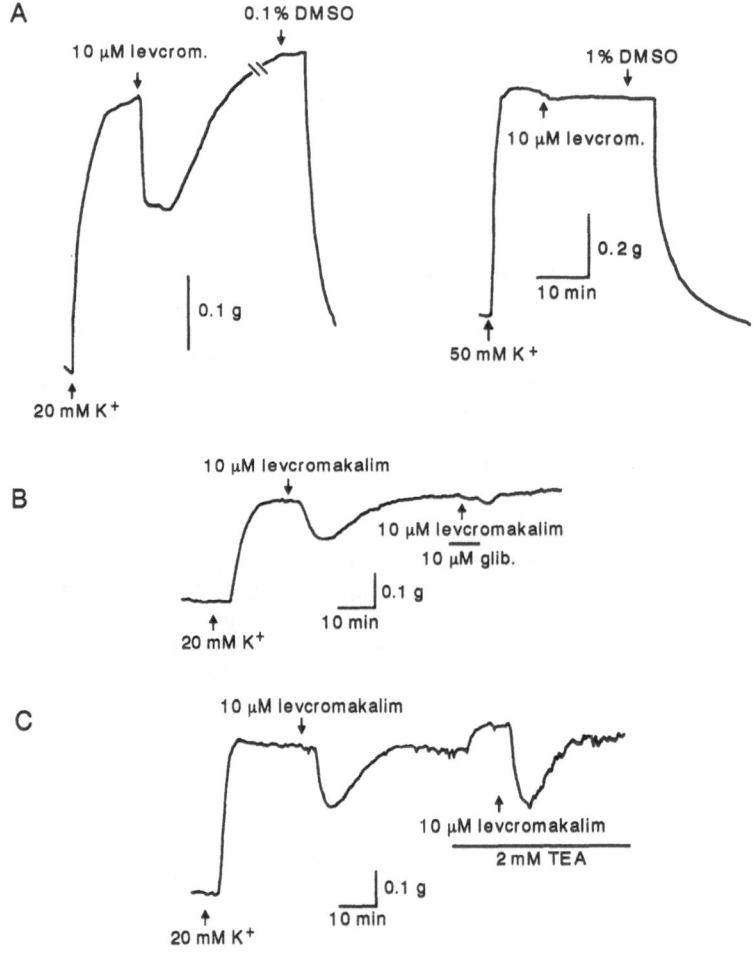

Figure 1. Vasodilation produced by levcromakalim in rabbit pulmonary artery involves activation of K^+ channels. (A) The effect of a 30 s application of levcromakalim (levcrom; 10 μM) on a muscle strip precontracted by elevating extracellular K^+ to 20 mM (left panel) and then to 50 mM (right panel). Dimethyl sulphoxide (DMSO) was without effect on tension at concentrations (up to 1%) greater than was used to make the final dilution of levcromakalim. The response to levcromakalim on strips precontracted with 20 mM K^+ is shown before and in the presence of either glibenclamide (glib., B) or tetraethylammonium ions (TEA; C). Parts B & C from Clapp and Gurney, 1993.

Effect of K+-channel Blockers on Levcromakalim Responses

Many studies have examined the effects of various K+-channel blockers on vasodilation evoked by KCOs in an attempt to identify the nature of the K+ channel giving rise to the hyperpolarization and relaxation produced by agents like cromakalim. One of the original observations that cromakalim-induced relaxations of smooth muscle strips could be reversed by gliblenclamide, led to the proposal that K_{ATP} channels were responsible for the effects of cromakalim (Quast and Cook, 1989; Standen et al., 1989). Similarly, glibenclamide is an effective blocker of the relaxant effects of levcromakalim in pulmonary artery (figure 1B; Clapp and Gurney, 1992). Other, less specific K+-channel blockers including phencyclidine and tetrapentylammonium ions, also inhibit relaxations to levcromakalim at concentrations as low as 1-10 μM (Clapp and Gurney, 1993; Langton et al., 1993b). The latter agent has been reported to block K_{ATP} channels in mesenteric artery (Kovacs et al., 1990). In contrast, tetraethylammonium (TEA) at 2 mM does not significantly affect levcromakalim-induced relaxations (figure 1C) despite blocking Ca-activated K+ current in pulmonary artery at this concentration (see later). This is consistent with other studies showing that both charybdotoxin and apamin, specific blockers of the large and small conductance Ca-activated K+ channels respectively, are not able to reverse the effects of cromakalim in vascular smooth muscle (see Longman and Hamilton, 1992). However, Ba^{2+}, which is effective against cromakalim in mesenteric artery does not block the response to levcromakalim in pulmonary artery (Clapp, unpublished observations), suggesting some heterogeneity in the K+ channel activated in arterial smooth muscle.

ELECTROPHYSIOLOGY

In the last couple of years there have been several patch-clamp studies on isolated smooth muscle cells describing the effects of KCOs both on single K+ channels and whole-cell K+ currents. One of the problems that has emerged from these single channel experiments is that KCOs can activate several different types of K+ channels, with varying conductances and sensitivities to Ca^{2+} and intracellular ATP (Standen et al., 1989: Gelband et al., 1989; Klockner et al., 1989; Kajioka et al., 1990; Miyoshi et al., 1992). Few studies have attempted to identify the nature of the K+ channel responsible for the membrane hyperpolarization induced by KCOs in the whole-cell. In order to address this issue, we have looked at the effects of various K+-channel blockers on whole-cell K+ currents and also looked at the effect of varying intracellular ATP concentration on membrane potential, current and the responsiveness to levcromakalim and K_{ATP} channel blockers.

Whole-cell Currents in Isolated Pulmonary Artery

Whole-cell voltage-clamp experiments have revealed that several types of K+ current exist in isolated smooth muscle cells from rabbit pulmonary artery. Typical voltage-clamp records, from a cell bathed in a physiological salt solution and dialysed with a K+-filled pipette solution containing 1 mM ATP, are illustrated in figure 2A. Under these conditions cells have a resting potential of -55 mV and a high input resistance (17 G Ω) with current not easily resolvable until the cell depolarizes to about -40 mV (Clapp and Gurney, 1991). Several current components can be distinguished on the basis of their voltage-sensitivity and differential sensitivities to the K+-channel blockers TEA and 4 aminopyridine (4AP). These include a transient, A-like current (I_{tran}), a Ca^{2+}-activated K+ current (I_{Kso}), a time-independent, background current (Clapp and Gurney, 1991) and a time and voltage-dependent current that is distinct from I_{Kso} and I_{tran}. At potentials positive to -20 mV, TEA (2 mM) is an effective blocker of the slowly activating, sustained current but not the transient current (figure 2B). The same TEA-sensitive current can also be inhibited by Cd^{2+} (Clapp and Gurney, 1991), suggesting this current is carried through large conductance Ca^{2+}-activated K^+ channels - the most prominent channel found in this preparation (Kirber et al., 1992). However, the time-dependent current activated below -20 mV is much less sensitive to TEA (figure 2C) being blocked by only 34% at -30 mV compared to 64% at +30 or +40 mV (Clapp and Gurney, 1991). The TEA-insensitive current is significantly inhibited by 4AP at a concentration of 2 mM (figure 2E), providing good evidence for the existence of another voltage-and time-dependent current. This current is most likely a delayed rectifier K^+ current

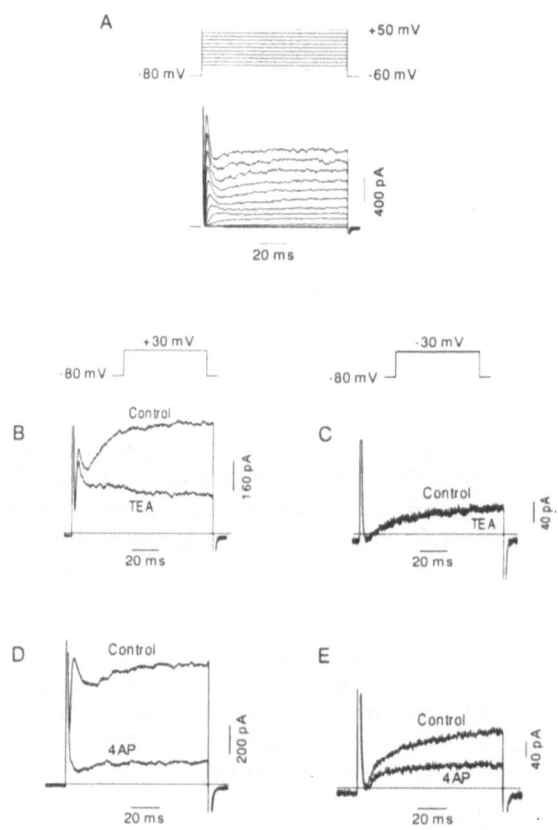

Figure 2. Whole-cell currents recorded from isolated pulmonary arterial cells bathed in a physiological solution. (**A**) A series of outward currents recorded with K^+-filled pipettes containing ATP and GTP. Currents were elicited by 100 ms depolarizing steps from a holding potential of -80 mV. Current first activated at -40 mV, with a clear separation of time-dependent currents seen above -10 mV. In Clapp and Gurney (1991), these were designated (I_{tran}) for the rapid transient component and I_{Kso}, for the more slowly activating sustained current at positive potentials. TEA blocks I_{Kso} but not I_{tran} or the time-dependent current seen at negative potentials. Currents in **B** & **C** were from the same cell and activated from -80 mV by a voltage step to either +30 mV or -30 mV in the absence (Control) and presence of 2 mM TEA. 4-aminopyridine (4AP) inhibits time-dependent current at all potentials, unmasking the presence of a quasi-instantaneous, background current. Currents in **D** & **E** were recorded from the same cell, and elicited by depolarizing steps to +30 mV and -30 mV in the absence and presence of 2 mM 4AP. (Modified from Clapp and Gurney, 1991).

which has previously been described in other vascular smooth muscles and reported to be inhibited by 4AP (Beech and Bolton, 1989; Gelband and Hume, 1992). At 2 mM, 4-AP markedly suppresses all time-dependent K^+ current, including I_{tran}, and I_{Kso}, unmasking the presence of a small time-independent, background current which is clearly evident at positive potentials (Figure 2D). The magnitude of this background current increases when cells are depleted of intracellular ATP, making its activation apparent at potentials as low as -70 mV (Clapp and Gurney, 1992). This current is not significantly affected by TEA (figure 2B) but is substantially blocked by higher concentrations (10 mM) of 4AP (Clapp and Gurney, 1991).

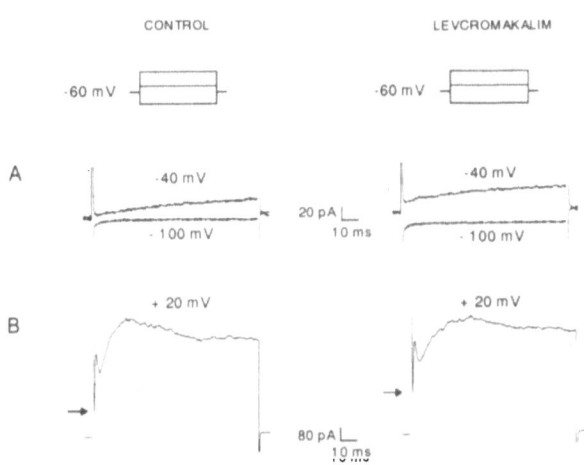

Figure 3. Levcromakalim increases the instantaneous, background K^+ current, but not I_{tran} or time-dependent K^+ currents. Records were obtained before (control) and during the application of levcromakalim (10 μM), which hyperpolarized the cell from -52 mV to -70 mV. The pipette contained 1 mM ATP, and currents were elicited by 100-ms voltage steps to -40 and -100 mV (A) and +20 mV (B) from a holding potential of -60 mV. Each record is an average of 5 consecutive traces. The reversal potential of the levcromakalim-induced current was -81 mV.

The Effect of Levcromakalim on Membrane Potential and Whole-cell Currents

Due to conflicting reports about the K^+ channel targeted by KCOs in different smooth muscle preparations, we have studied the effects of levcromakalim on membrane current in parallel with its effect on membrane potential. In isolated pulmonary artery cells dialyzed with 1 mM ATP, like in other smooth muscles, levcromakalim produces substantial membrane hyperpolarization which is fully reversed by glibenclamide (Clapp and Gurney, 1992; 1993). This hyperpolarization is associated with an increase in the instantaneous, background current. Levcromakalim does not however, alter the time-dependent current activated at negative potentials, or I_{tran} which is activated at positive potentials (figure 3A & B; Clapp and Gurney, 1993). The levcromakalim-induced current is carried by K^+ ions since it reverses direction at -80 mV (figure 3A), close to the calculated reversal potential for K^+ (E_K)

of -83 mV, and is inhibited by K_{ATP} channel blockers. Figure 3 shows that for cells displaying large hyperpolarizations in response to levcromakalim, the increase in steady-state current at -40 mV is < 20 pA. This suggests that the conductance of the activated channel under physiological conditions is probably small (see later). Although levcromakalim activates a background current at all potentials, it appears to have an additional effect at potentials above -20 mV which is not reversed by glibenclamide. The current increased by levcromakalim is smaller if measured at the end of a 100-ms command pulse compared to the beginning (figure 3B). At +40 mV there is no significant increase in this current (Clapp and Gurney, 1993), suggesting that levcromakalim has an opposing effect on a different current. This is consistent with other reports showing cromakalim or levcromakalim to decrease outward current in rat portal vein at positive potentials (Okabe et al., 1990; Noack et al., 1992a). However, this inhibitory effect is not likely to have any physiological consequence since rabbit pulmonary artery does not fire action potentials under normal conditions and does not depolarize much beyond -45 mV in the presence of noradrenaline (Casteels et al., 1977b). Furthermore, since the ability of levcromakalim to produce muscle relaxation depends almost entirely on whether it can produce hyperpolarization, activation of the background current must be the important mechanism underlying the relaxant effect of levcromakalim.

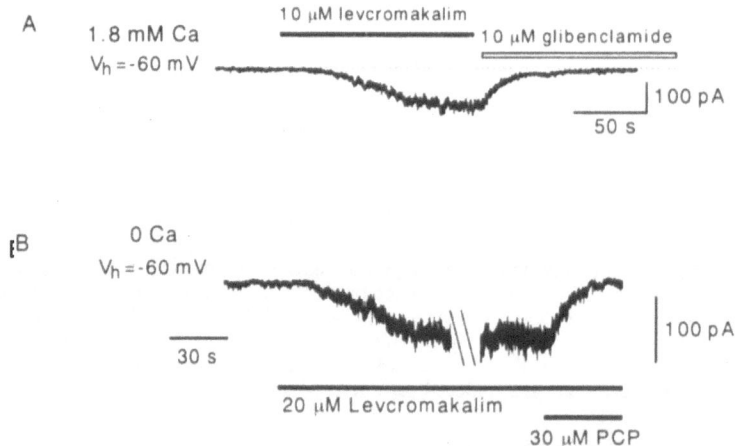

Figure 4. The effect of levcromakalim on whole-cell current recorded with 1 mM ATP in the pipette, in the presence (**A**) and absence (**B**) of 1.8 mM extracellular Ca^{2+}. The internal and external solutions contained 143 mM K^+. In each case, the cell was voltage-clamped at -60 mV, with the dashed line indicating zero current. The levcromakalim-induced current could be blocked by either glibenclamide or phencyclidine (PCP). Part (**A**) from Langton et al., 1993a.

Previously it has been shown that the effects of cromakalim on membrane potential and current in mesenteric artery or portal vein were markedly reduced in the absence of extracellular Ca^{2+} or when intracellular Ca^{2+} was buffered to low levels with EGTA (Nakeo et al., 1988; Okabe et al., 1990). Similarly, single-channel experiments have revealed that KCOs can activate a variety of K^+ channels sensitive to either intracellular or extracellular Ca^{2+} (Gelband et al., 1989; Kajioka et al., 1990; Miyoshi et al., 1992). However, in rabbit pulmonary artery, the effects of levcromakalim on membrane current and potential were observed when the pipette solution contained 1 mM EGTA, where the free intracellular Ca^{2+} concentration in the pipette solution would be < 15 nM according to the calculations of Fabiato (1991). Furthermore, the effect of levcromakalim on membrane current is essentially unaltered if extracellular Ca^{2+} is removed (figure 4), strongly suggesting that levcromakalim does not activate Ca^{2+}-sensitive K^+ channels in this preparation and that Ca^{2+}-sensitive K^+ channels do not contribute to the hyperpolarization or muscle relaxation produced by levcromakalim. This is also consistent with the lack of effect of TEA on levcromakalim-induced relaxation and on the background current.

We have used experiments like the one illustrated in figure 4 to analyse the current noise associated with the K^+ channels opened by levcromakalim. The cells were bathed in symmetrical K^+ solutions with 143 mM K^+ on each side of the membrane and voltage-clamped at -60 mV to minimise activation of other voltage-dependent K^+ channels. In the presence of levcromakalim, the current trace became increasingly noisy but declined when either glibenclamide or phencyclidine was added. Measurements of the mean whole-cell current and its variance were used to estimate the unitary amplitude (i) and the number of single channels (N) present. From a series of these experiments i was estimated to be -1.2 pA and N to be an average of only 360 channels per cell (see Langton et al., 1993a,b). This corresponds to a single channel conductance of 19.5 pS with 143 mM symmetrical K^+, which suggests that the conductance under physiological conditions is likely to be very small. Obviously the low density of these channels will make it hard to study the channel in isolated patches, and it probably explains why these channels have previously been hard to find. In rat portal vein, levcromakalim has also been shown to induce a voltage-independent K^+ current with an underlying conductance in a physiological K^+ gradient of between 10 and 20 pS (Noack et al., 1992a). Again the estimate of the number of channels involved was low at ~500 channels per cell.

ATP-depletion Experiments

Since the initial observations that K_{ATP} channels could be activated by KCOs and other hyperpolarizing vasodilators (Standen et al., 1989; Nelson et al., 1900), we have sought to identify whether the background current underlying the hyperpolarization produced by levcromakalim resulted from the opening of K_{ATP} channels, which, in the presence of physiological concentrations of intracellular ATP, would normally have a low probability of opening. Recently we have shown that depleting cells of intracellular ATP selectively increases the time-independent, background current, which had similar properties to the current activated by levcromakalim (Clapp & Gurney, 1992; 1993). In the presence of millimolar ATP, this current was small and hard to resolve. However, in the absence of ATP, the background current became quite prominent and could be clearly separated from the time-dependent currents. We also observed that the magnitude of this current correlated well with how negative the resting potential of the cell was. The average resting potential in ATP-depleted cells became -70 mV compared to -55 mV in the presence of ATP, and the input resistance declined 4-fold (see figure 5 and Clapp and Gurney, 1992). This would be predicted if removal of ATP opened K^+ channels normally held closed. Our results were recently criticised by Noack et al. (1992b), who claimed that the block of the background current we observed by raising ATP could have resulted from ATP chelating any free intracellular Ca^{2+}. This kind of effect might explain the ATP-block of the BK_{Ca} channel described by Silberberg and van Breemen (1990) (see Klöckner and Isenberg, 1992). However, it cannot explain our results which were obtained in conditions where the internal solution contained both EGTA (1 mM) and Mg^{2+} with no added Ca^{2+}. Thus ATP added to the pipette solution would bind Mg^{2+} in preference to Ca^{2+}, and EGTA would maintain the Ca^{2+} at a very low (nM) concentration, whether ATP was present or not. According to Fabiato (1991), we calculate that no change in free Ca^{2+} occurs under these conditions. Noack et al. (1992b) supported their argument by claiming mistakenly, that our data (figure 1, Clapp & Gurney, 1992) showed a reduction in the Ca^{2+}-dependent current when ATP was added. This is not the case. As we stated in our original paper, the amplitude of

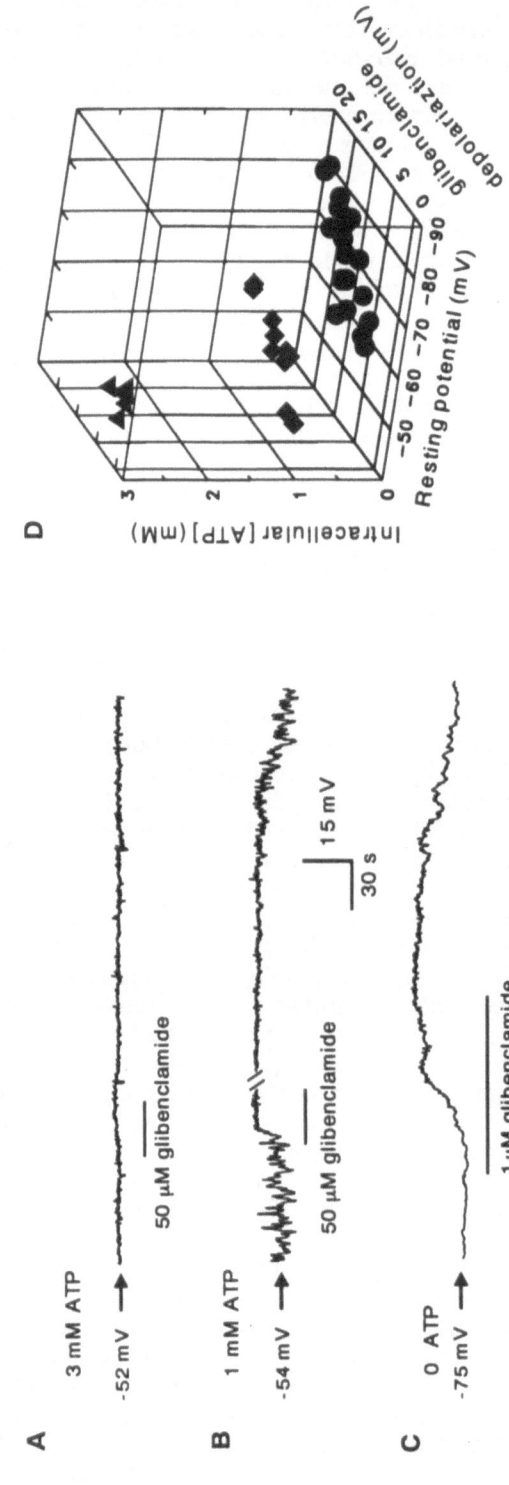

Figure 5. Membrane responses to glibenclamide depend on the intracellular ATP concentration. Effects of glibenclamide in 3 different cells dialysed with either 3 mM (A), 1 mM (B) or zero ATP (C) in the recording pipette. (D) The amplitudes of the glibenclamide depolarization from a series of experiments are plotted as a function of the resting potential and the intracellular ATP concentration. Recordings were made with either zero ATP (●), 1 mM ATP (♦) or 3 mM ATP (▲). Parts A, B, C from Clapp and Gurney, 1992.

the time-dependent current (partly Ca^{2+}-sensitive) varied from cell to cell and its size was not related to whether ATP was present or not. Clearly this current is not reduced when ATP is re-introduced into the cell (see figure 4, Clapp & Gurney, 1992).

Further evidence for K_{ATP} channels being responsible for the hyperpolarized resting potentials and effects of levcromakalim in pulmonary artery cells have come from experiments looking at the response to glibenclamide in cells dialysed with various ATP concentrations (Figure 5). In the presence of 1 mM ATP, glibenclamide produces a small but significant depolarization of the resting potential (figure 5B), associated with a small decrease in the background current. However, when the ATP concentration is raised to 3 mM, glibenclamide has essentially no effect on membrane potential (figure 5A), suggesting that under these conditions all the K_{ATP} channels are closed by intracellular ATP. In contrast, when ATP is removed from the pipette, pulmonary arterial cells are more hyperpolarized and glibenclamide becomes a potent depolarizing agent (figure 5C). The correlation between intracellular ATP, resting potential and the magnitude of the glibenclamide depolarization is clear from figure 5D. Furthermore, blockade of the background current by glibenclamide was significantly enhanced by ATP-depletion (figure 6). It is evident that the ATP concentration, by determining the resting conductance, governs the resting potential and the response to glibenclamide. Furthermore, at -60 mV, the normal resting potential of intact rabbit pulmonary artery (Casteels et al., 1977b), our results would suggest that K_{ATP} channels would contribute to the resting conductance and could therefore have an influence on vascular tone as recently demonstrated in the coronary vasculature (Samaha et al., 1992).

The most convincing evidence that the effects of ATP-depletion are due to changes in the activity of K_{ATP} channels, comes from experiments utilizing the technique of flash-photolysis to generate rapid concentration jumps of ATP. The inactive precursor of ATP (caged ATP) can be loaded into cells via the patch pipette which contains no added ATP. Under these conditions photo-release of ATP, but not

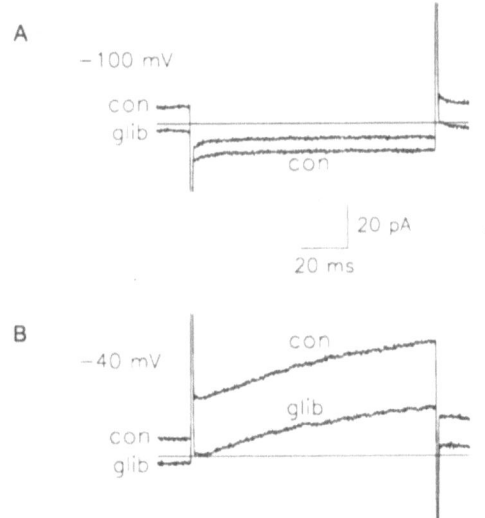

Figure 6. Blockade by glibenclamide of a time-independent K^+ current with similar properties to the levcromakalim-induced current. Records were obtained in the absence of ATP in the patch pipette, from a cell that had a resting potential of -80 mV before drug application. Currents were elicited by voltage steps to -100 mV (**A**) and -40 mV (**B**) from a holding level of -60 mV before (con) and during $10 \mu M$ glibenclamide (glib). Each record represents an average of 5 consecutive current traces. From Clapp and Gurney, 1992.

light alone, causes rapid depolarization ($<1s$) associated with block of the background current (Clapp and Gurney, 1992). The rapidity of the response to ATP suggests that ATP has its effects by directly interacting with channels in the membrane, although further experiments are required to determine more precisely the time course and to rule out that channel phosphorylation is required for ATP to have its effects.

CONCLUSION

The evidence is now clear that K_{ATP} channels exist in pulmonary artery and play an important role in the regulation of resting membrane potential. These channels have a relatively small unitary conductance and show little sensitivity to Ca^{2+}. Furthermore, the pharmacology of the ATP- and levcromakalim-sensitive K^+ current in the isolated cells correlates well with the pharmacology of levcromakalim responses in the intact artery, leading to the conclusion that K_{ATP} channels are the target for KCOs like levcromakalim. Given the sensitivity of these channels to intracellular ATP, they could well respond to changes in cell metabolism, and play a role under conditions of hypoxia. Future research should be aimed at identifying how these channels are regulated physiologically by intracellular messengers and endogenously circulating hormones. This may provide important insight into how selective therapeutic agents for the vasculature might be developed.

ACKNOWLEDGEMENTS
Supported by the British Heart Foundation, Royal Society and Wellcome Trust. LHC is a Wellcome Research Fellow.

REFERENCES

Ashcroft, S.J.H., and Ashcroft, F.M., 1990, Properties and function of ATP-sensitive K-channels, *Cellular Signalling* 2:197.

Beech, D.J., and Bolton, T.B., 1989, Properties of the cromakalim-induced potassium conductance in smooth muscle cells isolated from the rabbit portal vein, *Br. J. Pharmacol.* 98:851.

Casteels, R., Kitamura, K., Kuriyama, K., and Suzuki, H., 1977a, The membrane properties of the smooth muscle cells of the rabbit main pulmonary artery, *J. Physiol. Lond.* 271:41.

Casteels, R., Kitamura, K., Kuriyama, K., and Suzuki, H., 1977b, Excitation-contraction coupling in the smooth muscle cells of the rabbit main pulmonary artery. *J. Physiol. Lond.* 271:63.

Clapp, L.H., and Gurney, A.M., 1991, Outward currents in rabbit pulmonary artery cells dissociated with a new technique. *Exper. Physiol.* 76:677.

Clapp, L.H., and Gurney, A.M., 1992, ATP-sensitive K^+ channels regulate the resting potential of pulmonary arterial smooth muscle cells. *Am. J. Physiol.* 262: (*Heart Circ. Physiol.* 31) H916.

Clapp, L.H., and Gurney, A.M., 1993, ATP-sensitive K^+ channels mediate the vasodilation by lemakalim in rabbit pulmonary artery, *Am. J. Physiol. (Heart Circ. Physiol.) In press.*

Cook, N.S and Quast, U., 1990, Potassium channel pharmacology, *in*: "Potassium Channels: Structure, Classification, Function and Therapeutic Potential," N.S. Cook, ed., Ellis Horwood Ltd., Chichester, U.K., pp 209.

Escande, D. Thuringer D., Le Guern, S., Courteix, J., and Cavero, I., 1989, Potassium channel openers act through an activation of ATP-sensitive K^+ channels in guinea-pig cardiac myocytes, *Pflügers Arch.* 414:669.

Fabiato, A., 1991, Buffering: Computer programs and simulations, *in*: "Cellular Calcium, a Practical Approach", J.G. McCormack, and P.H. Cobbold, eds., Oxford University Press, New York, pp 159.

Fan, Z., Nakayama, K., and Hiraoka M., 1990, Pinacidil activates ATP-sensitive K^+ channels in inside-out and cell-attached patches of guinea-pig ventricular myocytes, *Pflügers Arch.* 415:387.

Gelband, C.H., and Hume, J.R., 1992, Ionic currents in single smooth muscle cells of canine renal artery, *Cir. Res.* 71:745.

Gelband, C.H., Lodge, N.J., and van Breemen, C., 1989, A Ca^{2+}-activated K^+ channel from rabbit aorta: Modulation by cromakalim. *Eur. J. Pharmacol.* 167:201.

Kajioka, S., Oike, M., and Kitamura, K., 1990, Nicorandil opens a calcium-dependent potassium channel in smooth muscle cells of the rat portal vein, *J. Pharmacol. Exp. Ther.* 254:905.

Kirber, M.T., Ordway, R.W. Clapp, L.H., Walsh, J.V. Jr., and Singer, J.J., 1992, Both membrane stretch and fatty acids directly activate large conductance Ca^{2+}-activated K^+ channels in vascular smooth muscle cells, *FEBS Let.*, 297:24.

Klöckner, U., and Isenberg, G., 1992, ATP suppresses activity of Ca^{2+}-activated K^+ channels by Ca^{2+} chelation, *Pflügers Arch.* 420:101.

Klöckner, U., Trieschmann, U., and Isenberg, G., 1989, Pharmacological modulation of calcium and potassium channels in isolated vascular smooth muscle cells, *Arzneim. Forsch. Drug Res.* 39:120.

Kovacs, R.J., Huang, Y., Brayden, J.E., and Nelson, M.T., 1990, Block of cromakalim action and arterial smooth muscle ATP-sensitive K^+ channels by tetrapentylammonium ions (Abstract), *Circ.* 82:III-341.

Langton, P. D., Clapp, L.H., Dart, C., Gurney, A.M., and Standen, N.B., 1993a, Whole-cell K^+ current activated by lemakalim in isolated myocytes from rabbit pulmonary artery: estimate of unitary conductance and density by noise analysis (Abstract), *J. Physiol. Lond.* 459:245P.

Langton, P. D., Clapp, L.H., Gurney, A.M., and Standen, N.B., 1993b, Lemakalim activated K_{ATP} channels in isolated myocytes of rabbit pulmonary artery occur at low density and have a small conductance (Abstract), *Biophys. J.* 64:A149.

Longman, S.D., and Hamilton, T.C., 1992, Potassium channel activator drugs: Mechanism of action, pharmacological properties, and therapeutic potential, *Medicinal Res. Rev.* 12:73.

Nelson, M.T., Huang, Y., Brayden, J.E., Heschler J.K., and Standen, N.B., 1990, Arterial dilations in response to calcitonin gene-related peptide involve activation of K^+ channels, *Nat. Lond.* 344:770.

Miyoshi, Y., Nakaya, Y., Wakatsuki, T., Nakaya, S., Fujino, K., Saito, K., and Inoue, I., 1992, Endothelin blocks ATP-sensitive K^+ channels and depolarizes smooth muscle cells of porcine coronary artery, *Cir. Res.* 70:612.

Noack, T., Edwards, G., Deitmer, P., and Weston, A.H., 1992a, Potassium channel modulation in rat portal vein by ATP depletion: a comparison with the effects of levcromakalim (BRL 38227), *Br. J. Pharmacol.* 107:945.

Noack, T., Deitmer, P., Edwards, G., and Weston, A.H., 1992b, Characterization of potassium currents modulated by BRL 38227 in rat portal vein, *Br. J. Pharmacol.* 106:717.

Noma, A., 1983, ATP-regulated K^+ channels in cardiac muscle. *Nat. Lond.* 305: 147.

Okabe, K., Kajioka, S., Nakao, K., Kitamura, K., Kuriyama, K. and Weston, A.H, 1990, Actions of cromakalim on ionic currents recorded from single smooth muscle cells of the rat portal vein, *J. Pharmacol. Exper. Ther.* 252:833.

Quast, U., And Cook, N.S., 1989, Moving together: K^+ channel openers and ATP-sensitive K^+ channels, *Trends Pharmacol. Sci.* 10:431.

Samaha, F.F., Heineman, F.W., Ince., C., Fleming, J., and Balaban, R.S., 1992, ATP-sensitive potassium channel is essential to maintain basal coronary vascular tone *in vivo*, *Am. J. Physiol.* 262: (*Cell Physiol.* 31) C1120.

Silberberg, S.D., and van Breemen, C., 1990, An ATP, calcium and voltage sensitive potassium channel in porcine coronary artery smooth muscle cells, *Biochem. Biophys. Res. Commun.* 17:517.

Standen, N.B., Quayle, J.M., Davies, N.W., Brayden J.E., Huang, Y., And Nelson, M.T., 1989, Hyperpolarizing vasodilators activate ATP-sensitive K^+ channels in arterial smooth muscle, *Sci. Wash.* 245:177.

Weir, S.W., and Weston, A.H., 1986, The effects of BRL 34915 and nicorandil on electrical and mechanical activity and on ^{86}Rb efflux in rat blood vessels, *Br. J. Pharmacol.* 88:121.

ELECTROPHYSIOLOGICAL ACTIONS OF EXTRACELLULAR ATP APPLICATION

C. D. Benham

SmithKline Beecham Pharmaceuticals
The Pinnacles
Harlow
Essex, CM19 5AD
England

INTRODUCTION

Considerable progress has been made in defining the receptors activated by adenosine, using selective agonists and antagonists. Our understanding of P_2 receptor subtypes, where ATP or another phosphorylated nucleoside is thought to be the endogenous ligand is much less precise (1). Highly selective agonists and antagonists are not available. This makes it particularly difficult to investigate the specific effects of receptor activation when cells express more than one receptor sub-type or in multi-tissue preparations where several receptors may again be involved in purinergic responses. A further complication in these preparations is the presence of ectoenzymes that rapidly hydrolyse ATP (2). The metabolites ADP, AMP and adenosine each have their own individual profiles of activity. The most useful antagonist compound to date may be suramin. This compound blocks P_{2X} & P_{2Y} mediated effects so it has little utility in sub classifying ATP receptors, but it is selective for ATP over the other fast transmitter activated conductances. This has allowed the identification of purinergic neuron-neuron synaptic currents in the coeliac ganglion (3). It seems to be a valuable ligand to identify purinergic mechanisms in whole tissues.

Studies of cell types that apparently express only one type of P_2 receptor have provided considerable detail of the transduction mechanisms activated by ATP. In single cell preparations ectoenzyme degradation is minimised which has also helped to consolidate receptor classification based on agonist pharmacology.

In many cell types, purinergic receptor activation results in the opening of cell membrane ion channels. This type of response can be divided into two broad classes. The first consists of responses where agonist-receptor binding results in direct activation of ion channels that carry a depolarising, non-selective cation current. As this cation conductance is in most cases Ca^{2+} permeable, secondary activation of Ca^{2+} dependant channels can complicate the electrophysiological signals generated. A second natural grouping is that working through G-protein coupled receptors to indirectly activate membrane ion channels. A second messenger pathway is involved which in most cases is the inositol trisphosphate - Ca^{2+} store release transduction system . This stimulated rise in $[Ca^{2+}]_i$ secondarily gates the opening of Ca^{2+} selective conductances in the cell membrane. Additionally, other G-protein coupled conductances may also be activated where there maybe a more direct link through receptor, G-protein and ion channel (5).

DIRECTLY COUPLED ION CHANNELS

The receptor channel system of which we have the most complete characterisation is the cation current activated by P_{2X} receptor activation. This type of response is seen in vascular smooth muscle cells from rabbit ear artery (6) and in smooth muscle from urinary bladder (7) and vas deferens (8). A very similar conductance is activated in sensory neurones (9). In all these preparations, electrophysiological studies suggest that only one receptor pathway is involved in the excitatory action of ATP.

Pharmacological studies in whole tissues have shown that AMPCPP is one of the most potent full agonists and more potent than ATP (10). Limited studies on single cells showed much more potent activity of ATP itself with a Kd of 2.3 μM in guinea-pig urinary bladder compared to 10.4 μM for AMPCPP (7). This higher potency probably reflects the reduced hydrolysis of ATP in the single cell preparation. The susceptibility of agonists to hydrolysis has caused a general problem in trying to compare agonist potency across tissues. This makes it unreliable to subdivide responses on this basis alone (11). However, differences are also seen at the single channel level which suggests that there may be a family of closely related ATP-gated cation channels (11) analogous to the sub types of nicotinic receptor.

Application of micromolar concentrations of ATP to smooth muscle cells from ear artery held under voltage clamp evoked a transient inward current that rapidly desensitises. The conductance had a reversal potential close to zero mV and ion replacement experiments demonstrate that it is a relatively non-selective cation current, permeable to monovalent and divalent cations. The first evidence that the ATP gated current carried significant amounts of Ca^{2+} is illustrated in Fig. 1. In panel A the response of a single cell bathed in normal tyrode is shown. ATP evoked transient inward current at -40mV followed by a more sustained outward current measured at

+5mV. In panel B in another cell the outward current response was not seen in the absence of extracellular Ca^{2+}.

Experiments on isolated outside out membrane patches show that unitary currents of about 20 pS conductance are activated by ATP in normal ionic gradients (6). This data, taken together with the short latency of activation in whole cell recording (a few

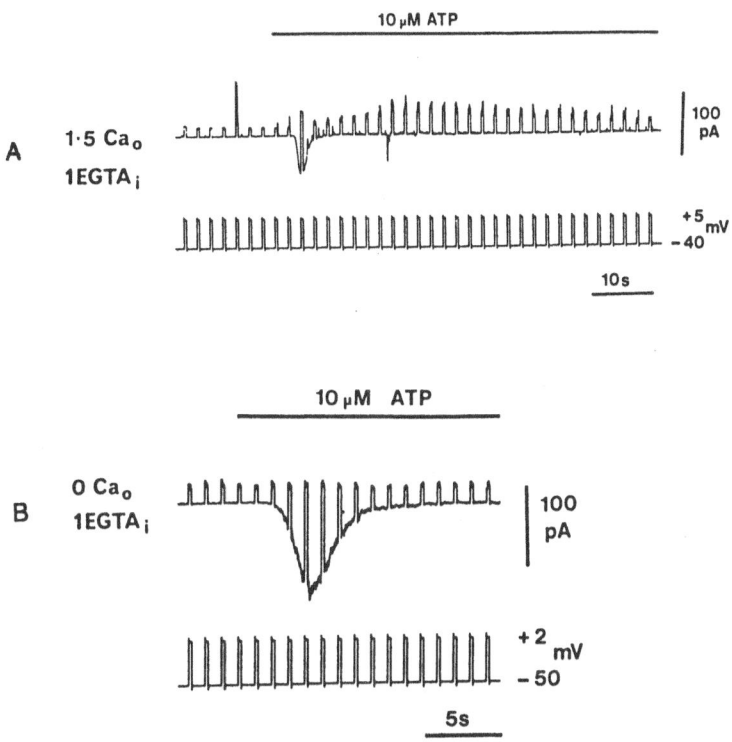

Fig. 1. Effect of bath application of 10μM ATP to single smooth muscle cells from rabbit ear artery.
A. In 1.5mM Ca^{2+} containing tyrode ATP evokes an inward current at -40mV followed by an outward current at +5mV.
B. Only an inward current is seen in the absence of Ca^{2+}. Both cells wre filled with a KCl based pipette solution also containing 1mM EGTA.
(reprinted with permission from Journal of Physiology (4)).

milliseconds , (4)) suggests that receptor activation and channel opening are closely coupled. The most parsimonious explanation is that the P_{2x} receptor is analogous to the cholinergic receptor at the neuromuscular junction where receptor and channel are the same pentameric protein (12).

Further work investigating the Ca^{2+} permeability of these channels has lead to the conclusion that they might admit functionally significant amounts of Ca^{2+}. At least this is suggested from experiments on single cells where cytoplasmic Ca^{2+} ($[Ca^{2+}]_i$) and membrane currents have been measured simultaneously and a contribution from voltage gated Ca^{2+} channels can be ruled out (13; 14). Fig 2A shows simultaneous records of membrane current and $[Ca^{2+}]_i$ from a single smooth muscle cell isolated from rabbit ear artery. Under voltage-clamp ATP application evokes an inward current and coincident rise in $[Ca^{2+}]_i$. As the cell membrane is held under voltage-clamp an influx of Ca^{2+} through voltage gated Ca^{2+} channels could be ruled out. The rise in $[Ca^{2+}]_i$ seemed to be due to Ca^{2+} influx, as in the absence of extracellular Ca^{2+}, no rise in $[Ca^{2+}]_i$ was seen following ATP application although a cation current was still evoked (13).

There seems to be little involvement of intracellular release in this $[Ca^{2+}]_i$ rise as shown in fig. 2B where an outward cation current response to ATP was recorded at positive holding potential. As expected, there was no change in $[Ca^{2+}]_i$ as there was no inward Ca^{2+} driving force at this potential. This result ruled out the possibility of a parallel pathway releasing intracellular stores following ATP receptor activation. 1 μM nifedipine was present in both A and B to block voltage gated Ca^{2+} channels.

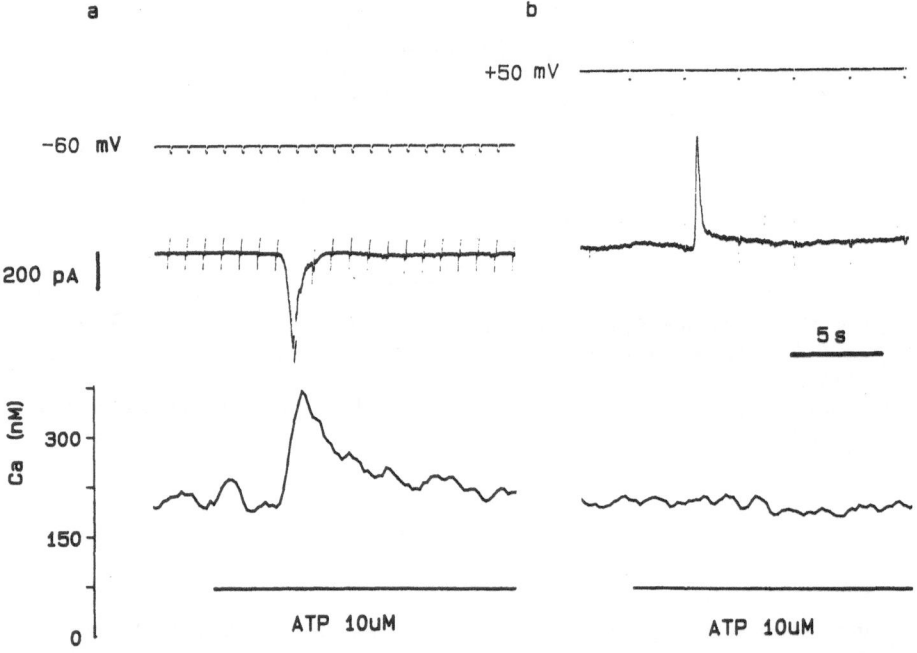

Fig. 2. Voltage dependence of $[Ca^{2+}]_i$ changes evoked by ATP application to single smooth muscle cells from rabbit ear artery. Cells bathed in 1.5 mM Ca^{2+} saline containing 1 μM nifedipine. Pipette and bath solutions are K^+ free. a, At -60 mV an inward ATP gated current is associated with a rise in $[Ca^{2+}]_i$. b) At +50 mV an outward current is seen but no rise in $[Ca^{2+}]_i$. (published in the Journal of Physiology, (13)).

In spite of this evidence for direct Ca^{2+} entry in single cells, the data from whole tissue experiments would suggest that the main role of these channels was to depolarise the tissue to a membrane potential range where voltage-gated Ca^{2+} channels could open. The evidence for this is that contractions to exogenous ATP or to nerve stimulation are invariably completely antagonised by concentrations of dihydropyridines that block voltage-gated Ca^{2+} channels without affecting the ATP gated current (e.g. 15).

Thus, the role of the direct Ca^{2+} influx is still uncertain. An increasing awareness that there is precise localisation of membrane, cytosolic and sarcoplasmic reticulum components of Ca^{2+} homeostatic mechanisms suggests a possible subtle physiological role for the ATP pathway. It offers a Ca^{2+} influx route that is spatially distinct from entry through voltage gated Ca^{2+} channels and so it is possible to envisage Ca^{2+} entry through this pathway being preferentially used to refill Ca^{2+} stores without interacting with contractile proteins. These thoughts are speculative, but the number of other receptor coupled channels that are Ca^{2+} permeable shows that this is not a potential transduction process that is unique to ATP. The permeability of the P_{2X} gated channels to Ca^{2+} is compared to other cation channels in smooth muscle in table 1.

The cholinergic and noradrenergic coupled conductances differ in that there is evidence that they are not directly coupled to the receptor as is the ATP gated channel but all three are significantly permeable to Ca^{2+}. Under reasonably comparable conditions the cation currents activated by acetylcholine in smooth muscle from guinea-pig ileum (16) and by noradrenaline in rabbit portal vein (17) have somewhat greater Ca^{2+} permeability. Both these conductances are one branch of a number of G-protein mediated second messenger pathways. The cation channels coupled to muscarinic receptors are membrane potential sensitive and regulated by $[Ca^{2+}]_i$ (16, 18). The combined effect of these properties is to provide a very effective positive feedback mechanism that maintains depolarisation and Ca^{2+} influx following action potential discharge. Although studied less comprehensively, the cation current activated by noradrenaline also seems to be modulated by intracellular factors (17).

ATP clearly functions as a co-transmitter in many neuro-effector synapses in the autonomic nervous system (19). The fast depolarisation mediated by the P_{2X} receptor-channels results in voltage gated Ca^{2+} entry and contraction. It may play a similar

TABLE 1. PERMEABILITY RATIOS FOR AGONIST GATED CURRENTS. SEE TEXT FOR REFERENCES

Agonist	Tissue	E_{rev} (mV)	P_{div}/P_{mon}	divalent$_o$
ATP	ear artery	+5.7	2.7	110 mM Ca^{2+}
Ach	ileum	+19.9	6.2	110 mM Ca^{2+}
Norad	portal vein	+17.6	6.6	89 mM Ba^{2+}

excitatory role in the central nervous system. The recent demonstration of excitatory post-synaptic potentials evoked by ATP at a neuro-neuronal synapse suggests a more general role for ATP as a fast transmitter in the nervous system (20). Up to now its transmitter role has been only clearly identified at sympathetic neuroeffector junctions.

In the pulmonary artery, nerve stimulation results in a relatively slow depolarisation reaching a peak after 10-15s. This response was completely blocked by phentolamine. There was no evidence of a fast non-adrenergic response indicating no involvement of purinergic receptors (21). Similar studies on the innervation of smaller resistance vessels in this vascular bed have not been reported. Evidence from other arterial beds would suggest that it is in the smaller vessels that the purinergic component is most important. So a role for P_{2X} receptors cannot be ruled out in this tissue.

2ND MESSENGER COUPLED ION CHANNELS

The classical P_{2Y} purinoceptor responses in smooth muscle cells of taenia coli and in hepatocytes are mediated by receptor activation of phospholipase C to produce inositol 1,4,5 trisphosphate which triggers Ca^{2+} store release. The effects on membrane potential are a consequence of this rise in $[Ca^{2+}]_i$ activating Ca^{2+} dependent ion channels. In taenia coli ATP hyperpolarises the smooth muscle cells as Ca^{2+} dependent K^+ channels are activated (22). This response is blocked by suramin. In many other smooth muscle cell types the membrane currents evoked by ATP application are complicated to interpret due to multiple receptor activation. Smooth muscle cells from portal vein express both P_{2X} and P_{2Y} receptors. In addition, the P_{2Y} receptor mediated events seemed to involve more than just Ca^{2+} store release. A fast transient cation current (P_{2X}) similar to that seen in rabbit ear artery was followed by a sustained, GTP-dependent inward current that was permeable to Na^+ and Ca^{2+} ions (5). As this sustained current was seen under conditions of high Ca^{2+} buffering it seems that this may be part of a separate G-protein mediated, pertussis toxin sensitive pathway independent of Ca^{2+} store release.

In the pulmonary circulation the most important source of P_{2Y} receptors is those found on endothelial cells. P_{2Y} receptor activation in these cells results in release of prostacyclin and nitric oxide as vasorelaxant messengers (2). This then represents an indirect pathway for vasodilation, but is the most important response to ATP released into the circulation where endothelial ectonucleotidases will limit the amount of ATP reaching the vascular smooth muscle cells.

ACKNOWLEDGEMENT

Thank J. Pitkin for help in preparing this manuscript.

REFERENCES

1. **Kennedy, C. & Burnstock, G.** Evidence for two types of P_2 purinoceptor in longitudinal muscle of the rabbit portal vein. Eur. J. Pharmacol. **111**, 49-56. 1985

2. **Gordon,** Extracellular ATP: Effects, sources and fate. Biochemical J. 233, 309-319, 1986.

3. **Evans, R. J., Derkach, V. & Surprenant, A.** ATP mediates fast synaptic transmission in mammalian neurons. Nature, 357, 503-505. 1992.

4. **Benham, C.D., Bolton, T.B., Byrne, N.G. & Large, W.A.,** 1987. Action of extracellular adenosine triphoshate in single smooth muscle cells dispersed from the rabbit ear artery. J. Physiol. 387, 473-488.

5. **Xiong, Z., Kitamura, K., & Kuriyama, H.**: ATP activates cationic currents and modulates the calcium current through GTP-binding protein in rabbit portal vein. J. Physiol, **440**, 143-165, 1991.

6. **Benham, C.D. & Tsien, R.W.** Receptor-operated, Ca-permeable channels activated by ATP in arterial smooth muscle. Nature, **328**: 275-278. 1987.

7. **Inoue, R. & Brading, A.F.** The properties of the ATP induced depolarisation and current in single cells isolated from the guinea-pig urinary bladder. Br. J. Pharmacol.**100**, 619-625, 1990.

8. **Friel, D.D.** An ATP sensitive conductance in single smooth muscle cells from the rat vas deferens. Journal of Physiology, **401**: 361-380, 1988.

9. **Krishtal, O.A., Marchenko, S.M. & Pidoplichko, V.I.** Receptor for ATP in the membrane of mammalian sensory neurones. Neuroscience Letters, **35**: 41-45   1983.

10. **Burnstock , G. & Kennedy, C.** Is there a basis for distinguishing two types of P_2-purinoceptor? Gen Pharmac. 16, 433-440 1985.

11. **Bean, B.P.** Pharmacology and electrophysiology of ATP-activated ion channels. T.I.P.S. 13, 87-90. 1992.

12. **Colquhoun, D., Ogden, D.C. & Mathie, A.** 1987. Nicotinic acetylcholine receptors of nerve and muscle: functional aspects. T.I.P.S. 8, 465-472.

13. **Benham, C.D.** ATP-activated channels gate calcium entry in single smooth muscle cells dissociated from rabbit ear artery. Journal of Physiology, **419**: 689-701, 1989.

14. **Schneider, P., Hopp, H.H., & Isenberg, G.** Ca^{2+} influx through ATP-gated channels increments $[Ca^{2+}]_i$ and inactivates I_{Ca} in myocytes from guinea-pig urinary bladder. J. Physiol. **440**,

15. **Katsuragi, T., Usune, S. & Furukawa, T.** Antagonism by nifedipine of contraction and Ca^{2+} influx evoked by ATP in guinea-pig urinary bladder. Br. J. Pharmacol. 100, 370-374. 1990.

16. **Inoue, R. & Isenberg, G.** Effect of membrane potential on acetylcholine induced inward current in guinea-pig ileum. J. Physiol. 424, 57-71. 1990a.

17. **Wang, Q. & Large, W.A.** Noradrenaline evoked cation conductance recorded with the nystatin whole cell method in rabbit portal vein cells. J. Physiol. 435, 21-39. (1991).

18. **Inoue, R. & Isenberg, G.** Intracellular calcium ions modulate acetylcholine-induced inward current in guinea-pig ileum. J. Physiol. 424, 73-92. 1990b.

19. **Sneddon, P., Westfall, D.P. & Fedan, J.S.** Co-transmitters in the motor nerves of the guinea-pig vas deferens: electrophysiological evidence. Science, **218**: 693-695 1982 **91**: 1-27.

20. **Edwards, F.A., Gibb, A.J., Colquhoun, D.** ATP receptor mediated synaptic currents in the central nervous system. Nature 359, 144-147. 1992.

21. **Suzuki, H.** An electrophysiological study of excitatory neuromuscular transmission in the guinea-pig main pulmonary artery. J. Physiol. 336, 47-59. 1983.

22. **Den Hertog, A. Nelemans,A. and Van Den Akker, J.** The inhibitory action of suramin on the P_2-purinoceptor response in smooth muscle cells of guinea-pig taenia caeci. Eur. J. Pharmacol.166, 531. 1989.

AN ATP-ACTIVATED POTASSIUM CHANNEL
IN SMOOTH MUSCLE CELLS FROM THE PULMONARY ARTERY

Sulayma Albarwani and Piers Nye

University Laboratory of Physiology
Parks Road
Oxford OX1 3PT, UK

INTRODUCTION

Pulmonary vascular smooth muscle and the carotid body chemoreceptors are both excited by hypoxia, but our understanding of the mechanism responsible for excitation of the former has lagged behind, and this is largely because the muscle cells have only recently been studied using patch (Post et al., 1992; Robertson et al., 1992a) and fluorescence (Robertson et al., 1993b) techniques. It seems to us that, just as the carotid body field is being advanced rapidly by the performance of experiments on single cells, and relating the results from these to more traditional work on the discharge of chemoreceptors, so experiments on single smooth muscle cells from pulmonary vessels, backed up by complementary experiments on isolated vessels and lungs, are likely to be the most successful way forward.

Direct observation of the length of single smooth muscle cells from the pulmonary artery (Madden et al., 1992) shows that these are constricted by hypoxia. They are therefore almost certainly the site of sensation of hypoxia as well as the site of its action. We have used the patch clamp technique to study the behaviour of single channels in the membranes of pulmonary vascular smooth muscle cells. Post et al. (1992) have provided evidence from whole-cell patch clamp experiments that hypoxia may act by closing a calcium-activated potassium (K_{Ca}) channel which may depolarise the cell and lead to the influx of Ca^{2+} through voltage-gated Ca^{2+} channels. This idea is supported by the single channel patch clamp experiments of Robertson et al., (1992a), in which there is no influence from cytoplasmic constituents. We have continued to study the behaviour of channels in isolated patches of membrane because it gives more control over the environment of the channels and therefore allows their essential characteristics - single channel currents and open probabilities - to be identified and studied in detail.

By far the most abundant channel that is recorded from patches of pulmonary arterial smooth muscle cells is a novel large conductance (240pS) K_{Ca} channel opened by intracellular MgATP. This was positively identified in 65 out of 75 patches, many of which contained at least 7 identical channels. We refer to it as a $K_{Ca,ATP}$ channel. Given its abundance and large conductance it seems not unreasonable to propose that the activity of this channel may be the most important factor determining membrane potential in these cells and that a reduction in the level of ATP immediately beneath the membrane may therefore link hypoxia to the process of excitation-contraction coupling. Technical problems have so far prevented us from testing the response of the channels to true hypoxia, but we are somewhat reassured by the observation of Ohe et al. (1986) that the ATP content of rings of pulmonary artery falls greatly when exposed to a Po2 of 11Torr.

This chapter describes our preliminary characterisation of $K_{Ca,ATP}$ channels and relates some of our results to parallel studies on isolated vessels. It also airs some of the problems involved in attempting to incorporate a Ca^{2+}-activated and voltage-activated channel into a mechanism responsible for the hypoxic excitation of pulmonary vascular smooth muscle.

CHARACTERISATION OF THE BEHAVIOUR OF $K_{Ca,ATP}$ CHANNELS IN INSIDE-OUT PATCHES

Cells were isolated from the pulmonary arteries of rats anaesthetized by 4% halothane by digesting short lengths of artery for 15min at 34 to 37°C in collagenase (725units/ml), elastase (0.25mg/ml), protease (0.1mg/ml) and bovine serum albumin (2.5mg/ml) titrated to pH 7.3 by NaOH in the presence of (mM) $NaHCO_3$ (4.2), Na_2HPO_4 (0.35), KH_2PO_4 (0.45), HEPES (10) glucose (10) and sucrose (2.9). The isolated cells were placed on polylysine-coated glass cover slips.

Single-channel currents were recorded from inside-out membrane patches using pipettes of 10-20MΩ resistance and an Axopatch 1-C amplifier. Traces were high-frequency filtered at 2kHz and digitized at 5kHz.

The electrodes contained (mM): KCl (140), $MgCl_2$ (1), $CaCl_2$ (1) and HEPES (10) taken to pH 7.4 by KOH. The intracellular aspect of the patch was bathed in the same KCl, $MgCl_2$, HEPES concentrations buffered to pH 7.3-7.4 with KOH. The free Ca^{2+} in the bath solution was varied by adding $CaCl_2$ and EGTA. All experiments were performed at room temperature 21-24°C.

All the patches used in this study contained two to eight $K_{Ca,ATP}$ channels activated by both Ca^{2+} and Mg-ATP applied to the intracellular membrane surface (Figure 1). This activation was prompt, developing with a time constant (ca. 10s) indistinguishable from that required to change the composition of the bath. The channels were voltage-sensitive, being activated markedly by depolarization. With symmetrical K^+ (140mM) on either side of the patch membrane their conductance was $245\pm6pS$, a value typical of large conductance K_{Ca} (BK_{Ca}) channels (Robertson et al., 1992a).

At a membrane potential of +40mV the EC_{50} for Ca^{2+} was $0.5\mu M$ (6 patches) and that for Mg-ATP was $50\mu M$ (16 patches). Calcium did not have to be present for ATP to activate the channels. The 5mM EGTA used to buffer all Ca^{2+} concentrations was also used for the zero Ca^{2+} solution to ensure chelation of any residual free calcium. This makes it surprising that ATP appeared to have some ability to activate the channel in the absence of Ca^{2+} (Figure 2). The activation by Ca^{2+} in the nominal absence of ATP could be the result of residual ATP levels that remained from a previous exposure. Raising the concentration of free Ca^{2+} from 0 to $10\mu M$ markedly increased sensitivity to ATP without affecting the $EC_{50,ATP}$.

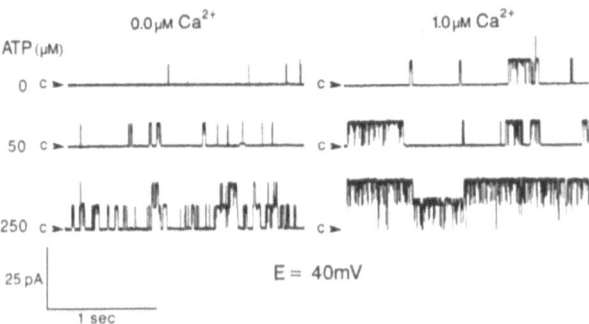

Figure 1. Recordings from typical patches each containing two $K_{Ca,ATP}$ channels. Ca^{2+} and ATP activate the channel. Both bath and electrode contained 140mM KCl.

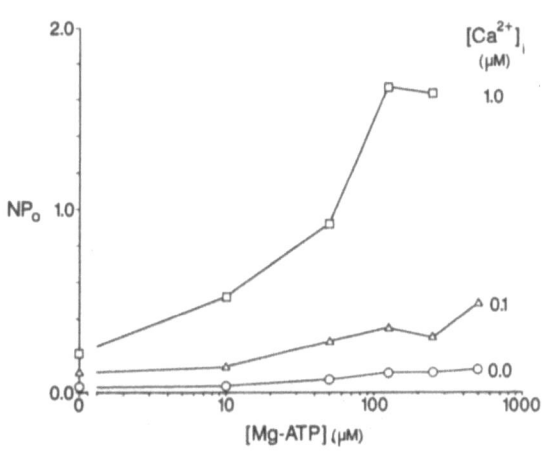

Figure 2. Dose-response curves in which each point represents the average from five or six different patches. NP_o is the open probability times the number of channels open. The stimulatory effects of Ca^{2+} and ATP interact multiplicatively.

Recruitment of Channels by ATP

The increase in channel activity that is brought on by the addition of ATP appeared to result more from the recruitment of channels that had previously been inactive than it did from an increase in the open probability of each channel. This was shown by binomial analysis which revealed that, at moderate and low stimulus intensities, there were fewer incidences of two or more channels being simultaneously open than would be expected if all channels had been active with the same open probability (Robertson et al., 1992a).

Activation Requires Phosphorylation by Mg-ATP

The activation of channels by ATP did not occur when Mg^{2+}, the presence of which is required for all kinase reactions, was omitted from the bath. The non-hydrolysable ATP analogue Mg-AMP-PNP was a feeble activator of the channels and Mg-ATP$_\gamma$S, which resists dephosphorylation, also resulted in only feeble activation (Robertson et al., 1992a). These observations strongly suggest that activation of the channel requires continuous phosphorylation by ATP. We imagine that the small effects of the ATP analogues may have arisen because of contaminating traces of ATP in the samples.

$K_{Ca,ATP}$ Channels are Insensitive to Drugs that Open or Close the K_{ATP} Channels of the Heart and Pancreas

It has previously been shown that drugs which modulate the activity of the K_{ATP} channels of the heart and endocrine pancreas can have important effects on hypoxic pulmonary vasoconstriction of isolated lungs and pulmonary vessels (Yuan et al., 1990, Wiener et al., 1991). It is important to emphasize here that cardiac and pancreatic K_{ATP} channels, which have also been reported to occur in the resistance vessels of the systemic circulation (Standen et al., 1989), are very different from the $K_{Ca,ATP}$ channels that are the primary subject of this chapter. The former are closed, not opened, by ATP and they are insensitive to Ca^{2+}.

The results of experiments with drugs selective for K_{ATP} channels, e.g. glibenclamide, have been interpreted to show that such channels are also found in the vessels of the lung, but that they are not directly involved in the mechanism responsible for hypoxic vasoconstriction (Hasunuma et al., 1991). They are closed in normoxia and also closed in hypoxia of an intensity sufficient to cause maximal vasoconstriction (Robertson et al., 1992b). It is only when hypoxia is so severe that it results in a secondary dilatation that these channels reveal their presence by opening. So they must first be opened by severe hypoxia, or by a drug such as cromakalim or diazoxide, before glibenclamide has an effect, and this effect is to constrict (Wiener et al., 1991).

Tolbutamide, which like other sulphonylureas such as glibenclamide, closes the K_{ATP} channels of the heart and pancreas, does not behave like glibenclamide in isolated lungs (Robertson et al., 1992b). It paradoxically relaxes pulmonary vessels as if it were activating the channels that glibenclamide closes. This idea is reinforced by the observation that glibenclamide restores constriction in hypoxic lungs that have been dilated by low concentrations of tolbutamide (Robertson et al., 1992b). It was therefore of interest to discover whether or not glibenclamide, tolbutamide and the potassium channel openers affected the activity of $K_{Ca,ATP}$ channels. Glibenclamide and the potassium channel openers cromakalim and diazoxide had no effect on the activity of the channels, and tolbutamide had variable insignificant effects. It seems therefore that unless tolbutamide acts exclusively on the external aspect of $K_{Ca,ATP}$ channels (all patches were inside-out) we must look elsewhere, perhaps to an effect on cAMP levels (Levey et al., 1974) if we are to explain the tolbutamide paradox.

$K_{Ca,ATP}$ Channels are Blocked by Barium, and 4-Aminopyridine but are Poorly Sensitive to Tetraethylammonium

Almost all potassium channels are blocked by barium, and many are blocked by 4-aminopyridine (4-AP). Both substances greatly reduced or abolished the opening of $K_{Ca,ATP}$ channels (Robertson et al., 1993a). However 4-AP does not affect the majority of BK_{Ca} channels; we can find only one reference to its doing so in single channel recordings (Cook and Quast, 1990). $K_{Ca,ATP}$ channels were also poorly sensitive to tetraethylammonium (TEA), requiring 100mM for a halving of unitary conductance. The effects of 4-AP and TEA are consistent with those reported by Okabe et al. (1987) whose experiments on the whole-cell current of pulmonary arterial smooth muscle cells showed that they contain many K_{Ca} channels. They also showed that the whole-cell current is about three orders of magnitude more sensitive to 4-AP than it is to TEA. These characteristics may therefore distinguish the present K_{Ca} channel from its relatives.

EXPERIMENTS ON SMALL ARTERIES ISOLATED FROM THE LUNG

Experiments on single small pulmonary vessels are useful because they can be used to relate observations made on channels to the behaviour of a tissue which is under more physiological conditions. When the results of such parallel experiments tally they reinforce the likelihood of their being of genuine significance rather than artefacts caused by the freak conditions experienced by naked, half-digested cell membranes stretched across the tip of a patch electrode. Experiments on vessels can also help in the selection of substances that would be worth studying at the level of the single channel.

Experiments on isolated blood vessels are not directly comparable to those on patches of cell membrane because the vessels contain many cell types, the metabolic machinery of their cells is intact, and substances are always applied to the external surface of cells. None of these conditions apply to experiments on channels in the inside-out patch clamp configuration, so it is perhaps surprising that good correlation between the responses of the two preparations is found. The next paragraph shows that it is.

4-Aminopyridine Mimics the Effects of Hypoxia but Tetraethylammonium Does Not

4-aminopyridine which powerfully blocks the $K_{Ca,ATP}$ channel in patches also constricts vessels, while TEA is a much less powerful blocker of the channel and is a less effective constrictor of vessels. An even more striking difference between the effects of these two drugs is apparent when one looks at the interaction of their effects and that of hypoxia. TEA never, in 34 trials made over a wide range of concentrations (1 to 20mM), enhanced the amplitude of hypoxic constriction (e.g. Figure 3A), while low concentrations of 4-AP, which often failed to elicit any contraction in normoxia, always increased the amplitude of a subsequent response to hypoxia (as has been shown in isolated lungs by Hasunuma et al., 1991) and did so in a dose-dependent way that suggested it was substituting for hypoxia (Figure 3B). High concentrations of 4-AP constricted the vessels and reversed the response to hypoxia (Figure 4) as if they had saturated the hypoxic constrictor mechanism to reveal an underlying inhibitory effect (Albarwani et al., 1993). These observations are consistent with the idea that 4-AP, but not TEA, acts via the same route that is responsible for the upwardly curving response to increasing hypoxia.

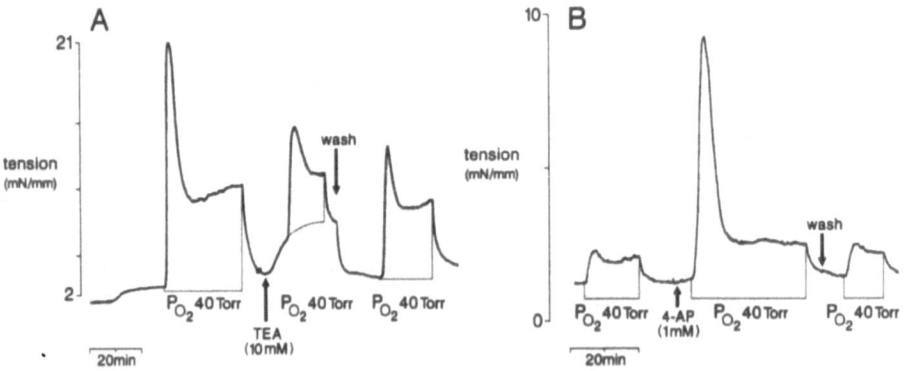

Figure 3. A. Traces from a small isolated pulmonary artery (ca. 400μm diameter) responding to three hypoxic challenges (shaded) given before, in the presence of, and after the removal of TEA. B. As for A but with a low dose of 4-AP instead of TEA. TEA constricted the vessel but reduced the amplitude of the response to hypoxia. TEA was *never* observed to enhance constriction by hypoxia. This dose of 4-AP (1mM) failed to constrict the vessel but *always* enhanced the response to hypoxia.

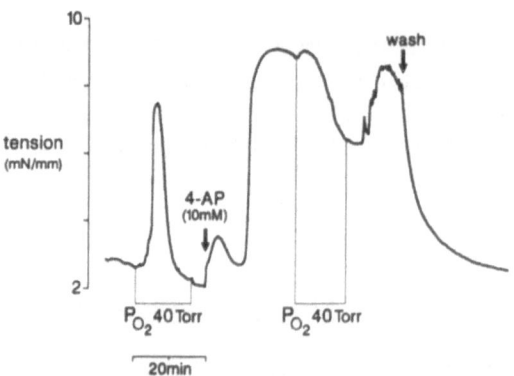

Figure 4. Responses of a small pulmonary artery (450μm diameter) to a control hypoxic challenge and then to one after constriction by a high concentration (10mM) of 4-AP. At this concentration 4-AP gives a triphasic constriction after which hypoxia relaxes the vessel and normoxia constricts it.

Potassium Channel Openers Abolish the Constriction of Isolated Vessels by Hypoxia

Hypoxically constricted isolated pulmonary vessels are relaxed by potassium channel openers in a way that is indistinguishable from that of whole lungs (Yuan et al., 1990; Robertson et al., 1992b). Figure 5 shows this happening in vessels preconstricted by prostaglandin F2α and then exposed to a quite modest hypoxia of 40Torr. In this example diazoxide (1μM) curtailed an ongoing constriction and abolished the subsequent one. Cromakalim (2μM) had a similar effect. We know that cromakalim and diazoxide have no effect on $K_{Ca,ATP}$ channels (see above), so if these drugs are acting only by opening potassium channels on the cell membrane of smooth muscle cells, there must be another channel type whose opening carries sufficient current to resist the depolarisation associated with hypoxia (Harder et al., 1985). The obvious candidate here is the K_{ATP} channel (Clapp and Gurney, 1991, 1992). This observation suggests to us that the response to hypoxia is unlikely to involve a series of purely intracellular events like those hypothesized for the carotid body by Duchen and Biscoe (1992). If the mechanism responsible for hypoxic constriction had been all inside the cell, hyperpolarisation of the cell membrane after the constriction had started would not be expected to cut short the initial transient component of the contraction.

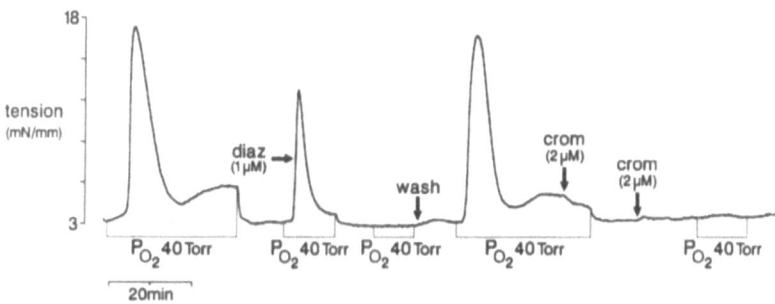

Figure 5. Trace from a small isolated pulmonary artery (480μm diameter) exposed to five consecutive hypoxic challenges. The potassium channel openers diazoxide (diaz) and cromakalim (crom) abolish both the transient and sustained components of the response to hypoxia even if given after the process has started.

CONCLUSION

The smooth muscle cells of the pulmonary artery possess potassium channels whose activity is reduced when intracellular Mg-ATP falls to below about 100μM (Figures 1 and 2). This is a low value for intracellular Mg-ATP, which is normally in the millimolar range, but it may not be low for the immediate environment of channels embedded in a metabolically active membrane. Local metabolic activity might also explain the observed recruitment of channels as Mg-ATP rises (Robertson et al., 1992a), with those channels that happen to be furthest from the nearest site of intense metabolism being the first to be activated by ATP.

The most awkward problem confronting the suggestion that $K_{Ca,ATP}$ channels may be central to the mechanism responsible for hypoxic pulmonary vasoconstriction is the calcium- and voltage-sensitivity of the channels. Both these influences would tend to return the channels to their open state soon after they had been closed by a fall in Mg-ATP. The physiological levels of Ca^{2+} and Mg-ATP beneath the cell membrane are not known, but cytosolic Ca^{2+} is about 100nM in resting pulmonary arterial smooth muscle and might rise to 400nM (Somlyo and Himpens, 1989) during the hypoxically induced calcium-dependent action potentials described by Harder et al. (1985). Figure 2 shows that within this range Ca^{2+} has large effects on open probability. There are also the activating effects of depolarisation to be considered, so the role of $K_{Ca,ATP}$ channels is far from resolved. Suffice it to say that hypoxia depolarises the smooth muscle cells of pulmonary vessels and results in the generation of action potentials (Harder et al., 1985) and that there is good evidence that it depolarises by reducing the calcium-activated potassium current (Post et al., 1992). There is also good evidence that the influx of calcium through voltage-gated membrane channels is required for hypoxic excitation (Archer et al., 1985; Rodman et al., 1989), so it may be that a reduction in Mg-ATP contributes to the initial triggering of this excitation and that a process involving action potentials carried by Ca^{2+} sustains it.

Acknowledgements. The published patch clamp experiments referred to above were performed by Blair Robertson and Peter Corry with much guidance from Chris Peers and Roland Kozlowski. RK also provided valuable advice on patch clamp work which is published here for the first time. Niya Xia performed many of the experiments on isolated vessels. We are very grateful to the British Heart Foundation and the Wellcome Trust for funding us.

REFERENCES

Albarwani, S., Xia, N. & Nye, P.C.G. 1993, Excitation of small isolated pulmonary blood vessels from the rat by 4-aminopyridine and hypoxia. *J Physiol Proceedings.* Queen Mary & Westfield Meeting, London. December 1992.

Archer, S.L., Yankovich, R.D., Chesler, E. & Weir, E.K. 1985, Comparative effects of nisoldipine, nifedipine and bepridil on experimental pulmonary hypertension. *J Pharmacol Exp Ther.* 233:12.

Clapp, L.H. & Gurney, A.M. 1991, Outward currents in rabbit pulmonary artery cells dissociated with a new technique. *Experimental Physiol.* 76:677.

Clapp, L.H. & Gurney, A.M. 1992, ATP-sensitive K+ channels regulate resting potential of pulmonary arterial smooth muscle cells. *Am J Physiol Heart Circ Physiol.* 262:H916.

Cook, N.S. & Quast, U. 1990, Potassium channel pharmacology, *in: "Potassium Channels,"* N.S.Cook, ed., Ellis Horwood, Chichester.

Duchen, M.R. & Biscoe, T.J. 1992, Mitochondrial function in type I cells isolated from rabbit arterial chemoreceptors. *J Physiol.* 450:13.

Harder, D.R., Madden, J.A. & Dawson, C. 1985, Hypoxic induction of Ca^{2+}-dependent action potentials in small pulmonary arteries of the cat. *J Appl Physiol.* 59:1389.

Hasunuma, K., Rodman, D.M. & McMurtry, I.F. 1991, Effects of K+ channel blockers on vascular tone in the perfused rat lung. *Am Rev Respir Dis.* 144:884.

Levey, G.S., Lasseter, K.C. & Palmer, R.F. 1974, Sulphonylureas and the heart. *Ann Rev Med* 25:69.

Madden, J.A., Vadula, M.S. & Kurup, V.P. 1992, Effects of hypoxia and other vasoactive agents on pulmonary and cerebral artery smooth muscle cells. *Am J Physiol.* 263:L384.

Ohe, M., Mimata, T., Haneda, T. & Takishima, T. 1986, Time course of pulmonary vasoconstriction with repeated hypoxia and glucose depletion. *Respir Physiol.* 63:177.

Okabe, K., Kitamura, K. & Kuriyama, H. 1987, Features of 4-aminopyridine sensitive outward current observed in single smooth muscle cells from the rabbit pulmonary artery. *Pflugers Arch.* 409:561.

Post, J.M., Hume, J.R., Archer, S.L. & Weir, E.K. 1992, Direct role for potassium channel inhibition in hypoxic pulmonary vasoconstriction. *Am J Physiol.* 262:C882.

Robertson, B.E., Corry, P.R., Kozlowski, R.Z. & Nye, P.C.G. 1993a, Responses of ATP-activated potassium channels of pulmonary arterial smooth muscle of the rat to potassium channel modulators. *J Physiol Proceedings.* Queen Mary & Westfield Meeting, London. December 1992.

Robertson, B.E., Corry, P.R., Nye, P.C.G. & Kozlowski, R.Z. 1992a, Ca^{2+} and Mg-ATP activated potassium channels from rat pulmonary artery. *Pflugers Arch.* 421:94

Robertson, B.E., Kozlowski, R.Z. & Nye, P.C.G. 1992b, Opposing actions of tolbutamide and glibenclamide on hypoxic pulmonary vasoconstriction. *Comp Biochem Physiol.* 102C:459.

Robertson, T.P., Priest, R., Leach, R.M., & Ward, J.P.T. 1993b, Chronic hypoxia results in depolarisation of the resting membrane potential of isolated small pulmonary arteries of the rat. *J Physiol Proceedings.* Queen Mary & Westfield Meeting, London. December 1992.

Rodman, D.M., Yamaguchi, T., O Brien, R.F. & McMurtry, I.F. 1989, Hypoxic contraction of isolated rat pulmonary artery. *J Pharmacol Exp Ther.* 248:952.

Somlyo, A.P. & Himpens, B. 1989, Cell calcium and its regulation in smooth muscle. *FASEB J.* 3:2266.

Standen, N.B., Quayle, J.M., Davies, N.W., Brayden, J.E., Huang, Y. & Nelson, M.T. 1989, Hyperpolarizing vasodilators activate ATP-sensitive K+ channels in arterial smooth muscle. *Science.* 245,177-180.

Wiener, C.M., Dunn, A. & Sylvester, J.T. 1991, ATP-dependent K+ channels modulate vasoconstrictor responses to severe hypoxia in isolated ferret lungs. *J Clin Invest.* 88:500.

Yuan, X-J., Tod, M.L., Rubin, L.J. & Blaustein, M.P. 1990, Contrasting effects of hypoxia on tension in rat pulmonary and mesenteric arteries. *Am J Physiol.* 259:H281.

THE EFFECT OF PROSTACYCLIN ANALOGS ON VASCULAR K CHANNELS

G. Siegel[1], M. Bostanjoglo[1], K. Rückborn[2], F. Schnalke[1], J. Mironneau[3], and G. Stock[4]

[1]Institute of Physiology, Biophysical Research Group, The Free University of Berlin, D-1000 Berlin 33, Germany
[2]Institute of Physiology, University of Rostock, D-2500 Rostock 1, Germany
[3]Laboratoire de Physiologie Cellulaire et Pharmacologie Moléculaire, Université de Bordeaux II, F-33076 Bordeaux, France
[4]Cardiovascular Pharmacology, Research Laboratories of Schering AG, D-1000 Berlin 65, Germany

INTRODUCTION

Drugs such as pinacidil, nicorandil, minoxidil sulphate and cromakalim effect vasodilatation by membrane hyperpolarization of the vascular smooth muscle cells, which in some tissues raises the membrane potential to a value close to the K^+ equilibrium potential [35]. Different K^+ channels are usually subdivided according to their mode of activation. Some are activated strictly voltage-dependently, others by a variation in the intracellular Ca^{2+} concentration, and some by the internal concentration of ATP, Na^+, cyclic nucleotides etc. The heterogeneous group of K^+ channel openers may be a potential therapy for hypertension, asthma, peripheral vascular disease, and diseases of the heart and nervous system. The central starting point of their physiological mode of action is the hyperpolarization of the smooth muscle cells which leads to relaxation by closing L- and/or T-type Ca^{2+} channels without, in the classical sense, the participation of cAMP or cGMP [34]. In the present contribution, it is convincingly demonstrated that prostacyclin analogs should be included into the class of K^+ channel agonists although cAMP seems to take part in the relaxation process. This may be justified by the fact that cyclic nucleotides also elicit a membrane hyperpolarization in the vascular smooth muscle.

MATERIALS AND METHODS

Experimental Preparations and Solutions

Intracellular potential recordings, measurements of isometric tension and radioisotope efflux were performed on the vascular smooth muscle of common carotid and coronary arteries [27,31]. The carotid arteries were surgically removed from sacrificed dogs or rabbits within 3-4 min. In order to examine the participation of G-proteins in the receptor-mediated K^+ channel opening, rabbits were treated in a test series with pertussis toxin (PTx) for 4-5 days. PTx was first injected intravenously in a single dose of 2.5 $\mu g/kg$ body weight [8]. Human coronary arteries were taken from heart transplant patients who had been suffering from a dilative cardiomyopathy or extensive atherosclerosis. There was a clear differentiation between normal and atherosclerotic vessels where the preparations had been examined by light and electron microscopy. All arterial segments were equilibrated in a Krebs solution of the following composition (mmol/l): Na^+ 151.16; K^+ 4.69;

Ca^{2+} 2.52; Mg^{2+} 1.1; Cl^- 145.4; HCO_3^- 16.31; $H_2PO_4^-$ 1.38; glucose 7.77 (temp. 37°C; pH 7.35). The solution was usually aerated with a 95% O_2 - 5% CO_2 gas mixture (carbogen). Blood substitute solutions were equilibrated with $O_2/CO_2/N_2$ mixtures containing different oxygen concentrations, but a constant concentration of carbon dioxide (5ml/dl) to vary the oxygen tension. Microelectrodes consisting of platinum cathodes were used for continuous oxygen tension measurements in the Krebs solution close to the blood vessel strip [7].

The PGI_2 analogs, iloprost (Schering, ZK 36374; mol. wt. 360.5 g/mol; 16R,S-methyl-18,18,19,19-tetradihydro-6a-carbaprostaglandin I_2) and cicaprost (Schering, ZK 96480; mol. wt. 374.5 g/mol; 5-{(E)-(1S,5S,6S,7R)-7-hydroxy-6-[(3S,4S)-3-hydroxy-4-methyl-nona-1,6-diinyl]-bicyclo[3.3.0]octan-3-ylidene}- 3-oxapentanoic acid), were used in the concentration steps 10^{-10}, 10^{-9}, 10^{-8}, 10^{-7}, 10^{-6}, $3 \cdot 10^{-6}$, and 10^{-5} mol/l. The dihydropyridine derivative, cicletanine (IHB-IPSEN; mol. wt. 298.17 g/mol; D-L-1,3-dihydro-3-(4-chlorophenyl)-7-hydroxy-6-methyl-(3,4-c)-furopyridine hydrochloride), was first dissolved in dimethylsulfoxide (Sigma, Deisenhofen) at a concentration of 0.671 mol/l (stock solution) [2]. From this, the concentration steps 10^{-6}, 10^{-5}, 10^{-4}, and 10^{-3} mol/l were obtained in a Krebs solution. The corresponding DMSO concentrations were 0.021, 0.21, 2.1, and 21 mmol/l. The effects of cicletanine were measured as the relative change to DMSO control experiments. Aqueous garlic extract from 0.0002, 0.002, 0.02 and 0.2 g powder/l was used, each solution being freshly prepared. For extraction, 3 g garlic powder (Kwai®, Lichtwer Pharma, Berlin) was elutriated with 9 ml H_2O and centrifuged (6,000 × g) after 20 min. The extraction was repeated twice. Circa 27 ml supernatant was freeze-dried (temp. -5°C) and resuspended in 15 ml Krebs solution. HPLC investigations of the freshly prepared suspension yielded a concentration of 3.58 mmol/l allicin and of 0.184 mmol/l ajoene, respectively. All diluted concentration steps were produced from this stock solution [30].

Noradrenaline (Arterenol, Hoechst) was used in the concentration step of $3 \cdot 10^{-6}$ mol/l, which has a half-maximal effect on vascular smooth muscle membrane potential and tone [29]. Indomethacin (Vonum, Econerica) was added at a concentration of 10^{-5} mol/l and 6-hydroxy-dopamine (Aldrich-Chemie, Steinheim) at a concentration of $1.8 \cdot 10^{-3}$ mol/l. Each solution was freshly prepared.

Mechanical Recording

4-5 mm long cylinders from the carotid or coronary arteries were cut lengthwise. Since the muscle cells of arteries are arranged circularly, the folded-out vessel rings, 3-15 mm long, were attached at the cut ends to an isometric tension measuring device (KWS 522.C, K 52 C; Hottinger-Baldwin, Darmstadt, Germany) [32]. The specimens were superfused with the Krebs solution at 10 ml/min for 30 min at an initial tension of 2 g. After this time, the tension was in equilibrium and amounted from 1.116 ± 0.243 g (n=6) (human coronary arteries) to 2.016 ± 0.097 g (n=8) (canine carotid arteries). The solutions with different iloprost concentrations were then applied consecutively for 30 min each, those with different cicaprost concentrations for 10 min each, first at rising, then at falling concentrations (reversibility). Cicletanine and aqueous garlic extract were also applied in a cumulative manner and the reversibility was tested [23,30,32]. The incubation time was 15 min for all concentration steps. In the experiments with varying oxygen tension, solutions with first decreasing and then increasing oxygen partial pressures were applied for 15 min each. After these periods the mechanical tension had reached a steady-state value for each concentration step [7,25,32]. The flow rate was 50-60 ml/min in order to bring about quick changes in P_{O_2} and reach low pressures. Voltage and tension values were averaged for identical effector concentrations.

Electrical Recording

Glass capillary microelectrodes filled with 3 mol/l KCl were used to record intracellular membrane potential. These recordings and the mechanical tension registration were performed simultaneously. The electrode resistances ranged from 60-100 MΩ and the tip potentials from -40 to -80 mV. The electrodes were shielded to just below the tips; otherwise conventional recording techniques were used. The microelectrode was inserted into the arterial muscle cell from the intimal surface [25,29]. Arteries in a normal Krebs solution with membrane potentials between -50 and -80 mV were selected for the final averaging. Membrane potentials more positive than -50 mV (20% of all impalements) had to be discarded because, according to our own measurements and to those of other authors, they stem from endothelial cells [17].

Isotope Flux Measurements

Carotid segments from the dog were saturated for at least 3 hs in a radioactive Krebs solution containing $^{24}Na^+$ and $^{42}K^+$. In addition, the test solutions contained iloprost at a concentration of 10^{-6} mol/l. Washout of the radioisotopes was performed in the corresponding non-radioactive solution. The activity in the preparation was measured using VIP series coaxial Ge(Li) detector systems with a peak height to Compton plateau height ratio of 30:1 (Ortec, Oak Ridge, TN, USA). This permitted adequate discrimination of the γ-lines (system resolution 2.2 keV using 1.333 MeV photons) of ^{24}Na at 1.37 MeV and ^{42}K at 1.52 MeV by means of stabilized single-channel analyzers (SCA-N-3; Elscint, Haifa, Israel) [22]. The combined γ-spectra from the preparation were measured continuously over 40 min, accumulated at 10 s intervals, and transferred to a PDP 11/23 Plus (Digital, Maynard, CA, USA). During the measurements, a strictly constant and laminar flow of the washout medium was maintained. The fluid in the measurement chamber was completely exchanged 1.2 times per second (chamber volume 2.3 ml, flow rate 160 ml/min). Experiments with dye indicators confirmed the favourable flow conditions. Thus, the accumulation of tracer ions in the preparation chamber, interstitium, or on the surface of the specimen was avoided. After each experiment the content of Na, K, Mg, Ca and Cl in the preparation as well as in the incubation and washout solution was determined by atomic absorption spectroscopy.

Patch-Clamp Recording

Portal veins from Wistar rats (150 g) were dissected free of connective tissue and single cells were obtained using a dispersal procedure similar to that previously described [15]. Whole-cell membrane currents were measured using a standard patch-clamp technique. Patch pipettes had resistances of 2-3 MΩ. The data were analyzed with a Plessey 6220 microcomputer and the illustrations were drawn with an X-Y plotter. The series resistance, measured at 10 kHz, was 4-6 MΩ and no correction for errors in measurements of potential was applied [26].

The extracellular medium (reference solution) contained: NaCl 130, CsCl or KCl 5.6, CaCl$_2$ or BaCl$_2$ 5, MgCl$_2$ 0.24, glucose 11, HEPES 8 mmol/l (pH 7.4; temp. 30°C). The pipettes were filled with solutions containing: CsCl or KCl 130 mmol/l, Cs- or K-pyruvate, Cs- or K-succinate, Cs- or K-oxaloacetate, HEPES 10 mmol/l (pH 7.3 with CsOH or KOH). The stimulation frequency was 0.05 Hz.

Determination of Cyclic Nucleotide Levels

Mesenteric arteries of the rabbit were cut into segments 5-7 mm in length (6-11 mg wet wt.). The segments were suspended in a Krebs-Ringer bicarbonate buffer containing indomethacin (10^{-5} mol/l) and stored in an ice bath. After one hour, the segments were blotted dry and suspended in a Krebs-Ringer bicarbonate buffer at 37°C, containing indomethacin (10^{-5} mol/l), 3-isobutyl-1-methyl-xanthine (10^{-3} mol/l, Janssen, Neuss) and papaverine ($1.1 \cdot 10^{-5}$ mol/l) [32]. After 2 min, iloprost, dissolved in ethanol, was added in concentrations as given in Fig. 1. The reaction was halted after 5 min with 500 μl trichloroacetic acid. The segments were homogenized and centrifuged at 3,000 r.p.m. for 5 min at room temperature. After the trichloroacetic acid had been extracted with ethyl ether, the cAMP content of the supernatant was assayed by radioimmunoassay (RIA test kits of NEN, Dreieich).

RESULTS

K^+ Channel Opening and Membrane Hyperpolarization

The activation of K^+ channels in the membrane of vascular smooth muscle cells implies their hyperpolarization. Hyperpolarization effects relaxation via L- and/or T-type Ca^{2+} channels which are closed voltage-dependently between 0 and -80 mV [16,26,31,32]. Which Ca^{2+} channels are affected by this closure is dependent on the actual membrane potential before hyperpolarization. The consequence is a fall in intracellular Ca^{2+} activity and thus vasodilatation. Numerous effector influences can evoke such a hyperpolarization.

Prostacyclin (PGI_2), a main but unstable metabolite of arachidonic acid cascade in vascular tissue, is often considered to be a vasorelaxing autacoid in hypoxic dilatation. In our studies we were interested to see whether the vasodilatation observed has an electrophysiologic correlative. Because of the chemical instability of natural PGI_2, we used iloprost, a stable prostacyclin analog. As can be seen in Fig. 1, a concentration-dependent hyperpolarization of membrane potential was found under iloprost. Iloprost hyperpolarized the membrane by 7.4 mV maximally, whereas this effect was reduced with very high doses. Part B of the figure shows the dose-dependent relaxation of the vascular strip. The half-maximal effect on membrane potential and tone was observed with a $2 \cdot 10^{-8}$ mol/l concentration [22,26,32].

Using cicaprost, another stable prostacyclin analog, we observed a concentration-dependent hyperpolarization of the membrane and a corresponding vasorelaxation (Fig. 2). The smooth muscle cell membrane was hyperpolarized from -60.1 ± 0.4 mV (n=5) to -71.2 ± 2.1 mV (n=5; P < 0.001) and the vascular strip relaxed from 1.450 ± 0.092 g (n=6) to 0.873 ± 0.103 g (n=6; P < 0.002). The half-maximal effect on membrane potential and tone was recorded at a $1.54 \cdot 10^{-8}$

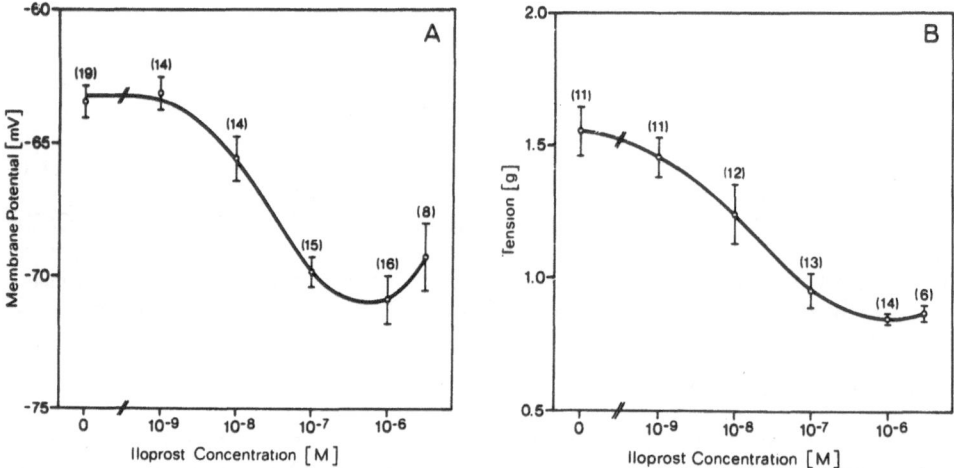

Figure 1. Membrane potential (A) and tension (B) of isolated carotid segments as a function of the iloprost concentration in the Krebs solution. The preparations were incubated for 30 min at each iloprost concentration step.

mol/l concentration. Only very high doses of cicaprost (10^{-5} mol/l) slightly attenuated the hyperpolarisatory and vasodilatory efficacy by -5% [cf. 24]. On the other hand, a cicaprost concentration of as low as 10^{-10} mol/l led to a significant 2.3 mV hyperpolarization and 0.1 g relaxation.

Vasodilatation, effected by a decrease in oxygen partial pressure or by a rise in blood flow, is due to membrane hyperpolarization. The cause of this hyperpolarization is quite different in each case, however, the end result is always elevated K^+ channel opening. The following experiments were designed to elucidate the role of the vasodilators prostacyclin (PGI_2) and endothelium-derived hyperpolarizing factor (EDHF) in oxygen deficiency. From Fig. 3 one can see how the lowering of O_2 partial pressure influences membrane potential and tension of vascular smooth muscle cells in human coronary arteries taken from heart transplant patients. There is a continuous hyperpolariza-

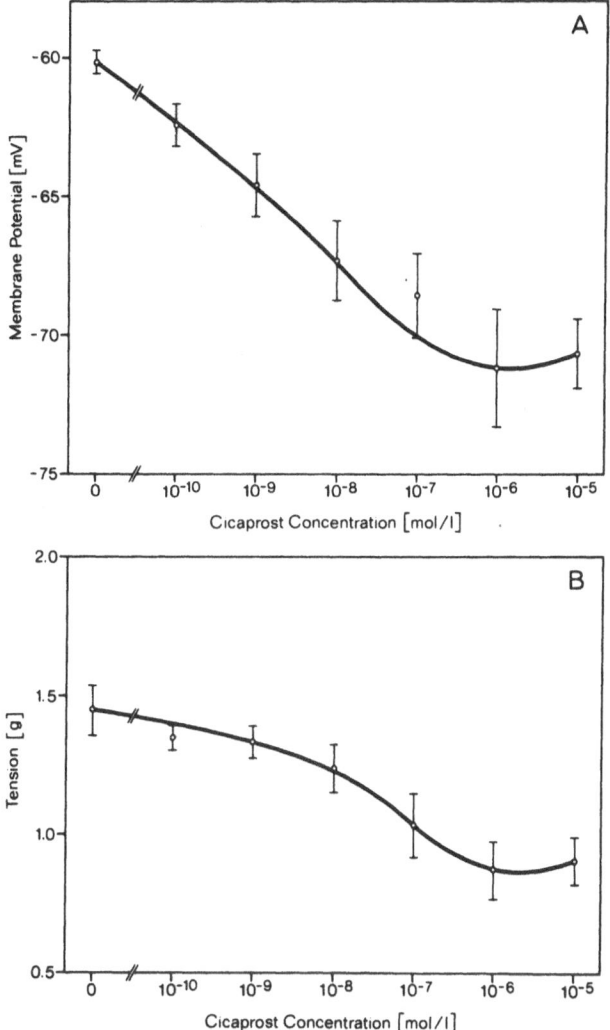

Figure 2. Membrane potential (A) and tension (B) of isolated carotid segments as a function of the cicaprost concentration in the Krebs solution. The preparations were incubated for 10 min at each cicaprost concentration step.

tion and relaxation with a reduction of P_{O_2} from 550 mm Hg to 35 mm Hg, and a depolarization and contraction between 35 and 0 mm Hg. It is striking that in the case of atherosclerosis, vascular smooth muscle cells are more depolarized and have more tone [28,32]. This is true for all O_2 tensions. The endothelial function is possibly limited in this tissue. Iloprost, in a concentration of 10^{-7} mol/l, is able to induce additional hyperpolarization and relaxation at all O_2 partial pressures, even with maximal hyperpolarization and dilatation at about 35 mm Hg P_{O_2}. In normal and in atherosclerotic blood vessels, the hypoxic depolarization and contraction with oxygen pressures below 35 mm Hg is prevented by iloprost. These results emphasize that also with an oxygen deficiency, vasodilatation can be attained by K^+ channel opening.

The compound cicletanine, a dihydropyridine derivative, was also investigated for possible K^+ channel opening properties. As cicletanine is virtually insoluble in blood substitute solutions, the influence of the solvent (DMSO) alone was contrasted to the effect of cicletanine plus solvent, illustrated in Fig. 4. While DMSO depolarized and contracted the vascular smooth muscle cells, a concentration-dependent hyperpolarization and relaxation appeared under cicletanine [24,31,32]. The hyperpolarization and relaxation elicited by this compound becomes amplified in a vessel precontrac-

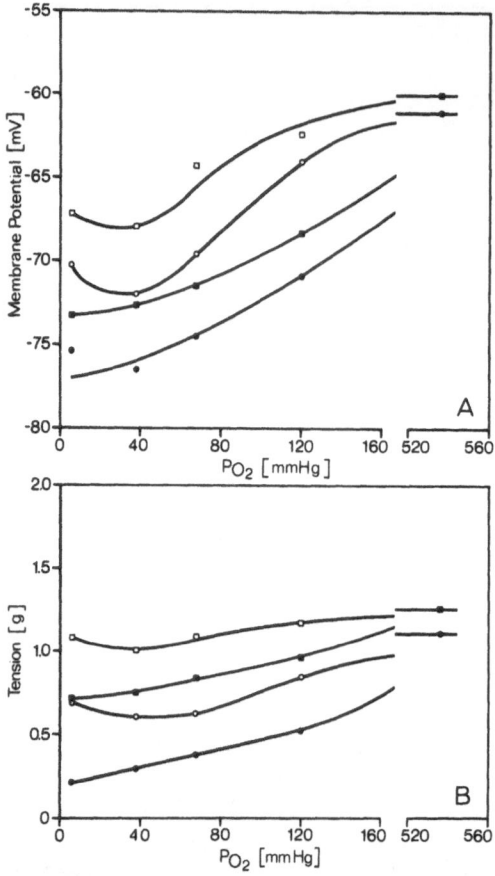

Figure 3. Membrane potential (A) and tension (B) in isolated vascular strips of human coronary arteries dependent on the oxygen partial pressure of the Krebs solution. Mean values from five experiments each. (O) Normal coronary arteries with intact endothelium, (●) plus iloprost 10^{-7} mol/l; (□) arteriosclerotic coronary arteries, (■) plus iloprost 10^{-7} mol/l.

ted with noradrenaline. Therefore, in this experimental series, the membrane of the vascular smooth muscle cells was predepolarized by noradrenaline and the vascular strips were precontracted over and above their tension (2 g stretch), before cicletanine was applied.

The net effect of cicletanine, i.e. the difference between the curves for cicletanine + DMSO-concentration vs. those for the DMSO-concentration alone, can be rated as follows. Noradrenaline depolarized the membrane from -60.7 ± 1.0 mV (n=12) to -53.3 ± 0.4 mV (n=12; P < 0.0001), while the membrane repolarized and hyperpolarized to -66.9 ± 2.0 mV (n=6; P < 0.0001) under cicletanine (10^{-3} mol/l). The force increased from 1.383 ± 0.112 g (n=12) to 3.970 ± 0.375 g (n=12; P < 0.0001) under noradrenaline. As the concentration rose, cicletanine reduced the tone step by step to 2.214 ± 0.346 g (n=5; P < 0.02). The strong hyperpolarization by ΔV = 13.6 mV suggests that cicletanine has K^+ channel opening properties. An experimental series with indomethacin should clarify whether cicletanine effects the membrane hyperpolarization and the decrease in tension observed via a release of prostacyclin. Fig. 5 shows that indomethacin did not cause any essential changes in the course of membrane potential and tone when the cicletanine concentration increased. A striking finding is, however, that a further depolarization and contraction occurred with application of the lowest DMSO and DMSO + cicletanine concentration (10^{-6} mol/l). This is an intrinsic indomethacin effect [cf. 25].

Fig. 6 illustrates the comparison of the membrane hyperpolarization and vasorelaxation between indomethacin-treated and untreated preparations. There is no significant difference in hyperpolariza-

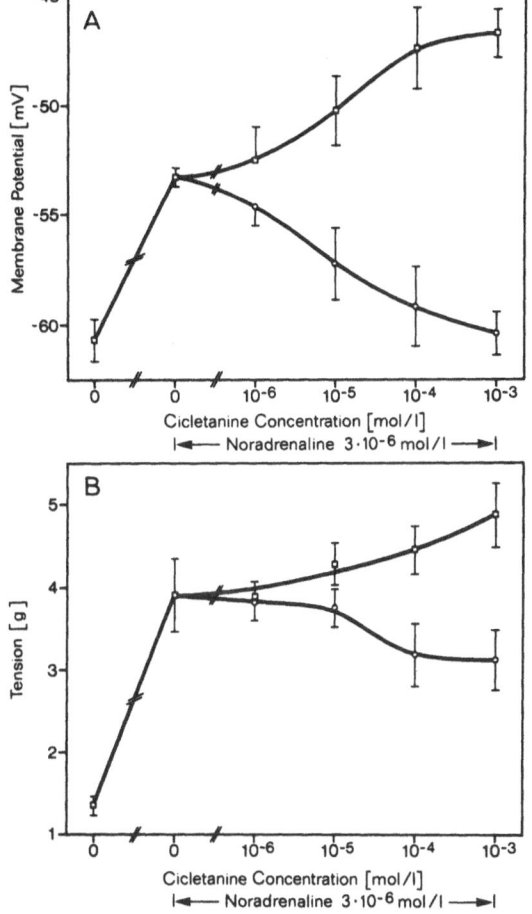

Figure 4. Membrane potential (A) and mechanical force development (B) of noradrenaline-depolarized and -contracted vascular strips of the canine carotid artery as a function of the cicletanine + DMSO-concentration (O), and of the DMSO-concentration alone (□), in the Krebs solution. The preparations (n=5) were incubated for 15 min at each concentration step.

tion and relaxation at all cicletanine concentrations. In the former case, the hyperpolarization amounted to 15.7 mV and the decrease in tone amounted to 1.992 g with 10^{-3} mol/l cicletanine, whereas a hyperpolarization by 13.6 mV and a relaxation by 1.756 g was measured without indomethacin. Cicletanine lowers blood pressure but, contrary to earlier expectations, does not stimulate prostacyclin synthesis.

Besides being an effective platelet anti-aggregating substance, aqueous garlic extract has been reported to relax a variety of vascular preparations [23,30,31,36], to reduce blood pressure [9,12], and to inhibit cholesterol biosynthesis [6]. Similarities in the mode of action with PGI_2 are indications for prostacyclin being the mediator of the formerly mentioned effects [cf. 12,13]. When aqueous garlic extract was applied at concentrations of 0.0002 to 0.2 g/l, the membrane was hyperpolarized from -59.9 $\pm$ 0.6 mV (n=6) to a maximum of -63.3 $\pm$ 1.0 mV (n=6; P < 0.02) in a concentration-dependent manner. Passing this curve consecutively from low to high garlic extract concentrations with a 15 min application time for each, the values depicted in Fig. 7A were obtained. Fig. 7B shows the effects of garlic extract on the mechanical force. In the concentration range of 0.0002 to 0.2 g/l, garlic extract caused a dose-dependent relaxation of the muscle cells. The tension decreased from 1.472 $\pm$ 0.107 g (n=5) to 1.280 $\pm$ 0.068 g (n=5; P < 0.18). The normal stress relaxation of the preparations has already been subtracted. Thus, hyperpolarization of the

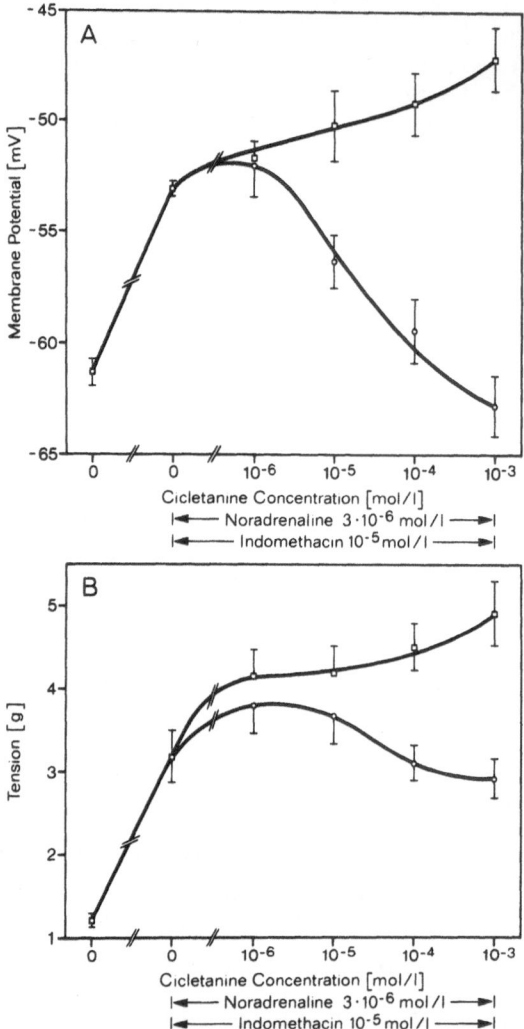

Figure 5. Membrane potential (A) and tension (B) of noradrenaline-depolarized and -contracted and indomethacin-pretreated vascular strips of the canine carotid artery as a function of the cicletanine + DMSO-concentration (O), and of the DMSO-concentration alone (□), in the Krebs solution. The preparations (n=7) were incubated for 15 min at each concentration step.

membrane and inhibition of tension in function of the garlic extract concentration run parallel. The maximum hyperpolarization of the cells amounted to 3.5 mV, the maximum relaxation to 0.192 g.

Prostacyclin, a K$^+$ Channel Activator

As has been shown, various substances lead to vascular relaxation via membrane hyperpolarization. How can the changes in membrane potential be explained? First of all the effects of iloprost. Fig. 8 shows a semilogarithmic plot of the ^{42}K$^+$ efflux with and without iloprost. The much steeper course of the K$^+$ decay under iloprost is immediately recognizable. After the exponential analysis and computation of a compartment model based on five experiments, we found an average increase of 250% in passive K$^+$ efflux with iloprost [22,26]. We calculated a rise in K$^+$ permeability of 340% causing the hyperpolarization. In spite of this enormous increase in K$^+$ permeability under iloprost, both flux curves in Fig. 8 start from the same K$^+$ concentration [32]. This is a clear

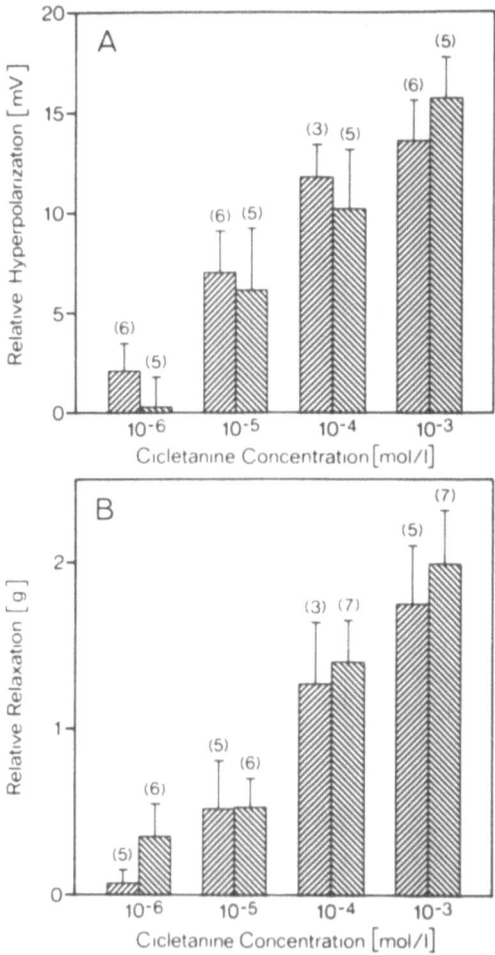

Figure 6. Relative hyperpolarization (A) and relative relaxation (B) of isolated canine carotid segments dependent on the cicletanine concentration in the Krebs solution. *Left columns:* preparations stretched by 2 g, and depolarized and contracted by noradrenaline $(3 \cdot 10^{-6}$ mol/l); *right columns:* preparations stretched by 2 g, depolarized and contracted by noradrenaline $(3 \cdot 10^{-6}$ mol/l) and pretreated by indomethacin $(10^{-5}$ mol/l). The effects of cicletanine were measured in comparison to the DMSO control series. Incubation with cicletanine for 15 min at each concentration step.

indication that the active K^+ inward transport also has to be raised in order to maintain the internal K^+ concentration. A quantification of the membrane physiological changes in vascular smooth muscle treated with iloprost is given in Table 1. There were no changes in Na^+ and K^+ equilibrium potentials, but transmembrane Na^+ net flux and permeability were modestly increased. On the contrary, K^+ net flux and permeability were strongly increased. The rise in Na^+ permeability by 40% and in K^+ permeability by 340% with iloprost demonstrates once again that, following treatment with this drug, the permeability of the cell membrane increases drastically and almost selectively for K^+ ions [22,32].

This property is again emphasized by patch-clamp experiments. When the Ca^{2+} channel current was tested by whole cell patch-clamping in the rat portal vein with Ba^{2+} ions acting as charge carriers, it was increased by merely 10% [24,26]. However, measuring the K^+ currents in either Ba^{2+} or Ca^{2+} solutions had drastic effects. In a Ba^{2+} solution, depolarizing voltage pulses produced an initial inward current followed by an outward current. Iloprost $(10^{-6}$ mol/l) strongly enhanced the outward current, thereby totally suppressing the net inward current. In a Ca^{2+}

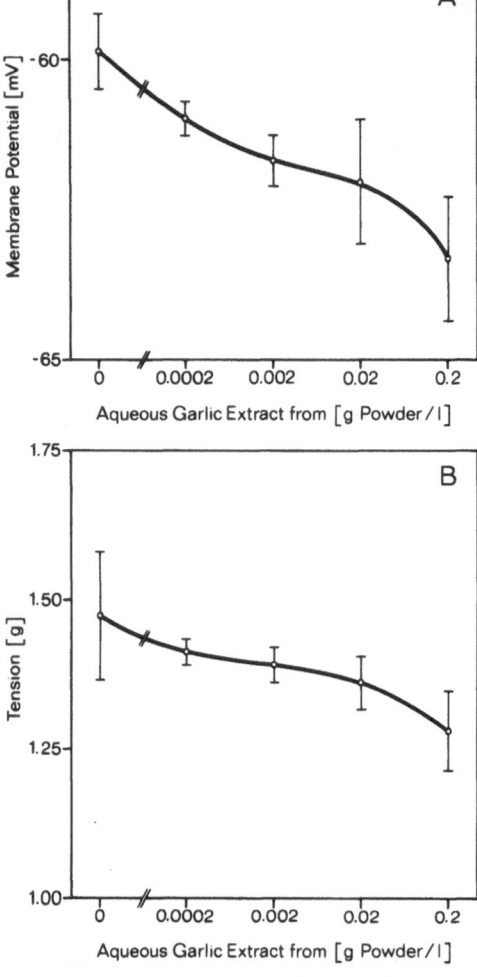

Figure 7. Membrane potential (A) and tension (B) of isolated carotid segments as a function of the aqueous garlic extract concentration in the Krebs solution. The preparations were incubated for 15 min at each garlic extract concentration step.

solution, depolarizing pulses only led to an outward current which was strongly enhanced by iloprost. Considering the relationship between voltage pulse and peak outward current, iloprost caused the latter to increase over the whole voltage range -60 mV < V < +30 mV, and by about 90% at a test potential of +30 mV (Fig. 9). According to these findings, iloprost may be characterized as a K$^+$ channel opener.

In order to arrive at a molecular description of the iloprost response, we first considered the PGI$_2$ receptor to which iloprost also binds. We assumed that the K$^+$ channel opening is induced via a G-protein. Experiments with pertussis toxin (PTx) resulted in membrane hyperpolarization and vasorelaxation with iloprost, being even more accentuated in the carotid musculature of PTx-pretreated animals than in the untreated control group (Fig. 10). This series of experiments indicates that cAMP may play a part in iloprost-induced vasorelaxation and that proteins of the G$_i$-type are also involved. After blocking the G-proteins in question with PTx, cAMP may well have increased intracellularly. This figure demonstrates that an additional hyperpolarization is combined with this supposed rise in cAMP [31,32].

The differing behaviour of the vessels treated with high external K$^+$ concentrations compared with the classical K$^+$ channel opener cromakalim implies cAMP participation. While iloprost causes

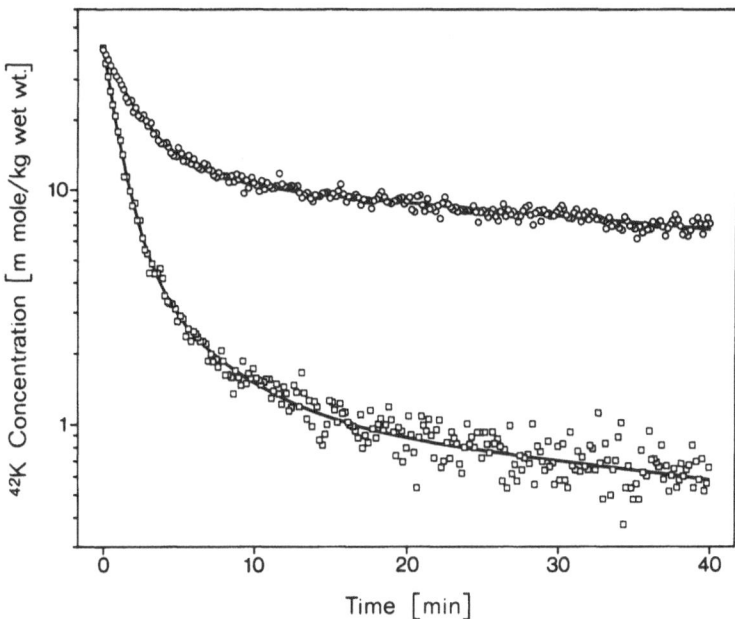

Figure 8. $^{42}K^+$ efflux of the carotid artery of the dog in Krebs solutions without (○) or with 10^{-6} mol/l iloprost (□). The graph shows the time course of the radioactive K^+ decay within the preparations at intervals of 10 s in single experiments. The superimposed lines represent the optimal double-exponential functions found by computer fitting.

Table 1. Membrane potentials (V), equilibrium potentials (E), net fluxes (ϕ) and permeabilities (P) in canine carotid vascular smooth muscle.

	Control	+Iloprost (10^{-6} mol/l)	
V	- 63.4 ± 0.6 (19)	- 70.8 ± 0.9 (16)	[mV]
E_{Na}	+35.5	+38.6	[mV]
E_K	- 86.3	- 85.3	[mV]
ϕ_{Na}	- 13.41 ± 2.15 (5)	- 20.71 ± 2.60 (5)	[pmol/cm²/s]
P_{Na}	$3.48 \cdot 10^{-8}$	$4.89 \cdot 10^{-8}$	[cm/s]
ϕ_K	9.04 ± 2.25 (5)	23.07 ± 5.34 (5)	[pmol/cm²/s]
P_K	$0.54 \cdot 10^{-6}$	$2.40 \cdot 10^{-6}$	[cm/s]

a dose-dependent relaxation of the mesenteric artery even in 70 mmol/l $[K^+]_o$, this can no longer be observed with cromakalim [20]. Fig. 11 shows that iloprost does effect an augmentation of the cAMP concentration. After 5 min incubation, the course of cAMP reflects the concentration-response curves for membrane potential and muscle tone (cf. Fig. 1).

When carrying out a more exact time analysis of the K^+ outward current in the whole cell voltage-clamp investigations, a rapid increase of the voltage-dependent K^+ current in a Ba^{2+} solution can be demonstrated (Fig. 12). However, this current together with the Ca^{2+}-activated K^+ current in a Ca^{2+} solution only rises after a delay. Apparently, cAMP plays a part in the increase of the Ca^{2+}-activated K^+ outward current. As suggested by Sadoshima et al. [19], the cAMP-dependent protein kinase seems to increase the Ca^{2+} sensitivity of the Ca^{2+}-activated K^+ channel. Obviously, one or two types of K^+ channels (perhaps the Ca^{2+}-activated K^+ channel and the delayed rectifier K^+ channel) are stimulated by phosphorylation with cA-PK [33]. The hyperpolarizing and relaxing effect of iloprost may be explained by cA-PK interventions at the Ca^{2+}-activated K^+ channel, the sarcolemmal Ca^{2+} pump, and the $I_{Ca(s)}$. In particular, cAMP-induced hyperpolari-

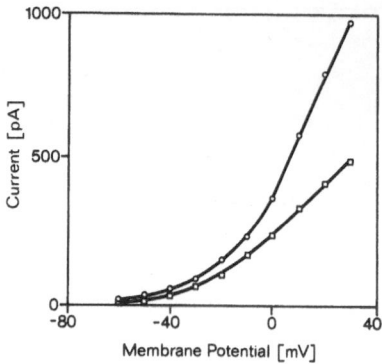

Figure 9. Current-voltage relationships established for the peak outward K^+ current, obtained in 5 mmol/l Ca^{2+} solution. The pipette solution contained KCl. Reference solution (□) and after the addition of 1 μmol/l iloprost for 7 min (O). Data shown are from a typical experiment. Similar estimates were obtained from four separate experiments.

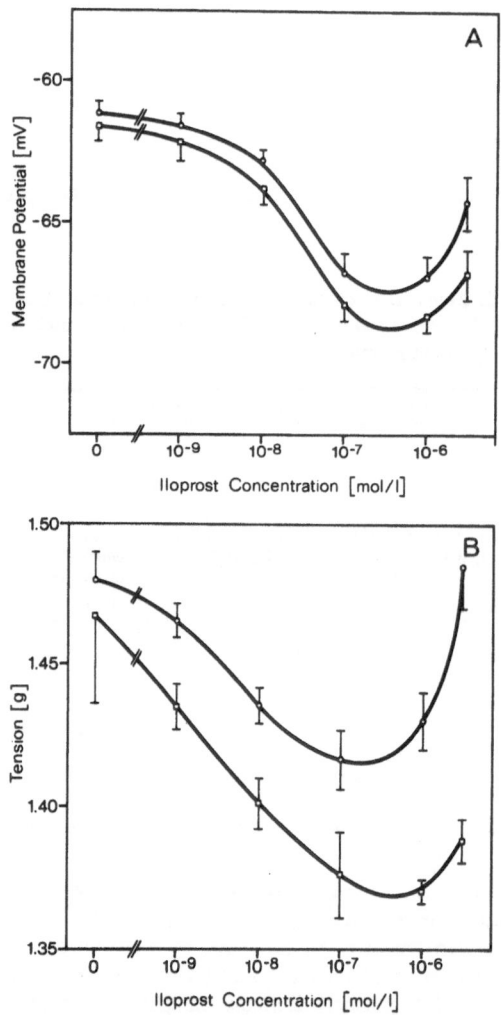

Figure 10. Membrane potential (A) and tone (B) in vascular strips of the carotid artery of untreated (O) and pertussis toxin-pretreated rabbits (□) as a function of the iloprost concentration in the Krebs solution. The initial pre-tension of the preparations was 2 g. Application of iloprost for 10 min at each concentration step.

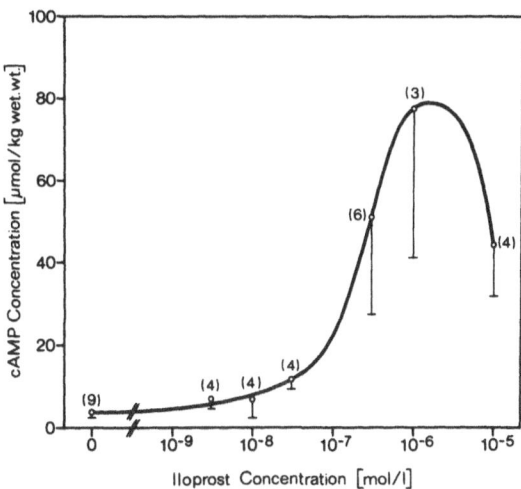

Figure 11. Concentrations of cAMP in vascular segments of the rabbit mesenteric artery in response to the iloprost concentration in the Krebs solution. Incubation with iloprost for 5 min at each concentration step.

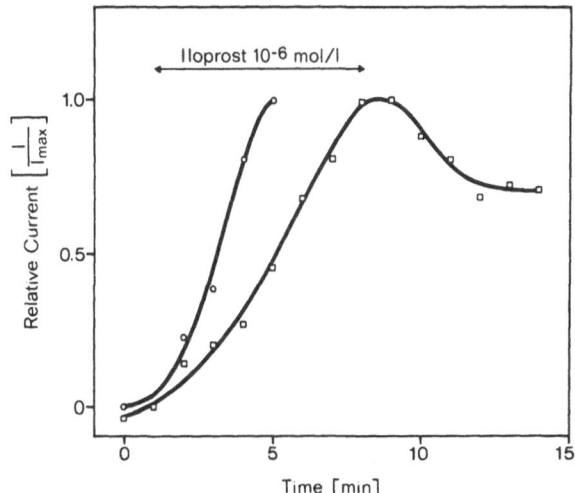

Figure 12. Time analysis of the K^+ outward current I/I_{max} under iloprost (10^{-6} mol/l) in patch-clamp investigations. (O) Ba^{2+} solution; (□) Ca^{2+} solution.

zation due to an increase in g_K cannot be distinguished from a hyperpolarization possibly elicited by a more direct effect of the drug on K^+ channels.

Fig. 13 shows the stationary activation curve 'tension vs. membrane potential' of the canine carotid artery smooth muscle. A number of different effector parameters are at the root of this curve [29]. Since in the investigations mentioned, the effector influences on membrane potential and vascular tone had been measured simultaneously, the elimination of the former parameters from the dose-response curves was possible [21]. Plotting the developed tension as a function of the membrane potential at each identical effector concentration produced the depolarized and hyperpolarized portions of the stationary activation curve. The point marked on the curve represents the membrane potential and tension under normal conditions. The range of effectivity of a hyperpolarization, caused by so-called K^+ channel opening, is in the lower linear and the nonlinear, hyperpolarized section of the curve. From the sigmoid, stationary activation curve and the values

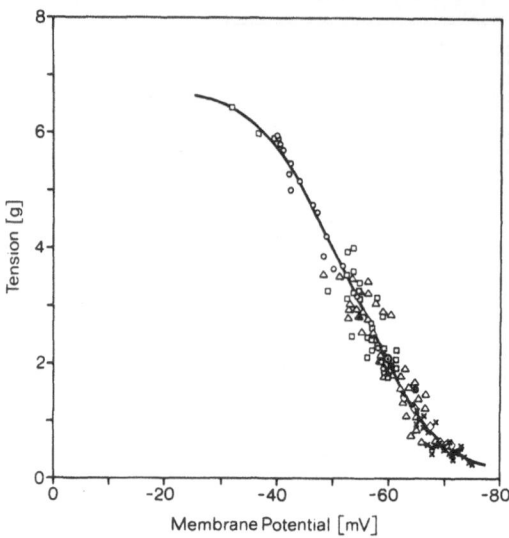

Figure 13. Dependency of mechanical tension on the membrane potential in isolated carotid arteries (stationary activation curve). The membrane potential was changed by varying the extracellular concentration of H^+ ($\triangle$), K^+ ($\square$), Ca^{2+} ($\bullet$), norepinephrine (O) or prostacyclin ($\Diamond$) or by lowering the oxygen tension ($\times$). The point marked on the curve (O) indicates membrane potential and mechanical force under control conditions.

measured, it is clear that the same change in tone occurs with a defined change in membrane potential, no matter from which effector influence the latter originates [21,29,31]. Thus, this curve provides strong evidence for the existence of electromechanical coupling in vascular smooth muscle, at least in the carotid artery.

DISCUSSION

A membrane hyperpolarization of vascular smooth muscle cells is generally a prerequisite for vasodilatation under physiological conditions. Such a hyperpolarization can simply be brought about through an increase in the open probability of K^+ channels. A variety of K^+ channels has also been described for smooth muscle which differ through their activability and single channel conductance. Different K^+ channel types, with their specific biophysical properties and pharmacological sensitivities, come about because of genetically coded channel proteins with slightly different amino-acid sequences. G-proteins, Ca^{2+}, and a number of kinases participate in their regulation. The clue to the initiation of vasorelaxation is a tight electromechanical coupling [21]. The strong voltage-sensitivity of the Ca^{2+} channel open probability provides an explanation for this [16].

Various compounds can produce a more or less pronounced membrane hyperpolarization. How much the smooth muscle cells relax depends on, among other things, their membrane potential before the intervention. A fall in the O_2 partial pressure [7,25,26,28,29,31,32], the application of iloprost [22,26,31,32] or cicaprost, cicletanine [24,31,32] or aqueous garlic extract [23,30,31] are only a selection from the many possibilities which lead to hyperpolarization.

The reasons for the hyperpolarization with prostacyclin (iloprost, cicaprost) and with a reduction in O_2 partial pressure are complicated. Iloprost induced a drastic increase in both the voltage-dependent and Ca^{2+}-activated K^+ outward current over the whole potential range in patch-clamp investigations. The rise in Na^+ permeability by 40%, and in K^+ permeability by 340%, as revealed by tracer studies, demonstrates once again that following treatment with this drug, the permeability of the cell membrane increases drastically and almost selectively for K^+ ions. The slight increase in Na^+ and Ca^{2+} permeability, as found in the flux and voltage-clamp studies, may signal the opening of an unspecific channel for small cations. Since this rise in permeabilities occurs to a noticeable extent only with high iloprost concentrations, it may explain the reversal of membrane hyperpolarization and vasorelaxation found in the same concentration range. This reduction in hyperpolarization and relaxation was blunted with cicaprost and only seen at very high concentrations (10^{-5} mol/l). Obviously, this compound does not have the prostaglandin E_1-like side effects of iloprost [24,32]. Finally, the studies with PTx, the studies of the time course of the increasing K^+ outward current, and the determination of cAMP resulted in the following effector chain of PGI_2. The hyperpolarizing and relaxing effect of prostacyclin analogs may be explained by cA-PK interventions at the Ca^{2+}-activated K^+ channel, the sarcolemmal Ca^{2+} pump and the $I_{Ca(s)}$. In principle, the effect of endothelium-derived hyperpolarizing factor (EDHF) in O_2 deficiency seems to be the same, but its effector chain proceeds via cG-PK [33]. This is in contrast to cardiac muscle. We presume that an ATP-sensitive channel, especially with respect to the effect of prostacyclin is not responsible for the K^+ channel opening resulting in hypoxic hyperpolarization, but the Ca^{2+}-activated K^+ channel.

Comparing the effect of hypoxia on membrane potential and tension of normal and arteriosclerotic human coronary arteries which were taken from heart transplant patients reveals a distinct difference. During oxygen deficiency, control preparations showed a maximal hyperpolarization of 10.7 mV and a maximal relaxation of 0.466 g. Arteriosclerotic arteries, however, became hyperpolarized by merely 6.6 mV and relaxed by 0.258 g. The relation was quite similar for a carbogen Krebs solution (resting, control conditions). In comparison to normal coronary arteries (V = -64.9 mV; T = 0.757 g) arteriosclerotic vessels were depolarized (V = -61.4 mV) and more contracted (T = 1.142 g) at a physiological P_{O_2} of 95 mm Hg. Thus, it may be concluded that in arteriosclerotic blood vessels synthesis and release of vasodilators are diminished. The beneficial effect of a compensation of the PGI_2 lacking was demonstrated by the application of iloprost (10^{-7} mol/l). This prostacyclin analog 'transformed' arteriosclerotic coronaries into normal ones as far as their membrane potential and tone response to a reduction in O_2 partial pressure is concerned. Above all, membrane depolarization and vasoconstriction failed to appear with O_2 pressures below 35 mm Hg even in arteriosclerotic vessels. This is a very important result because such low P_{O_2} values were found during coronary angiography in the large coronary blood vessels of man with extensive arteriosclerosis. Therefore, even lower values may be expected in peripheral arteries and arterioles.

Cicletanine and aqueous garlic extract led to vascular relaxation via membrane hyperpolarization as did iloprost (cicaprost). Assuming that patients have a plasma level of $3.1 \cdot 10^{-5}$ mol/l cicletanine for some time, and 90% of this compound is bound to proteins [4], the effective concentration is $3.1 \cdot 10^{-6}$ mol/l. Under sympathetic tone, this cicletanine concentration hyperpolarized the smooth muscle cell membrane by 4.2 mV; the relaxation of the vascular strip amounted to 0.238 g. This effect can be explained by closure of L-type Ca^{2+} channels with hyperpolarization. We observed a similar behaviour under iloprost [22,26]. These experiments support the idea of including cicletanine in the group of K^+ channel openers. Nevertheless, only detailed binding studies and flux experiments can validate whether K^+ channel opening or Ca^{2+} antagonistic properties exist. In our preparations, prostacyclin did not participate in the effect of cicletanine since unchanged dose-response curves were recorded under indomethacin. We could not confirm the participation of the eicosanoid system in the mechanism of the hypotensive action of cicletanine [3].

Dispersing the average allicin content of a garlic clove (9.3 mg) throughout the extracellular space in man (3.4 μmol/l), a 0.2 g/l garlic extract concentration reflects the allicin concentration in that space. This calculation is based on HPLC measurements of the allicin (3.58 mmol/l) and ajoene (0.184 mmol/l) concentrations in a 200 g/l garlic extract [31]. Further, applying a single therapeutic dose (e.g. 200 mg garlic powder) would correspond to an even fifteen times lower concentration of garlic extract, namely 0.012 g/l. Therefore, we recorded the effect of an aqueous garlic extract prepared from 0.0002 to 0.2 g powder/l on membrane potential and mechanical force development (Fig. 7). A dose-dependent membrane hyperpolarization and vasorelaxation were observed like in the high dose range [cf. 30,31]. Again, assuming that 0.01 - 0.2 g/l is a realistic dose range for man, our studies have proven a relevant hyperpolarization and inhibition of tone. As for nicorandil [14] and iloprost [31], the hyperpolarizing effect of 3.5 mV of aqueous garlic extract, could also be attributed to an increase of the K^+ permeability in the smooth muscle cells of the carotid artery. This effect is not surprising because essential constituents of garlic oil can be seen as being structurally related to arachidonic acid derivatives and behave in a similar fashion as these to membrane lipids. In future experiments it has to be proven whether prostacyclin participates in the effects of aqueous garlic extract.

How the hyperpolarization effects relaxation of the blood vessel is explained by the activation curve [21,29,31]. Every hyperpolarization leading to vasodilatation closes L- and/or T-type Ca^{2+} channels after all. The deciding factor is that a certain Ca^{2+} channel open probability exists with a resting potential of about -60 mV in the vascular smooth muscle cells [5,16]. In the linear range of the activation curve, a potential change of 5.1 mV corresponds to an alteration in tone of 1 g. This linear part comprises about 20 mV and thus 70% of the force developed. The activation curve can be shifted along the potential axis, for example under the influence of cyclic nucleotides or IP_3, or steepened in its degree of coupling [1,10,18]. A classical example is the effect of noradrenaline, which releases Ca^{2+} ions from intracellular stores, if present, and finally augments the affinity to the contractile filaments. On no account does this involve an abolition of electromechanical coupling. The force development is still voltage-dependent; only the system has switched to a shifted activation curve. The theory is confirmed by earlier observations in which a voltage-dependence of ROC-linked processes has been described [1]. Finally, the tight electromechanical coupling expressed by the stationary activation curve is supported through measurements in the vascular smooth muscle by Hermsmeyer [11], demonstrating a very tight electromechanical coupling, and through investigations of the strong voltage dependency of the Ca^{2+} channel found by Nelson et al. [16], who observed a reduction in tension of 50 % with a 2 mV hyperpolarization. Our coupling curve establishes a 50% vasorelaxation for 2.5 mV hyperpolarization.

SUMMARY

1. Numerous compounds shift the membrane potential of vascular smooth muscle to more negative values. The consequence is a vasodilatation because Ca^{2+} channels are closed. K^+ channel opening frequently causes the hyperpolarization.

2. Prostacyclin has a 20-30% share, and EDHF a 70-80% share, in hypoxic vasodilatation. Experiments with iloprost and cicaprost (PGI_2 analogs) confirmed the K^+ channel opening properties of these drugs. A voltage-dependent K^+ channel and a Ca^{2+}-activated K^+ channel, via the influence of cA-PK or cG-PK, are responsible for the hyperpolarization with iloprost/cicaprost and with oxygen deficiency.

3. Cicletanine and garlic extract cause a concentration-dependent membrane hyperpolarization and are potent vasodilators. A cicletanine concentration, which is attained by the dosage given to patients, is sufficient to produce these effects. Aqueous garlic extract exerts a hyperpolarizing and vasodilating influence even in a concentration which may occur in the extracellular space by the administration of a single garlic clove or a single therapeutic dose.

4. The stationary activation curve 'developed force vs. membrane potential' satisfactorily explains the effects of K^+ channel openers. The tight electromechanical coupling expressed by this curve comprises a 50% vasorelaxation for a 2.5 mV hyperpolarization. In the linear part of the curve, the coupling ratio is 5.1 mV/g.

5. In the vascular smooth muscle, vasorelaxation can be evoked by membrane hyperpolarization which is linked to a simultaneous increase in K^+ outward current and $^{42}K^+$ efflux. In the case of substances whose influence is solely or partially receptor-mediated, cyclic nucleotides may be involved in vasorelaxation. Since cyclic nucleotides also hyperpolarize through an increase in K^+ conductance, the resulting dilatation often cannot be divided into its single components. Therefore, it is sensible not to give the term "K^+ channel opener" too fine a definition. The term should be applied to all substances and changes in physical states which predominantly increase the open probability of K^+ channels finally via a conformational change in the cell membrane. Which K^+ channel and which single channel conductance is concerned in a particular case, and which 'mediator' may participate, become secondary questions.

ACKNOWLEDGEMENTS

The authors thank Mrs. I. Krukenberg for her excellent technical assistance and Mr. H. Ewald from the mechanical workshop for his constant help. We are grateful to Mrs. M. Krawczynski for her outstanding work in preparing the illustrations and Mr. P. Holzner for the photographs. We are indebted to Mrs. A. Scheuermann for the translation and editorial elaboration of the manuscript.

REFERENCES

1. Bolton, T. B. Mechanisms of action of transmitters and other substances on smooth muscle. *Physiol. Rev.* 59: 606-718, 1979.
2. Braquet, P., P. Guinot, and T. Tarrade. Cicletanine: biology, pharmacology and clinical sciences. *Drugs Exp. Clin. Res.* 14: 71-230, 1988.
3. Calder, J. A., M. Schachter, and P. S. Sever. Acute relaxant effect of cicletanine in human subcutaneous resistance arteries. *Blood Vessels* 28: 279, 1991.
4. Fredj, G. Clinical pharmacokinetics of cicletanine hydrochloride. *Drugs Exp. Clin. Res.* 14: 181-188, 1988.
5. Ganitkevich, V. Ya., and G. Isenberg. Contribution of two types of calcium channels to membrane conductance of single myocytes from guinea-pig coronary artery. *J. Physiol. (Lond.)* 426: 19-42, 1990.
6. Gebhardt, R. Multiple Wirkungen von Knoblauchextrakten auf die Cholesterin-Biosynthese. *Med. Welt* 7a: 12-13, 1991.
7. Grote, J., G. Siegel, K. Zimmer, and A. Adler. The influence of oxygen tension on membrane potential and tone of canine carotid artery smooth muscle. *Adv. Exp. Med. Biol.* 222: 481-487, 1988.
8. Haeusler, G., J.-E. de Peyer, and G. Schultz. Vascular effects of α_1- and α_2-adrenoceptor agonists in vitro and in hypertensive rats. *J. Cardiovasc. Pharmacol.* 10, Suppl. 4: 15-18, 1987.
9. Harenberg, J., C. Giese, and R. Zimmermann. Effect of dried garlic on blood coagulation, fibrinolysis, platelet aggregation and serum cholesterol levels in patients with hyperlipoproteinemia. *Atherosclerosis* 74: 247-249, 1988.
10. Hashimoto, T., M. Hirata, T. Itoh, Y. Kanmura, and H. Kuriyama. Inositol 1,4,5-trisphosphate activates pharmacomechanical coupling in smooth muscle of the rabbit mesenteric artery. *J. Physiol. (Lond.)* 370: 605-618, 1986.
11. Hermsmeyer, K. High shortening velocity of isolated single arterial muscle cells. *Experientia* 35: 1599-1602, 1979.
12. Jacob, R., M. Ehrsam, T. Ohkubo, and H. Rupp. Antihypertensive und kardioprotektive Effekte von Knoblauchpulver (Allium sativum). *Med. Welt* 7a: 39-41, 1991.
13. Jung, F., H. Kiesewetter, G. Pindur, E. M. Jung, C. Mrowietz, and E. Wenzel. Thrombozytenfunktionshemmende Wirkung von Knoblauch. *Med. Welt* 7a: 18-19, 1991.
14. Kajiwara, M., G. Droogmans, and R. Casteels. Effects of 2-nicotinamidoethyl nitrate (nicorandil) on excitation-contraction coupling in the smooth muscle cells of rabbit ear artery. *J. Pharmacol. Exp. Ther.* 230: 462-468, 1984.
15. Loirand, G., P. Pacaud, C. Mironneau, and J. Mironneau. Evidence for two distinct calcium channels in rat vascular smooth muscle cells in short-term primary culture. *Pflügers Arch.* 407: 566-568, 1986.
16. Nelson, M. T., J. B. Patlak, J. F. Worley, and N. B. Standen. Calcium channels, potassium channels, and voltage dependence of arterial smooth muscle tone. *Am. J. Physiol.* 259: C3-C18, 1990.
17. Northover, B. J. The membrane potential of vascular endothelial cells. *Adv. Microcirc.* 9: 135-160, 1980.
18. Rüegg, J. C. *Calcium in Muscle Activation.* Berlin, Heidelberg, New York, London, Paris, Tokyo: Springer-Verlag, 1986.
19. Sadoshima, J.-I., N. Akaike, H. Kanaide, and M. Nakamura. Cyclic AMP modulates Ca-activated K channel in cultured smooth muscle cells of rat aortas. *Am. J. Physiol.* 255: H754-H759, 1988.
20. Schröder, G., and G. Graichen. Comparison of the vasorelaxing effect of iloprost (PGI$_2$ analogue) and cromakalim (K-channel activator) in isolated vascular preparations of the rabbit. *Clin. Pharmacol.* 7: 97-105, 1990.
21. Siegel, G. Membranphysiologische Grundlagen der peripheren Gefäßregulation. *Physiol. akt.* 1: 31-52, 1986.
22. Siegel, G., A. Carl, A. Adler, and G. Stock. Effect of the prostacyclin analogue iloprost on K^+ permeability in the smooth muscle cells of the canine carotid artery. *Eicosanoids* 2: 213-222, 1989.
23. Siegel, G., J. Emden, F. Schnalke, A. Walter, K. Rückborn, and K. G. Wagner. Wirkungen von Knoblauch auf die Gefäßregulation. *Med. Welt* 7a: 32-34, 1991.

24. Siegel, G., J. Emden, K. Wenzel, J. Mironneau, and G. Stock. Potassium channel activation in vascular smooth muscle. *Adv. Exp. Med. Biol.* 311: 53-72, 1992.
25. Siegel, G., J. Grote, F. Schnalke, and K. Zimmer. The significance of the endothelium for hypoxic vasodilatation. *Z. Kardiol.* 78, *Suppl.* 6: 124-131, 1989.
26. Siegel, G., J. Mironneau, F. Schnalke, G. Schröder, B.-G. Schulz, and J. Grote. Vasodilatation evoked by K^+ channel opening. *Prog. Clin. Biol. Res.* 327: 299-306, 1990.
27. Siegel, G., H. Roedel, J. Nolte, H. W. Hofer, and O. Bertsche. Ionic composition and ion exchange in vascular smooth muscle. In: *Physiology of Smooth Muscle*, edited by E. Bülbring, M. F. Shuba. New York: Raven Press, 1976, p. 19-39.
28. Siegel, G., F. Schnalke, J. Schaarschmidt, J. Müller, and R. Hetzer. Hypoxia and vascular muscle tone in normal and arteriosclerotic human coronary arteries. *J. Vasc. Med. Biol.* 3: 140-149, 1991.
29. Siegel, G., A. Walter, M. Bostanjoglo, A. W. H. Jans, R. Kinne, L. Piculell, and B. Lindman. Ion transport and cation-polyanion interactions in vascular biomembranes. *J. Membrane Sci.* 41: 353-375, 1989.
30. Siegel, G., A. Walter, F. Schnalke, K. Rückborn, J. Emden, and K. G. Wagner. Knoblauch und Senkung des Gefäßtonus. *Vasomed* 4: 8-12, 1992.
31. Siegel, G., A. Walter, F. Schnalke, A. Schmidt, E. Buddecke, G. Loirand, and G. Stock. Potassium channel activation, hyperpolarization, and vascular relaxation. *Z. Kardiol.* 80, *Suppl.* 7: 9-24, 1991.
32. Siegel, G., K. Wenzel, F. Schnalke, J. Mironneau, G. Schultz, G. Schröder, E. Schillinger, O. Grauhan, and R. Hetzer. Prostacyclin analogues as K^+ channel openers. *Clin. Pharmacol.* 7: 72-96, 1990.
33. Sperelakis, N., and Y. Ohya. Cyclic nucleotide regulation of Ca^{2+} slow channels and neurotransmitter release in vascular muscle. *Prog. Clin. Biol. Res.* 327: 277-298, 1990.
34. Weston, A. H. The pharmacology of smooth muscle potassium channels. *Clin. Pharmacol.* 7: 1-18, 1990.
35. Weston, A. H., and A. Abbott. New class of antihypertensive acts by opening K^+ channels. *Trends Pharmacol. Sci.* 8: 283-284, 1987.
36. Wolf, S., M. Reim, and F. Jung. Effect of garlic on conjunctival vessels: a randomised, placebo-controlled, double-blind trial. *Br. J. Clin. Practice* 44, *Suppl.* 69: 36-39, 1990.

POTASSIUM CHANNEL MODULATED BY HYPOXIA AND THE REDOX
STATUS IN GLOMUS CELLS OF THE CAROTID BODY

Alberto R. Benot, María D. Ganfornina, and José López-Barneo

Departamento de Fisiología y Biofísica
Facultad de Medicina, Universidad de Sevilla
Sevilla, E-41009
Spain

INTRODUCTION

The mammalian carotid bodies are arterial chemoreceptors that mediate the reflex hyperventilation observed in response to physiological and pathophysiological conditions presenting a decrease in the oxygen tension (PO_2) of the blood. Although the chemosensory function of these organs were already known almost a century ago (3, 11), the mechanism involved in the transduction of the hypoxic stimulus has remained elusive. In the last decade there has been a general consensus that glomus, or type I, cells (the most numerous in the carotid body) are the key elements in chemotransduction. These cells secrete several transmitters in response to membrane depolarization and hypoxia and establish morphologically well-defined synapses with afferent endings of the sinus nerve which convey the sensory information to the central nervous system (6, 7).

Direct proofs for the chemoreceptive properties of glomus cells have come, however, from recent electrophysiological experiments. Research in a number of laboratories has shown that these cells, which are of neuroectodermal origin, have voltage-dependent Na^+, Ca^{2+} and K^+ currents and that, as other electrically excitable cells, they can fire action potentials repetitively (5, 19). Furthermore, it has been also demonstrated that the K^+ current is selectively and reversibly attenuated by low PO_2 (4, 13, 18). In the adult rabbit, hypoxia influences the macroscopic K^+ current by inhibiting the activity of a specific O_2-sensitive K^+ channel, referred to as KO_2 channel (8, 9). These findings have provided a framework for understanding the basic mechanisms underlying sensory transduction in the carotid body: inhibition of the O_2-sensitive K^+ channels under hypoxic conditions leads to Ca^{2+} influx, enhanced transmitter release, and activation of the afferent fibers of the sinus nerve. O_2-sensitive K^+ channels are surely involved in other physiological responses to hypoxia (such as pulmonary vasoconstriction, 16) and, thus, they may have a broad physiological significance. Here we describe the major characteristics of the KO_2 channel and of its modulation by changes in PO_2. We also illustrate that the gating of this K^+ channel can be influenced by sulfhydryl-preserving antioxidative agents, such as glutathione (GSH) and dithiothreitol (DTT), which suggests that the redox state of thiol groups may participate in the regulation of the ion channels by O_2 tension.

Ion Flux in Pulmonary Vascular Control, Edited by
E.K. Weir *et al.*, Plenum Press, New York, 1993

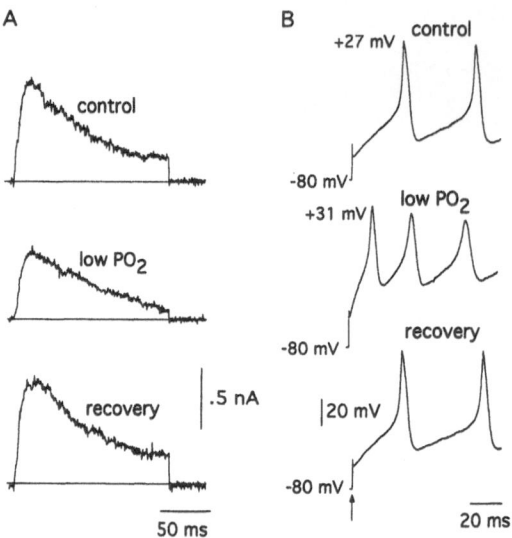

Figure 1. Changes in the electrical properties of glomus cells induced by modifications in O_2 tension. A. Whole-cell currents recorded during step depolarizations to 0 mV from a holding potential of -80 mV in a normoxic external solution (control and recovery, PO_2= 150 mmHg) and during exposure to hypoxia (low PO_2 ~10 mmHg). Solutions composition were, in mM, : external (140 NaCl, 2.7 KCl, 2.5 CaCl$_2$, 2 MgCl$_2$, 10 Hepes, 5 glucose, 1 μM tetrodotoxin, pH=7.3); internal (30 KCl, 80 K-glutamate, 20 KF, 10 Hepes, 10 EGTA, 2.5 MgCl$_2$, pH= 7.2). B. Action potential firing in a glomus cell during the first 100 ms after switching from voltage- to current-clamp in normoxic and hypoxic solutions. Holding potential was -80 mV and the instant of switching to current-clamp is indicated by the vertical arrow. Overshoot potentials in control and low PO_2 traces are also indicated. Solutions as in A, but the internal solution contained 3 mM Mg-ATP and 0.1 μM free Ca^{2+}. Temperature in all experiments was 22 to 25 °C.

POTASSIUM CURRENT INHIBITION BY HYPOXIA ENHANCES THE EXCITABILITY OF GLOMUS CELLS

After blockade of other voltage-gated ionic conductances, glomus cells dispersed from rabbit carotid bodies exhibit a typical macroscopic K$^+$ current that has an activation threshold at ~ -40 to -50 mV and inactivates almost completely in the course of 200-300 ms (19). The amplitude of this current is reversibly attenuated during exposure to solutions with low PO_2 (13). This phenomenon is illustrated in Fig. 1A with representative sweeps obtained from a cell bathed in normoxic (control and recovery, PO_2 = 150 mmHg) and hypoxic (PO_2 ~10 mmHg) solutions of identical ionic composition. In cells subjected to this low PO_2 value, the average reduction of the peak K$^+$ current is of approximately 40 %. The modulation of the K$^+$ current by the PO_2 levels is selective (with similar treatment the voltage-dependent Na$^+$ and Ca^{2+} currents remain unaltered) and can be elicited without attenuation several times in a given cell. It occurs roughly with the time course of bath exchange and is directly related to the decrease of PO_2 in the range between 150 and 50 mmHg (13, 14). In glomus cells subjected to whole-cell recording, switching from voltage- to current-clamp produces a sudden depolarization and the repetitive firing of action potentials (Fig. 1B). Following this protocol, hypoxia elicits a reversible increase in the depolarization rate of pacemaker potentials, which shortens the interspike intervals and leads to an augmentation of the cell's firing frequency (compare the low PO_2 trace with the control and recovery traces in Fig. 1B). Thus, the functional result of the attenuation of the repolarizing K^+ conductance induced by low PO_2 is an increase in the electrical excitability of glomus cells; changes in ambient O_2 tension are translated into a membrane electrical response which is the signal that encodes the output information sent to the central nervous system.

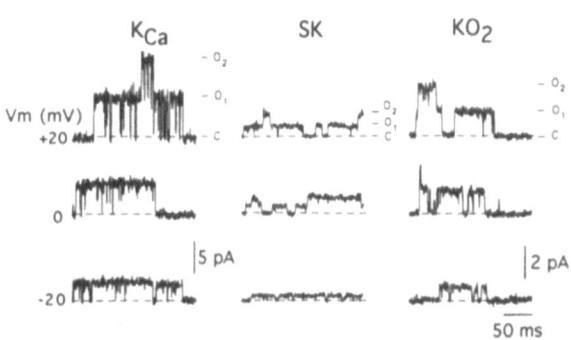

Figure 2. Potassium channel types recorded from enzimatically dispersed glomus cells. Data obtained during depolarizations at the indicated membrane potentials in three different inside-out excised patches with two channels each. K_{Ca}, Ca^{2+}-activated K^+ channels. SK, small conductance channels. KO_2, oxygen-regulated channels. Upward deflections from the zero current level (c) indicate outward current. Solutions as in Fig. 1A. K_{Ca} channels were recorded with an internal solution containing 1 µM free Ca^{2+}. Modified from ref. 9.

CHEMOTRANSDUCTION IS MEDIATED BY OXYGEN-SENSITIVE POTASSIUM CHANNELS

As in other excitable cells (12, 15), the macroscopic K^+ current of glomus cells reflects the activity of different classes of K^+ channels. We have recently characterized in excised membrane patches from rabbit cells three clearly distinguishable K^+ channel types differing in their voltage- and Ca^{2+}-dependence, single-channel conductance value, and O_2-sensitivity (8, 9). To facilitate comparison, single-channel events representative of the various channel types are illustrated in Fig. 2. The records are from patches where, by chance, the different channel classes were observed in isolation, but in most patches the various channels were found intermingled. In the three experiments there were two functional channels expressed in the patch and open probability was favored by membrane depolarization. However, in each case channel openings appeared as steps of clearly different amplitude. The abbreviation K_{Ca} denotes the large Ca^{2+}-activated K^+ channels; SK refers to channels of small conductance; and KO_2 stands for a channel population whose open probability is selectively regulated by ambient PO_2 (see below). With solutions containing asymmetrical K^+ concentrations (130 mM K^+ at the internal face of the membrane and 2.7 mM K^+ in the external milieu), the average values of the single-channel conductance of each channel species are 84, 6, and 20 pS respectively (9).

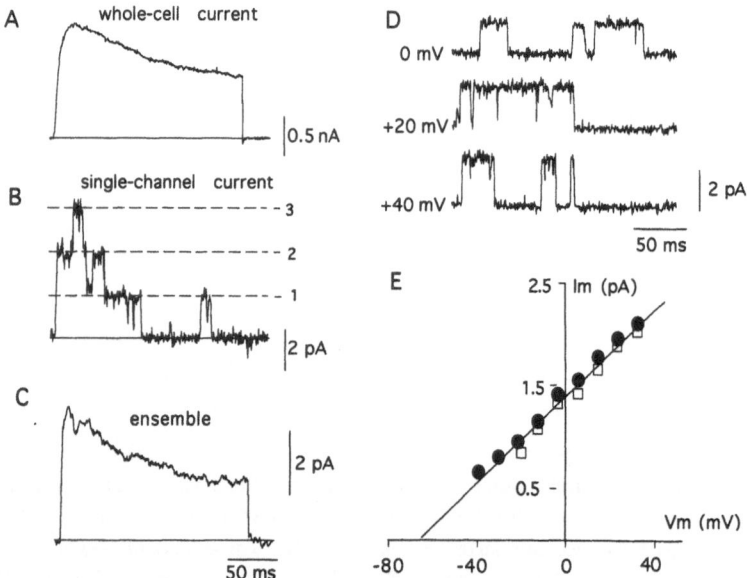

Figure 3. Time course of the macroscopic K^+ current and the O_2-sensitive channels of glomus cells. A. Whole-cell current elicited during a 200-ms pulse to +20 mV. B. Single-channel currents recorded with the same protocol after excision of an outside-out membrane patch that had at most three active channels. C. Average current of 23 consecutive sweeps recorded from the same patch with similar experimental protocol. D. Single-channel currents elicited during 200-ms step depolarization from -80 mV to the indicate membrane potentials in a different outside-out patch with at most one open channel. E. Single-channel current amplitude (Im) versus membrane potential (Vm). Filled symbols are average values from 33 patches recorded with normoxic solutions (PO_2 = 150 mmHg). Open symbols are average values from 15 patches exposed to hypoxia ($PO_2 \sim$ 5-10 mmHg). Solutions in all experiments as in Fig. 1A. Modified from ref. 8.

Although K_{Ca} and SK channels can be easily found in patches excised from glomus cells, the most numerous are the KO_2 channels, which are also the major contributors to the macroscopic O_2-sensitive K^+ current. The general properties of the KO_2 channels and their relationship with the macrosocpic K^+ current are summarized in Fig. 3. The typical time course of the whole-cell current, recorded during a depolarization to +20 mV, is shown in panel A. Panel B illustrates the single-channel events (steps of ~1.8 pA) generated by the same pulse protocol after excision of the membrane patch.The three channels of the patch open preferentially at the beginning of the pulse and progressively enter an inactivated state. The ensemble at the bottom (C), averaged from 37 consecutive single-channel current sweeps, makes clearly evident channel inactivation with a time course almost identical to the kinetics of the macroscopic current. Unitary currents recorded from a patch with at most one open K+ channel are shown in Fig. 3D, which also illustrates the increase in single-channel amplitude with membrane depolarization as the electrochemical driving force for K+ ions increases. Single-channel current amplitude (I_m) versus membrane potential (V_m) is plotted in Fig. 3E with average data obtained from patches exposed to normoxic (filled symbols) and hypoxic (open symbols) solutions. In these ionic conditions (2.7 mM external K+ and 130 mM internal K+), the current-voltage curve is almost linear in the range between ±40 mV and independent of O_2 tension. The relationship between the size of the macroscopic K+ current and the single-channel current amplitude can be used to estimate the density of the KO_2 channels which is of approximately 700 per cell. Thus, in the adult rabbit, glomus cells express a specific population of oxygen-sensitive K+ channels, referred to as KO_2 channels, which are responsible for their chemoreceptive properties.

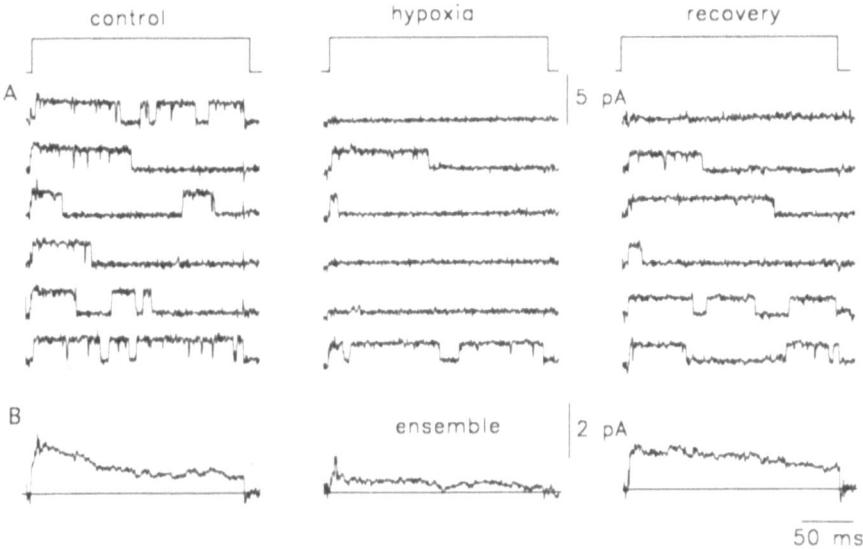

Figure 4. Modulation of KO_2 channels by O_2 tension. A. Representative sweeps obtained by step depolarization from -80 to +20 mV in an outside-out excised patch that contained at most one open channel. Recordings were obtained in the control external solution (equilibrated with air, PO_2= 150 mmHg), in low PO_2 (switching from 150 to 80 mmHg), and after returning to the solution with normal O_2 tension. Pulses were applied every 5 s. B. Ensemble averages of 15 to 30 consecutive sweeps in the various experimental conditions. Solutions as in Fig. 1A. Modified from ref. 8.

We have investigated the modulation of the K+ channels by PO_2 in excised membrane patches containing one, or in some cases two channels. In these experiments we could estimate several kinetic parameters derived from the distributions of open and closed intervals during exposure to control (PO_2= 150 mmHg) and hypoxic solutions. The most easily appreciable effect of low PO_2 is a reversible decrease in channel open probability. This is illustrated in Fig. 4A by the three sets of current sweeps generated in an outside-out patch during voltage steps to +20 mV in a normoxic solution (control and recovery) and on exposure to low PO_2. The average of 15 to 30 consecutive sweeps recorded in the different experimental conditions are shown in Fig. 4B. Average open probability of the channel ($p_0=0.61$), obtained by dividing the time spent in the open state by the total length of the pulse, decreased markedly after switching to the low PO_2 solution ($p_0=0.28$) and returned to a higher value ($p_0=0.74$) once the normal PO_2 was restored. In the range between 0 and +30 mV exposure to low PO_2 (~10 mmHg) leads to a 1.5 to 2-fold change in open probability (n=33). The recordings also show that single-channel current amplitude is unaffected by changes in O_2 tension. The kinetic analysis of single PO_2 channels recorded in normal and low PO_2 solutions indicates that hypoxia influences specific gating properties. Low PO_2 slows down the activation latency, decreases the number of bursts per trace, and favors the occurrence of traces without openings (perhaps due to inactivation from closed states). The open state and the transitions to adjacent closed or inactivated states are unaltered by hypoxia (10). Thus, at low PO_2 the number of channels that open in response to a depolarization decreases, and those channels that follow the activation pathway open more slowly and inactivate faster.

MODULATION OF OXYGEN SENSITIVE CHANNELS BY SULFHYDRYL-REDUCING AGENTS

Because low PO_2 modifies well-defined kinetic properties of the channels, it can be hypothesized that O_2 influences the structure of specific domains of the KO_2 channel molecule. O_2 tension does not alter either the conducting properties of KO_2 channels or the final transitions in the opening pathway, which are biophysical characteristics depending on structural domains probably not very accesible to ligands or enzymes. Changes in O_2 tension may regulate the channels by influencing the redox status of a few accesible amino acids. Along with this idea it has been shown that in some types of cloned K^+ channels inactivation time course is strongly accelerated by reduced glutathione (GSH) which prevents oxidation of sulfhydryl groups of cysteine residues in the amino terminus (17). The N-methyl-D-aspartate-receptor channel complex has also a redox modulatory site consisting of thiol groups that may be vicinal and capable of forming disulfide bonds (1).

Based on these ideas we have tested the effect of antioxidants on the KO_2 channel. The comparative effects of hypoxia (PO_2 ~10 mmHg) and GSH (1 mM) on a same inside-out patch containing at least six functional KO_2 channels are illustrated in Fig. 5. In the standard normoxic solution (A, control) the channels open and inactivate during depolarizing pulses with a time course that, as illustrated by the ensemble average of the figure, is similar to the kinetics exhibited by the O_2-sensitive K^+ current (see above). Low PO_2 leads to a marked reduction of channel activity and, thus, to a drastic attenuation of the ensemble average current. After recovery of channel activity in a normoxic solution (B, control), addition of GSH to the solution facing the internal side of the membrane produces a reduction of channel open probability without affecting the single-channel conductance. In most patches treated with GSH (between 1 and 5 mM) normal channel activity was restored within 5 min after wash-out of the metabolite. The inhibitory action of GSH on the KO_2 channel activity is mimicked by other reducing agents, such as DTT which is an artificial sulfhydryl-preserving compound used to prevent oxidation of structural proteins and enzymes. The effect of DTT (1 mM) added to the internal face of the membrane is shown in Fig. 6 with recordings from a macropatch containing at least six active K^+ channels. DTT leads to a decrease in the number of open channels without altering single-channel current amplitude and K^+ selectivity. In the concentration range of 1 to 5 mM, the effect of DTT was also reversible in most of the patches studied. Therefore, GSH (at concentrations similar to those found in living cells) and other sulfhydryl-preserving agents appear to have an effect on the KO_2 channels similar to the action of hypoxia. The interaction of antioxidants and the K^+ channels has not been studied in detail yet, nevertheless our preliminary data suggest that the modulation of the KO_2 channel by PO_2 could be exerted through the regulation of the redox status of the

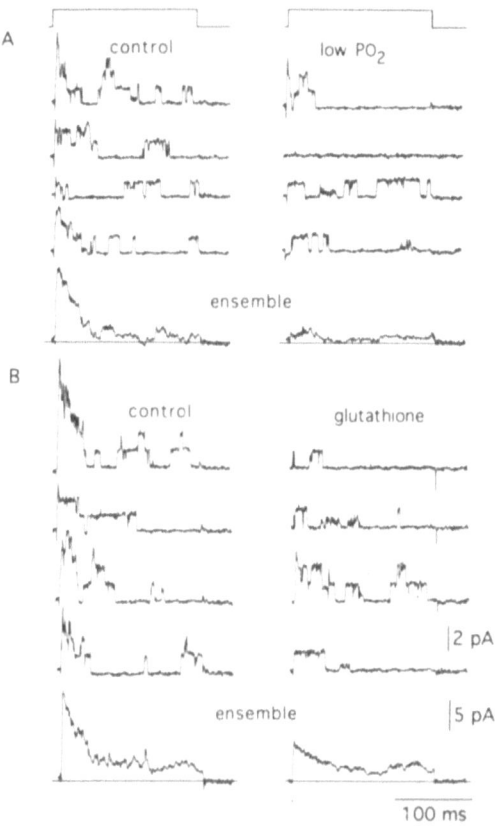

Figure 5. Exposure of an inside-out multichannel patch with at most six simultaneously active KO_2 channels to hypoxia (PO_2 ~10 mmHg) and glutathione (1 mM). A. Representative recordings of single-channel currents during depolarizations to +20 mV from -80 mV in normoxic (control) and low PO_2 solutions. Ensemble average currents are from 10 consecutive sweeps in each experimental condition. Note the inhibition of channel activity at low PO_2. B. Reversibility of the effect of hypoxia after returning to the control solution and inhibition of channel open probability on exposure of the same patch to an internal solution with glutathione. Ensemble average currents are from 10 and 12 sweeps respectively. Voltage pulses were applied at 30 s intervals. Solutions as in Fig. 1A.

channel molecule. Interestingly, sulfhydryl reagents, which are known long time ago to modify arterial vascular tone in the lung (20), can also regulate the activity of O_2-sensitive channels recently identified in smooth muscle cells dispersed from the pulmonary artery (16, Post et al., this volume).

INTERACTION BETWEEN OXYGEN AND THE KO₂ CHANNEL

Because the reversible inhibition of the macroscopic and single-channel K⁺ currents by low PO₂ can be repeatedly observed in dialyzed cells or excised membrane patches regardless the concentration of internal Ca^{2+}, ATP, or GTP-γ-S, it has been suggested that the O_2-K⁺ channel interaction occurs in the plasma membrane, without the participation of a soluble cytosolic mediator, perhaps through an intrinsic O_2 sensor which is a part of, or is closely

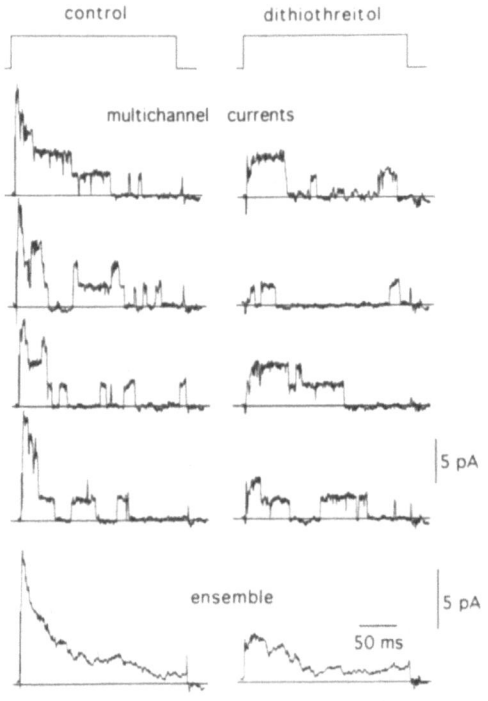

Figure 6. Activity of an inside-out multichannel patch with at most six simultaneously active KO₂ channels during step depolarizations to +20 mV from -90 mV in a standard internal solution (control) and after addition of dithiothreitol (1 mM). The ensemble averages are from 12 and 14 single sweeps in each case. Voltage pulses were applied at 30 s intervals. Solutions as in Fig. 1A.

associate with, the channel molecule (8). The mechanisms whereby O_2 could interact with the K⁺ channels are summarized in Fig. 7. A possibility is that O_2 could interact with a specific domain of one or several of the monomers forming the characteristic tetrameric structure of the K⁺ channel molecules and, thus, modify its function (Fig. 7A). In cloned K⁺ channels point mutation, even involving the replacement of closely related amino acids, are known to produce drastic changes in kinetics and open probability, hence the O_2-sensitivity could be the result of subtle structural modifications. The molecular lineage of the KO₂ channel is unknown but this hypothesis could be tested once the channel is cloned and its primary sequence is known.

Modulation of O_2-Sensitive Channels

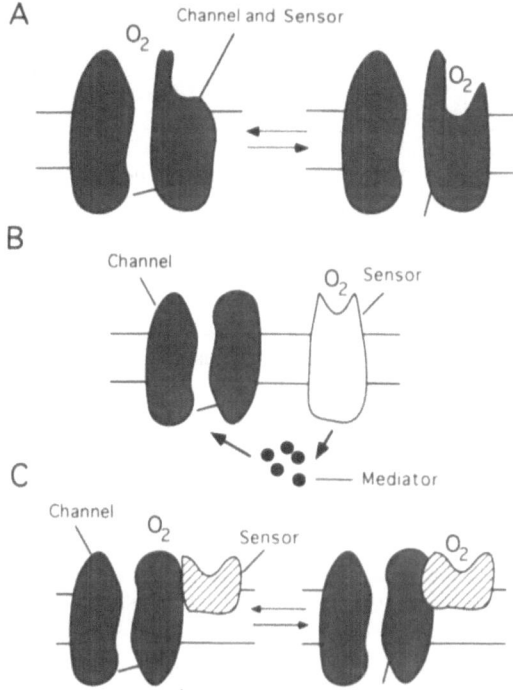

Figure 7. Possible mechanisms of interaction of O_2 and the O_2-sensitive K^+ channels. See text for explanation.

An alternative possibility is the existence of O_2 sensors expressed in the plasma membrane of glomus cells which could detect the changes in O_2 tension (Fig. 7 B and C). In this case, the sensor-channel interaction is likely to be direct (as depicted in frame C) because water soluble mediators, washed out in excised patches, or membrane difusible G proteins, blocked by GTP-γ-S, do not seem to be required for the KO_2 channel modulation (8, 9). The molecular nature of the O_2 sensor is unknown but it could belong to a family of heme-linked proteins (as for example the cytochrome-containing, membrane-bound, NADPH-oxidase, 2) that associated to the main channel subunits determine the redox state of thiol groups in the channel molecule. This basic scheme is attractive because it permits to envision the KO_2 channel as a member of a more broadly distributed family of O_2-sensitive channels susceptible to being co-expressed with an O_2-sensing subunit constituting part of a single oligomeric macromolecule. Thus, in different cells, or in a same cell at various developmental stages, the K^+ channel type being O_2-sensitive could be different. In smooth muscle cells from the pulmonary artery, where the macroscopic K^+ current is also inhibited by hypoxia (16, Blaustein et al., this volume), Ca^{2+}-activated K^+ channels appear to be O_2-sensitive (Post et al., this volume), whereas in glomus cells from adult rabbits channels of similar biophysical characteristics (such as the K_{Ca} channel described above) are unaffected by changes in O_2 (9).

Acknowledgments

Research was supported by grants from the Dirección General de Investigación Científica y Técnica.

REFERENCES

1. **Aizenman, E., S. A. Lipton, and R. H. Loring.** Selective modulation of NMDA responses by reduction and oxidation. *Neuron* 2: 1257-1263, 1989.

2. **Cross, A. R., L. Henderson, O. T. G. Jones, M. A. Delpiano, J. Hentschel, and H. Acker.** Involvement of an NAD(P)H oxidase as a PO_2 sensor protein in the rat carotid body. *Biochem. J.* 272: 743-747, 1990.

3. **De Castro, F.** Sur la structure et l'innervation de la glande intercarotidienne (glomus caroticum) de l'homme et des mamiféres, et sur un nouveau systeme d'innervation autonome du nerf glossopharyngien. *Trab. Lab. Invest. Biol. Univ. Madrid* 24: 365-432, 1926.

4. **Delpiano, M. A., and J. Hescheler.** Evidence for a PO_2-sensitive K^+ channel in the type I cell of the rabbit carotid body. *FEBS Lett.* 249: 195-198, 1989.

5. **Duchen, M.R., K. W. T. Caddy, G. C. Kirby, D. L. Patterson, J. Ponte, and T. J. Biscoe.** Biophysical studies of the cellular elements of the rabbit carotid body. *Neuroscience* 26: 291-311, 1988.

6. **Fidone, S.J., and C. González.** Initiation and control of chemoreceptor activity in the carotid body. In: *Handbook of Physiology: The Respiratory System II*. Bethesda, MD: American Physiological Society, 1986, p. 313-362.

7. **Fishman, M. C., W.L. Greene, and D. Platika.** Oxygen chemoreception by carotid body cells in culture. *Proc. Natl. Acad. Sci. USA.* 82: 1448-1450, 1985.

8. **Ganfornina, M.D., and J. López-Barneo.** Single K^+ channels in membrane patches of arterial chemoreceptor cells are modulated by O_2 tension. *Proc. Natl. Acad. Sci. USA.* 88: 2927-2930, 1991.

9. **Ganfornina, M.D., and J. López-Barneo.** Potassium channel types in arterial chemoreceptor cells and their selective modulation by oxygen. *J. Gen. Physiol.* 100: 401-426, 1992.

10. **Ganfornina, M.D., and J. López-Barneo.** Gating of O_2-sensitive K^+ channels of arterial chemoreceptor cells and kinetic modifications induced by low PO_2. *J. Gen. Physiol.* 100: 427-455, 1992.

11. **Heymans, C., J. J. Bouckaert, and L. Dautrebande.** Sinus carotidien et reflexes respiratoires. II. Influences respiratoires reflexes de l'acidose, de l'alcalose, de l'anhydride carbonique, de l'ion hydrogene et de l'anoxemie: sinus carotidiens et echanges respiratoires dans le poumons et au dela des poumons. *Arch. Internat. Pharmac. Ther.* 39: 400-408, 1930.

12. **Hoshi, T., and R. W. Aldrich.** Voltage-dependent K^+ currents and underlying single K^+ channels in pheochromocytoma cells. *J. Gen. Physiol.* 91: 76-106, 1988.

13. **López-Barneo, J., J. R. López-López, J. Ureña, and C. González.** Chemotrandsduction in the carotid body: K^+ current modulated by PO_2 in type I chemoreceptor cells . *Science Wash. DC.* 242: 580-582, 1988.

14. **López-López, J.R., C. González, J. Ureña, and J. López-Barneo.** Low PO_2 selectively inhibits K^+ channel activity in chemoreceptor cells of the carotid body. *J. Gen. Physiol.* 93: 1001-1015, 1989.

15. **Marty, A. and E. Neher.** Potassium channels in cultured bovine adrenal chromaffin cells. *J. Physiol. Lond.* 367: 117-141, 1985.

16. **Post, J. M., J. R. Hume, S. L. Archer, and E. K. Weir.** Direct role for potassium channel inhibition in hypoxic pulmonary vasoconstriction. *Amer. J. Physiol. 262 (Cell Physiol. 31):* C882-C890, 1992.

17. **Ruppersberg, J. P. M. Stocker, O. Pongs. S. H. Heinemann, R. Frank, and M. Koenen.** Regulation of fast inactivation of cloned mammalian $I_k(A)$ channels by cysteine oxidation. *Nature, Lond.* 352: 711-714, 1991.

18. **Stea, A., and C. A. Nurse.** Whole-cell and perforated patch recordings from O_2-sensitive rat carotid body cells grown in short- and long-term culture. *Pfluegers Arch.* 418: 93-101, 1991.

19. **Ureña, J., J. R. López-López, C. González, and J. López-Barneo.** Ionic currents in dispersed chemoreceptor cells of the mammalian carotid body. *J. Gen. Physiol.* 93: 979-999, 1989.

20. **Weir, E. K., J. A. Will, L. J. Lundquist, J. W. Eaton, and E. Chesler.** Diamide inhibits pulmonary vasoconstriction induced by hypoxia or prostaglandin $F_{2\alpha}$. *Proc. Soc. Exp. Biol. Med.* 173: 96-103, 1983.

REDOX REGULATION OF K^+ CHANNELS AND HYPOXIC PULMONARY VASOCONSTRICTION

J.M. Post*, E.K. Weir[+], S.L. Archer[+] and J.R. Hume*

Department of Physiology*
University of Nevada School of Medicine
Reno, NV 89557

Department of Medicine[+]
University of Minnesota
Minneapolis Veterans Administration Medical Center
Minneapolis, MN 55417

INTRODUCTION

The pulmonary vasculature is unique in that hypoxia causes vasoconstriction; whereas, most systemic vessels dilate (Daut et al., 1990). Hypoxic pulmonary vasoconstriction (HPV) serves as an adaptive mechanism by which blood flow is diverted from poorly ventilated to better ventilated regions of the lung to optimize ventilation/perfusion matching (Archer and Weir, 1989a; Cutaia and Rounds, 1990).

The mechanism by which pulmonary vascular smooth muscle cells sense oxygen tension has remained a mystery; however, recent studies provide evidence that links oxygen tension to the gating of ion channels (Ganfornina and López-Barneo, 1991; Post et al., 1992). HPV involves an increase in intracellular Ca^{2+} and can be inhibited by antagonists of voltage-dependent calcium channels (McMurtry et al., 1976; Archer et al., 1985). It is not known if hypoxia directly or indirectly activates Ca^{2+} channels in pulmonary artery smooth muscle. It has been demonstrated that K^+ channel inhibitors simulate HPV by increasing tension in pulmonary artery rings and pulmonary artery pressure in isolated lungs (McMurtry, 1984, Post et al., 1992). In addition, reduction in oxygen tension to hypoxic levels reduces K^+ channel activity and depolarizes the resting membrane potential in enzymatically dispersed pulmonary artery smooth muscle cells (Post et al., 1992). These studies which implicate involvement of Ca^{2+} and K^+ channels in HPV suggest the following hypothesis:

$$\text{Hypoxia}$$
$$\downarrow$$
$$K^+ \text{ channel inhibition}$$
$$\downarrow$$
$$\text{Depolarization of resting membrane potential (RMP)}$$
$$\downarrow$$
$$\text{Activation of voltage-dependent } Ca^{2+} \text{ channels}$$
$$\downarrow$$
$$Ca^{2+} \text{ influx}$$
$$\downarrow$$
$$\text{Pulmonary vasoconstriction}$$

Redox Hypothesis

Several theories have been proposed to explain HPV. Although each theory has its proponents, it appears that HPV is not secondary to the production of a chemical mediator, or depletion of ATP (for review see Archer and Weir, 1989d). As stated above, HPV may be initiated by K^+ channel inhibition. We have suggested that the redox status of the K^+ channel may in part determine its gating (Archer et al., 1986). The redox hypothesis proposes that alveolar oxygen tension regulates the production of oxygen radicals (e.g. superoxide anion, H_2O_2) in the lung (Freeman et al., 1982; Turrens et al., 1982; Archer et al., 1989b). These oxygen radicals produced during normal oxidative metabolism, regulate the redox state of glutathione, NADPH, and other intracellular constituents which may affect ion channel gating. Under normoxic conditions, oxygen or non-oxygen based radicals promote vasorelaxation and a normotensive state.

Archer, McMurtry and Weir (1989a) have suggested criteria and discussed the redox hypothesis. Evidence which supports the redox hypothesis includes: 1) Oxygen radicals are vasoactive in the lung (Weir et al., 1985; Archer et al., 1989b), 2) Oxygen radicals such as superoxide anion and H_2O_2 are produced in the lung in proportion to oxygen tension (Turrens et al., 1982; Archer et al., 1989b); 3) The sulfhydryl redox status of the cell may modulate both Ca^{2+} (Schmid et al., 1986) and K^+ channel activity (Meury and Kepes, 1982; Ruppersburg et al.,1991); 4) Oxygen tension regulates the ratio of reduced to oxidized glutathione (GSH\GSSG) and pyridine nucleotides (Patterson et al., 1985; White et al., 1986); 5) Hypoxia inhibits K^+ channel activity in pulmonary artery smooth muscle cells (Post et al., 1992).

The present chapter summarizes recent studies which test the validity of the redox hypothesis in HPV, as it applies to K^+ channel regulation. The results show that pulmonary vasoconstrictors such as hypoxia and doxapram inhibit K^+ channel activity. The reducing agent N-acetyl-L-cysteine (NAC; a membrane permeable sulfhydryl donor) also inhibits K^+ channel activity, whereas, sulfhydryl oxidizing agents like diamide, stimulate K^+ channel activity.

METHODS

Single canine pulmonary artery smooth muscle cells were enzymatically dissociated using a method described previously (Post et al., 1992). Membrane currents were recorded using either the whole-cell, cell-attached or inside-out configuration of the patch clamp technique. During whole-cell recordings, cells were dialyzed using 3-6 MΩ electrodes filled with a solution containing (in mM): 110 potassium gluconate, 20 KCl, 0.5 $MgCl_2$, 1.5 ATP (diK), 2.0 phosphocreatine, 5.0 N-2-hydroxyethylpiperazine-N'-2-ethanesulfonic acid (HEPES), 1.0 ethylene glycol-bis(β-aminoethyl ether)-N,N,N',N'-tetraacetic acid (EGTA), pH 7.2. Cells were bathed in an external solution containing (in mM): 130 NaCl, 10 $NaHCO_3$, 4.2 KCl, 1.2 KH_2PO_4, 0.5 $MgCl_2$, 1.5 $CaCl_2$, 5.5 glucose, 10 HEPES, pH 7.3. Whole-cell experiments were performed at 35 ± 1 °C. Normoxic solutions ($P_{O2} \approx 130$ mmHg) were obtained by aeration with a 20% O_2-5% CO_2-balance N_2 gas mixture. Hypoxic solutions ($PO_2 \approx 40$ mmHg) were obtained by aeration with a 5% CO_2-balance N_2 gas mixture. Experiments were performed using cell-attached and detached patches (inside-out) at room temperature using symmetrical K^+ solutions containing (in mM): 140 KCl, 10 HEPES, 0.1 EGTA, 0.05 $CaCl_2$, 5.5 glucose, pH 7.2. Estimates of free $[Ca^{2+}]_i$ were calculated with a computer program developed by Robertson and Potter (1984).

RESULTS AND DISCUSSION

Hypoxic Inhibition of K^+ Channels

We used the whole-cell patch clamp technique to directly measure the effect of changing oxygen tension on K^+ currents in enzymatically dispersed canine pulmonary artery smooth muscle cells. Whole-cell K^+ currents can be identified in these cells based upon their sensitivity to changes in the K^+ gradient and inhibition by K^+ channel blockers like tetraethylammonium (TEA) and 4-aminopyridine (Post et al., 1992). Fig. 1A shows an example of whole-cell K^+ currents which can be activated by the application of depolarizing voltage ramps. Outward currents begin to activate near -40 mV and increase in amplitude as the ramp voltage becomes more positive. There usually is an increase in current noise at positive membrane potentials. These K^+ currents are reduced considerably by organic Ca^{2+} channel blockers, suggesting that part of the total outward current can be attributed to activation of Ca^{2+}-dependent K^+ channels (Post et al., 1992). A reduction in oxygen tension from a normoxic ($P_{O2} \approx 130$ mmHg) to hypoxic level ($P_{O2} \approx 40$ mmHg) resulted in a decrease in K^+ current (Fig. 1A). The sensitivity of whole cell K^+ currents to hypoxia suggests the possible involvement of an O_2-sensitive K^+ channel in the regulation of HPV. Interestingly, inhibition of K^+ current by hypoxia was prevented if $[Ca^{2+}]_i$ was reduced by either strong buffering with BAPTA (Fig.1B) or blockade of voltage dependent Ca^{2+} channels by nisoldipine (Fig. 1C; Post et al., 1992). The observation that EGTA is much less effective in eliminating Ca^{2+}-dependent K^+ currents when compared with BAPTA is consistent with similar observations made in chromaffin cells (Marty and Neher, 1985).

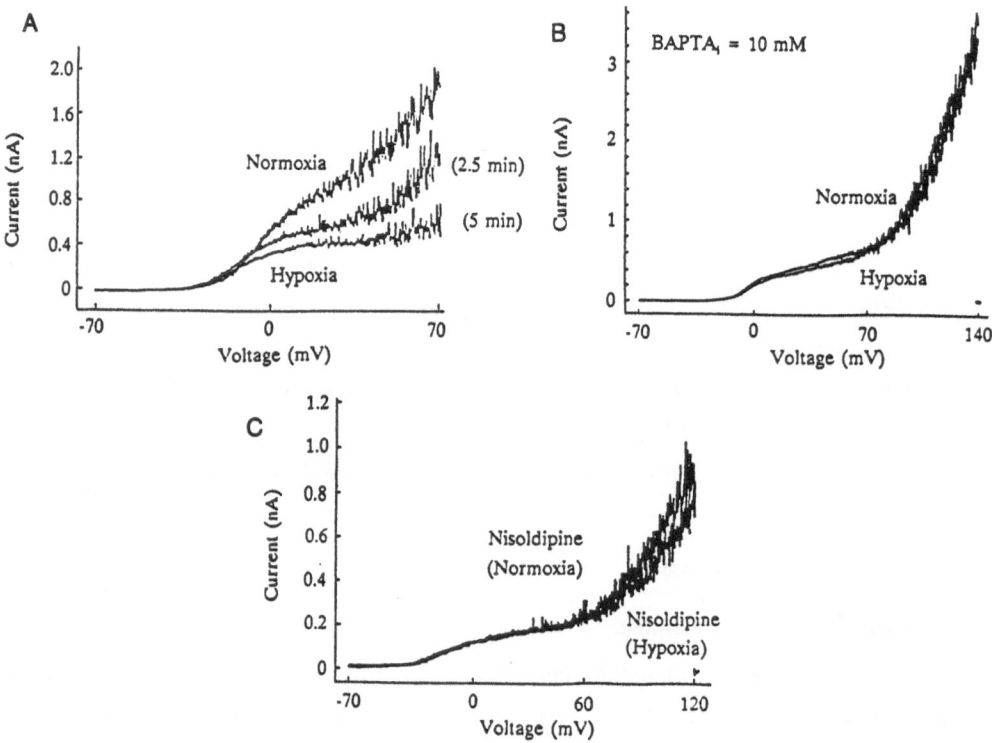

Figure 1 Effect of hypoxia on voltage dependent K^+ currents in pulmonary artery smooth muscle cells. A: Membrane currents activated by voltage ramps from -70 to 70 mV are shown for normoxic conditions and after 2.5 and 5.0 minutes exposure to hypoxia ($[EGTA]_i$ 0.1 mM; estimated $[Ca^{2+}]_i$ 8.5×10^{-9}M). B: Membrane currents activated by voltage ramps from -70 to 140 mV; hypoxia failed to inhibit K^+ current in cell dialyzed with 10 mM 1,2-bis(2-aminophenoxy)ethane-N,N,N',N'-tetraacetic acid (BAPTA)(estimated free $[Ca^{2+}]_i$ 4.6×10^{-14}M). C: Membrane currents activated by voltage ramps from -70 to 120 mV; hypoxia failed to inhibit K^+ currents in cell exposed to 1 μM nisoldipine ($[EGTA]_i$ 5 mM; control free $[Ca^{2+}]_i$ 1.6×10^{-10}M). Modified from Post et al. (1992) with permission.

These experiments suggest that hypoxic inhibition of K^+ currents in pulmonary artery cells involves inhibition of a Ca^{2+}-sensitive component of K^+ current.

Carotid body type 1 cells are similar to pulmonary artery cells in that they contain pO_2-sensitive K^+ channels which are reversibly inhibited by hypoxia. There is, however, disagreement whether the K^+ channel affected by hypoxia in these cells is Ca^{2+} sensitive (Peers, 1990) or Ca^{2+} insensitive (Lopez-Barneo et al., 1988; Delpiano and Hescheler, 1989).

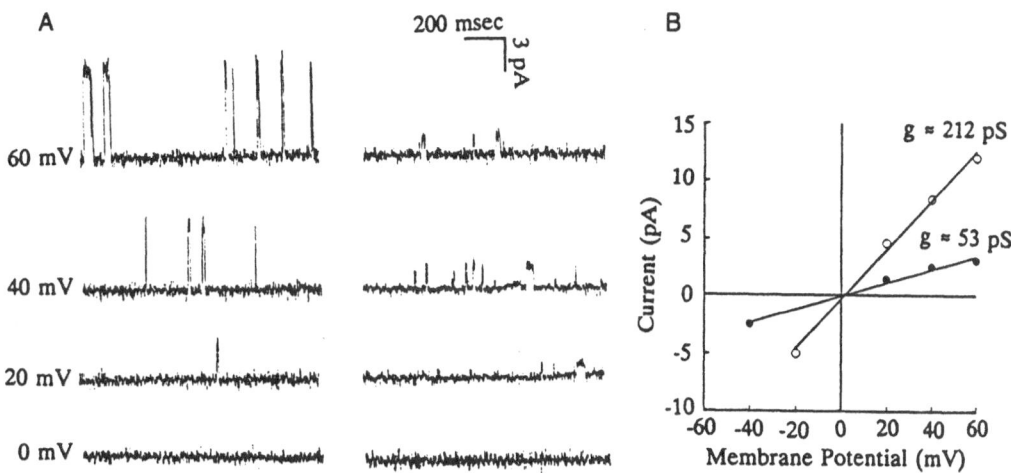

Figure 2 Single channel currents in cell-attached patches with a pipette $[K^+]$ of 140 mM. A: Both large and small conductance K^+ channels are present in canine pulmonary artery smooth muscle cells. B: Current-voltage relationships for the large (●) and small (○) conductance K^+ channels. Slope conductance was 212 and 53 pS for the large and small channels respectively.

Few characteristics are known about K^+ channels in pulmonary artery smooth muscle cells. Okabe et al.(1987), reported that the net outward K^+ current in rabbit pulmonary artery cells was sensitive to 4-aminopyridine and to tetraethylammonium. Cell-attached patches from canine pulmonary artery cells were used to measure unitary K^+ channel currents. These patches contain at least two types of K^+ channels with single channel conductances of approximately 215 and 55 pS in symmetrical K^+ solution (Fig.2). The large conductance K^+ channel is the predominant channel observed in these experiments and is activated by changes in Ca^{2+} at the cytoplasmic surface in inside-out detached membrane patches (Fig.3). This channel may be similar to a 245 pS Ca^{2+}-sensitive K^+ channel recently reported in rat pulmonary artery smooth muscle cells (Robertson et al., 1992).

Respiratory Stimulants Inhibit K$^+$ Channels

Doxapram is a respiratory stimulant which has been advocated for therapy of respiratory failure resulting from chronic pulmonary disease and drug-induced respiratory

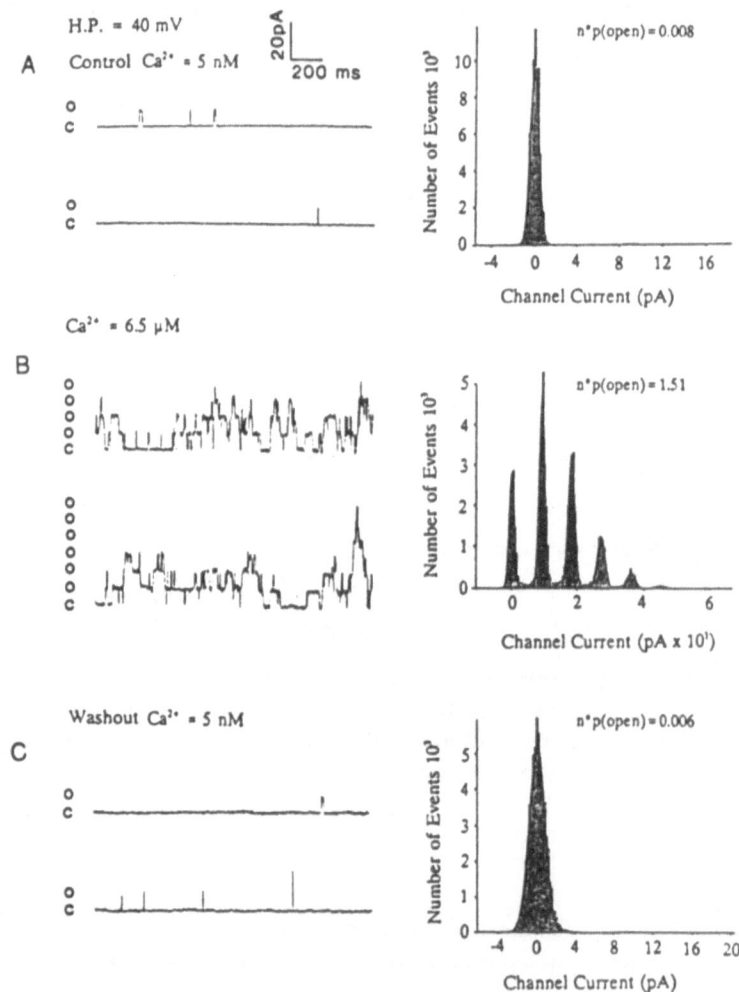

Figure 3 Effect of Ca^{2+} on the large conductance K$^+$ channels (inside-out patch). Membrane potential was zeroed with isotonic KCl in the pipette and bath solutions. Ca^{2+} added to the bath reversibly activated K$^+$ channel activity at a patch potential of 40 mV. Right column illustrates the single channel current amplitude histograms constructed from 1 minute recordings for each experimental condition.

depression. One mechanism by which doxapram increases ventilation and blood pressure is via stimulation of carotid chemoreceptors which in turn stimulate medullary inspiratory units (Mitchell and Herbert, 1975; Cote et al., 1992). Naeije et al. (1990), found that doxapram was able to restore HPV in dogs with a naturally absent hypoxic pulmonary pressor response and postulated that doxapram may exert its effect directly on the pulmonary artery. We tested the effect of doxapram on single pulmonary artery smooth muscle cells and found that doxapram (1 mM) significantly inhibits macroscopic K^+ currents by 65.5 ± 5.9 % at 60 mV (n=4) in pulmonary artery smooth muscle cells (Fig.4A). Cell-attached patch experiments demonstrate that the reduction in K^+ current probably results from inhibition of the large conductance Ca^+-activated K^+ channels (Fig.4B). This suggests

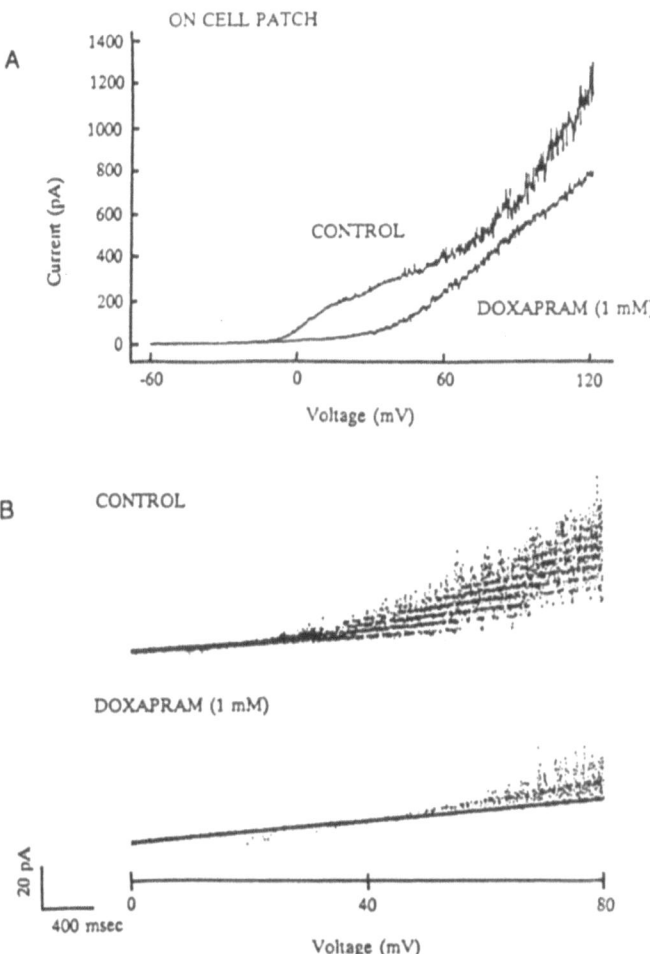

Figure 4 Effect of the respiratory stimulant doxapram on whole-cell and single channel (cell-attached) K^+ currents. A: Doxapram (1 mM) inhibited outward K^+ currents elicited by a ramp depolarization from -60 to 120 mV. B: Superimposed single channel currents activated by 5 ramp depolarizations applied from -60 to 120 mV before and after exposure of the cell to doxapram.

that respiratory stimulants, like hypoxia or doxapram, may initiate their vascular pressor response through direct inhibition of K^+ channels in pulmonary artery smooth muscle.

The observation that pharmacologic K^+ channel blockers increase pulmonary artery pressure (PAP) and pulmonary artery tension is consistent with the postulated role of K^+ channels in HPV (McMurtry, 1984; Post et al., 1992). Unfortunately, many of these blockers are not specific for a given type of K^+ channel and therefore it is difficult to discern the type of K^+ channel active during normoxia, which upon being blocked would depolarize the cell's resting membrane potential sufficiently to activate voltage-activated Ca^{2+} channels. In addition, although the previous experiments suggest that the K^+ channel serves as a sensor of oxygen tension in pulmonary artery cells, the question remains as to how the signal is transmitted to the channel protein. Does oxygen tension directly affect the channel protein, intracellular messengers, neurohormones, endothelial products and/or some other variable? Although vascular endothelial products are known to be important modulators of HPV, they do not appear to be essential since HPV occurs in endothelial denuded vessels (Burke-Wolin and Wolin, 1989; Yuan et al., 1990). In addition, since hypoxia inhibits K^+ channels in isolated pulmonary artery smooth muscle cells (Post et al., 1992) it would appear that oxygen tension either directly modulates the channel protein, or alters some membrane-bound substrate closely associated with the channel. The fact that hypoxia inhibits K^+ channels in well dialyzed single smooth muscle cells argues against the possible involvement of a soluble second messenger.

Modulation of K^+ Channels by Oxidants and Antioxidants

It is possible that the ability of oxygen tension to directly influence the redox state of the cell is involved in HPV. Oxygen radicals generated by xanthine /xanthine oxidase or glucose/glucose oxidase inhibit HPV in isolated lung models (Burghuber et al., 1984; Archer et al., 1989c). Allopurinol, an inhibitor of xanthine oxidase prevents this effect (Weir and Will, 1982). Similarly, the oxygen radical t-butylhydroperoxide, a membrane permeable analog of hydrogen peroxide inhibits HPV (Weir and Will, 1982). The manner by which t-butylhydroperoxide inhibits HPV may be associated with a decrease in GSH/GSSG and/or NADPH/NADP redox levels or the formation of hydroxyl radicals (Elkow et al., 1984).

Diamide is a sulfhydryl oxidant which rapidly and reversibly decreases the ratio of cellular GSH/GSSG and NADPH/NADP (Elkow et al., 1984). Although diamide itself does not produce oxygen radicals, it compromises the cell's ability to scavenge oxygen radicals by oxidizing GSH. If HPV is a result of inhibition of K^+ channels, then one might hypothesize that this is a consequence of increased levels of GSH and NADPH, since diamide inhibition of HPV (Weir et al., 1985) can be prevented by N-acetyl-L-cysteine (a membrane permeable sulfhydryl donor) or GSH (Archer et al., 1986). Therefore, we have examined the effects of diamide on whole-cell and single channel K+ currents in canine pulmonary artery cells. Diamide (10 μM) increases macroscopic K^+ currents (Fig. 5A) and increases the opening probability of large conductance Ca^{2+}-activated K^+ channels in cell-

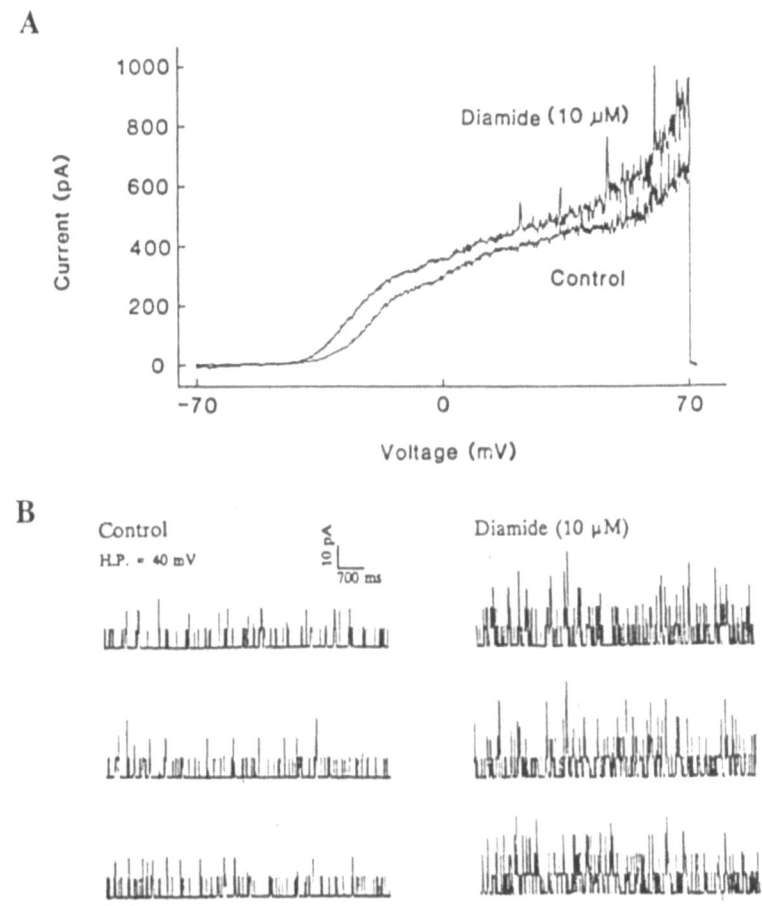

Figure 5 Effect of the sulfhydryl oxidant diamide on whole-cell and single channel K$^+$ currents. A: Diamide (10 μM) increased outward K$^+$ current during ramp depolarization from -70 to 70 mV. B: Diamide activated large conductance Ca^{2+}-activated K$^+$ channels in inside-out patch experiments. n*p(open) increased from 0.175 (control) to 0.584 (diamide 10 μM) over a 1 minute recording period.

197

detached membrane patches from pulmonary artery cells (Fig 5B). In contrast, the antioxidant N-acetyl-L-cysteine inhibits macroscopic K^+ currents in canine pulmonary artery smooth muscle cells (Fig. 6).

These studies further strengthen the hypothesis that K^+ channel activity is modulated by the redox state of the cell. Oxidizing agents augment K^+ channel activity, whereas, reducing agents inhibit K^+ channel activity. The observation that diamide augmented Ca^{2+}-activated K^+ channels in inside-out patches using an intracellular solution devoid of GSH suggests that diamide may directly oxidize the K^+ channel protein. This does not infer that reduced GSH is not normally important for channel function, on the contrary, GSH/GSSG

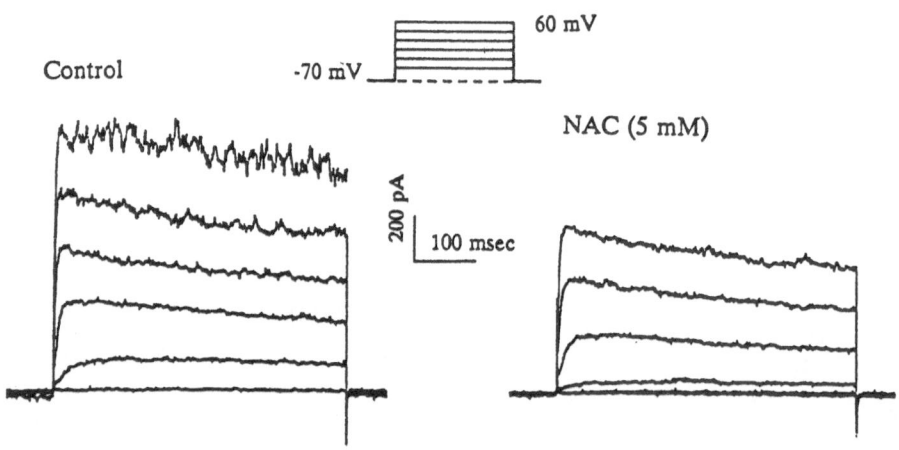

Figure 6 Effect of the antioxidant N-acetyl-L-cysteine (NAC) on K^+ current. NAC (5 mM) inhibited outward K^+ current during step depolarizations using the protocol shown (holding potential = -70 mV). NAC inhibited outward current at 60 mV by 36 ± 5 % (n=4).

may be a key intracellular mediator which regulates the redox state of cysteine residues involved with the K^+ channel gating mechanism.

An interesting corollary may exist in rat brain with a rapidly inactivating K^+ channel ($I_K(A)$) (Ruppersberg et al., 1991). These channels lose their fast inactivation characteristics when changing from a whole-cell to an inside-out patch configuration, but it can be restored if the intracellular surface of the patch is exposed to the reducing agents GSH or dithiotreitol (DTT). In addition, studies using site directed mutations show that replacement of the sulfhydryl containing cysteine residue in the N-terminal ball domain of the cloned RCK4 K^+ channel with a serine residue prevents the loss of fast inactivation when going from the whole-cell to inside-out configuration, suggesting that inactivation requires a reduced cysteine residue on the N-terminal domain (Ruppersberg et al., 1991). Although K^+ channels have yet to be cloned in smooth muscle, Atkinson et al. (1991) have deduced the amino acid sequence of Ca^{2+}-activated K^+ channels encoded by the drosophila slow poke (slo) locus. These Ca^{2+}-activated K^+ channels exhibit a large unitary conductance and charybdotoxin sensitivity (Elkins et al., 1986). Interestingly, this channel protein also contains cysteine residues in its N-terminal region. It is possible to speculate that K^+ channels in pulmonary artery may have a similar N-terminal inactivation gate which requires cysteine in its reduced form to function. If this cysteine is reduced by hypoxia, N-acetyl-cysteine or GSH, this may preferentially promote inactivation or closure of the K^+ channel, whereas oxidation of this cysteine residue, for example by diamide, may remove inactivation and enhance K^+ channel openings. Such speculation, however, can only be directly confirmed once a full length clone of the O_2-sensitive K^+ channel from pulmonary artery becomes available.

Redox Regulation of Other Channels

There is considerable evidence which suggests that HPV is closely linked to the activation of voltage-dependent Ca^{2+} channels. Ca^{2+} channel agonists augment (McMurtry, 1985; Tolins et al., 1986) and Ca^{2+} channel blockers inhibit (McMurtry et al., 1976, Harder et al., 1985) HPV respectively. Whether hypoxia directly activates Ca^{2+} channels or indirectly activates Ca^{2+} channels via inhibition of K^+ channels is not known. Electrophysiological studies designed to assess the effects of oxygen on Ca^{2+} channels in pulmonary artery smooth muscle cells have not been carried out. However, exposure of rat pulmonary artery smooth muscle to hypoxic solution (using sodium dithionite as an O_2 scavenger) increases intracellular Ca^{2+} within 30 seconds. Upon return to normoxic solution, $[Ca^{2+}]_i$ returned to control levels (Salvaterra and Goldman, 1991). Since this response was not inhibited by verapamil, it was postulated that Ca^{2+} influx may occur through verapamil-insensitive Ca^{2+} channels (Salvaterra and Goldman, 1991). A more recent study indicates that the initial rise in intracellular Ca^{2+} triggered by hypoxia may come from intracellular Ca^{2+} stores (Salvaterra and Goldman, 1993).

Circumstantial evidence in pancreatic cells (which have many similarities in terms of regulation of K^+ and Ca^{2+} channels to the pulmonary artery cells; Archer et al., 1986) suggests that the redox status of these cells may control Ca^{2+} uptake. The induction of a more reduced cytosolic redox state decreases K^+ conductance and increases cytosolic Ca^{2+} in pancreatic β-cells (Lebrun et al., 1983). This effect can be prevented by aminooxyacetate, an inhibitor of transamination reactions, which augments both ^{86}Rb and ^{45}Ca efflux. Furthermore, Ammon et al. (1977, 1983), found that methylene blue (an oxidant of NADPH), diamide and diazene (thiol oxidants) and t-butyl hydrogen peroxide (a substrate of GSH peroxidase) inhibits glucose-induced Ca^{2+} uptake (Ammon et al., 1983) and subsequent insulin release (Ammon et al., 1977) from pancreatic β-cells. It was postulated that a decrease in the GSH/GSSG ratio (Ammon et al., 1983) inhibits glucose-induced Ca^{2+}

uptake in β-cells. These studies and the present experiments with diamide suggest that oxidants may augment K^+ channel activity as a result of a decrease in the GSH/GSSG ratio and could explain why oxidants inhibit HPV.

Very little is known about Na^+ channels in vascular smooth muscle. Rabbit pulmonary artery cells have been reported to contain TTX-sensitive Na^+ channels (Okabe et al., 1988). It is interesting that GSH accelerates Na^+ channel inactivation in rat axon (Strupp et al., 1992). If GSH produced a similar effect in pulmonary artery smooth muscle cells then this would be expected to attenuate inward current and hence prevent depolarization and subsequent Ca^{2+} entry. This is opposite to what would be expected during HPV. However, in the absence of further evidence on the physiological role of Na^+ channels in pulmonary artery smooth muscle cells, it is premature to speculate on a role for Na^+ channels in HPV.

Although the present study indicates that Ca^{2+}-activated K^+ channels can be modulated by cellular redox status, careful analysis is still necessary to characterize possible redox regulation of other K^+ channels present in pulmonary artery smooth muscle. A recent study suggests that hypoxia may inhibit a different class of voltage-gated K^+ channels in cultured rat pulmonary arterial cells (Yuan et al., 1993), a conclusion which appears to be different from those of Post et al. (1992; see also Figure 1), which concluded that hypoxia inhibits a Ca^{2+}-sensitive K^+ channel in acutely isolated pulmonary arterial cells. Some of the discrepancy between these two studies might be explained by the recent demonstration that a rise in intracellular Ca^{2+}, which is known to occur early during the hypoxic vasconstrictor response (Salvaterra and Goldman, 1993), can cause inhibition of 4-aminopyridine-sensitive, voltage-dependent delayed rectifier K^+ channels in isolated pulmonary arterial cells (Gelband et al., 1993).

CONCLUSION

It appears that regulation of K^+ channel activity is of central importance in the regulation of pulmonary vascular tone. The present study suggests that changes in oxygen tension may modulate K^+ channel activity such that diminished oxygen radical production, reduced GSH, or a cytosolic environment which favors a reduced state, inhibits K^+ channels and that production of oxygen radicals and/or an environment which favors oxidation of intracellular substrates enhances K^+ channel activity. A diagram illustrating the potential role of K^+ channels in HPV with respect to the redox hypothesis is presented in figure 7. Whether the K^+ channel itself serves as the oxygen sensor or whether the redox status of intracellular substrates is essential for hypoxic inhibition of K^+ channels has yet to be determined. In either case, the redox hypothesis appears to be a valid hypothesis which may explain (at least in part) the HPV mechanism.

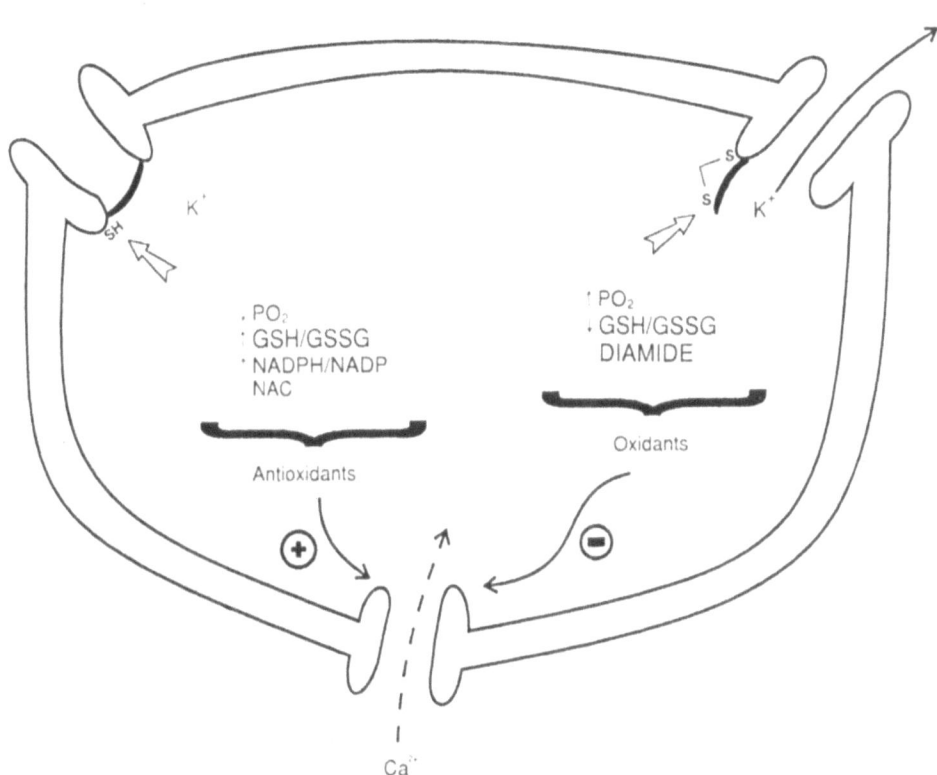

Figure 7 Hypothetical model which suggests that a cytosolic environment which favors a reduced state (e.g. hypoxia, GSH, N-acetyl-L-cysteine, NADPH) may inhibit K^+ channel activity; whereas, an environment which favors oxidation (e.g. normoxia, GSSG, NADP, diamide) may promote K^+ channel activity. If K^+ channels which regulate the resting membrane potential are inhibited by "antioxidants" then this might cause membrane depolarization and subsequent opening of voltage sensitive Ca^{2+} channels.

ACKNOWLEDGEMENTS

This work was supported by a National Research Service Awards postdoctoral fellowship (JMP), NIH grant HL49254 (JRH), NIH grant, HL45735-1 (SLA) and the Veterans Administration (SLA and EKW). The authors wish to thank Lan Xu for expert technical assistance. Canine pulmonary artery was graciously made available from a program project grant from the National Institute of Diabetes and Digestive and Kidney Diseases (DK-41315).

REFERENCES

Ammon, H.P.T., Akhtar, M.S., Niklas, H., and Hegner, D., 1977, Inhibition of p-chloromercuribenzoate- and glucose-induced insulin release in vitro by methylene blue, diamide, and tert-butyl hydroperoxide. *Mol. Pharm.* 13:598-605.

Ammon, H.P.T., Hägele, R., Youssif, N., Eujen, R., and El-Amri, N., 1983, A possible role of intracellular and membrane thiols of rat pancreatic islets in calcium uptake and insulin release. *Endocrinology* 112:720-726.

Archer, S.L., McMurtry, I.F., and Weir, E.K., 1989a, Mechanisms of Acute Hypoxic and Hyperoxic Changes in Pulmonary Vascular Reactivity, in: "Pulmonary Vascular Physiology and Pathophysiology," E.K. Weir and J.T. Reeves, eds., Marcel Dekker, Inc., New York p. 241-290.

Archer, S.L., Nelson, D.P. and Weir, E.K., 1989b, Simultaneous measurement of O_2 radicals and pulmonary vascular reactivity in rat lung. *J. Appl. Physiol.* 67:1903-1911.

Archer, S.L., Peterson, D., Nelson, D.P., DeMaster, E.G., Kelly, B., Eaton, J.W., and Weir, E.K., 1989c, Oxygen radicals and antioxidant enzymes alter pulmonary vascular reactivity in the rat lung. *J. Appl. Physiol.* 6: 102-111.

Archer, S.L., and Weir, E.K., 1989d, Mechanisms in Hypoxic Pulmonary Hypertension. *In*: "Pulmonary Circulation: Advances and Controversies. C.A. Wagenvoort and H. Denolin, eds., Elsevier Science Publishers, New York. p.87-113.

Archer, S.L., Will, J.A., and Weir, E.K., 1986, Redox status in the control of pulmonary vascular tone. *Herz* 11:127-141.

Archer, S.L., Yankovich, R.D., Chesler, E., and Weir, E.K., 1985, Comparative effects of nisoldipine, nifedipine and bepridil on experimental pulmonary hypertension. *J. Pharm. Exp. Ther.* 233:12-17.

Atkinson, N.S., Robertson, G.A., and Ganetzky, B., 1991, A component of calcium-activated potassium channels encoded by the drosophila slo locus. *Science Wash DC* 253:551-555.

Burghuber, O., Mathies, M.M., McMurtry, I.F., Reeves, J.T. and Voelkel, N.F. 1984, Lung edema due to hydrogen peroxide is independent of cyclooxygenase products. *J. Appl. Physiol.* 56:900-905.

Burke-Wolin, T., and Wolin, M.S., 1989, H_2O_2 and cGMP may function as an O_2 sensor in the pulmonary artery. *J.Appl. Physiol.* 66:167-170.

Côté, A., Blanchard, P.W., and Meehan, B., 1992, Metabolic and cardiorespiratory effects of doxapram and theophylline in sleeping newborn piglets. *J. Appl. Physiol.* 72:410-415.

Cutaia, M., and Rounds, S., 1990, Hypoxic pulmonary vasoconstriction. *Chest* 97:706-718.

Daut, J., Maier-Rudolph, W., von Bekerath, N., Mehrke, G., Günther, K., and Goedel-Meinen., L., 1990, Hypoxic dilation of coronary arteries is mediated by ATP-sensitive potassium channels. *Science Wash. DC* 247:1341-1344.

Delpiano, M.A., and Hescheler, J., 1989, Evidence for a P_{O2}-sensitive K^+ channel in the type-I cell of the rabbit carotid body. *FEBS Lett.* 249:195-198.

Eklöw, L., Moldéus, P., and Orrenius, S., 1984, Oxidation of glutathione during hydroperoxide metabolism. *Eur. J. Biochem.* 138:459-463.

Elkins, T., Banetzky, B., and Wu, C., 1986, A drosophila mutation that eliminates a calcium-dependent potassium current. *Proc. Natl. Acad. Sci. USA* 83:8415-8419.

Ganfornina, M.D., and López-Barneo, J., 1991, Single K^+ channels in membrane patches of arterial chemoreceptor cells are modulated by O_2 tension. *Proc. Natl. Acad. Sci. USA* 88:2927-2930.

Gelband, C.H., Ishikawa, T., Post, J.M., Keef, K.D. and Hume, J.R., 1993, Intracellular divalent cations block smooth muscle K^+ channels. *Circ. Res.* in press.

Lebrun, P., Malaisse, W.J., and Herchuelz, A., 1983, Impairment by aminooxyacetate of ionic response to nutrients in pancreatic islets. *Am. J. Physiol.* 245:E38-E46.

López-Barneo, J., López-López, J., Ureña, J., and González, C., 1988, Chemotransduction in the carotid body: K^+ current modulated by P_{O2} in type I chemoreceptor cells. *Science Wash. DC* 242:580-582.

Marty, A., and Neher, E., 1985, Potassium channels in cultured bovine adrenal chromaffin cells. *J. Physiol. Lond.* 367:117-141.

McMurtry, I.F., 1984, Angiotensin is not required for hypoxic constriction in salt solution-perfused rat lungs. *J.Appl. Physiol.* 56:375-380.

Mc Murtry, I.F., 1985, BAY K 8644 potentiates and A23187 inhibits hypoxic vasoconstriction in rat lungs. *Am. J. Physiol.* 249:H741-H746.

McMurtry, I.F., Davidson, A.B., Reeves, J.T., and Grover, R.F., 1976, Inhibition of hypoxic pulmonary vasoconstriction by calcium antagonists in isolated rat lungs. *Circ.Res.* 38:99-104.

Meury, J., and Kepes, A., 1982, Glutathione and the gated potassium channels of Escherichia coli. *EMBO J.* 1:339-343.

Mitchell, R.A., and Herbert, D.A., 1975, Potencies of doxapram and hypoxia in stimulating carotid-body chemoreceptors and ventilation in anesthetized cats. *Anesthesiology* 42:559-566.

Naeije, R., Lejeune, P., Vachiéry, J., Leeman, M., Mélot, Hallemans, R., Delcroix, M., and Brimioulle, S., 1990, Restored hypoxic pulmonary vasoconstriction by peripheral chemoreceptor agonists in dogs. *Am. Rev. Respir. Dis.* 142:789-795.

Okabe, K., Kitamura, K., and Kuriyama, H., 1987, Features of 4-aminopyridine sensitive outward current observed in single smooth muscle cells from the rabbit pulmonary artery. *Pflugers Arch.* 409:561-568.

Okabe, K., Kitamura, K., and Kuriyama, H., 1988, The existence of a highly tetrodotoxin sensitive Na channel in freshly dispersed smooth muscle cells of the rabbit main pulmonary artery. *Pfluegers Arch.* 411:423-428.

Patterson, C.E., Butler, J.A., Byrne, F.D., and Rhodes, M.L., 1985, Oxidant lung injury: intervention with sulfhydryl reagents. *Lung* 163:23-32.

Peers, C., 1990, Hypoxic suppression of K^+ currents in type I carotid body cells: selective effect on the Ca^{2+}-activated K^+ current. Neurosci. Letters 119:253-256.

Post, J.M., Hume, J.R., Archer, S.L., and Weir, E.K., 1992, Direct role for potassium channel inhibition in hypoxic pulmonary vasoconstriction. *Am. J. Physiol.* 262:C882-C890.

Robertson, B.E., Corry, P.R., Nye, P.C.G., and Kozlowski, R.Z., 1992, Ca^{2+} and Mg-ATP activated potassium channels from rat pulmonary artery. *Pflugers Arch* 421:97-99.

Robertson, S.P., and Potter, J.D., 1984, "Methods In Pharmacology," Plenum, New York p. 63-75.

Ruppersberg, J.P., Stocker, M., Pongs, O., Heinemann, S.H., Frank, R., and Koenen, M., 1991, Regulation of fast inactivation of cloned mammalian $I_K(A)$ channels by cysteine oxidation. *Nature* 352:711-714.

Salvaterra, C.G., and Goldman, W.F., 1991, Direct effects of hypoxia on apparent intracellular calcium levels in cultured pulmonary vascular smooth muscle cells. *Am. Rev. Resp. Dis.* 163:A373.

Salvaterra, C.G., and Goldman, W.F., 1991, Acute hypoxia increases cytosolic calcium in cultured pulmonary arterial myocytes. *Am. J. Physiol.* 264:L323-L328.

Schmid, A., Barhanin, J., Coppola, T., Borsotto, M., and Lazdunski, M., 1986, Immunochemical analysis of subunit structures of 1,4-dihydropyridine receptors associated with voltage-dependent Ca^{2+} channels in skeletal, cardiac, and smooth muscles. *Biochem.* 25:3492-3495.

Strupp, M., Quasthoff, S., Mitrović, and Grafe, P., 1992, Glutathione accelerates sodium channel inactivation in excised rat axonal membrane patches. *Pflugers Arch* 421:283-285.

Tolins, M., Weir, E.K., Chesler, E., Nelson, D.P., and From, A.H.L., 1986, Pulmonary vascular tone is increased by a voltage-dependent calcium channel potentiator. *J. Appl. Physiol.* 60:942-948.

Turrens, J.F., Freeman, B.A., Levitt, J.G., and Crapo, J.D., 1982, The effect of hyperoxia on superoxide production by lung submitochondrial particles. *Arch. Biochem. Biophys.* 217:401-410.

Weir, E.K., Eaton, J.W., and Chesler, E., 1985, Redox status and pulmonary vascular reactivity. *Chest* 88:249S-252S.

Weir, E.K., and Will, J.A., 1982, Oxidants: a new group of pulmonary vasodilators. *Clin Resp. Physiol.* 18:81-85.

White, R.E., Mimmack, R.F., and Repine, J.E., 1986, Accumulation of lung tissue oxidized glutathione (GSSG) as a marker of oxidant induced lung injury. *Chest* 89:111S-113S.

Yuan, X., Tod, M.L., Rubin, L.J., and Blaustein, M.P., 1990, Contrasting effects of hypoxia on tension in rat pulmonary and mesenteric arteries. *Am. J. Physiol.* 259:H281-H289.

Yuan, X., Goldman, W.F., Tod, M.L., Rubin, L.J. and Blaustein, M.P., 1993, Hypoxia reduces potassium currents in cultured rat pulmonary but not mesenteric arterial myocytes. *Am. J. Physiol.* 264:L116-L123.

THE SODIUM GRADIENT, POTASSIUM CHANNELS, AND REGULATION OF CALCIUM IN PULMONARY AND MESENTERIC ARTERIAL SMOOTH MUSCLES: EFFECTS OF HYPOXIA

Xiao-Jian Yuan[1-4], Carmen G. Salvaterra[2-4],
Mary L. Tod[1-4], Magdalena Juhaszova[1,4], William F. Goldman[1,4],
Lewis J. Rubin[1-4], and Mordecai P. Blaustein[1,3,4]

[1]Department of Physiology, [2]Division of Pulmonary and Critical Care Medicine,
[3]Department of Medicine, and [4]Hypertension Center
University of Maryland School of Medicine, Baltimore, MD 21201, U.S.A.

INTRODUCTION

Contraction in vascular smooth muscle (VSM) is generally initiated by the membrane excitation that triggers an increase in cytoplasmic free Ca^{2+} ($[Ca^{2+}]_i$) which then activates the contractile apparatus (19, 26). In general, $[Ca^{2+}]_i$ can be increased by i) Ca^{2+} influx, through voltage-gated Ca^{2+} channels by depolarization of the plasma membrane, and/or through receptor-operated Ca^{2+} channels by vasoconstrictive mediators; ii) Ca^{2+} release from sarcoplasmic reticulum (SR), mitochondrial, and other intracellular Ca^{2+} stores; iii) decreased Ca^{2+} extrusion (via Na-Ca exchange, Ca^{2+}-ATPase) and sequestration (via mitochondria, SR, Ca^{2+}-binding proteins); and iv) increased Ca^{2+} entry via Na-Ca exchange. Ca^{2+} influx through voltage-gated Ca^{2+} channels is controlled mainly by the membrane potential (E_m) (35) that is dominated by K^+ channel permeability and the transmembrane K^+ distribution (14). The smooth muscle cell membrane possesses a high membrane input resistance (13, 35, 56); thus, a small decrease in K^+ conductance should cause a relatively large depolarization, which should, in turn, open voltage-gated Ca^{2+} channels and thereby increase $[Ca^{2+}]_i$.

The mechanisms underlying hypoxic pulmonary vasoconstriction (HPV) have been investigated ever since 1876, when Lichtheim (29) first discovered that pulmonary arterial pressure rose during asphyxia. This phenomenon was further described by Bradford and Dean (1894) (10), and by Plumier (1904) (38), and studied extensively by von Euler and

Ion Flux in Pulmonary Vascular Control, Edited by
E.K. Weir *et al.*, Plenum Press, New York, 1993

Liljestrand (1946) (52). Physiologically, HPV is a mechanism for matching of ventilation (V_A) and perfusion (Q) in the lung (V_A/Q ratio). Persistent constriction of the pulmonary vasculature during hypoxia, however, may lead to pulmonary hypertension, as observed in patients with chronic obstructive pulmonary disorders and/or pulmonary heart disease, as well as in people residing at high altitude. Many investigators have demonstrated that Ca^{2+} is essential for the initiation of HPV (33, 54); however, the precise sequence of events leading from hypoxia to pulmonary vasoconstriction is still obscure.

Hypoxia elicits pulmonary vasoconstriction in isolated endothelium-denuded pulmonary arterial (PA) rings (5, 41, 42, 57), as well as in single PA myocytes (32, 34; Yuan et al., unpublished). This has directed attention to the intrinsic properties of the PA smooth muscle cells themselves. Bergofsky and Holtzman (6) first suggested that HPV may be due to depolarization of the vascular smooth muscle. Harder et al. (24) demonstrated that hypoxia depolarizes isolated cat small PA and induces Ca^{2+}-dependent action potentials, thereby causing these vessels to constrict. It appears that, no matter how hypoxia is sensed, membrane depolarization is involved in the transduction process that increases $[Ca^{2+}]_i$ and initiates HPV (6, 24, 48).

In VSM cells, at least three types of K^+ channels have been observed and characterized: Ca^{2+}-activated K^+ (K_{Ca}) channels, voltage-gated K^+ channels, and ATP-inhibited K^+ (K_{ATP}) channels (4, 13, 35, 50). K_{ATP} channels appear to be involved in hypoxia (or anoxia)-induced coronary (15, 51) and cerebral (8) vasodilation because of decreased ATP production. Post et al. (39) demonstrated recently that hypoxia inhibits K_{Ca} channels in canine PA smooth muscle, which thereby contribute to the depolarization induced by hypoxia, and to HPV. In rabbit carotid body type I cells, Ganfornina and Lopez-Barneo (21) identified an O_2-sensitive voltage-gated K^+ channel which was not regulated by intracellular Ca^{2+} and ATP. On exposure to hypoxia, the open-state probability of this O_2-sensitive K^+ channel was reversibly decreased without alteration of the single channel conductance. As described below, we directly investigated and compared the membrane potential and ionic currents during normoxia and hypoxia in primary cultured rat pulmonary artery (PA) and mesenteric artery (MA) myocytes. The effects of hypoxia on tonic (unstimulated) tension in isolated PA rings and the role of Na-Ca exchange in HPV were also investigated. The results of these studies lead to the conclusion that hypoxia closes some voltage-gated K^+ channels in PA, but not MA cells. This depolarizes the PA cells, promotes opening of voltage-gated Ca^{2+} channels and increases Ca^{2+} entry. As a consequence, $[Ca^{2+}]_i$ increases and raises tonic tension.

MATERIALS AND METHODS

Arterial Rings

Intra- (<400 μm, OD) and extra-pulmonary (~2 mm, OD) arterial rings from Sprague Dawley rats (240-270 g) were suspended in a muscle bath between tungsten or stainless steel wires, one of which was connected to an isometric transducer. Isometric tension was continuously monitored and recorded on a strip-chart recorder. The tissues were perfused with modified Krebs solution (MKS) in a small-volume (0.75 ml) chamber at a rate of 2.5 ml/min at 37°C (57).

Cell Preparation and Culture

Pulmonary (PA) and mesenteric (MA) arteries were dissected from Sprague-Dawley rats (125-250 g); the endothelium and adventitia were carefully removed by gentle digestion of the tissues with collagenase. The resulting smooth muscles were then digested with collagenase (2 mg/ml, Worthington Biochemical Co., Freehold, NJ) and elastase (0.5 mg/ml, Sigma Chemical, St. Louis, MO); the smooth muscle cells were separated and cultured in 10% fetal bovine serum culture medium for 3-6 days at 37°C before experiments (56). The cellular purity of cultures was monitored with specific monoclonal antibody (Boehringer Mannheim, Indianapolis, IN) raised against smooth muscle actin; a secondary antibody (Jackson ImmunoResearch, West Grove, PA) conjugated with fluorescein isothiocyanate was used to display fluorescent image. Virtually all cells in the cultures crossreacted with this actin antibody, indicating that > 99% of the primary cultured PA and MA cells were smooth muscle cells (Fig. 1).

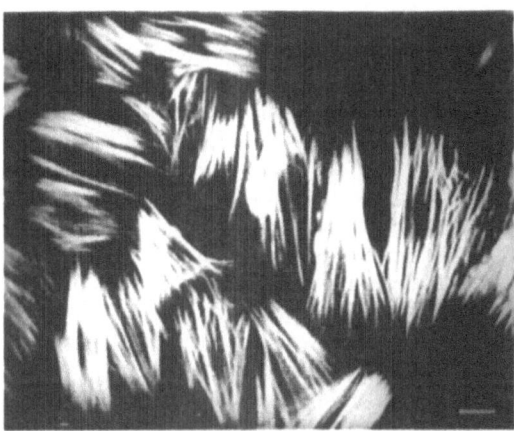

Figure 1. Primary cultured pulmonary myocytes stained with anti-smooth muscle actin antibody conjugated with fluorescein isothiocyanate. Bar = 10 μm.

Immunofluorescence Labeling

The primary cultured PA cells, attached to coverslips, were fixed by immersion in fixative at room temperature (22-24°C) for 45 min. The fixative consisted of 0.45% (w/v) formaldehyde, 75 mM cyclohexylamine (free base), 75 mM NaCl, 10 mM EGTA, 10 mM $MgCl_2$, 10 mM PIPES, buffered to pH 6.5 with HCl. The cells were then permeabilized in fresh fixative containing 0.5% Brij 58 (Sigma) for 10 min. The cultures were washed in 75 mM Tris-HCl, 10 mM $MgCl_2$, 10 mM NaN_3 (pH=6.7) for 30 min to remove and neutralize unreacted aldehyde. Nonspecific labeling was blocked by incubation in normal donkey serum (Jackson ImmunoResearch) diluted 1:6 in antibody buffer (containing 0.5 M NaCl, 10 mM $MgCl_2$, 20 mM Tris-HCl, 20 mM NaN_3, pH=7.4) for 4 hours. Affinity-purified anti-Na-Ca exchanger polyclonal antibody (27) was diluted 1:15 in antibody buffer and applied to the cultured PA cells for 17 hours at room temp. The cells were then washed in antibody buffer, 3 x 30 min, labelled with affinity purified Cy3-conjugated donkey anti-rabbit IgG (Jackson ImmunoResearch) diluted 1:100 in antibody buffer for 1 hr, and washed again in antibody buffer (3x30 min). The cells were mounted in 10% 1 M Tris-HCl, 90% glycerol (pH=8.5 containing 1 mg/ml p-phenylenediamine. The samples were examined with a Nikon Diaphot microscope (40x objective) and photographed on Kodak Ektachrome professional film (22).

Patch Clamp

Micro-hematocrit capillary glass tubes (VWR Scientific Inc., Bridgeport, NJ) were used to make high-resistance (2-6 GΩ) seals with the cell membrane for whole-cell voltage clamp studies (23). The pipettes had resistances of 2-4 MΩ when filled with high K^+ (125 mM) pipette solution. Whole-cell currents and membrane potential (I=0) were measured with the Axopatch-1D amplifier (Foster City, CA). The voltage protocols and data acquisition were carried out using a pClamp program, with a Tl-1 DMA digital interface coupled to an IBM compatible computer. Whole-cell currents were filtered at 2 kHz. Linear leakage current was subtracted using appropriately scaled, ensemble averaged current traces evoked by hyperpolarizing voltage steps from a potential 10 mV negative to the holding potential of -70 mV. A high membrane input resistance (R_m), 2-3 GΩ, was observed in most of the PA and MA cells used in these experiments, and series resistance (R_s) compensation was therefore used only in some experiments [(R_s/R_m) ≈ 0.3%] (13, 37). Values of R_s were calculated for cells from the decay time constant of the capacitive transient recorded in response to small hyperpolarizing voltage steps and from measurements of cell capacitance. On average R_s was 7.8±1.0 MΩ (n=24) and 7.4±1.9 MΩ (n=16) for primary cultured PA and MA cells, respectively.

Reagents and Solutions

For arterial ring experiments, the modified Krebs solution (MKS) had the following composition (mM): 138 NaCl, 4.7 KCl, 1.2 NaH_2PO_4, 1.2 $MgSO_4$, 5 N-2-hydroxyethylpiperazine-N'-2-ethanesulfonic acid (HEPES), 1.8 $CaCl_2$ and 10 glucose, buffered to pH 7.35-7.40 with 10 M NaOH. In K^+-rich solutions (e.g., 20 mM K^+-MKS), the NaCl in the MKS was replaced, mole-for-mole, by KCl. In Ca^{2+}-free MKS, the $CaCl_2$ was replaced by equimolar $MgCl_2$. In some instances (low Na^+ medium), all of the NaCl was replaced isosmotically by LiCl or N-methyl-glucamine (NMG) to reduce extracellular Na^+ ($[Na^+]_o$).

For patch clamp experiments, coverslips with cells were positioned in a recording chamber (≈0.75 ml) and superfused with external (bath) solution at a rate of 0.6-1.2 ml/min. The standard extracellular physiological salt solution (PSS) used for recording outward K^+ (I_K) currents contained (mM): 141 NaCl, 4.7 KCl, 1.8 $CaCl_2$, 1.2 $MgCl_2$, 10 HEPES, 10 glucose, buffered to pH 7.4 with 5 M NaOH. When the extracellular Ca^{2+} ($[Ca^{2+}]_o$) was varied between 0 (Ca^{2+}-free PSS) and 1.8 mM, the Mg^{2+} concentration was varied to compensate for surface charge and osmolarity. The internal (pipette) solution used for recording outward K^+ currents consisted of (mM): 125 KCl, 4 $MgCl_2$, 10 HEPES, 10 ethyleneglycol-bis-(β-aminoethyl ether)-N,N,N',N'-tetraacetic acid (EGTA), 5 Na_2ATP (or 5 MgATP), buffered to pH 7.2 with 1 M KOH.

Cromakalim (Beecham Pharmaceuticals, Epsom, Surrey, UK) and glibenclamide (Hoechst-Roussel Pharmaceuticals, Somerville, NJ) were both dissolved in ethanol to make stock solutions of 1 mM; aliquots of the stock solutions were diluted 1:250-500 into MKS or PSS. 4-Aminopyridine (4-AP, Aldrich Chemical Co., Milwaukee, WI) was dissolved directly in PSS immediately before experiments in which this agent was used.

Oxygen Tension

Normoxic conditions were established by bubbling the superfusion fluid with room air to achieve a PO_2 > 110 Torr (at 22-24°C). Hypoxic conditions were established by bubbling the superfusion MKS with 100% N_2 to achieve PO_2 < 40 Torr for PA ring experiments and by directly dissolving 1 mM sodium dithionite ($Na_2S_2O_4$, an oxygen scavenger, Sigma) (45) in extracellular PSS to achieve PO_2 < 74 Torr for patch clamp experiments, respectively. Oxygen tension was measured by using a pH/blood gas analyzer in arterial ring experiments and by an oxygen electrode (Microelectrodes Inc., Londonderry, NH) in single smooth muscle cell experiments. Sodium dithionite had no effect on outward K^+ currents (I_K) unless accompanied by a reduction in O_2 tension (55).

Measurement of $[Ca^{2+}]_i$

Rat PA cells grown on cover slips were loaded with fura-2 by incubation in culture medium containing 5 μM fura-2/AM for 45 min at 22°C. The cover slips were then transferred to the microscope stage and washed with PSS for 30 min before study. A quantitative fluorescence microscopy system, Photoscan M-series (Photon Technology International Inc., South Brunswick, NJ), was used to determine the average $[Ca^{2+}]_i$. Filtered light from a xenon lamp was passed to a rotating chopper and then alternately through either of two filters, centered at 340 nm and 380 nm, onto the cells under study. Fura-2 fluorescence (510 nm emission) from rat PA cells, as well as background fluorescence, were collected and monitored continuously using a photomultiplier tube. $[Ca^{2+}]_i$ was related to the ratio of fluorescence elicited at 340 and 380 nm, as previously described (22).

RESULTS

Hypoxia Constricts Isolated PA Rings

The effects of hypoxia on resting tension (without precontraction) in a small intra-(SPA) and a large extra-PA (LPA) ring are illustrated in Figures 2. Hypoxia (reducing PO_2 from 120 to 23 Torr) gradually and substantially increased resting tension in SPA; this contractile effect was abolished during Ca^{2+}-free MKS perfusion (Fig. 2A). In LPA, hypoxia also increased resting tension, as well as augmented the K^+-induced contractions (ΔT_K) at $[K^+]_o \leq 15$ mM, but attenuated ΔT_K at $[K^+]_o = 100$ mM (Fig. 2B). In contrast, the SMA rings did not contract in response to hypoxia, but sometimes relaxed slightly (57). These data demonstrate that acute hypoxia causes extrapulmonary arteries as well as intra-pulmonary arteries in the rat to contract in a Ca^{2+}-dependent fashion. Furthermore, prior administration of vasoconstrictors (5, 25, 42) is not required to evoke the hypoxic effect.

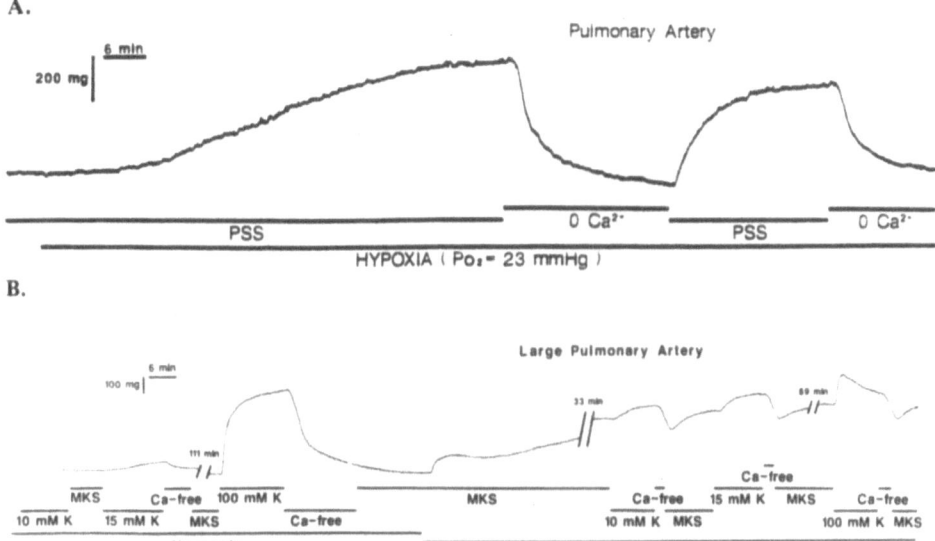

Figure 2. Effect of hypoxia on the resting tension and the K^+-induced contractions in isolated small (A) and large (B) PA rings.

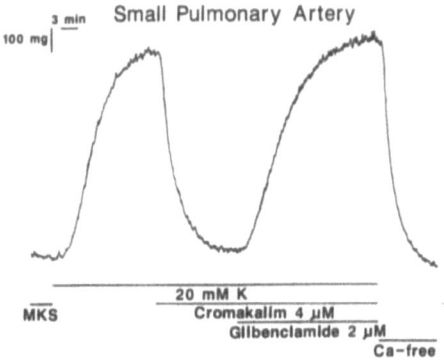

Figure 3. Effects of cromakalim and glibenclamide on 20 mM K^+-induced contraction in an isolated small pulmonary arterial ring.

The K_{ATP} channel activator, cromakalim, and the blocker, glibenclamide, were used to test whether K_{ATP} channels play a role in PA excitation-contraction regulation. Cromakalim (4 μM) had no effect on the baseline resting tension (57), but almost completely blocked the 20 mM K^+-induced contraction in SPA rings; these effects were reversed completely by 2 μM glibenclamide (Fig. 3). The hypoxia-induced increase in PA resting tension was also significantly inhibited by cromakalim, and this inhibitory effect was totally reversible by addition of glibenclamide (57). These data suggest that opening of K_{ATP} channels in PA rings should hyperpolarize the PA cells and thereby cause relaxation. Removal of endothelium from LPA did not prevent the hypoxia-induced increase in tension. In fact, the response to hypoxia was actually enhanced both in terms of magnitude and time course (57). Removal of the endothelium also did not alter the inhibitory effects of Ca^{2+}-free MKS or addition of verapamil and cromakalim during hypoxia (57). This indicates that the hypoxic vasoconstrictive response is an unique property of PA smooth muscle, and is not dependent upon the endothelium (32, 57). The implication is that freshly isolated or cultured PA myocytes may be useful preparations for studying the cellular mechanisms underlying hypoxia-induced pulmonary vasoconstriction.

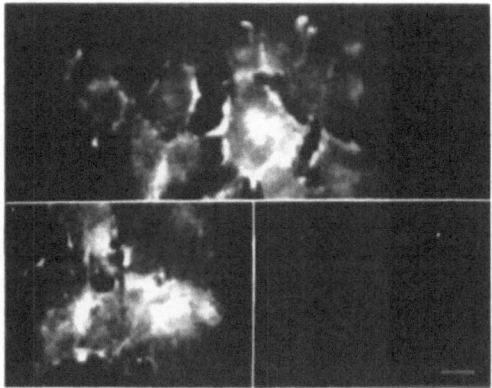

Figure 4. Distribution of Na-Ca exchanger in primary cultured PA smooth muscle cells with immunocytochemical methods. Bar = 10 μm.

The Na-Ca Exchanger in Rat Pulmonary Arteries

Identification of the Na-Ca Exchanger in Cultured PA Myocytes. One of the mechanisms that helps to regulate Ca^{2+} in VSM cells (7, 9), is the Na-Ca exchanger that is regulated by the transplasmalemmal Na^+ electrochemical gradient. Indeed, the exchanger may control the amount of Ca^{2+} in the SR stores, and may thereby control VSM contractility (7). We used polyclonal antibodies, raised against purified cardiac Na-Ca exchanger (1), to identify the Na-Ca exchanger in cultured rat PA myocytes with immunocytochemical methods (Fig. 4). This antibody cross-reacts specifically with 160, 120, and 70 kDa peptides from cultured rat VSM cells (Juhaszova et al., unpublished) as well as from rat and dog (1; Juhaszova et al., unpublished) heart. The distribution of the antibody (=distribution of Na-Ca exchanger) on the surface of primary cultured PA myocytes is shown in Figure 4. These cells were incubated with anti-Na-Ca exchanger antiserum followed by a Cy3-conjugated secondary antibody. This revealed a punctate labeling on the cell surface that was condensed at the cell edges. The labeling of the Na-Ca exchanger appears to form a reticular pattern, which is very similar to that in primary cultured rat mesenteric and aortic smooth muscle cells, but different from that in cultured cardiac myocytes (20, 27) and neurons (31). In controls (lower right panel in Fig. 4) treated with pre-immune serum and with Cy3 secondary antibody alone, labeling was observed only in the nuclear areas; this is due to the non-specific binding of secondary antibody.

Physiological Properties of the Na-Ca Exchanger in PA Rings: the Effect of Hypoxia. The operation of the Na-Ca exchanger can be tested by removing $[Na^+]_o$ (to inhibit Ca^{2+} extrusion and promote Ca^{2+} entry via the exchanger), or by treating the preparation with ouabain (to raise $[Na^+]_i$ and promote Ca^{2+} entry via the exchanger). The effects of replacing external Na^+ by Li^+ or NMG on resting tension in SPA and SMA rings are illustrated in Figures 5 and 6. Removal of $[Na^+]_o$ (replaced by either Li^+ or NMG) significantly increased the resting tension following K^+-induced contractions (Fig. 5A and B) or hypoxia-induced contractions (Fig. 6B, and Fig. 6) in PA, but not in MA. The relaxation rate (Fig. 6) was significantly slowed by low Na^+ MKS during normoxia and hypoxia (Figs. 6 and 7A); it is also noted from Figure 7A that hypoxia *per se* also slowed relaxation in SPA rings. Figure 7B shows that the hypoxia-induced increase of the resting tension in SPA was potentiated by replacing $[Na^+]_o$ with Li^+. These data demonstrate that the transmembrane Na^+ gradient also plays an important role in the regulation of $[Ca^{2+}]_i$ in PA via Na-Ca exchange; the inhibition of this mechanism during hypoxia modulates HPV, possibly by decreasing Ca^{2+} extrusion via Na-Ca exchange (44).

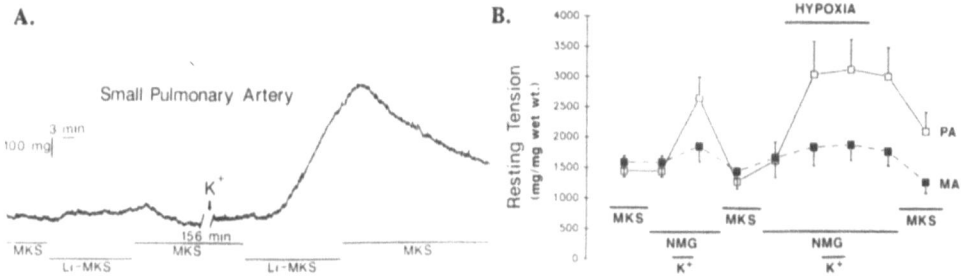

Figure 5. Effect of external Na^+ removal on resting tension in small PA (**A and B**) and MA (**B**) rings during normoxia and hypoxia. (**A**) Replacing external Na^+ by Li^+ did not increase SPA resting tension unless K^+-induced contractions (K^+) were applied. (**B**) Mean ($\pm$ SE) resting tensions of PA and MA rings perfused in normal (MKS) and low Na^+ (NMG) MKS during normoxia and hypoxia. A series of K^+ contractions (10-100 mM) were induced during NMG perfusion in both normoxia and hypoxia (K^+).

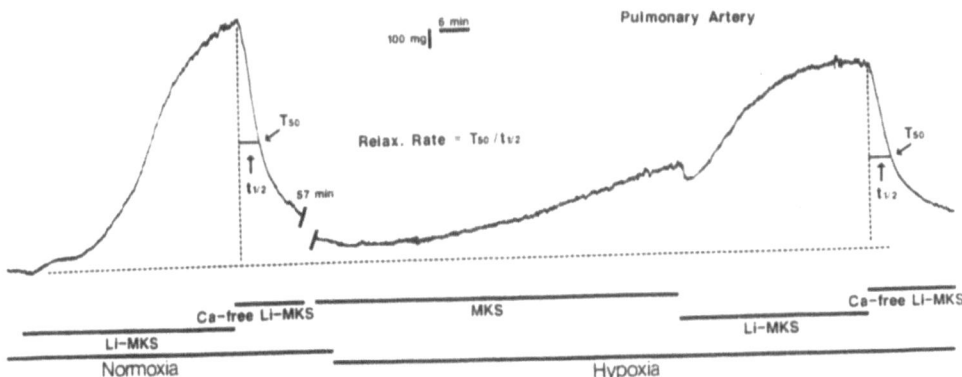

Figure 6. An original recording demonstrating the method used to calculate the relaxation rate, defined as one-half the peak tension amplitude (T_{50}) during a low Na^+ induced contraction divided by the time ($t_{1/2}$) to relax to this tension. Low Na^+-induced contractions were produced by perfusion with Li^+-MKS and relaxations were produced by removing external Ca^{2+} during normoxia and hypoxia in rat small PA.

A.   **B.**

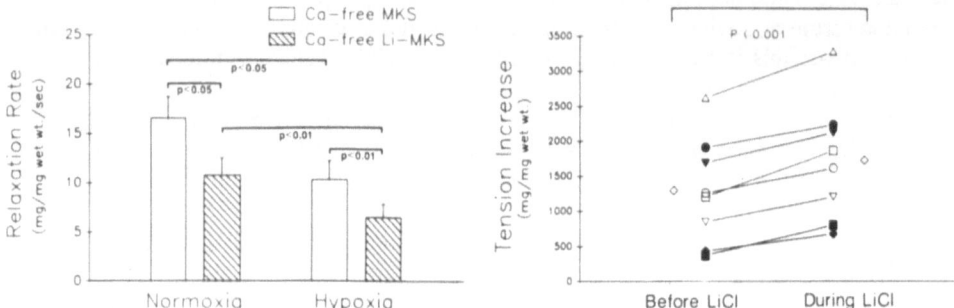

Figure 7. Effect of external Na^+ removal on the relaxation rate (**A**) and the resting tension (**B**) in small PA during normoxia and hypoxia. (**A**) Bars represent means $\pm$ SE of data from 10 experiments. (**B**) Replacement of external Na^+ with Li^+ potentiated the hypoxia-induced increase of PA resting tension.

Effects of Hypoxia on $[Ca^{2+}]_i$ in Primary Cultured PA Myocytes

We have shown that the HPV response in pulmonary arterial rings is Ca^{2+} dependent and is a function of the unique properties of PA cells (57). The effect of hypoxia on $[Ca^{2+}]_i$ was determined in cultured PA cells using fura-2 and quantitative fluorescence microscopy. Representative traces depicting the time course of the decline in superfusate PO_2 and the concomitant rise in $[Ca^{2+}]_i$ are shown in Figure 8A. Reduction in PO_2 from 150 to about 4 Torr was associated with a rise in $[Ca^{2+}]_i$ from approximately 90 nM to 215 nM. Peak levels of $[Ca^{2+}]_i$ were attained within 1 min and then declined until new steady state $[Ca^{2+}]_i$ were reached at 4 min (175-200 nM). Since the onset of the rise in $[Ca^{2+}]_i$ occurred when the PO_2 had decreased only minimally, the hypoxia-induced elevation appeared to result, at least in part, from the change in PO_2, rather than the absolute PO_2. The rise in $[Ca^{2+}]_i$ during hypoxia was attenuated by the Ca^{2+} channel blockers, nifedipine and verapamil (43). This hypoxia-induced elevation in $[Ca^{2+}]_i$ may explain the increase in PA tension and responsiveness observed in PA rings during hypoxia (Figs. 2 and 3). The following sections focus on possible mechanisms for this hypoxia-induced rise in $[Ca^{2+}]_i$.

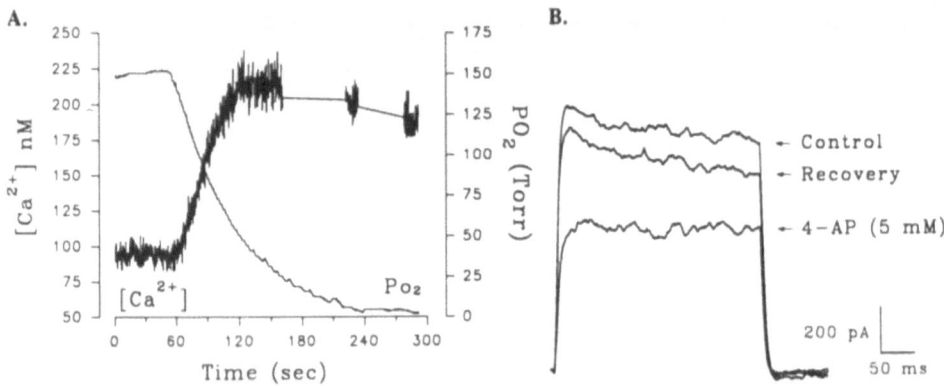

Figure 8. (**A**) Representative traces illustrating the effect of decline in PO_2 on $[Ca^{2+}]_i$ in primary cultured PA cells. PO_2 trace depicts the time-course of the decline in PO_2 following the onset of superfusion with sodium dithionite (0.8 mM; right-hand ordinate). Time-course of the hypoxia-induced elevation in $[Ca^{2+}]_i$ is depicted by the $[Ca^{2+}]$ trace (left-hand ordinate). Periods of data acquisition are connected by straight lines through periods when the shutter was closed. (**B**) Effect of 4-aminopyridine (4-AP, 5 mM) on I_{out} elicited by a test pulse of +60 mV from a holding potential of -70 mV in a primary cultured PA cell.

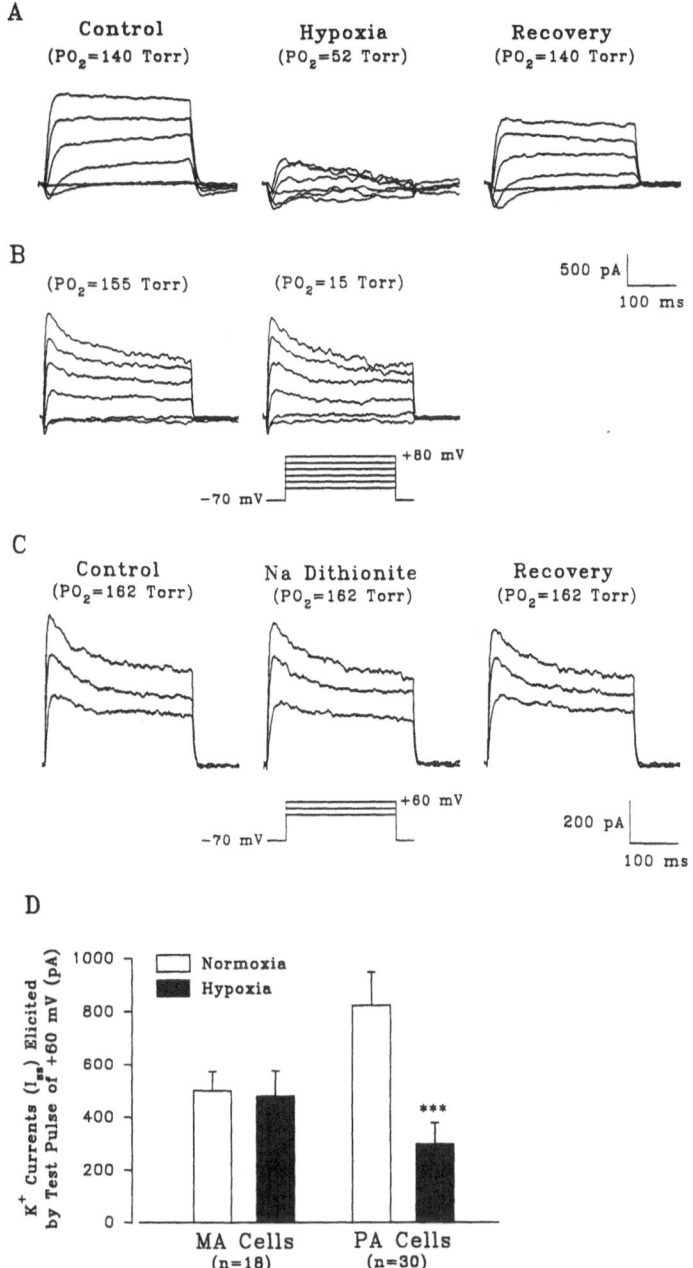

Figure 9. Effects of hypoxia on inward and outward currents in primary cultured rat PA and MA cells. The representative families of superimposed current records were elicited by depolarizing the cells to a series of test potentials between -20 and +80 mV for a PA cell (**A**) and a MA cell (**B**) during normoxia and hypoxia. (**C**) Effects of Na dithionite without accompanying hypoxia on outward K^+ currents in a PA cell. The superimposed current traces in (**C**) were elicited by depolarizing the cell to 20, 40, and 60 mV. The holding potential in all cases was -70 mV. PO_2 during normoxic and hypoxic conditions are shown in parentheses. Leakage currents were subtracted. (**D**) Mean ($\pm$SE) outward K^+ currents (I_{ss}, steady-state current) of PA and MA cells elicited by a test potential of +60 mV during normoxia (PO_2 was 142 ± 2 Torr for PA cells, and 145 ± 2 Torr for MA cells) and during hypoxia (PO_2 was 43 ± 5 Torr for PA cells, and 47 ± 5 Torr for MA cells). The numbers of cells are indicated in parentheses. ******* $P<0.001$, hypoxia *vs* normoxia (from Yuan *et al.*[55] with permission from *Am. J. Physiol.*).

A

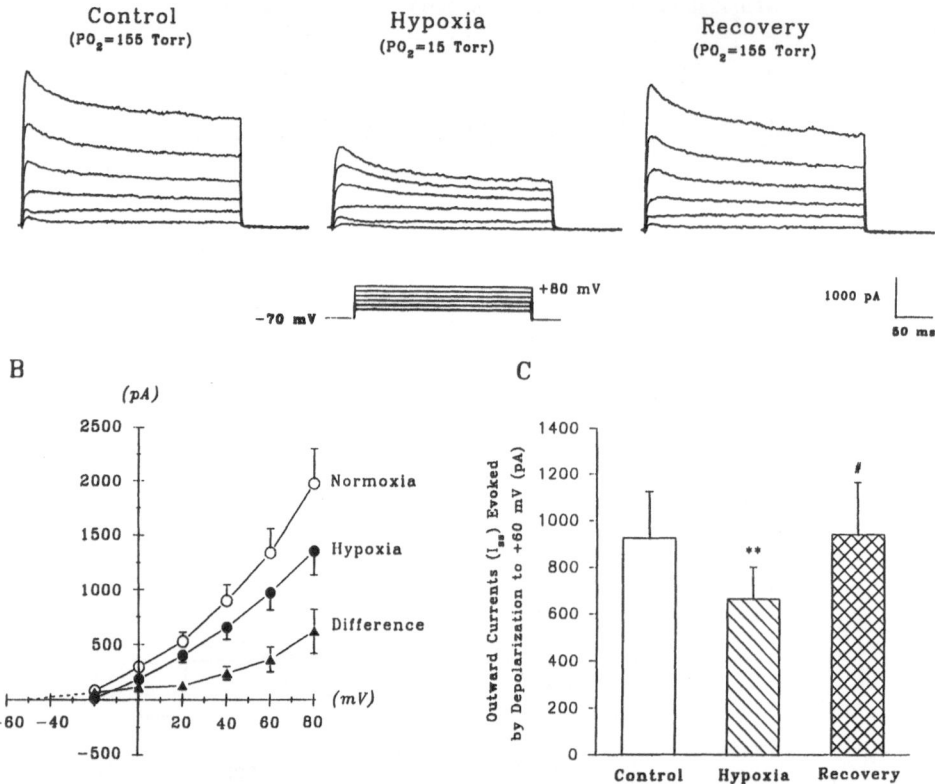

Figure 10. Effect of hypoxia on I_{out} in PA cells in the absence of $[Ca^{2+}]_o$. The Ca^{2+}-free PSS in all experiments contained 2 mM EGTA. **(A)** Representative current recordings elicited by depolarizing the PA cell to a family of test potentials from -20 to +80 mV during normoxia, hypoxia and normoxic recovery. Note that I_{in} was eliminated in the absence of $[Ca^{2+}]_o$. Linear leakage currents were subtracted. **(B)** Composite I-V curves of steady-state I_{out} (I_{ss}) were obtained from 7 PA cells bathed in Ca^{2+}-free PSS during normoxia and hypoxia. The hypoxia-sensitive component of I_{ss} (*difference* = I_{ss} during normoxia minus I_{ss} during hypoxia) is also shown. The broken line shows the linear extrapolation of the hypoxia-sensitive I_{ss} from 0 mV to the abscissa intercept (-52 mV). The I-V curve during hypoxia is significantly different from the I-V curve during normoxia ($P < 0.01$, *ANOVA*). **(C)** Mean I_{ss} evoked by repeated test pulses of +60 mV from a holding potential of -70 mV during normoxia ($PO_2 = 147 \pm 2$ Torr), hypoxia ($PO_2 = 42 \pm 9$ Torr), and normoxic recovery ($PO_2 = 145 \pm 3$ Torr). Data are means $\pm$ SE of I_{ss} values from 12 primary cultured PA cells. ** $P < 0.01$, hypoxia *vs* normoxia; # $P < 0.05$, recovery *vs* hypoxia (from Yuan *et al.*[55] with permission from *Am. J. Physiol.*).

Effects of Hypoxia on Outward K$^+$ Currents in Cultured PA and MA Myocytes

Both voltage-gated inward Ca^{2+} currents (I_{in}) and outward K$^+$ currents (I_{out}) are observed in primary cultured PA and MA cells (Fig. 9 left-hand column; 56). I_{out}, measured in standard PSS, using pipette solution containing Ca^{2+}-free media with 10 mM EGTA and 125 mM KCl (56), consists of three components, a rapidly inactivating (transient) component (I_{rt}, measured at the peak of I_{out}), a slowly inactivating component (I_{st}), and a non-inactivating (steady state) component (I_{ss}, measured at 250-290 ms). 4-Aminopyridine (4-AP, 5 mM), a voltage K$^+$ channel blocker (4, 37), substantially inhibited both I_{rt} and I_{ss} components of I_{out} in PA cell (Fig. 8B), but reduced I_{rt} to a greater extent; 4-AP had similar effects in MA cells (data not shown).

Reducing PO_2 in the bath solution from 140-155 Torr to $\leq$ 52 Torr, significantly and reversibly inhibited both I_{rt} and I_{ss} in PA cells (Fig. 9A), but did not affect I_{out} in MA cells (Fig. 9B) bathed in standard PSS that contained 1.8 mM Ca^{2+}. The apparent increase in I_{in} during hypoxia in PA cells was probably due to the decrease in I_{rt} rather than to a direct enhancement of I_{in}. Furthermore, when PA cells were superfused with normoxic solution (vigorously bubbled with room air) containing 1 mM Na dithionite, no reduction in I_{out} was observed (Fig. 9C). These findings demonstrate that the hypoxia, and not the Na dithionite, *per se*, caused I_{out} to decline in PA but not MA cells. The composite results from 30 PA cells and 18 MA cells showed that hypoxia inhibited I_{ss} in PA cells by about 60%, without affecting I_{ss} in MA cells (Fig. 9D).

The effects of removing extracellular Ca^{2+} on the hypoxia-induced inhibition of I_{out} were tested in PA cells bathed in Ca^{2+}-free PSS containing 2 mM EGTA (Fig. 10). Reduction of PO_2 from 155 to 15 Torr, still significantly and reversibly reduced both I_{rt} and I_{ss} components of I_{out} (Fig. 10A). The composite I-V curves obtained from 7 PA cells bathed in Ca^{2+}-free PSS during normoxia and hypoxia are illustrated in Figure 10B. The hypoxia-sensitive component of I_{ss}, obtained from the difference between I_{ss} during normoxia and I_{ss} during hypoxia, was activated at about -52 mV. This potential is more negative than the resting membrane potential (-41 mV) that we measured in these PA cells. These data support the hypothesis that HPV is due to depolarization that results from the inhibition of voltage-gated K^+ conductances (2, 39, 41, 53, 55, 57).

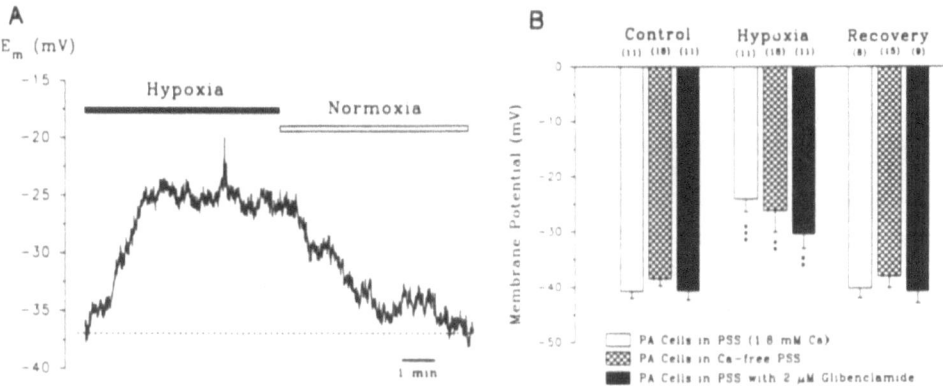

Figure 11. Effects of hypoxia on membrane potential (E_m) measured with current-clamp (I=0) in primary cultured PA cells. (A) E_m determined in the cell bathed in standard PSS during normoxia was about -45 mV. The cell depolarized reversibly during hypoxia (PO_2=15 Torr). (B) Mean ($\pm$ SE) membrane potentials for cells bathed in standard PSS (open bars), Ca^{2+}-free PSS (crosshatch bars), and glibenclamide-containing PSS (solid bars) during normoxia (PO_2 was 155$\pm$1, 145$\pm$2, and 132$\pm$1 Torr), hypoxia (PO_2 was 17$\pm$3, 16$\pm$4, and 3$\pm$1 Torr) and normoxic recovery (PO_2 was 151$\pm$2, 142$\pm$2, and 129$\pm$2 Torr), respectively. *** $P<0.001$, ** $P<0.01$, hypoxia *vs* control and *vs* recovery (modified from Yuan et al.[55]).

Effects of Hypoxia on Membrane Potential in Cultured PA Smooth Muscle Cells

The effect of hypoxia on the resting membrane potential (E_m) was tested in primary cultured PA smooth muscle cells. As illustrated in Figure 11A, hypoxia substantially and reversibly depolarized PA cells bathed in standard PSS; the same result was obtained from cells bathed in Ca^{2+}-free PSS containing 0.1 mM EGTA (Fig. 11B) (55). Hypoxia did not affect E_m in MA cells (data not shown). The composite results obtained from the PA cells bathed in standard PSS, Ca^{2+}-free PSS and glibenclamide-containing PSS, are summarized in Figure 11B. Reducing PO_2 from 155$\pm$1 Torr to 17$\pm$3 Torr decreased E_m by about 16 mV in PA cells bathed in PSS. Both perfusion of the PA cells with Ca^{2+}-free PSS and preincubation of the cells in 2 μM glibenclamide did not prevent the hypoxia-induced depolarization.

DISCUSSION

Hypoxic Pulmonary Vasoconstriction is an Intrinsic Property of PA Smooth Muscles

Pulmonary vasoconstriction is triggered by a rise in cytosolic Ca^{2+}, and can be elicited either by an increased Ca^{2+} influx, or by a decrease of Ca^{2+} extrusion and sequestration. Studies on systemic arterial smooth muscles showed that the removal of Ca^{2+} from the cytosol, required for relaxation, can be achieved by sequestering the Ca^{2+} into SR and/or extruding it across the sarcolemma (9). Rapid removal of Ca^{2+} from the cytosol can be accomplished by the SR Ca^{2+} pump as well as the Na-Ca exchanger. HPV requires elevation of $[Ca^{2+}]_i$ which can be initiated by membrane depolarization and Ca^{2+}(or second messenger)-induced Ca^{2+} release; however, the regulation of extrusion and sequestration of increased $[Ca^{2+}]_i$, via, e.g., Na-Ca exchange, may also play an important role in modulation of hypoxia-induced pulmonary vasoconstriction (18, 44, 49). Similar distribution of the Na-Ca exchanger in pulmonary (Fig. 4) and systemic VSM cells, provides direct evidence that the Na-Ca exchange mechanism also contributes to the regulation of $[Ca^{2+}]_i$ in PA cells.

Our data demonstrate that hypoxia substantially and reversibly increases the resting (unstimulated) tension in isolated intra- and extra-pulmonary arteries (Fig. 2). This effect is abolished by removal of extracellular Ca^{2+} and inhibited by verapamil, a voltage-gated Ca^{2+} channel antagonist, and cromakalim, a K_{ATP} channel activator (57). In contrast, comparable-size MA rings do not constrict in response to hypoxia (57). The Ca^{2+}-dependence of the hypoxia-induced increase in PA resting tension is consistent with either increased sensitivity of the intracellular biochemical machinery to the available Ca^{2+} or increased delivery of Ca^{2+} to the contractile apparatus. The latter can be achieved by an increase in Ca^{2+} influx through voltage-gated Ca^{2+} channels, or increased Ca^{2+} release (Fig. 12). In addition, enhancement of the hypoxia-induced increase in resting tension (Fig. 7B) along with the hypoxia-induced inhibition of the relaxation rate (Fig. 7A) by removal of extracellular Na^+ suggests that the Ca^{2+} extrusion via Na-Ca exchange may be inhibited during hypoxia; this might play a modulatory role in HPV, but it does not appear to be the primary cause of HPV.

Based on the data obtained from isolated PA rings, we hypothesized that HPV is an intrinsic mechanism of PA myocytes that is primarily related to hypoxia-induced depolarization resulting from either a decrease in K^+ conductance or an increase in Ca^{2+} conductance (2, 39, 41, 53, 57).

Effects of Hypoxia on I_{out} and E_m in Cultured PA and MA Smooth Muscle Cells

The resting membrane potential of most cells, including smooth muscle, is dominated by the relatively large K^+ permeability and the transmembrane K^+ distribution; other ions (e.g., Na^+ and Cl^-) generally make a much smaller contribution to the resting membrane potential (14, 35). Thus, hypoxia-induced inhibition of K^+ channels that are open at rest would lead to a gradual enhancement of the resting tension as we observed in isolated and endothelium-denuded pulmonary arterial rings (57). This hypothesis is also consistent with the report that some K^+ channel blockers (4-aminopyridine and tetraethylammonium, but not glibenclamide), cause vasoconstriction in the normoxic lung (25), suggesting that a background K^+ conductance helps maintain the low vascular resistance of the normoxic lung.

In our experiments, hypoxia (PO_2 decreased from ~150 to $\leq$ 74 Torr) significantly and reversibly attenuated the voltage-gated components (both I_n and I_{ss}) of the outward K^+ current in PA but not MA smooth muscle cells (Figs. 9 and 10). The selective effect of hypoxia on the PA cells implies that the mechanisms that mediate this hypoxic response are prevalent in PA but not MA cells; these differences are maintained when the cells are cultured. The implication is that smooth muscle cells in different vascular beds may have some fundamentally different electrophysiological properties consistent with their specific functions.

The membrane input resistance (R_m) of resting vascular smooth muscle cells is high, on the order of 1-10 $G\Omega$ (13, 35, 37, 56). Thus, even a small decrease of outward current through hypoxia-sensitive, voltage-gated K^+ channels should depolarize the cells, as we observed (Fig. 11) (55). These hypoxia-sensitive K^+ channels activate at a potential of about -52 mV (Fig. 12B), which is more negative than the resting potential of our PA smooth muscle cells (about -40 mV).

The current through voltage-gated Ca^{2+} channels is proportional to the time a Ca^{2+} channel spends in the open state which is strongly regulated by membrane potential (E_m). The steady-state fraction of time that a Ca^{2+} channel is open (open-state probability, P_{open}) increases exponentially with membrane depolarization. The exponential relationship between P_{open} and E_m can be simply expressed as the following approximation: $P_{open} \propto$ exp (E_m/k_s), where k_s is a steepness factor (when E_m ranges between -60 mV and -40 mV) that gives an indication of the E_m sensitivity of Ca^{2+} channels (35). This equation indicates that P_{open} will increase e-fold for a depolarization of k_s millivolts. A single open Ca^{2+} channel allows Ca^{2+} ions to flow passively down their electrochemical gradient at a rate of $> 10^7$ ions per sec. Thus, the hypoxia-induced depolarization as described above could cause opening of voltage-gated Ca^{2+} channels. This should increase Ca^{2+} entry and decrease Ca^{2+} exit via the voltage-sensitive Na-Ca exchanger (7, 9, 44), thereby increasing $[Ca^{2+}]_i$. and raising tension.

Is Hypoxia-induced Inhibition of I_{out} Dependent on Ca^{2+}?

Recently, Post et al. (39) demonstrated that hypoxia ($PO_2 \approx$ 40 Torr) inhibits K^+ currents in freshly dispersed canine PA cells. In their study, the hypoxia-induced attenuation of K^+ currents was substantially enhanced when the pipette solution contained 0.1 mM EGTA instead of 5 mM EGTA, and was completely prevented by either addition of nisoldipine (a Ca^{2+} channel blocker) to the bath solution, or replacement of EGTA by 10 mM [1,2-bis(2)aminophenoxy]ethane-N,N,N',N'-tetraacetic acid (BAPTA, another Ca^{2+} chelator) in the pipette solution. They therefore concluded that hypoxia-induced inhibition of outward K^+ current and hypoxia-induced depolarization were due mainly to block of Ca^{2+}-activated K^+ channels.

In contrast, in our primary cultured rat PA cells, I_{out} was still markedly and reversibly inhibited by hypoxia when the PA cells were bathed in Ca^{2+}-free PSS containing 2 mM EGTA while the Ca^{2+}-free pipette solution contained 10 mM EGTA (Fig. 10). Clearly, a large fraction of this hypoxia-induced decrease of I_{out} was not dependent upon extracellular or intracellular Ca^{2+}, and could not be attributed to block of K_{Ca} channels. Furthermore, the fact that hypoxia caused PA cells to depolarize even in the absence of external Ca^{2+}, and with 10 mM EGTA in the pipette solution (Fig. 11B), is further evidence that these hypoxic responses are Ca^{2+}-independent. Therefore, our findings suggest that voltage-gated K^+ channels play an important role in the hypoxia-induced inhibition of I_{out} and membrane depolarization in PA cells. Moreover, in these same primary cultured rat PA cells, hypoxia (PO_2=4-29 Torr) causes $[Ca^{2+}]_i$ to rise to a new, steady level. Such a rise in $[Ca^{2+}]_i$, as a result of membrane depolarization and the opening of voltage-gated Ca^{2+} channels (and perhaps release of Ca^{2+} from internal stores), could explain the HPV response (42), as diagrammed in Figure 12. A rise in $[Ca^{2+}]_i$ should, however, *activate* K_{Ca} channels (17, 58) and *hyperpolarize* the membrane (Fig. 12). Blocking these K_{Ca} channels (39) might then augment and prolong the contractions by inhibiting repolarization (11), but we would not expect block of K_{Ca} channels to *initiate* the HPV response.

Do ATP-sensitive K^+ Channels Play a Role in Hypoxia-induced Inhibition of I_{out}?

K^+ channels that are inhibited by intracellular ATP (K_{ATP} channels) have been identified in vascular smooth muscle cells (47). Activation of K_{ATP} channels in certain systemic arterial smooth muscle cells apparently plays a role in the vasodilator response to hypoxia or ischemia (15, 16, 51). Daut and his colleagues (15, 51) found that hypoxia-induced coronary vasodilation in guinea-pig hearts was mimicked by cromakalim, a K_{ATP} channel activator, and inhibited by glibenclamide, a K_{ATP} channel blocker. They proposed that activation of K_{ATP} channels, as a result of decreased ATP production during hypoxia, hyperpolarizes vascular smooth muscle cells and relaxes coronary arteries.

In the pulmonary circulation, however, hypoxia causes vasoconstriction. Furthermore, cromakalim has no effect on isolated rat PA rings during normoxia, but inhibits HPV in these rings; glibenclamide also has no effect during normoxia, but reverses the inhibitory effect of cromakalim on HPV (57). Cromakalim opens glibenclamide-sensitive K_{ATP} channels (35, 47). Therefore, we must assume either i) that the VSM cell membranes were sufficiently polarized during normoxia so that opening of these channels had no effect on membrane potential, and therefore did not influence $[Ca^{2+}]_i$ and contraction in the PA cells, or ii) that the ATP levels were sufficiently low during normoxia so that the K_{ATP} channels were already opened (which seems unlikely). If the latter possibility were true, then, to explain our observations (57), we would need to postulate that hypoxia paradoxically increases cytosolic ATP to a level sufficient to close the K_{ATP} channels in order to depolarize pulmonary arterial smooth muscle and induce contraction. Our data indicate, however, that glibenclamide has no effect on the resting membrane potential, nor does it prevent hypoxia-induced depolarization, in primary cultured PA cells.

The relationship between hypoxia and intracellular ATP levels in PA smooth muscle cells has been controversial. Buescher and colleagues (12) have reported that the ATP concentration in isolated, degassed pig lung is unchanged or even (paradoxically) increased during hypoxia and the induction of HPV. However, several other investigators (2, 36) have shown that HPV is initiated or augmented when the ATP level [and the phosphate potential, *i.e.* the {ATP/(ADP+Pi)} ratio] is reduced as a consequence of hypoxia and, in some studies, subsequent inhibition of glycolysis (46). The weight of evidence strongly suggests that hypoxia does not raise ATP levels and, therefore, that block of K_{ATP} channels is not involved in the initiation of the HPV response. However, it is clear that K_{ATP} channels exist in rat pulmonary arterial smooth muscle (Fig. 3), and potentially could have a modulatory effect on membrane potential and contraction in response to hypoxia. Recently, K^+ channels that, in contrast to the aforementioned "classic" K_{ATP} channels, are *activated* by intracellular ATP have been observed in rat hypothalamic neurones (3) and pulmonary arterial smooth muscle cells (40). Indeed, Robertson et al. (40) have suggested that hypoxia may depolarize pulmonary arterial smooth muscle cells and promote vasoconstriction by reducing the activity of these ATP-activated K^+ channels. How these ATP-activated K^+ channels relate to our voltage-gated K^+ channels is not known. We might speculate that some or all voltage-gated K^+ channels must be phosphorylated to be fully available; dephosphorylation may reduce the conductance or block the depolarization-induced activation of these channels (28).

Summary and Conclusions: Implications for the Mechanisms of Hypoxic Pulmonary Vasoconstriction

The observations described here show that i) HPV is principally related to the intrinsic properties of pulmonary smooth muscles *per se*; ii) some aspects of the hypoxia-induced responses in the pulmonary vasculature can profitably be studied in cultured PA smooth muscle cells (34), and iii) mechanisms that affect $[Ca^{2+}]_i$ homeostasis modulate HPV. Moreover, our results demonstrate that the electrophysiological responses of PA and MA cells to hypoxia are different, and are consistent with the different contractile responses of pulmonary and mesenteric arteries to hypoxia (57).

The proposed cellular mechanisms responsible for hypoxia-induced pulmonary vasoconstriction are diagrammed in Figure 12. The initiation of HPV appears to be due to the depolarization that results from block of voltage-gated K^+ channels. The *initiation* of the response cannot be explained by the block of either K_{Ca} or K_{ATP} channels in PA cells. However, the hypoxia-induced inhibition of K_{Ca} (39) along with the glibenclamide-induced blockade of K_{ATP} (13, 16, 35, 57) could secondarily modulate the hypoxia-induced depolarization, and HPV, by inhibiting membrane repolarization and thereby maintaining: i) the depolarization, ii) the elevated $[Ca^{2+}]_i$, and iii) the contraction (Fig. 12) (11, 35).

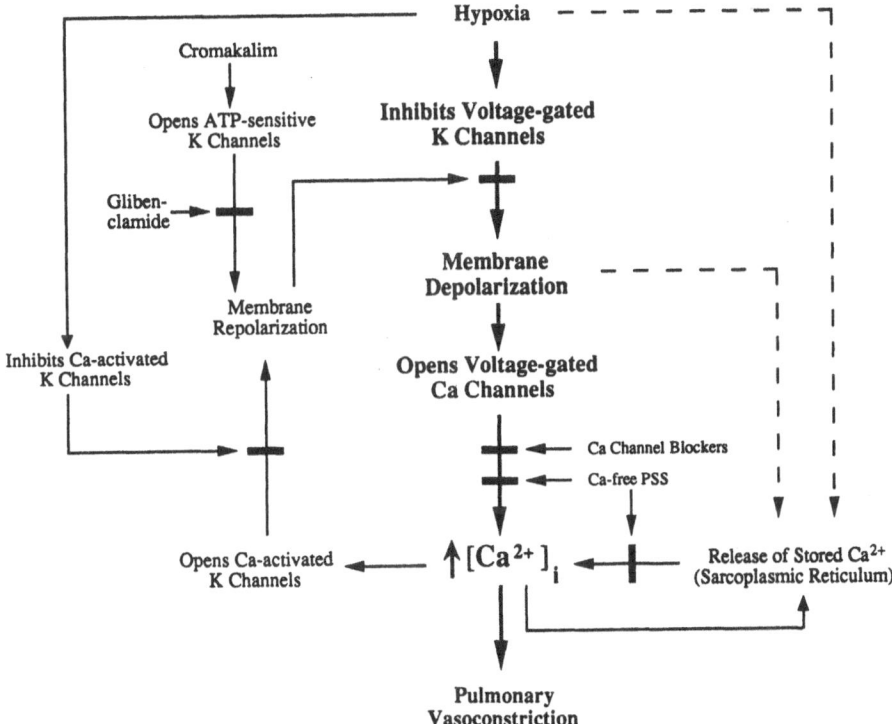

Figure 12. Schematic diagram depicting the proposed cellular mechanisms responsible for acute hypoxia-induced pulmonary vasoconstriction. The response appears to be initiated by blockade of the voltage-gated K^+ channels and the attendant depolarization and rise in $[Ca^{2+}]_i$. The response is maintained by hypoxia-induced blockade of the Ca^{2+}-activated K^+ channels, which inhibits repolarization, and by hypoxia-induced release of Ca^{2+} from sarcoplasmic reticulum stores.

Our results provide direct evidence that hypoxia selectively blocks voltage-gated K^+ channels (Figs. 9 and 10) in PA but not MA cells. These voltage-gated K^+ channels are activated at potentials more negative than the resting membrane potential. Some of these channels are probably open at rest and may therefore contribute to the resting membrane conductance and resting potential. Block of these hypoxia-sensitive, voltage-gated K^+ channels can be expected to depolarize the PA cells. This will, secondarily, open voltage-gated Ca^+ channels, thereby raising $[Ca^{2+}]_i$ and promoting contraction (Fig. 12). Whether these hypoxia-sensitive, voltage-gated K^+ channels are affected directly by the reduced O_2 level (21, 30), or by free radicals or another second messenger that is generated as a result of the hypoxia (2), remains to be determined.

ACKNOWLEDGMENTS

We gratefully acknowledge Dr. E.M. Santiago for his excellent assistance with cell preparation and culture, and Drs. A. Ambesi and G.E. Lindenmayer for providing us with anti-Na-Ca exchanger antibody.

This research was supported by a Research Fellowship from the AHA-Maryland Affiliate Inc. (X.-J.Y), by NHLBI grants HL-32276 (M.P.B), HL-43091 (W.F.G) and HL-43304 (M.L.T), and by research grants from the VA Research Services (L.J.R and C.G.S), and by Parker B. Francis Fellowship (C.G.S.). M.L.T is an Established Investigator of the AHA.

REFERENCES

1. **Ambesi, A., E.E. Bagwell, and G.E. Lindenmayer.** Purification and identification of the cardiac sarcolemma Na/Ca exchanger. *Biophys. J.* 59:138A, 1991.
2. **Archer, S.L., J.A. Will, and E.K. Weir.** Redox status in the control of pulmonary vascular tone. *Ht* 11(3):127-141, 1986.
3. **Ashford, M.L.J.** ATP-regulated K^+ channels in rat hypothalamic neurones. *J. Physiol. Lond.* in press, 1992.
4. **Beech, D.J. and T.B. Bolton.** Two components of potassium current activated by depolarization of single smooth muscle cells from the rabbit portal vein. *J. Physiol. Lond.* 418:293-309, 1989.
5. **Bennie, R.E., C.S. Packer, D.R. Powell, N. Jin, and R.A. Rhoades.** Biphasic contractile response of pulmonary artery to hypoxia. *Am. J. Physiol.* 261:L156-L163, 1991.
6. **Bergofsky, E.H. and S. Holtzman.** A study of the mechanisms involved in the pulmonary arterial pressor response to hypoxia. *Circ. Res.* 20:506-519, 1967.
7. **Blaustein, M.P., W.F. Goldman, G. Fontana, B.K. Krueger, E.M. Santiago, T.D. Steele, D.N. Weis, and P.J. Yarowsky.** Physiological roles of the sodium-calcium exchanger in nerve and muscle. *Ann. NY Acad. Sci.* 639:254-274, 1991.
8. **Bonnet, P., D. Gebremedhin, N.J. Rush, and D.R. Harder.** Effects of hypoxia on a potassium channel in cat cerebral arterial muscle cells. *Zeitschrift für Kardiologie* 80(Suppl. 7):25-27, 1991.
9. **Bova, S., W.F. Goldman, X.-J. Yuan, and M.P. Blaustein.** Influence of Na^+ gradient on Ca^{2+} transients and contraction in vascular smooth muscle. *Am. J. Physiol.* 259:H409-H423, 1990.
10. **Bradford, J.R. and H.P. Dean.** The pulmonary circulation. *J. Physiol. Lond.* 16:34-96, 1894.
11. **Brayden, J.E. and M.T. Nelson.** Regulation of arterial tone by activation of calcium-dependent potassium channels. *Science Wash. DC* 256:532-535, 1992.
12. **Buescher, P.C., D.B. Pearse, R.P. Pillai, M.C. Litt, M.C. Mitchell, and J.T. Sylvester.** Energy state and vasomotor tone in hypoxic pig lungs. *J. Appl. Physiol.* 70(4):1874-1881, 1991.
13. **Clapp, L.H. and A.M. Gurney.** ATP-sensitive K^+ channels regulate resting potential of pulmonary arterial smooth muscle cells. *Am. J. Physiol.* 262:H916-H920, 1992.
14. **Cox, R.H.** Potassium channel activators in vascular smooth muscle. *In: Cellular and Molecular Mechanisms in Hypertension*, edited by R.H. Cox. New York: Plenum Press, 1991, p. 27-43.
15. **Daut, J., W. Maier-Rudolph, N. von Beckerath, G. Mehrke, K. Guntherk and L. Goedel-Meinen.** Hypoxic dilation of coronary arteries in mediated by ATP-sensitive potassium channels. *Science Wash. DC* 247:1341-1344, 1990.
16. **Davies, N.W., N.B. Standen and P.R. Stanfield.** ATP-dependent potassium channels of muscle

cells:their properties, regulation, and possible function. *J. Bioenergetics and Biomembranes* 23(4):509-535, 1991.

17. **Duchen, M. R.** Effects of metabolic inhibition of the membrane properties of isolated mouse primary sensory neurones. *J. Physiol. Lond.* 424:387-409, 1990.

18. **Farrukh, I.S. and J.R. Michael.** Cellular mechanisms that control pulmonary vascular tone during hypoxia and normoxia. *Am. Rev. Respir. Dis.*, 145:1389-1397, 1992.

19. **Filo, R.S., D.F. Bohr, and J.C. Ruegg.** Glycerinated skeletal and smooth muscle: calcium and magnesium dependence. *Science Wash. DC* 147:1581-1583, 1972.

20. **Frank, J.S., G. Mottino, D. Reid, R.S. Molday, and K.D. Philipson.** Distribution of the Na^+-Ca^{2+} exchange protein in mammalian cardiac myocytes: An immunofluorescence and immunocolloidal gold-labeling study. *J. Cell Biol.* 117:337-345, 1992.

21. **Ganfornina, M.D. and J. Lopez-Barneo.** Potassium channel types in arterial chemoreceptor cells and their selective modulation by oxygen. *J. Gen. Physiol.* 100:401-426, 1992.

22. **Goldman, W.F., S. Bova, and M.P. Blaustein.** Measurement of intracellular Ca^{2+} in cultured arterial smooth muscle cells using fura-2 and digital imaging microscopy. *Cell Calcium* 11:221-231, 1990.

23. **Hamill, O.P., A. Marty, E. Neher, B. Sakmann, and F.J. Sigworth.** Improved patch-clamp techniques for high-resolution current recording from cells and cell-free membrane patches. *Pflügers Arch.* 391:85-100, 1981.

24. **Harder, D.R., J.A. Madden, and C. Dawson.** Hypoxic induction of Ca^{2+}-dependent action potentials in small pulmonary arteries of the cat. *J. Appl. Physiol.* 59(5):1389-1393, 1985.

25. **Hasunuma, K., D.M. Rodman, and I.F. McMurtry.** Effects of K^+ channel blockers on vascular tone in the perfused rat lung. *Am. Rev. Respir. Dis.* 144:884-887, 1991.

26. **Kamm, K.E., and J.T. Stull.** The function of myosin and myosin light chain kinase phosphorylation in smooth muscle. *Ann. Rev. Pharmacol. Toxicol.* 25:593-620, 1989.

27. **Kieval, R.S., R.J. Bloch, G.E. Lindenmayer, A. Ambesi, W.J. Lederer.** Immunofluorescence localization of the Na-Ca exchanger in heart cells. *Am. J. Physiol.* 263:C545-C550, 1992.

28. **Levitan, I.B., S. Chung, and P.H. Reinhart.** Modulation of a single ion channel by several different protein kinases. *Advances in Second Messenger and Phosphoprotein Research* 24:36-40, 1990.

29. **Lichtheim.** Die Stoerungen des Lungenkreislaufes und ihr Einfluss auf den. Blutdruck. *Inaug. Dissert.* Berlin, 1876.

30. **Lopez-Lopez, J., C. Gonzalez, J. Urena, and J. Lopez-Barneo.** Low pO_2 selectively inhibits K channel activity in chemoreceptor cells of the mammalian carotid body. *J. Gen. Physiol.* 93:1001-1015, 1989.

31. **Luther, P.W., R.K. Yip, R.J. Bloch, A. Ambesi, G.E. Lindenmayer, and M.P. Blaustein.** Presynaptic localization of sodium/calcium exchangers in neuromuscular preparations. *J. Neurosci.*, in press, 1992.

32. **Madden, J.A., M.S. Vadula, and V.P. Kurup.** Effects of hypoxia and other vasoactive agents on pulmonary and cerebral artery smooth muscle cells. *Am. J. Physiol.* 263:L384-L393, 1992.

33. **McMurtry, I.F., A.B. Davidson, J.T. Reeves, and R.F. Grover.** Inhibition of hypoxic pulmonary vasoconstriction by calcium antagonists in isolated rat lungs. *Circ. Res.* 38:99-104, 1976.

34. **Murray, T.R., L. Chen, B.E. Marshall, and E.J. Macarak.** Hypoxic contraction of cultured pulmonary vascular smooth muscle cells. *Am. J. Respir. Cell Mol. Biol.* 3:457-465, 1990.

35. **Nelson, M.T., J.B. Patlak, J.F. Worley, and N.B. Standen.** Calcium channels, potassium channels, and voltage dependence of arterial smooth muscle tone. *Am. J. Physiol.* 259:C3-C18, 1990.

36. **Ohe, M., T. Mimata, T. Haneda and T. Takishima.** Time course of pulmonary vasoconstriction with repeated hypoxia and glucose depletion. *Respir. Physiol.* 63:177-186, 1986.

37. **Okabe, K., K. Kitamura, and H. Kuriyama.** Features of 4-aminopyridine sensitive outward current observed in single smooth muscle cells from the rabbit pulmonary artery. *Pflügers Arch.* 409:561-568, 1987.

38. **Plumier, P.L.** La circulation pulmonaire chez le chien. *Arch. Int. Physiol.* 1:176-213, 1904.

39. **Post, J.M., J.R. Hume, S.L. Archer, and E.K. Weir.** Direct role for potassium channel inhibition in hypoxic pulmonary vasoconstriction. *Am. J. Physiol.* 262:C882-C890, 1992.

40. **Robertson, B.E., P.R. Corry, P.C.G. Nye, and R.Z. Kozloswski.** Ca^{2+} and Mg-ATP activated potassium channels from rat pulmonary artery. *Pflügers Arch* 421:94-96, 1992.

41. **Rodman, D.M. and N.F. Voelkel.** Regulation of vascular tone. *In: The Lung, Scientific Foundations.* R.G. Crystal, J.B. West, P.J. Barnes, N.S. Cherniack, and E.R. Weibel, editors. Raven Press, Ltd., New York, 1991, p. 1105-1119.

42. **Rodman, D.M., T. Yamaguchi, K. Hasunuma, R.F. O'Brien, and I.F. McMurtry.** Hypoxic contraction of isolated rat pulmonary artery. *J. Pharmacol. Exp. Ther.* 248:952-959, 1988.

43. **Salvaterra, C.G. and W.F. Goldman.** Acute hypoxia increases cytosolic calcium in cultured pulmonary arterial myocytes. *Am. J. Physiol.*, in press, 1992.

44. **Salvaterra, C.G., L.J. Rubin, J. Schaeffer, and M.P. Blaustein.** The influence of the transmembrane sodium gradient on the responses of pulmonary arteries to decreases in oxygen tension. *Am. Rev. Respir. Dis.* 139:933-939, 1989.

45. **See, K.L., I.J. Forbes and W.H. Betts.** Oxygen dependency of phototoxicity with hematoporphyrin derivative. *Biochem. Photobiol.* 39(5):631-634, 1984.

46. **Stanbrook, H.S. and I.F. McMurtry.** Inhibition of glycolysis potentiates hypoxic vasoconstriction in rat lung. *J. Appl. Physiol.* 55:1467-1473, 1983.

47. **Standen, N.B., J.M. Quayle, N.W. Davies, J.E. Brayden, Y. Huang, and M.T. Nelson.** Hyperpolarizing vasodilators activate ATP-sensitive K^+ channels in arterial smooth muscle. *Science Wash. DC* 245:177-190, 1989.

48. **Suzuki, H. and B.M. Twarog.** Membrane properties of smooth muscle cells in pulmonary hypertensive rats. *Am. J. Physiol.* 242:H907-H915, 1982.

49. **Voelkel, N.F., I.F. McMurtry, and J.T. Reeves.** Hypoxia impairs vasodilation in the lung. *J. Clin. Invest.* 67:238-246, 1981.

50. **Volk, K.A., J.J. Matsuda, and E.F. Shibata.** A voltage-dependent potassium current in rabbit coronary artery smooth muscle cells. *J. Physiol. Lond.* 439:751-768, 1991.

51. **Von Beckerath, N., S. Cyrys, A. Dischner, and J. Daut.** Hypoxic vasodilation in isolated, perfused guinea-pig heart: an analysis of the underlying mechanisms. *J. Physiol. Lond.* 442:297-319, 1991.

52. **Von Euler, U.S. and G. Liljestrand.** Observations on the pulmonary arterial blood pressure in the cat. *Acta Physiol. Scand.* 12:301-320, 1946.

53. **Yuan, X.-J.** The cellular mechanisms of hypoxia pulmonary vasoconstriction. *Progress in Physiol. Sci.* 20(4):301-306, 1989.

54. **Yuan, X.-J and Y.N. Cai.** Effects of calcium antagonists on pulmonary hypertension during acute hypoxia in rats. *Chinese J. Appl. Physiol.* 2(2):136-141, 1989.

55. **Yuan, X.-J., W.F. Goldman, M.L Tod, L.J. Rubin, and M.P. Blaustein.** Hypoxia reduced potassium currents in cultured rat pulmonary but not mesenteric arterial myocytes. *Am. J. Physiol.* in press, 1992.

56. **Yuan, X.-J., W.F. Goldman, M.L. Tod, L.J. Rubin, and M.P. Blaustein.** Ionic currents in rat pulmonary and mesenteric arterial myocytes in primary culture and subculture. *Am. J. Physiol.* in press, 1992.

57. **Yuan, X.-J., M.L. Tod, L.J. Rubin, and M.P. Blaustein.** Contrasting effects of hypoxia on tension in rat pulmonary and mesenteric arteries. *Am. J. Physiol.* 259:H281-H289, 1990.

58. **Yuan, X.-J., T. Sugiyama, W.F. Goldman, L.J. Rubin, and M.P. Blaustein.** A mitochondrial uncoupler, FCCP, increases K^+ current in rat pulmonary arterial myocytes. *Biophys. J.* in press, 1993.

THE IMPORTANCE OF CALCIUM IN THE REGULATION OF EDRF SYNTHESIS IN THE PULMONARY VASCULATURE

Stephen L. Archer, Vaclav Hampl,
James Huang, and Nancy Cowan

Minneapolis Veterans Administration Medical Center
Minnesota, 1 Veterans Drive
Minneapolis, MN., 55417 and Department of Medicine,
University of Minnesota, 55455

INTRODUCTION

The vascular endothelium and smooth muscle exist in close apposition yet they have opposing responses to stimuli which increase free cytosolic calcium concentration, $[Ca^{2+}]_i$. Stimuli which increase $[Ca^{2+}]_i$ in the endothelium initiate synthesis of vasodilators, such as prostacyclin and nitric oxide; increases in $[Ca^{2+}]_i$ elicits vasoconstriction of smooth muscle. The net response to a stimulus which elevates $[Ca^{2+}]_i$ in both the endothelium and smooth muscle is the sum of these divergent responses. This chapter focuses on the central role of the Ca^{2+} ion in regulating the synthesis of endothelial derived relaxing factor (EDRF), which is nitric oxide (NO) or a related substance. The importance of both intra- and extra-cellular Ca^{2+} to EDRF/NO synthesis is supported by the observation that agents which cause EDRF synthesis increase endothelial $[Ca^{2+}]_i$ [1-5]. Although elevation of intracellular $[Ca^{2+}]_i$ activates NO synthase, a continued influx of Ca^{2+} is necessary to sustain endothelium-dependent vasodilatation [1]. The chapter begins with a description of the NO hypothesis emphasizing the role of Ca^{2+} in the initiation and sustenance of NO synthesis. While endothelial $[Ca^{2+}]_i$ is an important signal for NO synthesis, changes in cytosolic calcium also initiate synthesis of other endogenous vasodilators (e.g. endothelium-derived hyperpolarizing factor (EDHF) and prostacyclin). The importance of age, species and vessel type in determining the diverse means by which the endothelium modulates vascular tone is considered, using the differences between EDRF in the pulmonary and systemic circuits as a paradigm. This is followed by a discussion of the regulation of Ca^{2+} influx into endothelial cells, emphasizing the

Ion Flux in Pulmonary Vascular Control, Edited by
E.K. Weir et al., Plenum Press, New York, 1993

importance of endothelial membrane potential as a determinant of the electrochemical gradient for Ca^{2+} influx. The chapter concludes with a discussion of the techniques for, and value of, measuring endothelial $[Ca^{2+}]_i$ using fluorescent fura microscopy. Examples of the value of this technique in determining mechanisms of endothelium dependent vasodilatation in the pulmonary circulation are provided.

EDRF and the Nitric Oxide Hypothesis

EDRF is a short-lived (biological half-life 6-45 seconds), endogenous dilator of vascular smooth muscle produced by the vascular endothelium [6,7]. EDRF

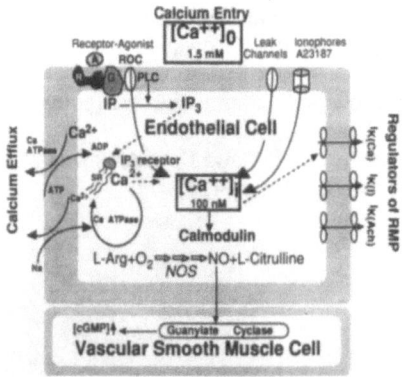

Figure 1. Ca^{2+} Homeostasis and Nitric Oxide Synthesis in the Endothelial Cell
This diagram illustrates some of the mechanisms for regulation of Ca^{2+} homeostasis in the endothelial cell which impact on nitric oxide (NO) synthesis. SR=sarcoplasmic reticulum, L-Arg=L-arginine. $I_{K(Ca)}$, $I_{K(i)}$, $I_{K(Ach)}$ are the Ca $^{2+}$ gated, inward rectifying and acetylcholine-sensitive K^+ channels, respectively. G=G protein, A=agonist (e.g. bradykinin), R=receptor, PLC=phospholipase C, ADP=adenosine diphosphate, IP=inositol phosphate, IP_3=inositol triphosphate. The Ca^{2+} gradient from the extracellular space $[Ca^{++}]_o$ to the cytosol $[Ca^{++}]_i$ is 1.5mM / 100nM. Nitric oxide synthase (NOS) is activated by association with its cofactor , calmodulin , which occurs when calmodulin responds to increases in cytosolic Ca^{2+}.

synthesis is dependent on extracellular Ca^{2+} [1,2]. EDRF-induced vasodilatation is associated with activation of soluble guanylate cyclase and accumulation of c-GMP [7]. The observation that NO and EDRF have similar biological activity suggested that EDRF is NO, an hypothesis further supported by detection of NO production in response to endothelial-dependent vasodilators [8,9]. The complex means by which changes in endothelial $[Ca^{2+}]_i$ and the availability of extracellular Ca^{2+} alter synthesis of NO is illustrated in Figure 1.

224

The Diversity of Endothelium Dependent Vasodilatation: Lessons from the Pulmonary Circulation

EDRF is one of many endogenous vasoactive substances that modulate pulmonary vascular tone (e.g. prostaglandins, endothelin, leukotrienes, neuropeptides, neural activity). The importance of EDRF, relative to these other factors, varies among vascular beds and even differs between larger "conduit" and smaller "resistance" vessels. The species and maturational stage of the individual are also important predictors of EDRF activity.

Prior to reviewing the evidence that $[Ca^{2+}]_i$ is an important second messenger in NO synthesis, it is useful to consider the diversity of mechanisms by which the endothelium regulates vascular tone. Even in situations where NO synthesis can occur, simply observing that an agonist increases endothelial $[Ca^{2+}]_i$ does not prove the agent's actions are solely or primarily due to NO. Many agents increase endothelial $[Ca^{2+}]_i$ even in the absence of extracellular Ca^{2+}; but EDRF/NO is only synthesized when adequate extracellular Ca^{2+} is present. Furthermore, the increase in endothelial $[Ca^{2+}]_i$ which elicits NO synthesis often signals the simultaneous synthesis of other vasodilators, such as prostacyclin. It is the net production of vasoactive substances which determines the integrated hemodynamic response to the endothelial dependent vasodilator.

Both systemic and pulmonary vascular beds have functional EDRF pathways (e.g. ionophores or agonists, such as bradykinin (BK), elevate $[Ca^{2+}]_i$, cause NO synthesis and result in vasodilatation in both circulations). However, when inhibitors of nitric oxide synthase (NOS) (e.g. N^G-monomethyl-L-arginine, L-NMMA) are administered to adult rats there is much less increase in basal blood pressure and resistance in the pulmonary , as compared to the systemic vasculature. To compare the role of EDRF in the regulation of basal tone between pulmonary and systemic vascular bed, we developed a model in which an isolated lung and kidney, from the same rat, are perfused in series (Figure 2). The organs are perfused with a Krebs-albumin solution containing an inhibitor of cyclooxygenase, meclofenamate, permitting direct comparison of the effects of endothelial dependent vasodilators and EDRF inhibitors while eliminating the effects of prostaglandins and blood as well as reducing the confounding effect of individual variation among rats. The pulmonary and renal circulations display opposing responses to hypoxia (vasoconstriction in the lung and vasodilatation in the kidney), as they do in vivo, supporting the circulatory relevance of the model (Figure 3).

The lung-kidney model was used to assess the contribution of basal NO synthesis to resting vascular tone in the lung, as compared to the kidney. In these experiments basal NO synthesis was inhibited by N^G-nitro-L-arginine methylester (L-NAME,1.5 mM). During constant-flow conditions, L-NAME did not significantly change mean pulmonary perfusion pressure (Δ -0.8 %) but more than doubled the perfusion pressure in the kidney (Δ+145 %). L-NAME had no effect on the pressure-flow (P/Q) relationship of the pulmonary vasculature but markedly increased the slope of the P/Q relationship in the kidney (Figure 4). L-arginine, the substrate for NOS, reversed the vasoconstrictor effects of L-NAME on the renal vasculature (Figure 4). The preferential constriction of renal, as compared to pulmonary circulations, caused by L-NAME also occurred with another inhibitor of NOS, N^G-nitro-L-arginine. These data suggest that EDRF is an important determinant of basal tone in the renal but not the pulmonary circulation, in this model. Despite the seemingly minor role of NO in regulation of basal pulmonary vascular tone, the pulmonary vascular endothelium responds to EDRF agonists with elevation of $[Ca^{2+}]_i$ and production of NO [5]; clearly the potential to modulate vascular tone using this mechanism is present, if not always biologically important.

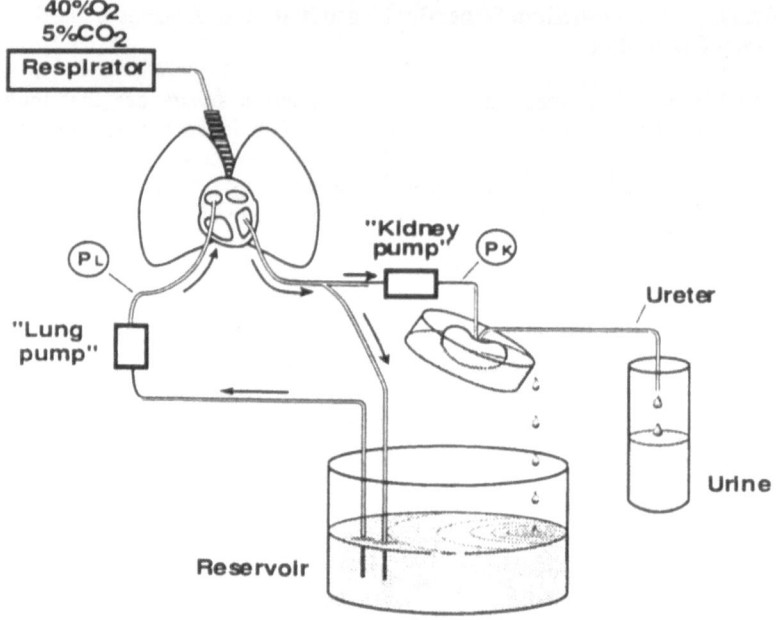

Figure 2. Lung-Kidney Model

The lung is suspended from the trachea in a humid chamber and ventilated at 65 breaths/minute with a tidal volume of 2.5 ml. In this constant flow model changes in perfusion pressure reflect changes in vascular resistance. The perfusate (Krebs-albumin solution with meclofenamate, 38°C) is pumped by the "lung pump" from reservoir into the pulmonary artery while perfusion pressure (P_L) is recorded. The outflow from the left atrium is divided into two portions; one part returns to the reservoir, while the other perfuses the kidney. The kidney is placed in a Petri dish above the reservoir and perfused through a cannula inserted into renal artery, using a "kidney pump". Renal perfusion pressure (P_K) is measured continuously. The renal vein is cut to allow free outflow to the Petri dish and back to reservoir. Urine is collected from the cannulated ureter.

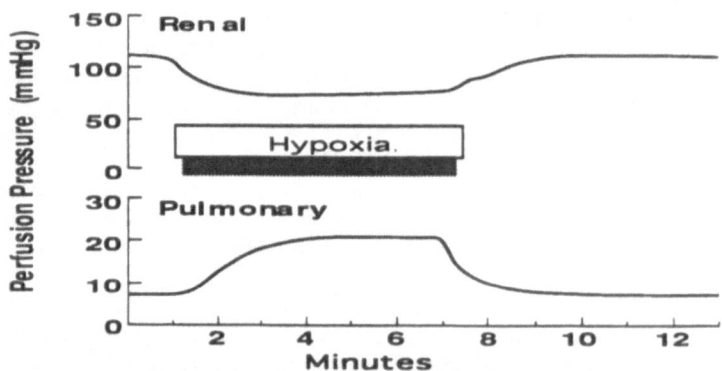

Figure 3. Hypoxia Causes Simultaneous Pulmonary Vasoconstriction and Renal Vasodilatation in the Isolated Lung-Kidney Model

This is a computer scanned image of an actual experimental trace in which the lung and kidney of an adult rat were perfused with Krebs-albumin solution at 15 and 7 ml/minute, respectively. At constant perfusion rates, hypoxic ventilation of the lung (2.5% inspired O_2, pO_2 40 mmHg) increased pulmonary artery pressure and decreased renal perfusion pressure (recorded simultaneously).

226

The conclusion that basal NO production is a less important determinant of resting tone in the pulmonary, as compared to the renal, circulation is consistent with studies where only one of the organs is perfused [10-16, 17] and is further supported by experiments in which NO synthesis is chronically suppressed. Adult rats given L-NAME (1.85 mM) in their drinking water for 3 weeks, developed systemic hypertension (mean systemic arterial pressure was 212 ± 10 mm Hg, compared to 159 ± 5 mm Hg in controls) but mean pulmonary arterial pressure was unchanged (13 ± 1 mm Hg, compared to control, 15 ± 1 mm Hg). Thus, chronic suppression of NO synthesis produces systemic but not pulmonary hypertension.

The lack of a consistent constrictor effect of the arginine analogs in the whole lung seems to contradict in vitro studies in which they consistently constrict conduit pulmonary arterial rings [18, 19]. Several potential explanations exist for this discrepancy. Firstly, isolated rings are usually "primed" with phenylephrine; without this pretreatment minimal vasoconstriction occurs in response to arginine analogs [17]. The mechanism by which phenylephrine sensitizes arterial rings to NOS-inhibitors, such as L-NAME, is unknown, but may relate to phenylephrine's ability to increase endothelial Ca^{2+} levels. Secondly, tone in the intact lung is a function of small resistance arteries, rather than the larger, more easily isolated, conduit arteries used in ring studies. Thus conduit rings may be the wrong model to evaluate the importance of basal EDRF synthesis to the integrated pulmonary circulation. There is precedent for the concept that EDRF activity differs between large and small vessels. Pulmonary vasodilator responses to acetylcholine (ACH) (in vitro [18, 19] and in vivo [20, 21]) are much less in small pulmonary arteries, as compared to large ones. Perhaps the constrictor response to arginine analogs observed in pulmonary conduit arteries is not biologically important to regulation of vascular resistance in the normal lung.

Species differences in the pulmonary vascular effects of the endothelium-dependent vasodilator, acetylcholine (ACH), are well known. While ACH causes pulmonary vasodilatation in humans [23], calves [24] and rats [5, 25], it has pulmonary vasoconstrictor properties in rabbits [26] and dogs [20, 21]. These differences reflect diversity in the amount and type of prostanoid synthesis elicited by ACH [26].

Diversity in the importance of EDRF exists even within the pulmonary circulation of adult animals. Arginine analogs do not affect resting pulmonary vascular tone in isolated rat lungs (Figure 4) [10, 12, 13, 16] and in conscious dogs [27]. On the other hand, pulmonary vascular resistance is increased by arginine analogs in anesthetized rabbits [28] and cats [22]. In addition to those factors which predict EDRF activity already discussed (e.g., vascular bed, vessel size, species) one must also consider the maturational status of the species in question. In contrast to the adult pulmonary vasculature, which conducts high flow at low resistance, the fetal pulmonary circulation has low flow and high pressure/resistance, similar to the systemic circulation. Furthermore the response to many vasoconstrictors is prominent in the fetal pulmonary vasculature [29]. Consequently, it is not surprising that the importance of EDRF appears similar in the fetal pulmonary and systemic circuits. The fetal pulmonary circulation , in several species, constricts in response to infusion of arginine analogs, similar to the response of systemic vessels [29-32] (see Abman's Chapter). It appears that circulations in which the resting tone is high are particularly likely to display basal EDRF activity, evident as constriction in response to administration of arginine analogs.

The role that EDRF plays in the pathogenesis of pulmonary hypertension is controversial. Two opposing hypotheses exist: either EDRF deficiency causes or contributes to pulmonary hypertension or EDRF synthesis is up-regulated in pulmonary hypertension, as a homeostatic mechanism. In chronic hypoxic pulmonary hypertension, reactivity to endothelium-dependent vasodilators has variously been reported to be reduced[33,34] or preserved[24,35,36]. ACH-induced vasodilatation was preserved in conscious, pulmonary hypertensive calves but absent in isolated, conduit, pulmonary arterial rings, from these same animals [35]. We demonstrated enhanced vasodilatation in response to substance P and NO in isolated lungs from adult rats with chronic hypobaric hypoxic pulmonary hypertension created by exposure to 0.45 atmospheres for 3 weeks [36]

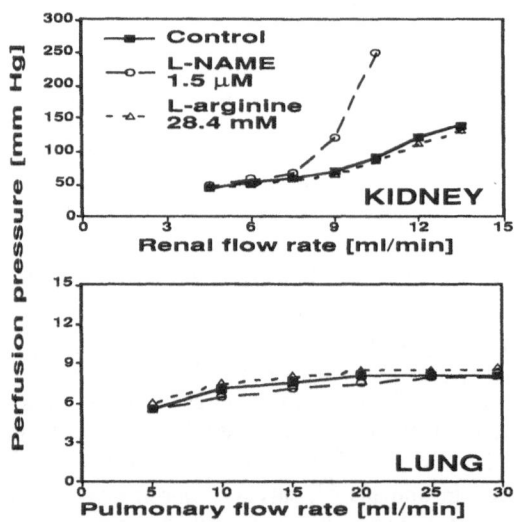

Figure 4. Acute Inhibition of Nitric Oxide Synthesis Increases the Slope of the Pressure-Flow Relationship in the Renal, but not the Pulmonary, Circulation

These data were obtained from the lung-kidney model perfused with Krebs-albumin solution, containing meclofenamate. Pressure/flow relationships were constructed by observing the plateau perfusion pressure achieved with increments in flow (5-15 ml in the kidney, 5-30 ml in the lung). In the presence of L-NAME renal flow could not be advanced beyond 10 ml/minute due to severe hypertension.

Upper panel: The slope of the pressure/flow (P/Q) relationship in the kidney is increased by the inhibitor of NO synthase, L-NAME. The renal P/Q relationship is normalized by giving an excess of the substrate for NO synthase, L-arginine.

Lower Panel: The pulmonary P/Q relationship, in lungs from the same rat, has a much flatter slope and is not significantly shifted by L-NAME, suggesting that either basal NO synthesis is a less important determinant of vascular tone in the lung than in the kidney. Alternatively, L-NAME may be less potent in the pulmonary circulation.

Even chronic inhibition of NO synthesis (accomplished by adding L-NAME, a "false substrate" for NOS, to the drinking water for 3-weeks) did not produce pulmonary hypertension in normoxic rats nor did it aggravate pulmonary hypertension in chronically hypoxic rats. However, chronic ingestion of L-NAME did reduce cardiac index, elevate systemic blood pressure and cause left ventricular hypertrophy in normoxic and hypoxic rats, indicating the loss of a significant basal NO synthesis in the systemic vasculature. Chronic inhibition of NO synthesis did not increase pulmonary artery pressure in rats with chronic hypoxic pulmonary hypertension (pulmonary artery pressure 28 ± 3 mm Hg in hypoxic rats ingesting L-NAME compared to 28 ± 2 mm Hg in control hypoxic rats). Although acute administration of L-NAME had little effect on normotensive lungs, a bolus of L-NAME acutely increased the slope of the pressure flow relationship lungs from rats with chronic hypoxic pulmonary hypertension. To confirm the hypothesis that NO

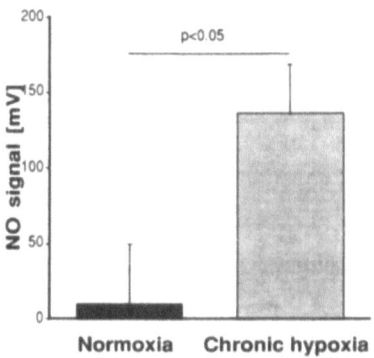

Figure 5. Basal Levels of NO Are Increased in Chronic Hypoxic Pulmonary Hypertension

Isolated lungs were perfused (0.04 ml/min/g) for 80 minutes with normoxic Krebs-4 % albumin solution containing meclofenamate. "NO signal" = the net chemiluminescence produced by NO, NO_2 and NO_3 in the perfusate as measured in a chemiluminescence assay (0.25 sec integration time, see ref [38].) in the presence of potassium iodide and HCl. The backgroud signal from the perfusate, due to nitrite contaminants in the albumin as supplied commercially, was subtracted to yield these values.

synthesis was up-regulated in rats with hypoxic pulmonary hypertension we measured nitrites/nitrates in perfusate from the isolated lung, using a chemiluminescence assay. The perfusate levels of nitrite were much greater in the hypertensive lungs than in controls, although both were perfused at the same flow rate for 80 minutes prior to the nitrite measurement (Figure 5). Furthermore, removal of extracellular Ca^{2+} from the perfusate eliminated NO synthesis, evident as a dramatic decrease in perfusate content of mixed nitrite/nitrate. These findings suggest that EDRF synthesis is preserved or up-regulated in the rat model of chronic hypoxic pulmonary hypertension. This is analogous to prostacyclin synthesis which is increased in monocrotaline-induced pulmonary hypertension, presumably in an incompletely successful effort to reduce pulmonary hypertension [37].

Interim Summary

Although the synthesis of NO by the endothelium can modulate vascular tone, its biological importance as a regulator of blood pressure is variable among species, vascular beds, and with disease states, analogous to the earlier recognized heterogeneity in the role of the prostaglandin pathway. In the adult rat NO synthesis seems to be a more important determinant of renal than of pulmonary vascular tone. The extent to which the diversity in the importance of NO reflects differences in the distribution and activity of NOS, alterations in regulation of endothelial $[Ca^{2+}]_i$ or variability in the synthesis of other endogenous vasodilators is unknown [2]. The multiple mechanisms by which the endothelium can mediate vasodilatation underscores the need to correlate hemodynamic changes with concurrent measurements of the effects of putative NOS agonists and antagonists on cell biochemistry (e.g. NO and c-GMP levels and $[Ca^{2+}]_i$) [38].

EDRF Bioactivity: Dependence on Extracellular Ca^{2+}

An important clue that Ca^{2+} is important as a signal for EDRF/NO synthesis is that most endothelial dependent vasodilators cause an immediate increase in endothelial $[Ca^{2+}]_i$ [1-5, 39, 40]. Furthermore, the Ca^{2+} ionophore, A23187, is a potent endothelial dependent vasodilator [1,9,41] and has been confirmed to elicit NO synthesis, measured using a chemiluminescence assay [5, 9, 41]. Furthermore, endothelium dependent vasodilatation is reduced in Ca^{2+} free media [1]. Several groups have shown that EDRF activity, although associated with a rise in intracellular $[Ca^{2+}]$ is more closely correlated with the availability of extracellular Ca^{2+} [1, 2]. The increase in $[Ca^{2+}]_i$ caused by agents such as bradykinin is rapid and consists of two components, an initial increase, dependent on the release of intracellular Ca^{2+} stores, and a later, more sustained, "plateau" phase which is dependent on entry of extracellular Ca^{2+} [2]. The initial release of intracellular Ca^{2+} is in part accomplished by the G protein, inositol triphosphate (IP$_3$) pathway (Figure 1). The plateau phase, which appears essential for sustained NO synthesis and vasodilatation, can be inhibited by removal of extracellular Ca^{2+} or by LaCL$_3$, a non-selective Ca^{2+} entry blocker. The plateau phase which occurs in response to receptor-agonist interaction (e.g. bradykinin) may involve influx of Ca^{2+} through receptor operated channels.

Many agonists which increase $[Ca^{2+}]_i$ elicit both NO and prostacyclin synthesis. A qualitative difference in the importance of extracellular Ca^{2+} is evident when one compares the effects of removal of perfusate Ca^{2+} on the synthesis of EDRF and prostacyclin. Although Ca^{2+}-free media reduces both EDRF and prostacyclin synthesis, EDRF synthesis is much more completely inhibited by removal of extracellular Ca^{2+} [2]. Furthermore, while bradykinin-induced synthesis of prostacyclin correlates closely with bradykinin-induced changes in endothelial $[Ca^{2+}]_i$, the correlation between changes in $[Ca^{2+}]_i$ and EDRF activity is weak [2].

Synthesis of the endothelium dependent hyperpolarizing factor (EDHF) is also associated with increases in intracellular Ca^{2+} [42], therefore, one cannot distinguish the reliance of a vasodilator on EDRF or EDHF on the basis of the Ca^{2+}-dependency of the response. Most, if not all, the stimuli which cause NO synthesis also cause at least transient membrane hyperpolarization. Adequate extracellular Ca^{2+} must be available and endothelial cytosolic Ca^{2+} levels must rise for either EDRF and EDHF activity to be expressed [42]. The ability of endothelial dependent vasodilators, such as ACH and A23187, to hyperpolarize vascular smooth muscle is at least partially Ca^{2+}-dependent [42].

Ca^{2+}-Calmodulin and the Regulation of NOS

Nitric oxide synthase (NOS) is a dioxygenase which combines nitrogen (from L-arginine) with molecular oxygen to form the short-lived radical NO and citrulline (Figure 1). NOS exists in constitutive and inducible forms. Constitutive forms of the enzyme (NOS$_c$), such as those in the vascular endothelium, produce transient bursts of NO synthesis in rapid response to vasodilators, such as bradykinin and the ionophore A23187. The activation of NOS$_c$ is dependent on extracellular Ca^{2+} entering the endothelial cell; intracellular stores of Ca^{2+} seem inadequate to cause or sustain NO synthesis. When Ca^{2+} binds calmodulin, this obligatory cofactor associates with NOS and the enzyme is activated [43] (Figure 1). In contrast, inducible forms of NOS (NOS$_i$), found in inflammatory cells, make large amounts of NO, over a sustained period, beginning several hours after priming with stimuli such as lipopolysaccharide. NOS$_i$ is probably not regulated by cytosolic Ca^{2+} because there already exists a tight association between calmodulin and the enzyme at normal cytosolic Ca^{2+} concentrations [43]. In endothelial and cerebellar cells the EC$_{50}$ of NOS$_c$ for calmodulin and Ca^{2+} range from 1-10 nM and 0.2-0.4 μM, respectively [43]. NOS$_i$'s activity is regulated at the transcriptional level, accounting for the long delay (hours) between stimulation and NO synthesis.

Routes of Ca^{2+} Entry in Endothelial Cells

Since extracellular Ca^{2+} concentrations exceed those in the cell by a factor of 10,000, it is evident that Ca^{2+} influx is tightly regulated. The primary route(s) by which the extracellular Ca^{2+} required for endothelium-dependent vasodilatation enters the cell remains controversial. Adams et al reviewed four mechanisms for Ca^{2+} entry into endothelial cells: 1) receptor operated channels (ROC) coupled to a second messenger, 2) "leak" channels 3) stretch activated, non-selective cation channels, and 4) Na$^+$/Ca^{2+} exchangers [44] (Figure 1). Receptor-agonist interaction triggers opening of receptor operated channels. These non-selective ion channels permit influx of a significant amount of extracellular calcium. Johns et al used patch clamping and ^{45}Ca to determine that the rise in Ca^{2+} seen in response to low doses of thrombin and bradykinin was due to influx of Ca^{2+} through a receptor operated channel [3]. The receptor operated channel allowed influx of 912 pmol/10^{6}cells/2 minutes, an amount which is probably adequate to elevate [Ca^{2+}]$_i$ to levels necessary to initiate NO synthesis.

At higher doses of agonist, intracellular release of Ca^{2+} contributes to the net increase in [Ca^{2+}]$_i$ [3]. In addition to opening a channel, the agonist-receptor interaction activates membrane bound guanosine triphosphate binding or "G " proteins. The G proteins in turn activate phospholipases, such as phospholipase C, which then cause the production of second messengers, such as IP$_3$, which further elevate [Ca^{2+}]$_i$ (Figure 1). Although the receptor operated channels may permit sufficient Ca^{2+} into the cell to sustain EDRF release, EDRF synthesis can be triggered by the ionophore A23187, which elevates [Ca^{2+}]$_i$ and causes NO synthesis without activating receptor operated channels. Thus it appears that entrance of adequate amounts of extracellular Ca^{2+}, regardless the route, is a necessary stimulus for NO synthesis.

Interaction Between Membrane Potential and [Ca^{2+}]$_i$

Despite the absence of voltage-gated Ca^{2+} channels in most endothelial cells[4] the resting membrane potential (RMP) is an important regulator of transmembrane Ca^{2+} flux by virtue of its ability to alter the electrical driving force for Ca^{2+}. In cultured pulmonary artery endothelial cells the membrane potential is directly related to the log of the extracellular KCl concentration so that a 10 fold increase in extracellular KCl causes a 52 mV depolarization of the cell [3]. Johns et al speculated that "resting intracellular calcium concentration and basal EDRF release is probably influenced by the activity of the inward rectifying K^+ channel and membrane potential of the endothelial cell" [3]. KCl decreases basal cytosolic Ca^{2+} levels and diminishes bradykinin-evoked elevations of $[Ca^{2+}]_i$ in endothelial cells [3, 44]. Depolarization-induced inhibition of Ca^{2+} signaling may explain the impaired ability of endothelial dependent vasodilators to reverse constrictions caused by drugs which alter RMP, such as KCl [45]. Laskey et al demonstrated that agonist-induced oscillations of cytosolic Ca^{2+} were mediated by influx of Ca^{2+} across the cell membrane [46]. This oscillatory or "wave-like" influx of Ca^{2+} is associated with, though not necessarily the result of, cyclic changes in the endothelial cell's membrane potential [46].

We compared the efficacy of bradykinin-induced vasodilatation in isolated rat lungs which were pre-constricted with hypoxia or KCl. In these experiments the isolated lungs of adult rats were perfused with a Krebs/4% albumin solution containing meclofenamate, to eliminate the confounding effects of prostaglandins. Bradykinin reduced hypoxic vasoconstriction but failed to reduce the larger KCl-induced vasoconstriction (Table 1). The ability of acetylcholine to relax second division (conduit) pulmonary artery rings from adult rats was also markedly reduced in KCl versus norepinephrine-constricted rings. The endothelium-dependent relaxation caused by 2 receptor mediated endothelial dependent vasodilators is reduced by KCl, presumably a consequence of depolarization.

The relationship between membrane potential and cytosolic calcium concentrations suggests that even if EDRF and EDHF are separate substances, there could be significant interaction between them. For example, if EDHF causes membrane hyperpolarization this promotes elevation of $[Ca^{2+}]_i$ and may concomitantly stimulate EDRF/NO synthesis. Mehrke and Daut found that bradykinin and adenosine triphosphate (ATP) have biphasic effects on membrane potential in cultured coronary endothelial cells, with an initial hyperpolarization followed by a sustained depolarization. Bradykinin-induced hyperpolarization was inhibited by high extracellular K^+ concentrations and by the putative inhibitors of the Ca^{2+}-dependent K^+ channel (iK_{Ca}), d-tubocurarine and apamin. Mehrke and Daut

Table 1. KCl Inhibits Endothelium Dependent Vasodilatation in Isolated Rat Lungs and Pulmonary Artery Rings

Isolated Rat Lungs Constrictor Stimulus	Δ Pulmonary Artery Pressure (mm Hg)	%Δ Pulmonary Artery Pressure BK (10^{-7}M)
HYPOXIA (n=5)	8±1	-17±2*
KCl (n=5)	29±5†	1±1

Pulmonary Artery Rings Constrictor Stimulus	ΔTension (mg)	%Δ Tension ACH (10^{-6}M)
NOREPINEPHRINE (n=6)	+551±37	-56±8*
KCl (n=6)	+674±27†	-13±4

ΔTension is the change in arterial ring tension above basal tension of 800 mg caused by norepinephrine or KCl. *p<0.001 vasodilatation is greater than in KCl constricted rings/lungs. † p <0.01 constriction caused by KCl (40mM) is greater than that caused by hypoxic ventilation (2.5% inspired O_2) or norepinephrine (3.1×10^{-8}M).

postulate that the transient hyperpolarization caused by bradykinin and ATP reflects activation of G proteins, enhanced inositol phosphate metabolism and thence increased $[Ca^{2+}]_i$. The elevation of $[Ca^{2+}]_i$ in turn promotes K^+ efflux through the Ca^{2+}-dependent K^+ channel and culminates in membrane hyperpolarization and vasodilatation [47]. Ca^{2+}-dependent K^+ channels have recently been demonstrated in freshly isolated endothelial cells (see Adams chapter).

How is RMP Determined in the Vascular Endothelium?

Voltage gated Na^+ and Ca^{2+} channels are thought to be absent in cultured endothelial cells [3]. The predominant current in cultured endothelial cells is the inwardly rectifying K^+ channel, although this channel is much less evident in freshly dispersed cells (Figure 1) [3]. The inward rectifying K^+ channel tends to stabilize the membrane potential by allowing influx of K^+ during hyperpolarization and permitting K^+ efflux under depolarized conditions. However, recent reports have identified Ca^{2+}-dependent K^+ channels both in cultured endothelial cells (Personal communication Andreas Luckhoff) and freshly isolated cells (Personal communication David Adams). This channel has the potential to modulate RMP, causing hyperpolarization in response to stimuli which increase endothelial $[Ca^{2+}]_i$. Activation of iK_{Ca} channels promotes hyperpolarization thereby preserving the electrochemical gradient for Ca^{2+} influx and perhaps permitting sustained Ca^{2+} entry and prolonged EDRF responses.

One of the difficulties in determining the primary regulatory mechanism for endothelial RMP is the endothelial cell's high input resistance, compared to larger cells such as the myocyte. In small cells, with high input resistance, activation or inactivation of a relatively small number of channels can greatly alter RMP. Further studies using patch clamp techniques and pharmacological inhibitors of K^+ channels will be necessary to determine the key regulators of endothelial RMP.

METHODOLOGY FOR MEASUREMENT OF CYTOSOLIC Ca^{2+}

The study of Ca^{2+} in cells began with the use of radioactive Ca^{2+} but the explosion of data regarding the regulation of cytosolic Ca^{2+} in endothelial cells followed the introduction of relatively sensitive and specific dyes which could be loaded into cells by incubation, rather than injection. The technology for measurement of $[Ca^{2+}]_i$ using the dye fura 2-AM will be briefly reviewed, as an appreciation of the strengths and limitations of the methodology assists critical interpretation of the data derived from pulmonary artery endothelial cells, using this technique.

Fura 2 has higher quantum efficiency and is less prone to bleaching than prior generations of Ca^{2+}-sensitive dyes. Perhaps the most important aspect of fura 2's profile is the ability to determine $[Ca^{2+}]_i$ based on the ratio of light emission at two stimulating wavelengths, rather than the absolute emission intensity at a single wavelength (as with Quin) [48]. Cell stimulation is accomplished by a computerized, system which rapidly alternates between stimulating wave lengths (340 and 380 nm) using a chopper and light emission is detected at 510 nm. Increases in $[Ca^{2+}]_i$ enhance the 510 nm light emission caused by 340 nm stimulation while depressing emission caused by 380 stimulation (Figures 6-8). The ability to quantitate $[Ca^{2+}]_i$ based on the *ratio* of 510 emission caused by stimulation at 340 and 380 nm, rather than the absolute emission intensity at a single wavelength reduces errors caused by autoluminescence and dye bleaching.

Endothelial cells are typically loaded for 30-40 minutes with 1-10 µM fura 2 pentaacetoxymethyl ester (AM),whereas the salt of fura is used for in vitro measurements and assay calibration. The ester group permits entry of fura 2 into the cytoplasm. The ester group cleaves from the fura 2 during a 20-30 minute incubation at 37⁰C, leaving 90% of the dye free in the cytoplasm. Extracellular fura is rinsed off the plate following incubation. Inadequate incubation time does not permit homogenous distribution of fura 2 in the cytosol; excessive incubation results in sequestration of the dye in organelles, such as the Golgi apparatus (particularly in certain cells e.g. alveolar epithelium). Visual inspection to ensure the homogeneity of the fluorescence can obviate the problem of excessive or inadequate incubation times.

Quantitative measurement of $[Ca^{2+}]_i$ is a highly technical endeavor with few clear references available to guide the novice. The following discussion will describe the calibration of a fluorescent microscopy system, explain measurement of Ca^{2+} in solutions and finally illustrate the application of these techniques to measurement of endothelial $[Ca^{2+}]_i$. The measurement of Ca^{2+} in cells employs a microscopic technique in which the signal from a few cells is transmitted from the microscope stage via fiber optic cables to photon counters. The microscope must be equipped with optics which will pass fluorescent light.

By performing the steps outlined in Tables 2 and 3 one can be sure that the batch of fura 2 is active, the light source is aligned on both 340 and 380nm monochrometers, the photon counters adequately sensitive, and that a correct dissociation constant for fura 2-Ca^{2+} (Kd) is used. This calibration procedure obviates most non-biological sources of error. If the wavelength scan is performed on the microscope stage one also evaluates the sensitivity of the microscope's optics.

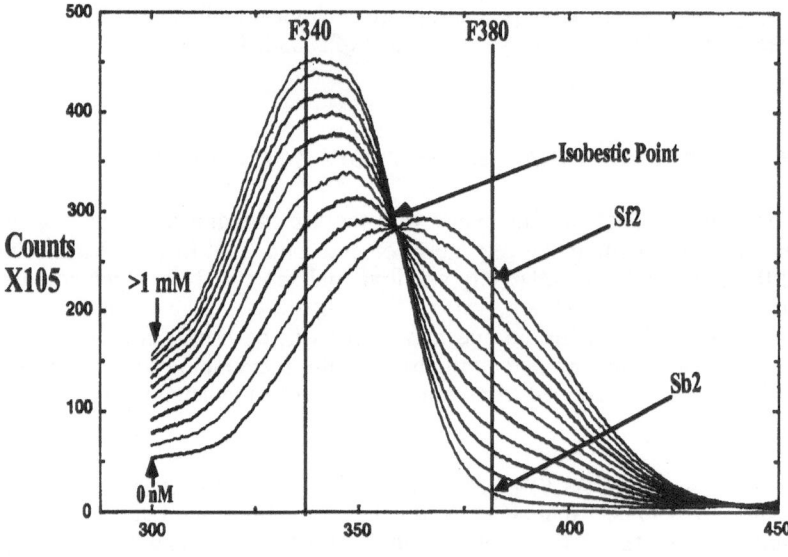

Wavelength (nm)

Figure 6. A Wavelength Scan of Ca^{2+} Standards Provides the Values Necessary to Calculate the Isobestic Point and Kd of Fura 2

The rationale behind, and methodology for, constructing this wavelength scan of 11 Ca^{2+} standards (0nM-1mM) is in Tables 2 and 3. The fluorescent emission caused by 340nm stimulation is divided by that of 380 nm , yielding R values. Sf2 and Sb2 are the emission intensity (at 380 nm) with Ca^{2+} free and >1mM Ca^{2+}. These variables are used to determine the Kd of fura-Ca^{2+} (Tables 2 and 3).

Table 2. Technical Aspects of Calibration of a Monochrometer-Based Spectrofluorometer and Measurement of Endothelial Cytosolic Calcium

Procedure	Technical Considerations
•**Prepare Ca²⁺ standards** Ca^{2+} standards (0 nM-1mM) are prepared by precise mixing of Ca^{2+} free solutions, containing EGTA and Ca^{2+} replete solutions. By stepwise replacement of aliquots of Ca^{2+}-free with equal volumes of Ca^{2+} replete solution, incremental Ca^{2+} levels are produced in accordance with the dissociation constant of Ca^{2+}-EGTA (K) (**Equation 1**) $$[Ca^{2+}] = K\frac{[Ca^{2+} - EGTA]}{[EGTA]}$$	**Preparation of standards** Prepare two 10 mM EGTA buffers (pH 7.0) containing 100 mM KCl, 10 mM K-MOPS and 1µM fura 2 pentapotassium salt. Add 10 mM $CaCl_2$ to one ("Ca^{2+}-EGTA solution") but not the other ("EGTA solution"). By stepwise replacement of aliquots of Ca^{2+}-free with equal volumes of Ca^{2+} replete solution, incremental Ca^{2+} levels are produced in accordance with the dissociation constant of Ca^{2+}-EGTA (K) (Equation 1). Based on the volume replacement protocol defined in Table 3 the $[Ca^{2+}]$ standards are = 380 x 1/9, 380 x 2/8, 380 x 3/7...or 42, 95, 163 nM., respectively.
•**Perform a wavelength scan of Ca²⁺ standard solutions** Determine Isobestic Point of fura	The superimposed excitation spectra obtained by scanning the range of Ca^{2+} standards produces a family of discrete curves which intersect at a single "isobestic" point ($\approx$360nm), indicating that the spectra are formed from two components in equilibrium (Figure 6).
•**Measure the effect of increasing [Ca²⁺] on the ratio of light emission at 340 and 380 nM**	The ratio F340/F380 light emission is termed R. The R in the Ca^{2+} free solutions is Rmin, and is subtracted from all subsequent R values. The R-Rmin is calculated for each standard (Table 3)
•**Calculate the dissociation constant for fura 2 and Ca²⁺ (Kd)** **Equation 2** $$Kd(M) = \frac{-1}{m \times \dfrac{Sf_2}{Sb_2}}$$	Plot (R-Rmin)/$[Ca^{2+}]$ (Y axis) over the range of 42-3420nM (R-Rmin 0-9). The slope of the fitted line (m) is used to calculate the Kd (Figure 7). The Kd (approximately 169-200 M) varies with temperature and other experimental factors and thus should be calculated for each laboratory. Sf2 and Sb2 are the emissions at 380 nm of the Ca^{2+}-free and bound samples, respectively (Sf2/Sb2 = 11.1 in Figure 6) .
Perform a time-based scan	Compare the nominal value of the Ca^{2+} standards to the measured Ca^{2+} values, calculated using Equation 3 (Figure 8)

Measurement of Cytosolic Calcium in Endothelial Cells (Figure 8) [49]

(Equation 3) [49]	
$$[Ca^{2+}]_i = Kd\frac{(R-Rmin)}{(Rmax-Rmin)} \times \frac{sf380}{sb380}$$	Now one can calculate Ca^{2+} levels in cells by determining the ratios of F340/380 caused by: • the experimental intervention (R, e.g. giving bradykinin) • saturating fura 2 with Ca^{2+} (Rmax, e.g. permeabilizing cells with ionomycin or lysing the cells using the detergent Triton (Sigma) in the presence of >1mmol Ca^{2+}, Figure 9) • depriving fura 2 of Ca^{2+} (Rmin e.g. give EGTA, 2 mmol/L, to Triton-lysed cells, Figure 9)

Table 3. Fluorescent Emission (F) at 340 and 380 nm from Ca^{2+} Standards, Measured in Vitro, are Used to Produce a Ratio (R) Which is Used in Calculating the Dissociation Constant of Fura 2 and Ca^{2+}

Volume Exchanged to Create Ca^{2+} Standards (ml)	Ca^{2+} Standards nM	F340 countsX10^5	F380 countsX10^5	R F340/F380	R-Rmin	R-Rmin / $[Ca^{2+}]$ X10^{-3}
0.000	0	1.95	2.54	0.77	-	-
0.300	42.2	2.32	2.34	1.00	0.23	5.34
0.333	95	2.61	2.08	1.25	0.48	5.09
0.375	162.85	2.96	1.86	1.59	0.82	5.05
0.429	253.33	3.24	1.61	2.01	1.24	4.89
0.500	380	3.51	1.35	2.60	1.83	4.82
0.600	570	3.75	1.13	3.33	2.56	4.48
0.750	880	3.96	.877	4.51	3.74	4.25
1.000	1520	4.17	.658	6.33	5.56	3.66
1.500	3420	4.39	.440	9.98	9.21	2.69
2.991	>0.1 mM	4.50	.230	19.57	18.80	-

Volume Exchanged refers to the amount of Ca^{2+} free solution which is removed and replaced by an equal volume of Ca^{2+} replete solutions to create the Ca^{2+} standards in the adjacent column. F=fluorescent intensity (counts) caused by 340 or 380 nm stimulation and measured at 510nm. R=F340/F380. Rmin is the light emission of the Ca^{2+} free solution. $[Ca^{2+}]$ is the nominal level of the Ca^{2+} standard.

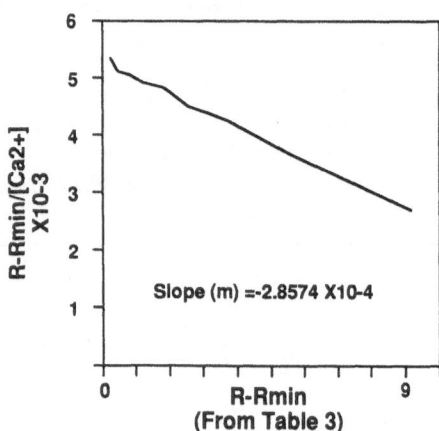

Figure 7. The Slope of the Line Created by Plotting R-Rmin/ $[Ca^{2+}]$ for each of the Ca^{2+} Standards is Used to Calculate the Dissociation Constant (Kd) for Fura-Ca^{2+}

The methods for constructing this graph are detailed in Table 2. The values for R-Rmin/ $[Ca^{2+}]$ originate from Figure 6 and are presented in Table 3.

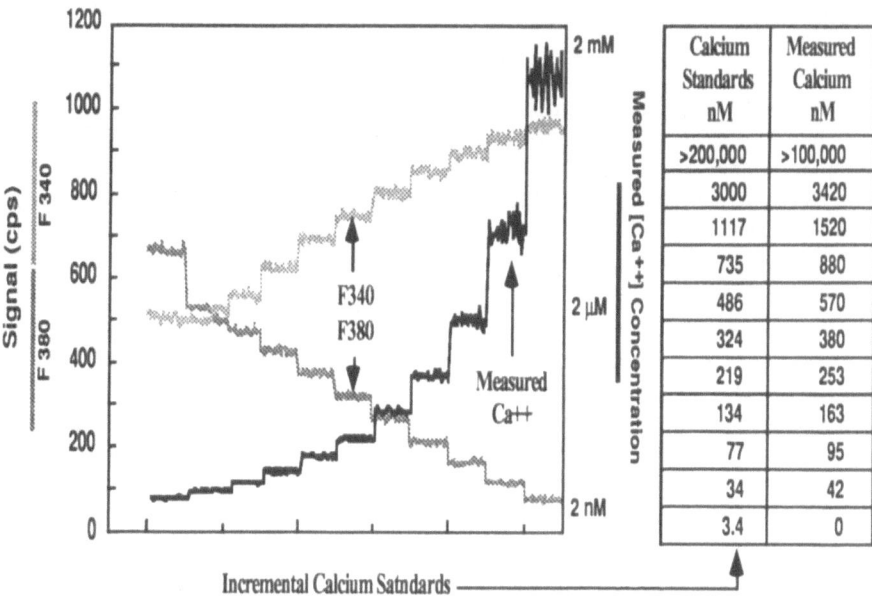

Figure 8. A Time Based Scan is Used to Validate the Assay and the Kd: Measured Ca^{2+} Values (Equation 3) are Compared to the Nominal Values of the Standard Solutions

The rationale behind and methodology for constructing this wavelength scan is described in Table 2. The ratio F340/F380 (R), and the Rmax and Rmin are used to calculate Ca^{2+} levels (using equation 3). The calculated Ca^{2+} values are close to those of the Ca^{2+} standards, indicating the equipment is properly calibrated, the Kd appropriate and that Equation 3 is valid in this experimental condition.

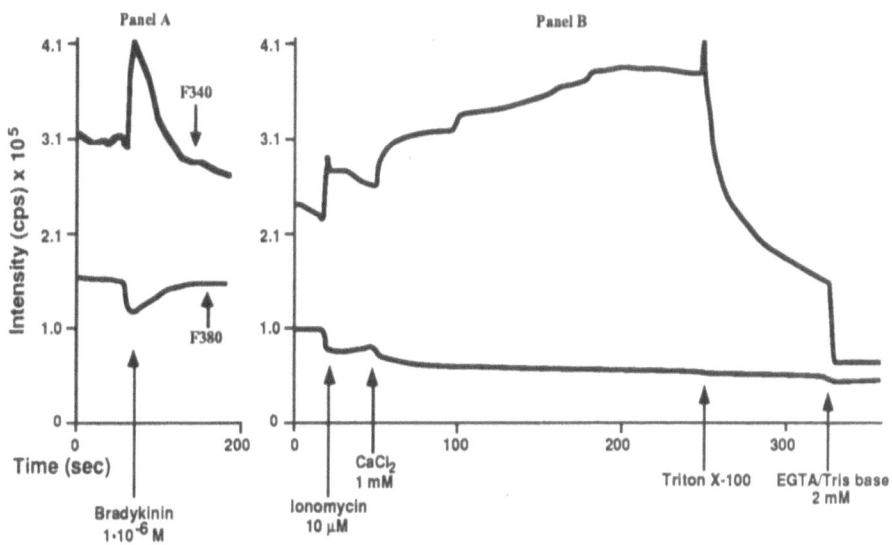

Figure 9. Measurement of R, Rmax and Rmin in Endothelial Cells

Panel A shows the change in the emission caused by 340 and 380 stimulation caused by bradykinin (BK). Panel B illustrates how the R, Rmax and Rmin were obtained. Ionomycin was given to permeabilize the bovine pulmonary artery endothelial cells. An excess of Ca^{2+} was then given to saturate intracellular fura 2 (Rmax). The cells were then lysed with the detergent Triton and Ca^{2+} was chelated by administration of EGTA (Rmin, Table 3).

237

Fura 2 Strategy

Several techniques can be used to establish the source of a Ca^{2+} transient in endothelial cells. Lanthanum chloride ($LaCl_3$) blocks the influx of Ca^{2+} from the extracellular space via all routes whereas the dihydropyridines only alter Ca^{2+} flux through the voltage-dependent Ca^{2+} channel. Increases in $[Ca^{2+}]_i$ that persist in the presence of $LaCl_3$ or when extracellular Ca^{2+} is removed are thought to result from release of intracellular Ca^{2+}.

A second approach to determining the source of elevations of $[Ca^{2+}]_i$ is based on the ability of manganese (Mn^{2+}) to quench fura-2 fluorescence [39, 40]. Mn^{2+}, like Ca^{2+}, enters the intact cell very slowly in the resting state; however, when the membrane is permeabilized (e.g. ionomycin), or a Ca^{2+} channel is opened, Mn^{2+} enters rapidly (Figure 10). Because fura-2 binds Mn^{2+} more than forty times as avidly as it does Ca^{2+}, fluorescence rapidly falls. A rise in $[Ca^{2+}]_i$ will only be associated with Mn^{2+} influx (and a fall in fluorescence) if it relies on extracellular Ca^{2+}. This technique also permits temporal resolution of the influx of Ca^{2+} in response to agents which may also mobilize intracellular Ca^{2+}. Since the influx of Mn^{2+} and its interaction with fura-2 are rapid, the time of onset of the quenching of fura-2 correlates fairly well with the time of onset of influx of extracellular Ca^{2+}.

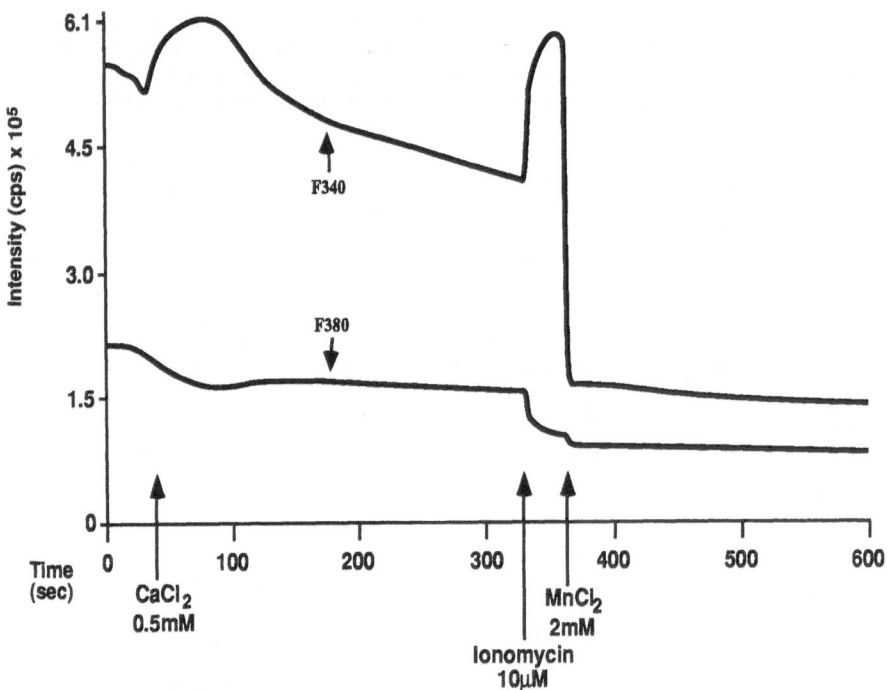

Figure 10. Manganese Rapidly Quenches Fura 2 Fluorescence When Ionomycin Initiates Influx of Ca^{2+}

Calcium influx was triggered in pulmonary artery endothelial cells by the ionophore ionomycin. Manganese was then able to enter through the membrane pores created by the ionophore and rapidly displaced Ca^{2+} from the fura 2, thereby quenching the fluorescence and demonstrating the source of increased F340/F380 caused by ionomycin was the extracellular Ca^{2+} pool, not release of intracellular stores.

MEASUREMENT OF [Ca^{2+}]$_i$ IN PULMONARY ARTERY ENDOTHELIUM

Measurement of [Ca^{2+}]$_i$ has been useful in evaluating endothelial dependent vasodilatation in the pulmonary circulation. This chapter will conclude with examples, from our laboratory, of how measurement of endothelial [Ca^{2+}]$_i$ can be used.

Correlation of Endothelial [Ca^{2+}]$_i$ and Nitric Oxide Synthesis in the Rat Lung

The pharmacological profile of acetylcholine does not resemble that of NO in the isolated rat lung. ACH causes more prolonged vasodilatation and displays tachyphylaxis which can be overcome by NO. In contrast, bradykinin causes briefer vasodilatation more reminiscent of the effects of bolus injection of NO [25]. Furthermore, the putative NOS inhibitor N^G-monomethyl-L-arginine (L-NMMA) inhibits bradykinin, but not ACH-induced pulmonary vasodilatation in isolated adult rat lungs [10]. Measurement of NO production in response to ACH, bradykinin and A23187 in isolated rat conduit pulmonary artery rings showed that A23187, and to a lesser extent bradykinin, stimulated NO synthesis. ACH was a more effective pulmonary vasodilator than either bradykinin or A23187 but did not stimulate NO synthesis [5]. Measurement of endothelial [Ca^{2+}]$_i$ confirmed that bradykinin, A23187 and ACH have different effects on Ca^{2+} homeostasis. Bradykinin and A23187 increase [Ca^{2+}]$_i$ in cultured bovine pulmonary artery endothelial cells; ACH does not alter or tends to reduce [Ca^{2+}]$_i$ [5]. The bovine pulmonary artery endothelial cells used in these experiments were in the 15th-21st passages in culture. Because endothelial cells are known to change many characteristics in culture we were concerned that the failure of ACH to increase [Ca^{2+}]$_i$ reflected the loss of the muscarinic receptor. However, [Ca^{2+}]$_i$ fell with addition of atropine, suggesting that the muscarinic receptor was present. This study utilized measurements of [Ca^{2+}]$_i$ to support the hypothesis that acetylcholine is an endothelium dependent pulmonary vasodilator which acts through a NO-independent mechanism.

Arginine Analogs Depress [Ca^{2+}]$_i$ in Pulmonary Artery Endothelial Cells

The widespread use of the arginine analogs, such as L-NAME, as probes for NOS activity has been very helpful and there is no question that arginine analogs inhibit EDRF and reduce NO synthesis. However, like most "specific-inhibitors" the perception of specificity diminishes as familiarity with the inhibitor increases [17, 50]. Several groups have demonstrated that the arginine analogs enhance the responses to a number of chemically unrelated pulmonary vasoconstrictors (e.g., hypoxia, phenylephrine, norepinephrine) [17, 51, 52]. This could indicate that all vasoconstrictors cause a secondary, counter-balancing increase in NO synthesis (revealed through the inhibitory properties of the arginine analogs). However, we have been unable to detect basal or vasoconstrictor-induced NO production by pulmonary arterial or aortic rings with a chemiluminescence assay [17, 53]. Perhaps basal or constrictor-induced NO levels fell below the detection threshold of our chemiluminescence assay (20 picomoles) [38]; although this is unlikely as the chemiluminescence nitrite/nitrate assay demonstrates minimal basal NO production by the normoxic lung (Figure 5). Consequently we evaluated the hypothesis that the arginine analogs can enhance vasoconstriction by means in addition to their effects on NOS [5]. This might involve a direct action of the analogs on the vascular smooth muscle or an inhibition of the production of other endothelial relaxing factors, in addition to NO. Since the arginine analogs do inhibit NO synthesis, a process initiated by elevation of endothelial [Ca^{2+}]$_i$, we hypothesized that they might diminish production of many types of endothelial vasodilator by interfering with agonist-induced increases in endothelial $[Ca^{2+}]_i$. Preliminary studies show that L-NMMA rapidly reduces $[Ca^{2+}]_i$ in cultured bovine pulmonary artery endothelial cells (Figure 11). The mechanism appears to be related primarily to the presence of arginine, since L- and D-arginine also reduce $[Ca^{2+}]_i$. In addition, L-NMMA reduces bradykinin-induced elevations of $[Ca^{2+}]_i$, suggesting these agents may interfere with endothelial Ca^{2+} signaling. Perhaps the effects of arginine and arginine analogs on endothelial $[Ca^{2+}]_i$ contribute to their ability to alter NO synthesis.

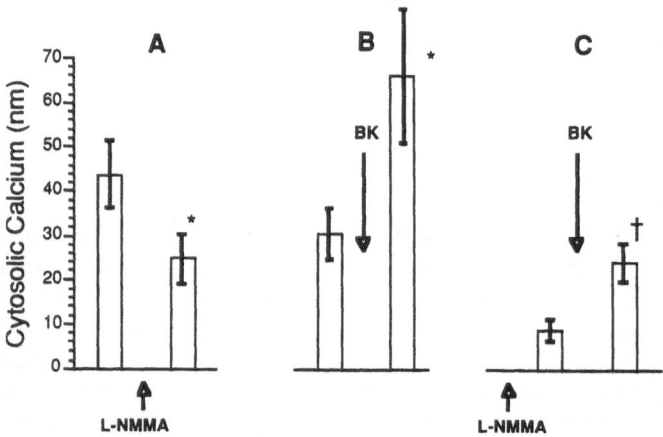

Figure 11. L-NMMA Reduces Basal [Ca2+]$_i$ and Diminishes Bradykinin-Induced Increases in [Ca2+]$_i$ In Pulmonary Artery Endothelial Cells

Values are the mean±SEM., n=5/group. * p<0.05 cytosolic Ca^{2+} is reduced by administration of L-NMMA (panel A) or increased by BK (Panel B), compared to appropriate control value. † p<0.05 L-NMMA reduces the BK-induced increase in $[Ca^{2+}]_i$ (panel B versus panel C). Panel A: L-NMMA $(10^{-4}M)$ lowers $[Ca^{2+}]_i$. Panel B, BK, $10^{-8}M$ increases $[Ca^{2+}]_i$. Panel C the peak BK response is blunted by L-NMMA pre-treatment .

It remains to be seen whether the interference with agonist-induced increases in cytosolic Ca^{2+} caused by arginine analogs impairs the production of EDHF or prostacyclin.

Acute Hypoxia Elevates [Ca^{2+}]$_i$ in Pulmonary Artery Endothelial Cells

While removal of the endothelium inhibits hypoxic constriction of isolated, conduit pulmonary artery rings [54], pharmacological inhibition of EDRF, by arginine analogs, enhances hypoxic pulmonary vasoconstriction (HPV) in isolated lungs [10, 16,12]. It is uncertain whether inhibition of NO synthesis augments HPV by removing basal NO synthesis or by blocking a compensatory increase in NO synthesis, which moderates HPV. The data presented in this chapter supports the notion that there is minimal basal NO synthesis in the lung and that NO production is increased in response to prolonged hypoxia.

Since elevation of $[Ca^{2+}]_i$ by vasodilators, such as bradykinin, initiates NO synthesis, we evaluated the hypothesis that hypoxia produces a signal (elevation of $[Ca^{2+}]_i$) which could elicit homeostatic increases in NO synthesis. The effects of acute hypoxia (pO_2 50 mm Hg for 3 minutes) on basal and bradykinin-stimulated $[Ca^{2+}]_i$ were determined in bovine pulmonary artery endothelial cells using fura-2 microscopy. Cells were studied in the presence and absence of extracellular Ca^{2+} (1 mM) at 37^oC. Acute hypoxia transiently increased $[Ca^{2+}]_i$, even in the absence of

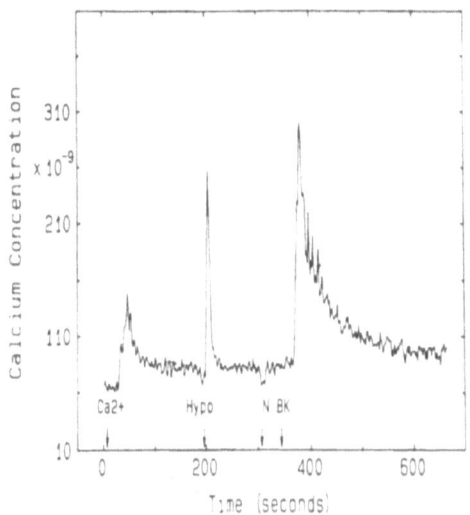

Figure 12. Acute Hypoxia Increases Cytosolic Ca^{2+} in Bovine Pulmonary Artery Endothelial Cells

This is a representative trace of endothelial cytosolic Ca^{2+} levels from pulmonary artery endothelial cells, studied in the presence of 1 mmol extracellular Ca^{2+}. Acute hypoxia (Onset at "Hypo") causes a brief increase in Ca^{2+} levels which was not sustained despite persistence of hypoxia for several minutes (from Hypo to "N "on the Time axis). This figure shows a normal response to bradykinin (BK $10^{-8}M$) which was given following resumption of normoxia (indicated by "N" on the time axis).

extracellular Ca^{2+} (Table 4, Figure 12). Bradykinin-induced increases in $[Ca^{2+}]_i$ were not reduced by acute hypoxia, but were smaller in the absence of extracellular Ca^{2+} (Table 4) Although hypoxia does not interfere with Ca^{2+} signaling in the endothelium, it is unknown whether it impairs distal steps in the NO synthesis pathway. The significance of the Ca^{2+} spike caused by acute hypoxia is unknown although it may represent a signal for synthesis of prostacyclin, NO or other vasodilator substances.

Table 4. Acute Hypoxia Increases $[Ca^{2+}]_i$ and does not Inhibit Bradykinin-Induced Increases in $[Ca^{2+}]_i$ in Bovine Pulmonary Artery Endothelial Cells

Group	Basal $[Ca^{2+}]_i$ During Normoxia nM	Peak $[Ca^{2+}]_i$ During Hypoxia nM	Peak $[Ca^{2+}]_i$ Following BK nM
1 Ca^{2+} Replete Normoxic during BK response	76±7	158±25	221±21
2 Ca^{2+} Replete Hypoxia during BK response	85±19	123±19	251±58
3 Ca^{2+} Free Normoxic during BK response	67±5	153±18	117±21*

Values are the mean±SEM, * $p<0.05$ (n=5/group) In these experiments the effects of hypoxia on cytosolic Ca^{2+} were studied in all groups (middle column). In group 1 the hypoxia ceased prior to administration of bradykinin (BK), as in figure 11. In group 2 the hypoxia was continued throughout the BK response, which did not alter the magnitude of the BK-induced increase in cytosolic Ca^{2+}. Group 3 cells were studied in Ca^{2+} free perfusate, but as for group 1, hypoxia ceased prior to administration of BK. Compared to Ca^{2+} replete cells the Ca^{2+}-free group had smaller responses to BK , but equivalent response to hypoxia.

Conclusion

Nitric oxide synthesis by the vascular endothelium is dependent on extracellular Ca^{2+} and is associated with a transient increase in endothelial $[Ca^{2+}]_i$. The route by which extracellular Ca^{2+} enters the cell is not the key determinant of whether NO synthesis will occur. Influx of Ca^{2+} , whether through receptor operated channels or nonspecific pores in the plasma membrane caused by ionophores, elicits NO synthesis and causes vasodilatation. Elevation of endothelial $[Ca^{2+}]_i$ increases the interaction of NOS_c with its co-factor, calmodulin, thereby facilitating NO synthesis and partially explaining the regulatory role of Ca^{2+} for the constitutive form of NOS. Measurement of $[Ca^{2+}]_i$ in the pulmonary artery endothelium, using fura 2 microscopy, suggests:

- Acetylcholine is an endothelium dependent, NO-independent pulmonary vasodilator
- Arginine and arginine analogs lower endothelial $[Ca^{2+}]_i$ and may interfere with bradykinin-induced increases in $[Ca^{2+}]_i$
- Acute hypoxia increases endothelial $[Ca^{2+}]_i$

Acknowledgments

Dr. Archer is supported by NIH grant 1R29-HL45735 and the Veterans Administration. Dr. Hampl is supported by a Proshek Foundation Fellowship in Medicine.

References

1 M. J. Peach, H. A. Singer, N. J. Izzo and A. L. Loeb, Role of calcium in endothelium-dependent relaxation of arterial smooth muscle, *Am. J. Cardiol.* 59: A35-A43 (1987)
2 A. Luckhoff, U. Pohl, A. Mulsch and R. Busse, Differential role of extra- and intracellular calcium in the release of EDRF and prostacyclin from cultured endothelial cells, *Br. J. Pharmacol.* 95: 189-196 (1988)

3 A. Johns, T. W. Lategan, N. J. Lodge, U. S. Ryan, C. vanBreeman and D. J. Adams, Calcium entry through receptor operated channels in bovine pulmonary artery endothelial cells, *Tissue&Cell.* 19: 733-745 (1987)

4 M. Colden-Stanfield, W. P. Schilling, A. K. Ritchie, S. G. Eskin, L. T. Navarro and D. L. Kunze, Bradykinin-induced increases in cytosolic calcium and ionic currents in cultured bovine aortic endothelial cells, *Circ. Res.* 61: 632-640 (1987)

5 S. L. Archer and N. J. Cowan, Acetylcholine causes Endothelium Dependent Vasodilatation but does not stimulate Nitric Oxide Production by Rat Pulmonary Arteries or Elevate Endothelial Cytosolic Calcium Concentrations, *Circ. Res.* 68: 1569-1581 (1991)

6 L. J. Ignarro, Endothelium-derived nitric oxide: actions and properties, *FASEB J.* 3: 31-36 (1989)

7 L. J. Ignarro, R. G. Harbison, K. S. Wood and P. J. Kadowitz, Activation of purified soluble guanylate cyclase by endothelium-derived relaxing factor from intrapulmonary artery and vein: stimulation by acetylcholine, bradykinin and arachidonic acid, *J. Pharm. Exp. Ther.* 237: 893-900 (1986)

8 L. J. Ignarro, G. M. Buga, R. E. Byrns, K. S. Wood and G. Chaudhuri, Endothelium-derived relaxing factor and nitric oxide possess identical pharmacological properties as relaxants of bovine arterial and venous smooth muscle, *J. Pharm. Exp. Ther.* 246: 218-226 (1988)

9 R. M. J. Palmer, A. G. Ferridge and S. Moncada, Nitric oxide release accounts for the biological activity of endothelium-derived relaxing factor, *Nature.* 327: 524-526 (1987)

10 S. L. Archer, J. P. Tolins, L. Raij and E. K. Weir, Hypoxic pulmonary vasoconstriction is enhanced by inhibition of the synthesis of an endothelium derived relaxing factor, *Biochem.Biophys.Res.Comm.* 164: 1198-1205 (1989)

11 J. Gardes, J.-M. Poux, M.-F. Gonzalez, F. Alhenc-Gelas and J. Menard, Decreased renin release and constant kallikrein secretion after injection of L-NAME in isolated perfused rat kidney, *Life Sci.* 50: 987-993 (1992)

12 K. Hasunuma, T. Yamaguchi, D. M. Rodman, R. F. O'Brien and I. F. McMurtry, Effects of inhibitors of EDRF and EDHF on vasoreactivity of perfused rat lungs, *Am J Physiol.* 260: L97-L104 (1991)

13 S. Liu, D. E. Crawley, P. J. Barnes and T. W. Evans, Endothelium-derived relaxing factor inhibits hypoxic pulmonary vasoconstriction in rats, *Am Rev Respir Dis.* 143: 32-37 (1991)

14 J. Radermacher, U. Forstermann and J. C. Frolich, Endothelium-derived relaxing factor influences renal vascular resistance, *Am J Physiol.* 259: F9-F17 (1990)

15 J. Radermacher, B. Klanke, H.-J. Schurek, H. F. Stolte and J. C. Frolich, Importance of NO/EDRF for glomerular and tubular function: studies in the isolated perfused rat kidney, *Kidney Int.* 41: 1549-1559 (1992)

16 B. E. Robertson, J. B. Warren and P. C. G. Nye, Inhibition of nitric oxide synthesis potentiates hypoxic vasoconstriction in isolated rat lungs, *Exp Physiol.* 75: 255-257 (1990)

17 S. L. Archer and V. Hampl, NG-Monomethyl-L-Arginine causes nitric oxide synthesis in isolated arterial rings:Trouble in paradise, *Biochem. Biophys. Res. Comm.* 188: 590-596 (1992)

18 T. M. Zellers and P. M. Vanhoutte, Heterogeneity of endothelium-dependent and independent responses among large and small porcine pulmonary arteries, *Pulmon Pharmacol.* 2: 201-208 (1989)

19 R. M. Leach, C. H. Twort, I. R. Cameron and J. P. T. Ward, A comparison of the pharmacological and mechanical properties in vitro of large and small pulmonary arteries of the rat, *Clin Sci.* 82: 55-62 (1992)

20 S. A. Barman, E. Senteno and A. E. Taylor, Acetylcholine's effect on vascular resistance and compliance in the pulmonary circulation, *J Appl Physisol.* 67: 1495-1503 (1989)

21 H. A. ElKashef, W. F. Hofman, I. C. Ehrhart and J. D. Catravas, Multiple muscarinic receptor subtypes in the canine pulmonary circulation, *J Appl Physiol.* 71: 2032-2043 (1991)

22 T. J. McMahon, J. S. Hood, J. A. Bellan and P. J. Kadowitz, N^{ω}-nitro-L-arginine methyl ester selectively inhibits pulmonary vasodilator responses to acetylcholine and bradykinin, *J Appl Physiol.* 71: 2026-2031 (1991)

23 H. W. Fritts, P. Harris, R. H. Clauss, J. E. Odell and A. Cournand, The effect of acetylcholine on the human pulmonary circulation under normal and hypoxic conditions, *J Clin Invest.* 37: 99-110 (1958)

24 J. H. K. Vogel, D. Cameron and G. Jamieson, Chronic pharmacological treatment of experimental hypoxic pulmonary hypertension. With observations on rate of change in pulmonary arterial pressure, *Am Heart J.* 72: 50-59 (1966)

25 S. L. Archer, K. Rist, D. Nelson, E. De Master, N. Cowan and E. K. Weir, Comparison of the hemodynamic effects of nitric oxide and endothelium dependent vasodilators in the lung, *J. Appl. Physiol.* 68: 735-747 (1990)

26 A. L. Hyman and P. J. Kadowitz, Influence of tone on responses to acetylcholine in the rabbit pulmonary vascular bed, *J Appl Physiol.* 67: 1388-1394 (1989)

27 K. Nishiwaki, D. P. Nyhan, P. Rock, P. M. Desai, W. P. Peterson, C. G. Pribble and P. A. Murray, N^{ω}-nitro-L-arginine and pulmonary vascular pressure-flow relationship in conscious dogs, *Am J Physiol.* 262: H1331-H1337 (1992)

28 M. G. Persson, L. E. Gustafsson, N. P. Wiklund, S. Moncada and P. Hedqvist, Endogenous nitric oxide as a probable modulator of pulmonary circulation and hypoxic pressor response in vivo, *Acta Physiol Scand.* 140: 449-457 (1990)

29 J.-K. Chang, P. Moore, J. R. Fineman, S. J. Soifer and M. A. Heymann, K^+ channel pulmonary vasodilation in fetal lambs: role of endothelium-derived nitric oxide, *J Appl Physiol.* 73: 188-194 (1992)

30 S. H. Abman, B. A. Chatfield, S. L. Hall and I. F. McMurtry, Role of endothelium-derived relaxing factor during transition of pulmonary circulation at birth, *Am J Physiol.* 259: H1921-H1927 (1990)

31 D. Davidson and A. Eldemerdash, Endothelium-derived relaxing factor: evidence that it regulates pulmonary vascular resistance in the isolated neonatal guinea pig lung, *Pediatr Res.* 29: 538-542 (1991)

32 J. R. Fineman, R. Chang and S. J. Soifer, EDRF inhibition augments pulmonary hypertension in intact newborn lambs, *Am J Physiol.* 262: H1365-H1371 (1992)

33 S. Adnot, B. Raffestin, S. Eddahibi, P. Braquet and P.-E. Chabrier, Loss of endothelium-dependent relaxant activity in the pulmonary circulation of rats exposed to chronic hypoxia, *J Clin Invest.* 87: 155-162 (1991)

34 A. T. Dinh-Xuan, T. W. Higenbottam, C. A. Clelland, J. Pepke-Zaba, G. Cremona, A. Y. Butt, S. R. Large, F. C. Wells and J. Wallwork, Impairment of endothelium-dependent pulmonary-artery relaxation in chronic obstructive lung disease, *N Eng. J Med.* 324: 1539-1547 (1991)

35 E. C. Orton, J. T. Reeves and K. R. Stenmark, Pulmonary vasodilation with structurally altered pulmonary vessels and pulmonary hypertension, *J Appl Physiol.* 65: 2459-2467 (1988)

36 T. Isaacson, E. K. Weir, V. Hampl, D. Nelson and S. Archer, Enhanced vasoconstriction in chronic hypoxic pulmonary hypertension is not associated with reduced responsiveness to EDRF or nitric oxide (abstract), *FASEB J.* 6: A1177 (1992)

37 A. Molteni, W. F. Ward, C.-H. Ts'ao, C. D. Port and N. H. Solliday, Monocrotaline-induced pulmonary endothelial dysfunction in rats, *Proc. Soc. Exp. Biol.* 176: 88-94 (1984)

38 S. L. Archer, Measurement of nitric oxide in biological models, *FASEB J.* 7: (in press) (1993)

39 T. J. Hallam, J. D. Pearson and L. A. Needham, Thrombin-stimulated elevation of human endothelial-cell cytoplasmic free calcium concentration causes prostacyclin production, *Biochem. J.* 251: 243-249 (1988)

40 T. J. Hallam, R. Jacob and J. E. Merritt, Evidence that agonists stimulate bivalent-cation influx into human endothelial cells, *Biochem J.* 255: 179-184 (1988)

41 N. Menon, A. Wolf, M. Zehetgruber and R. Bing, An improved chemiluminescence assay suggests non nitric oxide-mediated action of lysophosphatidylcholine and acetylcholine, *Proc. Soc. Exp. Biol. Med.* 191: 316-319 (1989)

42 G. Chen and H. Suzuki, Calcium dependency of the endothelium-dependent hyperpolarization in smooth muscle cells of the rabbit carotid artery, *J. Physiol.* 421: 521-534 (1990)

43 C. Nathan, Nitric oxide as a secretory product of mammalian cells, *FASEB J.* 6: 3051-3064 (1992)

44 D. J. Adams, J. Barakeh, R. Laskey and C. vanBreeman, Ion channels and regulation of intracellular calcium in vascular endothelial cells, *FASEB J.* 3: 2389-2400 (1989)

45 R. F. Furchgott, Role of the endothelium in the response of vascular smooth muscle to drugs, *Ann. Rev. Pharmacol.* 24: 175-197 (1984)

46 R. E. Laskey, D. J. Adams, M. Cannell and C. vanBreeman, Calcium entry-dependent oscillations of cytoplasmic calcium concentration in cultured endothelial cell monolayers, *Proc. Natl. Acad. Sci.* 89: 1690-1694 (1992)

47 G. Mehrke and J. Daut, The electrical response of cultured guinea-pig coronary endothelial cells to endothelium-dependent vasodilators, *J. Physiol. (London).* 430: 251-272 (1990)

48 R. Y. Tsien, T. J. Rink and M. Poenie, Measurement of cytosolic free Ca^{2+} in individual small cells using fluorescence microscopy with dual excitation wavelengths, *Cell Calcium.* 6: 145-157 (1985)

49 G. Grynkiewicz, M. Poenie and R. Y. Tsien, A new generation of Ca^{2+} indicators with greatly improved fluoresence properties, *J. Biol. Chem.* 260: 3440-3550. (1985)

50 D. A. Peterson, D. C. Peterson, S. Archer and E. K. Weir, The non specificity of specific nitric oxide synthase inhibitors, *Biochem. Biophys. Res. Comm.* 187: 797-801 (1992)

51 T. M. Cocks and J. A. Angus, Evidence that contraction of isolated arteries by L-NMMA and NOLA are not due to inhibition of basal EDRF release, *J. Cardiovasc. Pharmacol.* 17: S159-S164 (1991)

52 G. Thomas and P. Ramwell, Interaction of non-arginine compounds with the endothelium-derived relaxing factor inhibitor, NG-monomethyl L-arginine, *J. Pharm. Exp. Ther.* 260: 676-679 (1991)

53 S. L. Archer and N. J. Cowan, Measurement of endothelial cytosolic calcium concentration and nitric oxide production reveals discrete mechanisms of endothelium-dependent pulmonary vasodilation, *Circ Res.* 68: 1569-1581 (1991)

54 T. Yamaguchi, D. Rodman, R. O'Brien and I. F. McMurtry, Modulation of pulmonary artery contraction by endothelium-derived relaxing factor, *Eur J Pharmacol.* 161: 259-262 (1989)

THE ROLE OF ION CHANNELS IN VASCULAR ENDOTHELIUM

Andreas Lückhoff

Department of Pharmacology
Mayo Foundation
Rochester, MN 55905, USA

INTRODUCTION

In the past few years, the role of ion channels in several endothelial cell functions has been progressively elucidated. Particularly, we have learned how the membrane potential, controlled mostly by K^+ channels, exerts a crucial control function on Ca^{2+} influx (and thereby on Ca^{2+}-dependent endothelial reactions like release of nitric oxide), while Ca^{2+} influx, in turn, indirectly affects the membrane potential. Some pathways for Ca^{2+} influx have been studied on the level of single channel analysis, although the knowledge of these pathways is still far from complete. Furthermore, ion channels appear to be intimately involved in the response to shear stress and flow.

My own work is mostly concerned with the mechanisms of Ca^{2+} influx after receptor-dependent stimulation, and this will be one major topic of this review. Discussing ion channels, I will concentrate on results obtained with the patch-clamp method. Being aware that major differences may exist between endothelial cells from different vascular segments, species, and culture conditions, I should point out that all figures represent experiments performed on long-term cultured endothelial cells from bovine aorta.

ENDOTHELIAL POTASSIUM CHANNELS

Whole-cell measurements of unstimulated endothelial cells reveal as prominent current an inward-rectifying K^+ current (Fig. 1a). This current is not dependent on internal Ca^{2+} because it is regularly found with <10 nM Ca^{2+} in the internal (pipette) solution (Olesen et al., 1988b). The corresponding ion channel is shown in Fig. 1a. With high K^+ concentrations in the pipette, several of those channels are consistently found in cell-attached patches. Characteristic features include long open times (several hundred ms) and the absence of a voltage-dependency. The conductance is 24 pS with 120 mM K^+ as charge carrier. The inward rectification is striking; no reversal of the current can be found in the cell-attached mode; outside-out or inside-out patches rarely reverse.

Ion Flux in Pulmonary Vascular Control, Edited by
E.K. Weir *et al.*, Plenum Press, New York, 1993

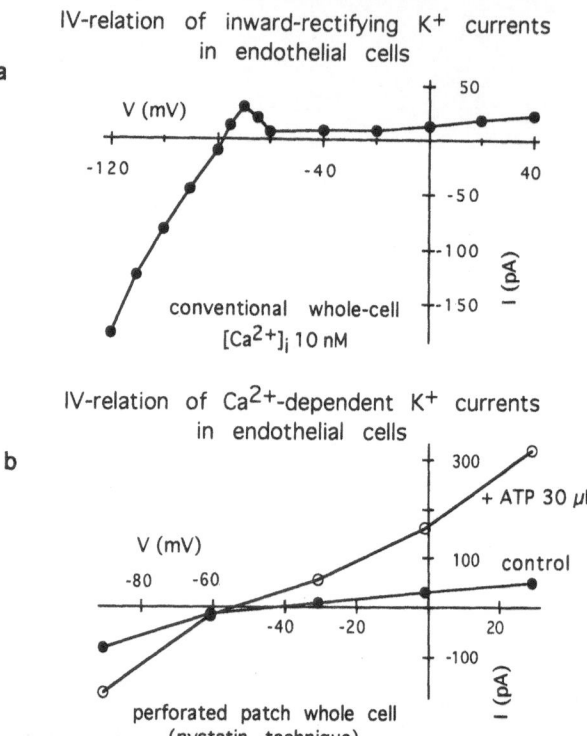

IV-relation of inward-rectifying K$^+$ currents in endothelial cells

conventional whole-cell
[Ca^{2+}]$_i$ 10 nM

IV-relation of Ca^{2+}-dependent K$^+$ currents in endothelial cells

+ ATP 30 μM

control

perforated patch whole cell
(nystatin technique)

Fig. 1. Current-voltage relation of endothelial K$^+$ currents. **a)** K$^+$ currents in a cell during a conventional whole-cell patch-clamp (Hamill et al., 1981) experiment. The pipette contained (in mM) KCl$_2$ 120, MgCl$_2$ 2, CaCl$_2$ 1, EGTA 10, ATP 1.5, GTP 0.3, HEPES 10, pH (NaOH) 7.2; [Ca^{2+}] <10 nM. The bath contained NaCl 140, KCl 5, CaCl$_2$ 1, MgCl$_2$ 2, HEPES 10, glucose 5, pH 7.4. Note that the outward current is largest close to the equilibrium concentration of K$^+$. **b)** K$^+$ currents in a cell held in voltage-clamp with the perforated-patch-clamp technique (Horn and Marty, 1988), before (control) and during application of ATP to the bath. The pipette contained KCl 40, K$_2$SO$_4$ 80, CaCl$_2$ 1, MgCl$_2$ 2, HEPES 5, nystatin 0.3 mg/ml, pH (NaOH) 7.4. In the bath, NaCl was substituted with choline chloride. The effect of ATP was fully reversible after washout (not shown).

In light of this rectification, one may question the physiological relevance of this channel type because K$^+$ inward currents do not normally occur in non-excitable cells. However, Fig. 1a reveals that when Ca^{2+} dependent currents are eliminated, there is a narrow voltage range where a marked K$^+$ *outward* current may be observed. This range, slightly more positive than the reversal potential for K$^+$, is so small that it may be easily overlooked during routine measurements.

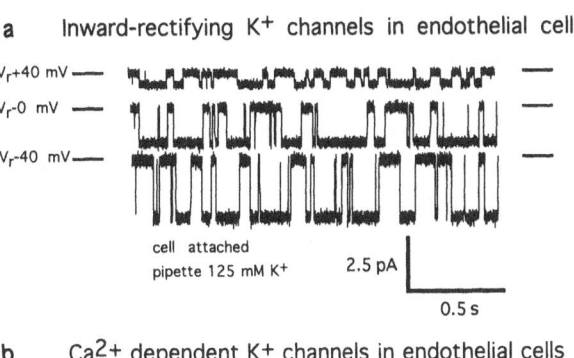

a   Inward-rectifying K$^+$ channels in endothelial cells

V$_r$+40 mV

V$_r$-0 mV

V$_r$-40 mV

cell attached
pipette 125 mM K$^+$

2.5 pA

0.5 s

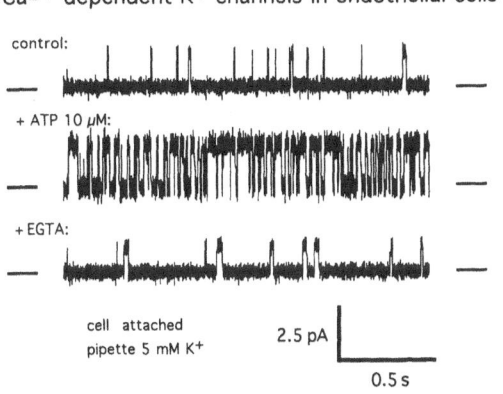

b   Ca^{2+} dependent K$^+$ channels in endothelial cells

control:

+ ATP 10 μM:

+ EGTA:

cell attached
pipette 5 mM K$^+$

2.5 pA

0.5 s

Fig. 2. K$^+$ channels in endothelial cells. Cell-attached patch-clamp experiments. The closed level of the channels is indicated by lines. **a)** Inward-rectifying K$^+$ channel. **b)** Ca^{2+} dependent K$^+$ channel before (control) and during bath application of ATP and after addition of EGTA (5 mM). Note that the channel conducts in the outward direction, that its increase in activity by ATP is reversed by chelation of external Ca^{2+} with EGTA, and that the channel amplitude changes, reflecting changes in the membrane potential (V$_{hold}$ 0 mV). Bath solution as in Fig. 1a; pipette solution in (b) equal to bath solution, in (a) equal to bath solution with NaCl substituted with 120 mM KCl.

These results indicate that inward-rectifying K$^+$ channels help to stabilize the membrane potential close to -60 mV but do not carry major outward currents when the cells are depolarized. This situation is similar to that in the myocardium (Josephson and Brown, 1986).

In order to analyze Ca^{2+} dependent whole-cell currents, internal Ca^{2+} buffering less rigorous than in Fig. 1a is desirable. The currently best method for this purpose appears to be the perforated-patch technique (Horn and Marty, 1988). Electrical access to the cell interior is not gained by a "hole in the membrane", as it is in conventional whole-cell experiments (Hamill et al., 1981), but rather via pores formed by the antibiotic nystatin or amphotericin B added to the pipette fluid. These pores are well permeable to monovalent cations, less permeant to monovalent anions, and practically impermeant to divalent cations such as Ca^{2+} and Mg^{2+}. Furthermore, the method avoids dialysis of the cell that inevitably results in rapid loss of soluble cytosolic factors.

Endothelial cells held in voltage-clamp with the perforated-patch method reveal resting currents (Fig. 1b) not much different from those found with the conventional whole-cell technique. However, after stimulation with substances such as ATP or bradykinin, large outward currents carried by K^+ are found (Fig. 1b). These currents do not occur when external Ca^{2+} is chelated with EGTA. These and other (Colden-Stanfield et al., 1987) experiments demonstrate that ATP and bradykinin do not activate ion channels directly; rather, channel activation is a consequence of the increases in $[Ca^{2+}]_i$ elicited by the stimuli.

In cell-attached patches, outward-conducting K^+ channels can sometimes be found, although not so regularly as inward-rectifying K^+ channels. Channel activity is markedly increased after bath application of ATP (Fig. 2b). This effect is reversible after addition of EGTA to the bath. These channels are likely identical to those studied by others as indicators of internal Ca^{2+} (Sauvé et al., 1988). Ca^{2+}-dependent K^+ channels in endothelial cells have a similar conductance (22 pS with 120 mM K^+ as charge carrier in inside-out patches) as the inward rectifyer, however they can be discriminated by their short open times (18 ms), and, in inside-out patches, by their Ca^{2+} dependency and less marked rectification. However, discrimination is complicated because many channels of either type are frequently present in single patches.

Under physiological conditions, activation of Ca^{2+} dependent outward K^+ currents leads to hyperpolarization. This hyperpolarization exerts a major control function on Ca^{2+} influx, which is also stimulated by ATP and bradykinin. It has been repeatedly shown (Cannell and Sage, 1989; Schilling, 1989; Schilling et al., 1989; Lückhoff and Busse, 1990b) that Ca^{2+} influx uses the membrane potential as driving force so that it is nearly abolished when the cells are depolarized whereas it is enhanced under experimental conditions leading to hyperpolarization. Moreover, release of nitric oxide (NO) from endothelial cells has been shown to be mainly determined by Ca^{2+} influx (Lückhoff et al., 1988). This finding explains why NO release is diminished or absent at high external K^+ concentrations (Furchgott, 1983).

It should be mentioned that endothelial K^+ currents may be increased by some K^+ channel activators such as pinacidil and cromakalim (Lückhoff and Busse, 1990a). These findings suggest that the vasodilating effects of K^+ channel activators may have an endothelium-dependent component. Indeed, this notion has recently been experimentally confirmed (Nakashima et al., 1992).

ENDOTHELIAL CHLORIDE CHANNELS

Chloride channels are frequently found during experiments with hyperosmolar pipette solutions when the whole-cell configuration is inadvertently reached from the cell-attached formation and an outside-out patch is formed by pulling back the pipette. A typical example of the chloride channels observed in such an outside-out patch is shown in Fig. 3. The channels' conductance is relatively big; it is ~320 pS with 150 mM Cl^- as charge carrier. Mean open time is 8 ms. As also seen in Fig. 3, there is no obvious voltage-dependency of channel activity.

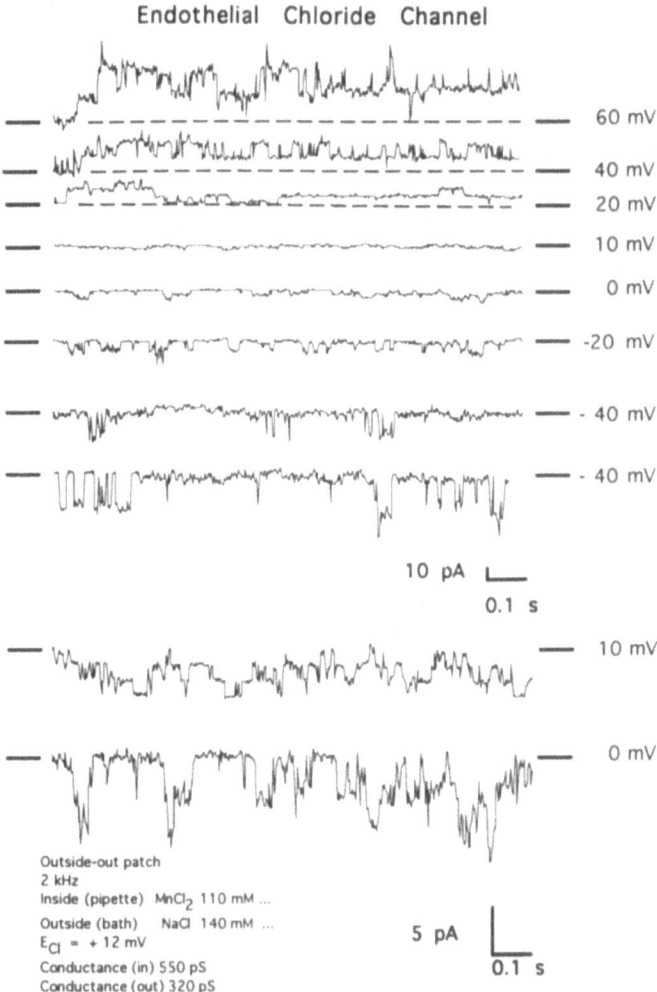

Fig. 3. Endothelial Cl⁻ channel. Outside-out patch with a hypertonic (330 mosmol/l) pipette solution (in mM): $MnCl_2$ 110, $CaCl_2$ 1, $MgCl_2$ 2, HEPES 10, pH 7.1. Bath as in Fig. 1a. From the IV-relation, a conductance of 550 pS in the outward direction and of 320 pS in the inward direction is calculated. The activity of this channel was irreversibly abolished by bath application of the chloride channel inhibitor DIDS (0.3 mM) (not shown).

The fact that these chloride channels are found after dialyzing the cell interior with hyperosmolar solutions suggests that cell swelling may be their physiological activator. Indeed, the same channels were observed with isosmotic Ringer's solution in the pipette when the cell were exposed to a hyposmotic bath (180 mosm/l)(unpublished). The channels are readily inhibited by the chloride channel blocker 4,4',diisothiocyanatostilbene-2,2'-disulfonic acid (DIDS, 0.3 mM).

However, I could not find any evidence for cAMP-dependent or voltage-regulated chloride channels in cultured bovine aortic endothelial cells.

ENDOTHELIAL CALCIUM-PERMEABLE CHANNELS

Endothelial cells are non-excitable cells and, in general, are not equipped with voltage-activated Ca^{2+} channels. The few studies (e.g. Bossu et al., 1989) that have contradicted this notion did not involve "typical" endothelial cells.

This review will mostly concentrate on endothelial Ca^{2+} pathways activated after receptor-dependent stimulation. Such stimulation, however, may be less important *in vivo* than stimulation by blood flow, exerting shear-stress on endothelial cells. Stretch-activated Ca^{2+}-permeable cation channels have been found (Lansman et al., 1987), but it is not known whether they may carry an appreciable whole-cell current in response to flow-induced shear stress. In other studies, shear-stress induced membrane currents (Olesen et al., 1988a; Schwarz et al., 1992b) and fluorometrically-determined calcium entry (Schwarz et al., 1992a) have been reported, but single-channel-analysis was not possible and the relation to the stretch-activated channels is not known.

After stimulation of endothelial cells with various agonists such as thrombin and histamine, Ca^{2+}-permeable cation channels have been demonstrated (Johns et al., 1987; Adams et al., 1989; Nilius, 1990; Nilius, 1991) that are probably of the same or a similar type. They are likely coupled to membrane receptors, possibly by an (so far unidentified) G-protein. However, gating by a cytosolic factor (but not by Ca^{2+}) is also possible. These channels are not specific for Ca^{2+}, so that under physiological conditions, they conduct mostly Na^+ and also K^+. Therefore, they lead to depolarization, counteracting the hyperpolarizing effect of Ca^{2+}-dependent K^+ channels. In spite of their non-selectivity, the Ca^{2+} current that they carry may account for the elevations in cytosolic Ca^{2+} observed after stimulation with the same agonists (Nilius, 1990).

However, no permeability of the channels to Mn^{2+} was found (B. Nilius, personal communication). This is surprising in light of the fact that Mn^{2+} entry can be used as a probe for Ca^{2+} entry in fura-2 studies (Hallam et al., 1989; Jacob, 1990). Furthermore, endothelial cells voltage-clamped with the perforated-patch clamp technique exhibit a Mn^{2+} current of several tens of pA (external $[Mn^{2+}]$ 6 mM) after stimulation with ATP (Lückhoff and Clapham, 1992).

Looking for a membrane channel that would account for this Mn^{2+} current, I found a channel that is equally permeant to Ca^{2+}, Mn^{2+}, and Ba^{2+} but not to Na^+ and K^+ (Lückhoff and Clapham, 1992). In cell-attached patches, this channel is activated by bath application of ATP, bradykinin, and inositol (Fig. 4). Its conductivity is very small, only 2.5 pS (with 110 mM Mn^{2+}; Fig. 5), which makes resolution of single channel events difficult. Fortunately, the mean open time of the channel (230 ms) is long enough to allow substantial filtering of the recordings.

The channel exhibits a marked run-down (i.e. loss of activity) over 1-3 min in inside-out patches. Yet this time is sufficient to reveal two important regulatory principles of the channel. First, its activity is virtually abolished at Ca^{2+} levels corresponding to that in resting cells. Maximal activation is found at Ca^{2+} concentrations of 10 µM-1 mM; at 0.5 µM, the activation is 0.3 of the maximum. Second, the channel activity is markedly (some 4-fold) increased (Fig. 6) by 1,3,4,5-inositol-tetrakisphosphate ($InsP_4$, 10-30 µM), a second messenger stimulating Ca^{2+} influx in some other systems (e.g. Morris et al., 1987. Changya et al., 1989). Again, there is an absolute requirement for Ca^{2+} (Fig. 7): when Ca^{2+} in the

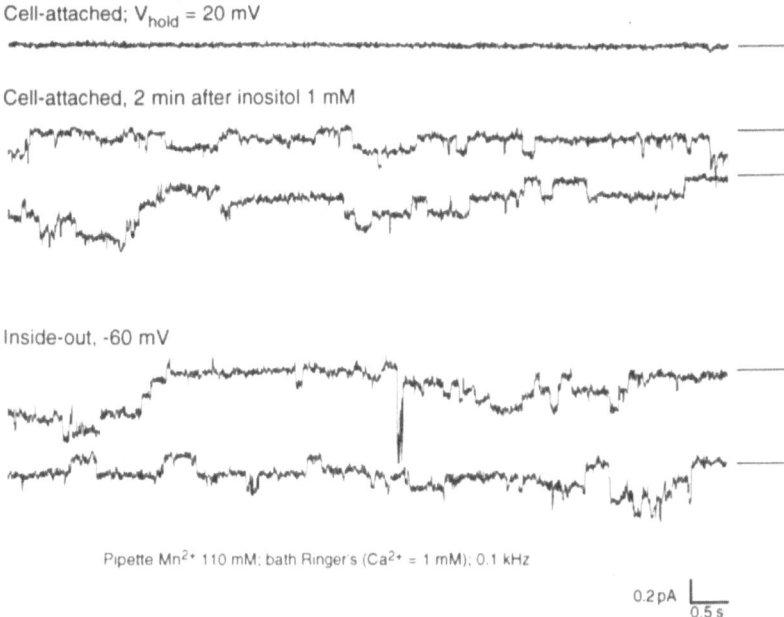

Cell-attached; V$_{hold}$ = 20 mV

Cell-attached, 2 min after inositol 1 mM

Inside-out, -60 mV

Pipette Mn^{2+} 110 mM; bath Ringer's (Ca^{2+} = 1 mM); 0.1 kHz

0.2 pA
0.5 s

Fig. 4. Mn^{2+} (and Ca^{2+}) permeable channel in endothelial cells. A pipette was sealed on to an endothelial cell in the cell-attached mode. No channels were visible under control conditions (top). 2 min after bath application of inositol (1 mM), channel activity started (middle). This channel activity was preserved when the patch was excised into the inside-out configuration (bottom). The pipette solution was (in mM) MnCl$_2$ 110, CaCl$_2$ 1, MgCl$_2$ 2, HEPES 10, pH 7.1. Bath as in Fig. 1a. Since endothelial cell typically have a membrane potential of -40 mV and the holding potential in the cell-attached mode was set to 20 mV, the transmembrane potential in the patch was ~ -60 mV.

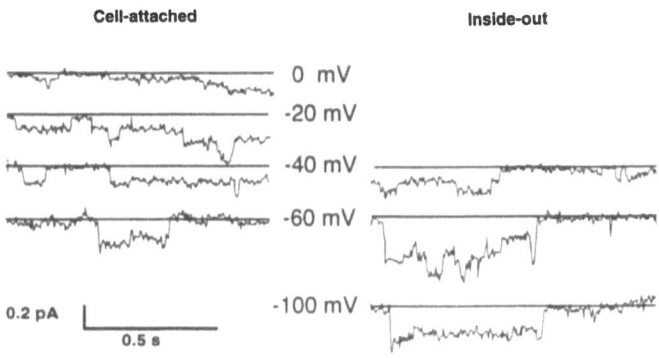

Cell-attached   Inside-out

0 mV
-20 mV
-40 mV
-60 mV
-100 mV

0.2 pA
0.5 s

Pipette 110 Mn^{2+}; bath Ringer's; 0.1 kHz

Fig. 5. Current-voltage-relation of endothelial Mn^{2+} permeable channel. Same experiment as in Fig. 4, after addition of inositol. The channel has a slope conductance of 2.5 pS.

Effect of InsP$_4$ on endothelial Ca^{2+} channels

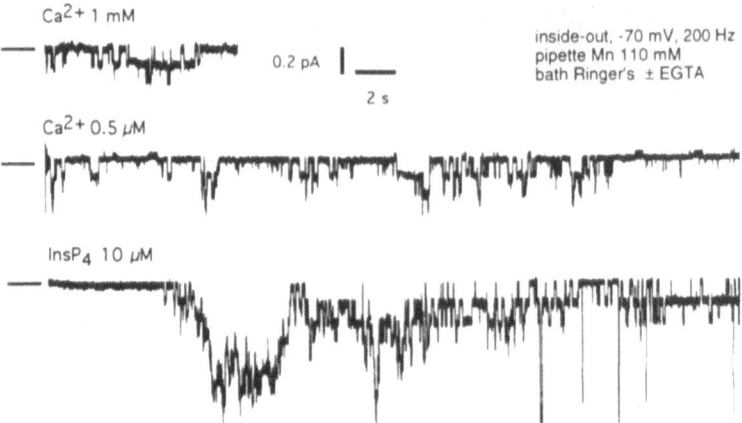

Ca^{2+} 1 mM

0.2 pA

2 s

inside-out, -70 mV, 200 Hz
pipette Mn 110 mM
bath Ringer's ± EGTA

Ca^{2+} 0.5 μM

InsP$_4$ 10 μM

Fig. 6. Effect of InsP$_4$ on the activity of endothelial Mn^{2+} permeable channels in the inside-out configuration. The patch was excised into standard bath solution ([Ca^{2+}] 1 mM) (top), then EGTA (1.17 mM) was added to lower [Ca^{2+}] to 0.5 μM (middle). Finally, InsP$_4$ (30 μM) was applied (bottom). Note that EGTA decreases channel activity to some 30 % whereas InsP$_4$ increases it about 6-fold.

Ca^{2+}-dependency of InsP$_4$ effects on endothelial Ca^{2+} channels

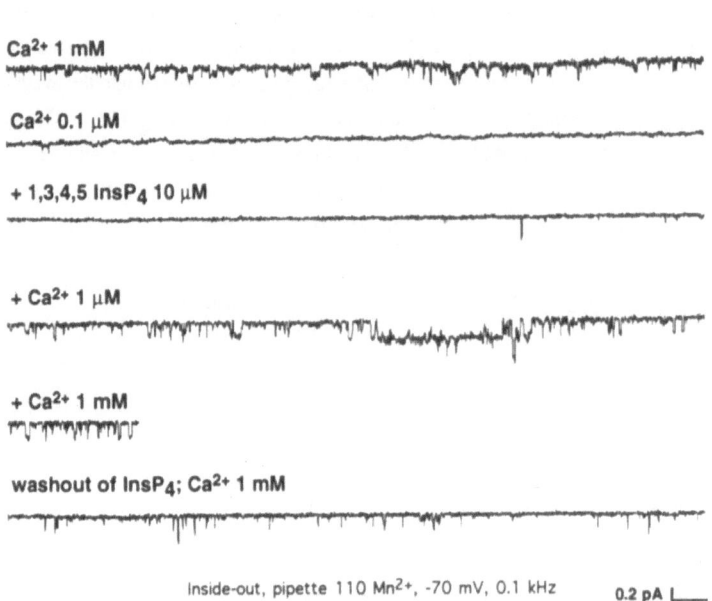

Ca^{2+} 1 mM

Ca^{2+} 0.1 μM

+ 1,3,4,5 InsP$_4$ 10 μM

+ Ca^{2+} 1 μM

+ Ca^{2+} 1 mM

washout of InsP$_4$; Ca^{2+} 1 mM

Inside-out, pipette 110 Mn^{2+}, -70 mV, 0.1 kHz

0.2 pA
0.5 s

Fig. 7. Ca^{2+} dependency of endothelial Mn^{2+} permeable channels. To an excised patch in standard bath solution, the following additions were made: a) EGTA (1.67 mM), lowering [Ca^{2+}] to 0.1 μM; b) InsP$_4$ (10 μM); c) CaCl$_2$ (0.55 mM), raising [Ca^{2+}] to 1 μM; d) CaCl$_2$ (1 mM). Finally, after washout, the original bath solution was restored. Note that low [Ca^{2+}] completely abolishes channel activity even in the presence of InsP$_4$, whereas at higher [Ca^{2+}], channel activity is markedly enhanced by InsP$_4$ and that this effect is reversible after washout.

bath solution of an inside-out patch is lowered from 1 mM to 0.1 μM, channel activity is completely inhibited, and no effect of $InsP_4$ is observed. Raising Ca^{2+} to 1 μM, still in the presence of $InsP_4$, enhances the activity markedly. The open probability under these conditions is higher than initially, even though some channel run-down has supposedly occurred and the Ca^{2+} concentration is below its optimum. After washout of $InsP_4$, activity is reduced considerably below the original control level, reflecting run-down.

The physiological role of this channel appears to be a positive feed-back mechanism that evokes Ca^{2+} influx under conditions where internal Ca^{2+} is elevated due to the $InsP_3$-induced release of Ca^{2+} from internal stores and inositol phosphate turnover is enhanced due to activation of phospholipase C. By this influx mechanism, cytosolic Ca^{2+} can be maintained at elevated levels when release from the stores alone is no longer sufficient.

However, there is evidence for yet another Ca^{2+} entry pathway in endothelial cells that is activated by depletion of internal calcium stores and does not require cytosolic Ca^{2+} or any known second messengers (Jacob, 1990; Schilling et al., 1992). Such a mechanism exists in many other cell types (for review see Putney, 1990). Electrophysiological evidence for the pathway has been obtained in lymphocytes (Lewis and Cahalan, 1989) and mast cells (Hoth and Penner, 1992). In the latter study, a Ca^{2+}-selective current (named I_{CRAC}, calcium-release activated calcium current) was induced after emptying the stores with $InsP_3$ or ionophore, or by blocking re-uptake of cytosolic Ca^{2+} into the stores with high cytosolic Ca^{2+} buffer and/or thapsigargin. I am currently characterizing a Ca^{2+} channel in A431 cells that may explain I_{CRAC} on the single-channel level. Especially, this channel is activated by thapsigargin under conditions where cytosolic Ca^{2+} levels, and levels of soluble cytosolic factors, are kept constant. Comparable studies in endothelial cells have not yet been possible.

CONCLUSIONS

There are at least two distinct types of endothelial K^+ currents, one through an inward-rectifying, Ca^{2+}-independent channel and one through a Ca^{2+}-activated channel. The former may adjust and stabilize the resting membrane potential, the latter leads to hyperpolarization after stimulation. Hyperpolarization increases the driving force for Ca^{2+} influx and thus contributes to Ca^{2+}-dependent cellular activation.

Endothelial cells are equipped with chloride channels that are activated by cell swelling. These channels may have a particular role in the microcirculation where osmotic challenges to endothelial cells occur.

Several pathways for Ca^{2+} entry exist, all of them are voltage-independent. One Ca^{2+} channel is activated by stretch. Others are receptor-operated and possibly linked to G-proteins. One further channel is dependent on internal Ca^{2+} and is activated by $InsP_4$. Evidence from fluorometric studies and from other cell types point to another Ca^{2+}-permeable channel type that is activated by the depletion of internal Ca^{2+} stores.

ACKNOWLEDGEMENT

I wish to thank Rudi Busse and David Clapham for their continuous support and encouragement. I am supported by a Heisenberg fellowship of the Deutsche Forschungsgemeinschaft.

REFERENCES

Adams, D.J., Barakeh, J., Laskey, R., and van Breemen, C., 1989, Ion channels and regulation of intracellular calcium in vascular endothelial cells, FASEB. J. 3:2389-2400.

Bossu, J.L., Feltz, A., Rodeau, J.L., and Tanzi, F., 1989, Voltage-dependent transient calcium currents in freshly dissociated capillary endothelial cells, FEBS Lett. 255:377-380.

Cannell, M.B., and Sage, S.O., 1989, Bradykinin-evoked changes in cytosolic calcium and membrane currents in cultured bovine pulmonary artery endothelial cells, J. Physiol. (London) 419:555-568.

F., Potter, B.V.L., and Petersen, O.H., 1989, Inositol 1,3,4,5-tetrakisphosphate is essential for the sustained activation of the Ca^{2+}-dependent K^+ current in single internally perfused mouse lacrimal acinar cells, J. Memb. Biol. 109:85-93.

Colden-Stanfield, M., Schilling, W.P., Ritchie, A.K., Eskin, S.G., Navarro, L.T., and Kunze, D.L., 1987, Bradykinin-induced increases in cytosolic calcium and ionic currents in cultured bovine aortic endothelial cells, Circ. Res. 61:632-640.

Furchgott, R.F., 1983, Role of endothelium in responses of vascular smooth muscle, Circ. Res. 53:557-573.

Hallam, T.J., Jacob, R., and Merritt, J.E., 1989, Influx of bivalent cations can be independent of receptor stimulation in human endothelial cells, Biochem. J. 259:125-129.

Hamill, O.P., Marty, A., Neher, E., Sakmann, B., and Sigworth, F.J., 1981, Improved patch-clamp techniques for high-resolution current recording from cells and cell-free membrane patches, Pflügers Arch. 391:85-100.

Horn, R., and Marty, A., 1988, Muscarinic activation of ionic currents measured by a new whole-cell recording method, J. Gen. Physiol. 92:145-159.

Hoth, M., and Penner, R., 1992, Depletion of intracellular calcium stores activates a calcium current in mast cells, Nature 355:353-356.

Jacob, R., 1990, Agonist-stimulated divalent cation entry into single cultured human umbilical vein endothelial cells, J. Physiol. (London) 421:55-77.

Johns, A., Lategan, T.W., Lodge, N.J., Ryan, U.S., van Breemen, C., and Adams, D.J., 1987, Calcium entry through receptor-operated channels in bovine pulmonary artery endothelial cells, Tissue Cell 19:733-745.

Josephson, I.R., and Brown, A.M., 1986, Inwardly rectifying single-channel and whole cell K^+ currents in rat ventricular myocytes, J. Memb. Biol. 94:19-35.

Lansman, J.B., Hallam, T.J., and Rink, T.J., 1987, Single stretch-activated ion channels in vascular endothelial cells as mechanotransducers?, Nature 325:811-813.

Lewis, R.S., and Cahalan, M.D., 1989, Mitogen-induced oscillations of cytosolic Ca^{2+} and transmembrane Ca^{2+} current in human leukemic T cells, Cell Regul. 1:99-112.

Lückhoff, A., Pohl, U., Mülsch, A., and Busse, R., 1988, Differential role of extra- and intracellular calcium in the release of EDRF and prostacyclin from cultured endothelial cells, Br. J. Pharmacol. 95:189-196.

Lückhoff, A., and Busse, R., 1990a, Activators of potassium channels enhance calcium influx into endothelial cells as a consequence of potassium currents, Naunyn-Schmiedebergs Arch. Pharmacol. 342:94-99.

Lückhoff, A., and Busse, R., 1990b, Calcium influx into endothelial cells and formation of EDRF is controlled by the membrane potential, Pflügers Arch. 416:305-311.

Lückhoff, A., and Clapham, D.E., 1992, Inositol 1,3,4,5-tetrakisphosphate activates an endothelial Ca^{2+}-permeable channel, Nature 355:356-358.

Morris, A.P., Gallacher, D.V., Irvine, R.F., and Petersen, O.H., 1987, Synergism of inositol triphosphate and inositol tetrakisphosphate in activating Ca^{2+}-dependent K^+ channels, Nature 330:653-655.

Nakashima, M., Akata, T., and Kuriyama, H., 1992, Effects on the rabbit coronary artery of LP-805, a new type of releaser of endothelium-derived relaxing factor and a K^+ channel opener, Circ. Res. 71:859-869.

Nilius, B., 1990, Permeation properties of a non-selective cation channel in human vascular endothelial cells, Pflügers. Arch. 416:609-611.

Nilius, B., 1991, Regulation of transmembrane calcium fluxes in endothelium, News Physiol. Sci. 6:110-114.

Olesen, S.P., Clapham, D.E., and Davies, P.F., 1988a, Haemodynamic shear stress activates a K^+ current in vascular endothelial cells, Nature 331:168-170.

Olesen, S.P., Davies, P.F., and Clapham, D.E., 1988b, Muscarinic-activated K^+ current in bovine aortic endothelial cells, Circ. Res. 62:1059-1064.

Putney, J.W., 1990, Capacitive calcium entry revisited, Cell Calcium 11:611-624.

Sauvé, R., Parent, L., Simoneau, C., and Roy, G., 1988, External ATP triggers a biphasic activation process of a calicum-dependent K^+ channel in cultured bovine aortic endothelial cells, Pflügers Arch. 412:469-481.

Schilling, W.P., 1989, Effect of membrane potential on cytosolic calcium of bovine aortic endothelial cells, Amer. J. Physiol. 257:H778-H784.

Schilling, W.P., Rajan, L., and Strobl-Jager, E., 1989, Characterization of the bradykinin-stimulated calcium influx pathway of cultured vascular endothelial cells. Saturability, selectivity, and kinetics, J. Biol. Chem. 264:12838-12848.

Schilling, W.P., Cabello O.A., and Rajan, L., 1992, Depletion of the inositol 1,4,5-trisphosphate-sensitive intracellular Ca^{2+} store in vascular endothelial cells activates the agonist-sensitive Ca^{2+}-influx pathway, Biochem. J. 284:521-530.

Schwarz, G., Callewaert, G., Droogmans, G., and Nilius, B., 1992a, Shear stress induced calcium transients in endothelial cells from human umbilical cord veins, J. Physiol. (London) (in press)

Schwarz, G., Droogmans, G., and Nilius, B., 1992b, Shear stress induced membrane currents in human vascular endothelial cells, Pflügers. Arch. (in press)

CALCIUM SIGNALLING IN VASCULAR ENDOTHELIAL CELLS:
Ca²⁺ ENTRY AND RELEASE

David J. Adams, Julius Rusko, and Glen Van Slooten

Department of Molecular and Cellular Pharmacology
University of Miami School of Medicine
Miami, FL 33101

INTRODUCTION

Endothelial cells serve both autocrine and paracrine functions within the cardiovascular system to modulate blood pressure and maintain tissue perfusion. Endothelial cells respond to a variety of humoral and physical stimuli to release endothelium-dependent vasodilators, such as prostacyclin (PGI_2) and endothelium-derived relaxing factor (EDRF), and vasoconstrictors such as endothelin (Furchgott and Vanhoutte, 1989). The synthesis and release of endothelium-derived vasodilators have been shown to be Ca^{2+}-dependent whereby the production of PGI_2 and EDRF is attenuated by the removal of extracellular Ca^{2+} (Singer and Peach, 1982; Long and Stone, 1985; Griffith et al., 1986; Lückhoff et al., 1988). It is now well established that an EDRF released from endothelial cells, both *in situ* and in culture, is nitric oxide (NO) or a nitroso compound that readily releases NO, and that NO is synthesized in endothelial cells by the oxidation of one of the two equivalent guanidino nitrogens of L-arginine via a cytoplasmic NADPH- and Ca^{2+}-dependent enzyme, termed NO synthase (see Moncada et al., 1991). Changes in the extracellular Ca^{2+} concentration around the physiological range have been shown to modulate the synthesis/release of NO by the vascular endothelium and consequently, vascular tone (Lopez-Jaramillo et al., 1990).

Both the entry of extracellular Ca^{2+} and the mobalization of Ca^{2+} from intracellular stores can contribute to the increase in the intracellular free Ca^{2+} concentration ($[Ca^{2+}]_i$) in endothelial cells, which is an essential step in the activation of NO synthase. Regulation of endothelial $[Ca^{2+}]_i$ is composed of activating mechanisms which supply Ca^{2+} to the cytoplasm and homeostatic mechanisms which maintain low $[Ca^{2+}]_i$ and remove cytoplasmic free Ca^{2+} after stimulation. Activation of Ca^{2+} entry from the extracellular space and Ca^{2+} release from intracellular stores (endoplasmic reticulum) occurs in response to humoral and physical stimuli. Although the importance of intracellular Ca^{2+} as a second messenger is unquestioned, its regulation in endothelial cells is not clearly delineated.

Ion Flux in Pulmonary Vascular Control, Edited by
E.K. Weir *et al.*, Plenum Press, New York, 1993

The activation of endothelial cell-surface receptors by vasoactive substances evoke a biphasic increase in the cytoplasmic free Ca^{2+} concentration (Hallam and Pearson, 1986; Hallam et al., 1988a). An initial transient component reflects the release of Ca^{2+} from intracellular stores by the intracellular second messenger, inositol 1,4,5-trisphosphate ($InsP_3$) (Derian and Moskowitz, 1986; Lambert et al., 1986; Pirotten et al., 1987; Freay et al., 1989), whereas a subsequent sustained elevation in $[Ca^{2+}]_i$ results from the influx of Ca^{2+} from the extracellular space (Hallam and Pearson, 1986; Morgan-Boyd et al., 1987; Hallam et al., 1988; Schilling et al., 1988). Endothelial cells often respond to prolonged agonist stimulation with repetitive spikes or oscillations in $[Ca^{2+}]_i$ superimposed on the plateau phase (Jacob et al., 1988; Sage et al., 1989; Laskey et al., 1990).

PLASMALEMMAL Ca^{2+} ENTRY

Extracellular Ca^{2+} may enter the cell via at least five different mechanisms: [1] a passive Ca^{2+} leak pathway dependent on the electrochemical gradient for Ca^{2+}; [2] receptor-mediated Ca^{2+} influx via receptor-operated or second messenger activated non-selective cation channels; [3] a "capacitative Ca^{2+} influx" pathway dependent on the state of the intracellular Ca^{2+} stores; [4] mechanosensitive (stretch-activated) cation channels, and [5] Na-dependent Ca^{2+} entry via a Na^+-Ca^{2+} exchanger. Electrophysiological and unidirectional ^{45}Ca flux measurements in cultured endothelial cells derived from large blood vessels indicate an absence of the L-type Ca^{2+} channels which are sensitive to dihydropyridines and activated by membrane depolarization (Colden-Stanfield et al., 1987; Johns et al., 1987; Takeda et al., 1987). In recent studies on freshly dissociated endothelial cells from rabbit aorta, no evidence for depolarization-activated Ca^{2+} influx via voltage-dependent Ca^{2+} channels was found (Sakai, 1990; Rusko et al., 1992a), however, such channels may be present in capillary endothelium (Bossu et al., 1989; Bossu et al., 1992).

Passive Ca^{2+} Leak Pathway

A passive Ca^{2+} "leak" influx driven by the electrochemical gradient for Ca^{2+} (E_m - E_{Ca}) and ubiquitous to all eukaryotic cells is present in vascular endothelial cells (Johns et al., 1987; Schilling et al., 1989). Johns et al (1987) observed a resting ^{45}Ca influx into unstimulated, cultured bovine pulmonary artery endothelial cells which was postulated to mediate basal EDRF release. Lanthanum ions, but not other calcium channel blockers such as verapamil and diltiazem, inhibited the passive ^{45}Ca uptake into cultured endothelial cells. Depolarization reduced the driving force for Ca^{2+} entry through the leak pathway by reducing the electrochemical gradient. The rate of basal ^{45}Ca leak, corresponding to 16 $pmol/10^6$ cells/s, was reduced by approximately 15% in an isotonic KCl solution. Changes in extracellular pH also modulate the Ca^{2+} leak. The resting ^{45}Ca influx increased progressively as extracellular pH was raised from 5.2 to 9.2 (S. Purkerson, D.J. Adams and C. van Breemen, unpublished observations). "Leak" channels which open at negative membrane potentials and are permeable to divalent cations have been described in vascular smooth muscle cells (Benham and Tsien, 1986) and Duchenne human and *mdx* mouse myotubes (Fong et al., 1990). The nature of the Ca^{2+} leak pathway in endothelial cells is unknown but may play an important physiological role in the basal release of EDRF and thus regulation of vascular tone and peripheral resistance.

Receptor-mediated Ca^{2+} Entry via Non-selective Cation Channels

Direct support for a receptor-mediated Ca^{2+} influx pathway in endothelial cells came from the finding that ^{45}Ca influx is enhanced by agonists such as bradykinin, histamine and thrombin (D'Amore and Shepro, 1977; Whorton et al., 1984; Johns et al., 1987). Thrombin and bradykinin evoked an inward nonspecific cation current in pulmonary artery endothelial cells which could account for the enhanced Ca^{2+} influx (Johns et al., 1987; Lodge et al., 1988). Recently, bradykinin-evoked inward currents carried by Na^+ and Ca^{2+} with amplitudes of 5-25 pA/cell were reported only in clusters of electrically coupled endothelial cells from bovine aorta (Mendelowitz et al., 1992). Histamine and A23187 (a calcium ionophore) also induced a $[Ca^{2+}]_i$ dependent inward current in human umbilical vein endothelial cells (Bregestovski et al., 1988). The existence of receptor-operated cation channels permeable to Ca^{2+} has been demonstrated in cultured endothelial cells from bovine pulmonary artery (Johns et al., 1987; Lodge et al., 1988), human umbilical vein (Bregestovski et al., 1988, Nilius, 1990; Nilius and Rienmann, 1990) and capillary endothelial cells (Popp and Gögelein, 1992). Bradykinin- and thrombin-activated single channels recorded in approximately 10% of cell-attached patches exhibited a slope conductance of ~ 15 pS with 100 mM Ba^{2+} (Lodge et al., 1988; Adams et al., 1989). A non-selective cation channel activated by histamine and thrombin was also described in cell-attached patches in cultured human umbilical vein endothelial cells and exhibited a single channel conductance of 26 pS with asymmetrical solutions (140 Na^+//140 K^+) and 8 pS with 110 mM Ca^{2+} pipette solution (Nilius, 1990; 1991). The permeation ratio for this receptor-operated non-selective cation channel is $P_K:P_{Na}:P_{Ca}$ = 1:0.9:0.2 (Nilius, 1990). Recently, a histamine-activated cation channel with a slope conductance of 22.5 pS in physiological solutions was reported in undispersed endothelial cells of the rat intrapulmonary artery (Yamamoto et al., 1992). The channel exhibited inward rectification, possibly due to block by intracellular Mg^{2+}, and the permeability ratios calculated from the reversal potentials were $P_K:P_{Na}:P_{Ca}$ = 1:1:15.7. The observation that apparently similar ion channels are activated by different agonists suggests a convergent intracellular messenger cascade between receptor activation and channel opening. Lückhoff and Clapham (1992) recently reported the activation of a Ca^{2+}-permeable channel by inositol 1,3,4,5-tetrakisphosphate (InsP$_4$) and Ca^{2+} applied to the cytoplasmic surface of inside-out membrane patches from bovine aortic endothelial cells. These channels were not activated by GTPγS or InsP$_3$, although the phosphorylation of InsP$_3$ to generate InsP$_4$ would be an obvious final step in a convergent signalling pathway. These channels appear to have a lower conductance than those described above, that is, 2.5 pS with 110 mM Mn^{2+} in the pipette solution. The data presently available, therefore, suggest the presence of two types of receptor-operated Ca^{2+} permeable cation channels, a low conductance channel activated by InsP$_4$ and Ca^{2+}, and a channel approximately 10 times larger.

Capacitative Ca^{2+} Influx Pathway

An alternative mechanism proposed for the regulation of agonist-stimulated Ca^{2+} entry involves a pathway controlled by the content of the intracellular stores, as first proposed for rat parotid acinar cells (Putney, 1986; Merritt and Rink, 1987). According to this mechanism, termed "capacitative Ca^{2+} entry", the Ca^{2+} influx across the plasma membrane is also activated by emptying the internal Ca^{2+} stores. This may account for the phenomenon of receptor-mediated divalent cation entry reported in stimulated endothelial cells despite the removal of agonist (Hallam et al., 1988,1989; Jacob, 1990). This model has gained support from observations made in a variety of non-excitable cells using either thapsigargin (TG), or 2',5'-di(*tert*-butyl)-1,4-benzohydroquinone(BHQ). Thapsigargin and BHQ are known to selectively inhibit the Ca^{2+}-ATPase in the endoplasmic reticulum and prevent Ca^{2+} reuptake thereby mobilizing intracellular Ca^{2+} by a mechanism independent of receptor stimulation and InsP$_3$ formation (Takemura et al., 1989; Kass et al., 1989).

In contrast to the rapidly developing, transient rise in $[Ca^{2+}]_i$ initiated by bradykinin, TG and BHQ have been shown to induce a slow rise and prolonged elevation in $[Ca^{2+}]_i$ in cultured endothelial cells from bovine aorta and pulmonary artery (Dolor et al., 1992; Schilling et al., 1992). Mobilization of intracellular Ca^{2+} with either inhibitor depleted intracellular Ca^{2+} stores and greatly reduced subsequent mobilization of the InsP$_3$-sensitive Ca^{2+} store by bradykinin. Although TG and BHQ had no effect on phosphoinositide hydrolysis, both agents stimulated ^{45}Ca uptake and divalent cation permeability, as measured by enhanced Mn^{2+} quenching of intracellular fura-2, suggesting that depletion of the agonist-sensitive store is sufficient to activate Ca^{2+} influx (Dolor et al., 1992; Schilling et al., 1992). However, in a separate study on pig aortic endothelial cells (Lückhoff and Busse, 1990a), BHQ increased $[Ca^{2+}]_i$ in prestimulated and resting cells without stimulation of Mn^{2+} entry. The TG- and BHQ-induced elevation in $[Ca^{2+}]_i$ required extracellular Ca^{2+} and was inhibited by extracellular Ni^{2+} or La^{3+}, membrane depolarization and the putative receptor-operated channel blocker, SKF 96365 (Dolor et al., 1992; Schilling et al., 1992). These results are identical to those obtained for the agonist-stimulated increase in the plateau value of $[Ca^{2+}]_i$ suggesting that the influx pathway activated by depletion of the agonist-sensitive internal Ca^{2+} store is similar to the agonist-activated Ca^{2+} influx pathway. A Ca^{2+} current that is activated by depletion of intracellular Ca^{2+} stores has been recently identified in mast cells (Hoth and Penner, 1992). This "Ca^{2+} release-activated Ca^{2+} current" observed only when cytoplasmic free Ca^{2+} concentration was buffered to very low values, was voltage-insensitive and highly selective for Ca^{2+} over Mn^{2+}. This Ca^{2+} entry pathway may contribute to Ca^{2+} homeostasis and replenish empty Ca^{2+} stores in electrically non-excitable cells.

The original capacitative Ca^{2+} entry model which featured a direct pathway for Ca^{2+} entry into the ER lumen, was modified primarily due to the lack of additivity of the effects of carbachol and TG on steady-state $[Ca^{2+}]_i$ in parotid acinar cells (Putney, 1990). Depletion of ER Ca^{2+} was postulated to signal the opening of plasmalemmal Ca^{2+}-permeable channels connecting the extracellular space to the cytoplasm, however, no plausible candidate for this signalling pathway has been identified.

Mechanosensitive Ca^{2+}-permeable Channels

Mechanosensitive (stretch-activated) ion channels in endothelial cells may serve as transducers in detecting changes in blood pressure or flow. Depending on their sensitivity to these stimuli and on their ionic selectivity, mechanosensitive ion channels could, for example, change endothelial cell membrane potential, and thus the driving force for passive Ca^{2+} entry, if they are K^+ permeable, or directly allow the entry of Ca^{2+}, if they are Ca^{2+} permeable. Indeed, stretch-activated, single-channel currents have been described in cultured endothelial cells from neonatal pig aorta (Lansman et al., 1987). With the application of suction pulses (10-20 mm Hg, 300 ms) to cell-attached membrane patches and physiological ion concentrations in the patch pipette and external media, the frequency of opening of inward unitary currents increased for negative membrane potentials. These currents displayed an ohmic I-V relationship with a slope conductance of 40 pS and an extrapolated reversal potential of 9 mV. They were impermeable to Cl⁻, but able to pass K^+ (elementary conductance of 56 pS with an isotonic KCl pipette solution) and Ca^{2+} (19 pS with isotonic CaCl$_2$). It was reported that the calculated P_{Ca}/P_{Na} ranged between 1.2 and 8.4. These suction-activated channels were not Ca^{2+}-sensitive. Stretch-activated unitary currents with similar characteristics have also been recently described in endothelial cells from porcine cerebral capillaries (Popp et al., 1992).

If membrane stretch is similar to wall shear stress generated by flowing blood, then activation of stretch-activated channels should elicit changes in membrane potential and

$[Ca^{2+}]_i$. In a more direct test of the effects of increased shear stress, a laminar flow-activated, whole-cell, inwardly rectifying K^+ current was described in cultured bovine aortic endothelial cells (Olesen et al., 1988). Half-maximal activation of this current in symmetrical K^+ solutions occurred at a flow rate producing wall shear stress of 0.7 dyn/cm^2, with saturation being reached at 15 dyn/cm^2. It was reported that flow caused cell hyperpolarization of 0 to 6 mV under physiological ionic conditions, where the average resting potential was -77 mV (E_K = -90 mV). This is in accord with another study where the flow-associated shift in the fluorescence of a potential-sensitive dye was consistent with membrane hyperpolarization (Nakache and Gaub, 1988). Shear stress-induced $[Ca^{2+}]_i$ transients have been described in cultured bovine aortic endothelial cells (Ando et al., 1988), however, this result has been attributed to a potentiation of the ATP-induced Ca^{2+} transient by shear stress (Mo et al., 1991). Simultaneous measurements of shear stress-induced membrane currents and $[Ca^{2+}]_i$ changes have been recently reported in cultured endothelial cells from human umbilical vein (Schwarz et al., 1992). In the presence of extracellular Ca^{2+} (10 mM), shear stress (~ 10 dyn/cm^2) evoked an inward current at a holding potential of 0 mV which was accompanied by a slow rise in $[Ca^{2+}]_i$. In the absence of extracellular Ca^{2+} no increase in $[Ca^{2+}]_i$ could be evoked and the reversal potential of the shear stress induced current was shifted by approximately -24 mV. A permeability ratio, P_{Ca}/P_{Cs} of 12.5 was calculated indicating that shear stress activated a Ca^{2+}-permeable channel in endothelial cells thereby inducing a Ca^{2+} influx and an increase in $[Ca^{2+}]_i$. Similarly, measurement of intracellular Ca^{2+} in cultured bovine aortic endothelial cells showed that upon the initiation of flow (shear stress 0.2 to 8 dyn/cm^2), $[Ca^{2+}]_i$ increased within 15-40 s to a peak and either declined to an elevated plateau that persisted for > 5 min (Geiger et al., 1992) or back to baseline within 40-80 s (Shen et al., 1992). Removal of extracellular Ca^{2+}, blockade of Ca^{2+} entry with La^{3+} or depolarization with high K^+ did not eliminate this $[Ca^{2+}]_i$ response (Geiger et al., 1992; Shen et al., 1992). Taken together with the reported increase in $InsP_3$ levels in endothelial cells exposed to flow (Nollert et al., 1990) these findings lend support to the hypothesis that initial flow effects may be mediated by the release of Ca^{2+} from intracellular stores.

At low shear stresses, the flow-activated K^+ current would hyperpolarize endothelial cells, whereas at higher shear stresses, stretch-activated channels might produce depolarization. The dependence of current direction on stress magnitude, the location of a particular endothelial cell within the circulatory system and the channel type expressed could combine to modulate the release of various endothelium-derived vasoactive factors. This may be a possible mechanism by which vascular endothelium in intact vessels regulates smooth muscle tone in response to haemodynamic stimuli (Rubanyi et al., 1990).

Na^+-dependent Ca^{2+} Influx

The removal of extracellular Na^+ has been shown to have little effect on the basal and bradykinin-activated levels of intracellular Ca^{2+} in cultured endothelial cells (Schilling et al., 1988; Laskey et al., 1990). Similarly, Ca^{2+}-dependent K channel activity, an indirect estimate of $[Ca^{2+}]_i$, is unaffected by the removal of extracellular Na^+ (Sauvé et al., 1988). This suggests that Na^+/Ca^{2+} exchange does not contribute either to Ca^{2+} extrusion from the cultured endothelial cells or to the supply of Ca^{2+} during the plateau phase of agonist mediated activation. However, if the endothelial cells were Na^+-loaded using monensin and then exposed to a Na^+-free solution, a large transient increase in $[Ca^{2+}]_i$ ensued (Sage et al., 1991). Pretreatment of Na^+-loaded cells with ouabain doubled the transient $[Ca^{2+}]_i$ response. Removal of extracellular Na^+ stimulated a large transient increase in $[Ca^{2+}]_i$ in these cells due to Ca^{2+} influx because the Na^+-Ca^{2+} exchanger was operating in reverse. This Ca^{2+} influx mediated by Na^+/Ca^{2+}-exchange was diminished as intracellular Na$^+$ was lost to the Na$^+$-free medium. Electrogenicity of the carrier was demonstrated by the observation that this Ca^{2+} influx was augmented by high external K$^+$ depolarization, as expected from the coupling by the exchanger of the transport of 3 Na$^+$ out for 1 Ca^{2+} in. Although the presence of the Na$^+$/Ca^{2+} exchanger could be demonstrated under these extreme experimental conditions, the lack of effects of Na$^+$ removal under physiological conditions leave its possible contribution to Ca^{2+} entry in doubt. A recent study in endothelial cells freshly dispersed from porcine arteries showed a > 4-fold increase in [Ca^{2+}]$_i$ when the Na$^+$ gradient was reduced to 48% of normal, while subcultured cells showed a varied response with no [Ca^{2+}]$_i$ increase by the 7th passage (Sturek et al., 1991).

Role of K Channels in Ca^{2+} Influx

The rate of Ca^{2+} entry through plasmalemmal ion pathways is modulated by the resting membrane potential (E$_m$), which is regulated by membrane K$^+$ conductance. Membrane depolarization by elevation of extracellular K$^+$ or under voltage clamp reduces the agonist-stimulated Ca^{2+} influx in vascular endothelial cells (Adams et al., 1989; Laskey et al., 1990; Lückhoff and Busse, 1990b). There are at least four types of potassium channels present in endothelial cells: [1] an inwardly rectifying K channel activated upon hyperpolarization (Johns et al., 1987; Takeda et al., 1987; Silver and DeCoursey, 1990) or by shear stress (Olesen et al., 1988); [2] a transient (A-type) K channel (Takeda et al., 1987; Silver and DeCoursey, 1990); [3] an ATP sensitive K channel (Janigro et al., 1992) and [4] a Ca^{2+}-dependent K channel activated by membrane depolarization and a rise in [Ca^{2+}]$_i$ (Sauvé et al., 1988; Colden-Stanfield et al., 1990; Rusko et al., 1992a). The activation of Ca^{2+}-dependent K channels concomitant with the receptor-mediated increase in [Ca^{2+}]$_i$ is most likely to underlie agonist-induced changes in E$_m$ of endothelial cells.

Agonist-induced changes in K$^+$ permeability, initially described by measuring the rate of ^{86}Rb efflux, were shown to be largely dependent on the presence of extracellular Ca^{2+} and inhibited by La^{3+} (Gordon and Martin, 1983; Schilling et al., 1988). Whole-cell outward currents evoked by agonists have been observed in cultured endothelial cells in the absence of significant intracellular Ca^{2+} buffering (Colden-Stanfield et al., 1987; Cannell and Sage, 1989; Lückhoff and Busse, 1990c; Takeda and Klepper, 1990). The agonist-induced outward K$^+$ current produces a transient hyperpolarization of the endothelial cell and has been observed in response to ACh in native endothelial cells from rabbit aorta (Busse et al., 1988; Sakai, 1990) and pig coronary artery (Chen and Cheung, 1992). Similarly, bradykinin, substance P, ATP and adenosine evoked a transient hyperpolarization in cultured endothelial cells from porcine coronary artery and bovine aorta (Brunet and Bény, 1989; Lückhoff and Busse, 1990b; Mehrke and Daut, 1990; Takeda and Klepper, 1990), as did histamine in cultured pig coronary artery and human umbilical vein endothelial cells (Mehrke and Daut, 1990). The peak amplitude of the agonist-induced hyperpolarizations in cultured endothelial cells from pig coronary artery is dependent on the extracellular concentrations of agonist and K$^+$ (Brunet and Bény, 1989; Mehrke and Daut, 1990). This transient hyperpolarization was attenuated by the removal of external Ca^{2+} and by blockers of Ca^{2+}-activated K$^+$ channels (Mehrke and Daut, 1990; Chen and Cheung, 1992).

The ACh-induced outward current observed in freshly dispersed rabbit aortic endothelial cells shows a similar dependence on extracellular [K$^+$], and is attenuated in Ca^{2+}-free solution (Busse et al., 1988; Sakai, 1990; Rusko et al., 1992a). ACh, bradykinin and ATP each evoke a biphasic increase in the open probability (P$_{open}$) of Ca^{2+}-activated K$^+$ channels, while the removal of external Ca^{2+}, both in the presence and absence of agonist stimulation, failed to inhibit Ca^{2+}-activated K$^+$ channel activity (Sauvé et al., 1988;

Sakai, 1990; Rusko et al., 1992b). These data suggest that endothelium-dependent vasodilators produce a large initial Ca^{2+} transient due to Ca^{2+} release from intracellular stores together with Ca^{2+} influx via receptor-operated ion channels. This elevated $[Ca^{2+}]_i$ activates Ca^{2+}-dependent K channels, increasing K^+ permeability and thereby hyperpolarizing the endothelial cell. This membrane hyperpolarization provides an electrochemical gradient for maintained Ca^{2+} entry during agonist stimulation.

Ca^{2+} RELEASE FROM INTRACELLULAR STORES

The dynamic equilibrium between Ca^{2+} entry and extrusion at the plasma membrane and Ca^{2+} release and reuptake by internal stores, may be altered by agonist stimulation. Numerous studies have led to the view of two phases of Ca^{2+}-mobilization following receptor activation. Direct measurement of $[Ca^{2+}]_i$, using fluorescent Ca^{2+} indicator dyes, show that the initial, rapid increase in $[Ca^{2+}]_i$ is largely independent of the presence of extracellular Ca^{2+} and is relatively insensitive to blockers of the plasmalemmal Ca^{2+} entry pathways, while the sustained elevation in $[Ca^{2+}]_i$ has an absolute requirement for extracellular Ca^{2+} (Hallam et al., 1988; Lückhoff et al., 1988; Schilling et al., 1988). In the absence of extracellular Ca^{2+}, bradykinin evokes a transient increase in $[Ca^{2+}]_i$ and ^{45}Ca efflux from cultured bovine pulmonary artery endothelial cells (Peach et al., 1987; Johns et al., 1987; Freay et al., 1989). These results clearly indicate an agonist-induced release of Ca^{2+} from an intracellular store.

A biphasic increase in the P_{open} of the Ca^{2+}-dependent K channels was observed upon stimulation of cultured endothelial cells from bovine aorta with either bradykinin or ATP (Sauvé et al., 1988; 1990) and native endothelial cells from rabbit aorta with either bradykinin, ATP or ACh (Sakai, 1990; Rusko et al., 1992a). This finding is reminiscent of the biphasic $[Ca^{2+}]_i$ responses observed in endothelial cells following stimulation by bradykinin (Morgan-Boyd et al., 1987), ATP (Hallam and Pearson, 1986) or ACh (Danthuluri et al., 1988; see **Figure 1A**). Tetraethylammonium ions (TEA), which inhibited Ca^{2+}-activated K^+ currents, attenuated the bradykinin-induced biphasic increase in P_{open} of the channels in native endothelial cells. Thus, in addition to direct measurement of $[Ca^{2+}]_i$ using fluorescent indicator dyes, agonist-activated Ca^{2+} store release can be monitored using ionic current measurement of Ca^{2+}-dependent K channel activity.

Unitary and spontaneous transient outward currents (STOCs) observed in native endothelial cells are believed to represent the sporadic release of Ca^{2+} from intracellular stores adjacent to TEA-sensitive, Ca^{2+}-activated K channels (Rusko et al., 1992a). Involvement of receptor-mediated Ca^{2+} release from intracellular stores was also demonstrated by inhibition of the ACh-evoked outward currents by caffeine, ryanodine and heparin, agents which disrupt Ca^{2+} uptake and release from intracellular stores (Sakai, 1990). STOCs recorded in native endothelial cells were characterized by a large amplitude varying between 50 and 150 pA (at 0 mV) representing the simultaneous activation of several Ca^{2+}-dependent K channels. STOCs were initially described in vascular and visceral smooth muscle cells of rabbit (Benham et al., 1986; Benham and Bolton, 1986) and were either reduced in amplitude or abolished on exposure to Ca^{2+}-free, EGTA-containing external solutions (Benham and Bolton, 1986). In marked contrast, the absence of extracellular Ca^{2+} evoked an increase in both P_{open} and the amplitude of STOCs in endothelial cells freshly dissociated from rabbit aorta (Rusko et al., 1992b). The lack of effect of Ca^{2+}-free external solutions on unitary outward currents and STOCs suggests that Ca^{2+} influx is not necessary for maintaining the activity of unitary currents and STOCs in native endothelial cells. Therefore, the intracellular stores must be an important source of Ca^{2+} for activation of Ca^{2+}-dependent K channels.

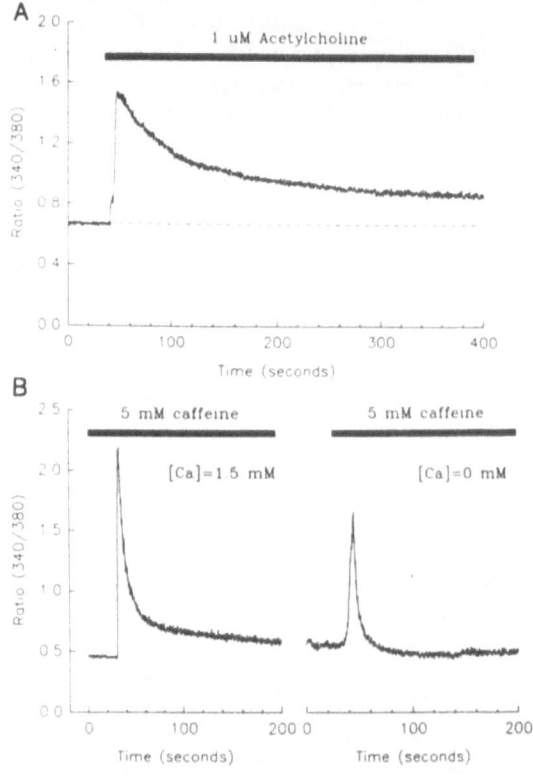

Figure 1. Acetylcholine- and caffeine-stimulated increases in $[Ca^{2+}]_i$ in endothelial cells freshly dissociated from rabbit aorta.
A. A biphasic $[Ca^{2+}]_i$ response to bath application of acetylcholine (1 μM) in physiological salt solution. Removal of extracellular Ca^{2+} abolished the $[Ca^{2+}]_o$-dependent plateau phase.
B. Caffeine-induced transient rise in $[Ca^{2+}]_i$ observed in the presence ($[Ca^{2+}]$ = 1.5 mM) and absence ($[Ca^{2+}]$ = 0 mM) of extracellular Ca^{2+}.

Ins(1,4,5)P$_3$-mediated Ca^{2+} release

In *myo*[³H]inositol-labelled cultured human umbilical vein endothelial cells, both thrombin- and histamine-induced increases in InsP$_3$ occurred in less than 15 sec and were temporally correlated with $[Ca^{2+}]_i$ increases (Lambert et al., 1986; Pollock et al., 1988). Bradykinin also transiently stimulated InsP$_3$ production (up at 15 s, down at 90 s) (Derian and Moskowitz, 1986; Lambert et al., 1986). InsP$_3$ is released into the cytosol following the hydrolysis of phosphatidylinositol 4,5-bisphosphate in response to cell-surface receptor-mediated activation of phospholipase C. InsP$_3$, in micromolar concentrations, binds to specific membrane receptors to rapidly release Ca^{2+} from a non-mitochondrial store in a variety of cells (see Berridge and Irvine, 1989). InsP$_3$ has been shown to be the second messenger for agonist-induced Ca^{2+} release from an intracellular store by opening Ca^{2+}-permeable channels in the ER (Ferris et al., 1989; Bezprozvanny et al., 1991). This messenger was effective in releasing the ER Ca^{2+} with a high affinity (K_d = 1 μM) but had no effect on the mitochondrial Ca^{2+} stores in cultured endothelial cells (Freay et al., 1989). Flash photolysis of 'caged' InsP$_3$ in voltage-clamped aortic endothelial cells evoked a rise in $[Ca^{2+}]_i$ with a delay and time to peak which decreased with increasing concentrations of InsP$_3$ over the range 0.2-5 μM (Carter and Ogden, 1992), consistent with a direct binding and gating action of InsP$_3$ on the ER Ca channel.

The release of Ca^{2+} from intracellular stores of endothelial cells was investigated in both intact and saponin-permeabilized cultured cells (Freay et al., 1989). Mg-ATP dependent Ca^{2+} uptake by the intracellular organelles exhibited two compartments, a high Ca^{2+} affinity (half saturation at 0.8 μM) low capacity (2.5 nmoles/10^6 cells) compartment and a low affinity (half saturation at >8 μM) high capacity (>14 nmoles/10^6 cells) compartment. The low affinity Ca^{2+} uptake was inhibited by azide and could therefore be identified as mitochondrial Ca^{2+} uptake. Only 74% of the high affinity Ca^{2+} compartment was releasable by InsP$_3$ with an EC$_{50}$ of 1 μM. The high-affinity Ca^{2+} compartment was identified as ER, allowing for the possibility that it is not homogeneous with respect to its Ca^{2+} release mechanisms. The [Ca^{2+}]$_i$ transient and efflux of ^{45}Ca were lost after a single exposure to a maximally effective bradykinin concentration in Ca^{2+}-free media and restored by addition of Ca^{2+} to the extracellular space (Morgan-Boyd et al., 1987; Freay et al., 1989). Once the ER store of Ca^{2+} has been discharged it must be refilled by extracellular Ca^{2+} although the precise Ca^{2+} entry pathway into the ER is not known. It appears that Ca^{2+} does not cycle between the ER and cytoplasm exclusively but follows a larger cycle involving the plasma membrane (see Capacitative Ca^{2+} Influx Pathway).

In addition to activating internal Ca^{2+} store release, InsP$_3$ has also been implicated in activating Ca^{2+} influx. Various mechanisms have been proposed to underlie InsP$_3$-mediated Ca^{2+} entry in non-excitable cells (see above). In most studies, direct application of InsP$_3$ to the plasma membrane failed to increase permeability to Ca^{2+} although InsP$_3$ has been reported to increase the plasmalemmal Ca^{2+} permeability in human T-lymphocytes (Kuno and Gardner, 1987). Exposure of mast cells to InsP$_3$ appeared to increase [Ca^{2+}]$_i$ by Ca^{2+} entry, but an InsP$_3$-activated Ca^{2+} channel could not be identified (Penner et al., 1988). Irvine and collaborators have suggested that the phosphorylated product of InsP$_3$ metabolism, InsP$_4$, plays a significant role in the regulation of Ca^{2+} entry (Irvine et al., 1988; Irvine, 1992). It was observed that alone neither InsP$_3$ nor InsP$_4$ in the patch pipette could evoke mobilization of Ca^{2+} in lacrimal cells. However, when added together a biphasic response was observed, which apparently resulted from Ca^{2+} release as well as entry (Morris et al., 1987; Changya et al., 1989).

The internal release of Ca^{2+} mediated by InsP$_3$ is from a discrete intracellular pool that probably does not include all of the non-mitochondrial sequestered Ca^{2+}. There is evidence that the size of this pool can be regulated in cells by a GTP-dependent mechanism (Mullaney et al., 1988; Ghosh et al., 1989). The organelle from which InsP$_3$ releases Ca^{2+} is probably a component of the ER and may also be an InsP$_3$-sensitive "calciosome" (Volpe et al., 1988). However, the heterogeneity of InsP$_3$ receptor distribution within cells (Ross et al., 1989) is inconsistent with a single specialized InsP$_3$-responsive organelle (the "calciosome") but rather suggests the occurrence of complex and dynamic interactions between InsP$_3$-sensitive and -insensitive organelles mediated by intracellular G proteins. Two non-mitochondrial Ca^{2+} compartments have been proposed to exist in rat pancreatic acinar cells (Streb et al., 1984). Similarly, evidence suggests that there are at least two functionally distinct non-mitochondrial Ca^{2+} compartments present in native endothelial cells: [1] an InsP$_3$-sensitive Ca^{2+} store, and [2] an InsP$_3$-insensitive Ca^{2+} store which may be coupled to the InsP$_3$-sensitive pool by a GTP-dependent pathway (see Freay et al., 1989).

Caffeine-induced Ca^{2+} Release

Caffeine translocates Ca^{2+} from intracellular storage sites, such as the sarcoplasmic reticulum in excitable cells, into the cytosol by an enhanced Ca^{2+}-induced Ca^{2+} release mechanism. There is considerable evidence to show that an intracellular Ca^{2+} store in muscle can be released by caffeine (Endo, 1985). In skinned smooth muscle cells of the guinea-pig taenia caeci, it has been suggested that InsP$_3$-sensitive Ca^{2+} stores do not completely overlap with caffeine-sensitive stores (Iino et al., 1988). In contrast, pretreatment of cells with caffeine abolished the carbachol-induced Ca^{2+}-dependent K$^+$ currents in smooth muscle cells of rabbit small intestine, suggesting that caffeine had released the carbachol-sensitive Ca^{2+} stores (Bolton and Lim, 1989). Alternatively, caffeine inhibition of InsP$_3$-mediated responses in *Xenopus* oocytes has been attributed to caffeine antagonism of the binding of InsP$_3$ to its intracellular receptor (Parker and Ivorra, 1991).

In cultured bovine aortic endothelial cells, caffeine (5 mM) caused a small increase in [Ca^{2+}]$_i$ which was abolished following incubation in Ca^{2+}-free medium (Buchan and Martin, 1991). Furthermore, a recent study of the effects of caffeine on bradykinin-induced fluctuations of [Ca^{2+}]$_i$ by monitoring the activity of Ca^{2+}-dependent K channels in bovine aortic endothelial cells of confluent monolayers (Thuringer and Sauvé, 1992), provides evidence for the co-existence of InsP$_3$-sensitive and caffeine-sensitive Ca^{2+} stores. Caffeine-induced [Ca^{2+}]$_i$ transients which are abolished in the presence of ryanodine (10 μM) demonstrate the presence of functional ryanodine-sensitive Ca^{2+} release channels in the ER of freshly isolated endothelial cells (see **Figure 1B**). Caffeine (5-20 mM) also stimulated a dose-dependent increase in unitary current activity superimposed on a large, prolonged transient outward current. In the continued presence of caffeine, unitary and spontaneous outward currents were observed to occur at both a higher frequency and larger amplitude (see **Figure 2A**). External TEA inhibited unitary currents and STOCs but failed to completely inhibit the large, long-lasting transient outward current evoked by caffeine (Rusko et al., 1992b) suggesting the presence a TEA-insensitive component of the caffeine-induced outward current in native endothelial cells.

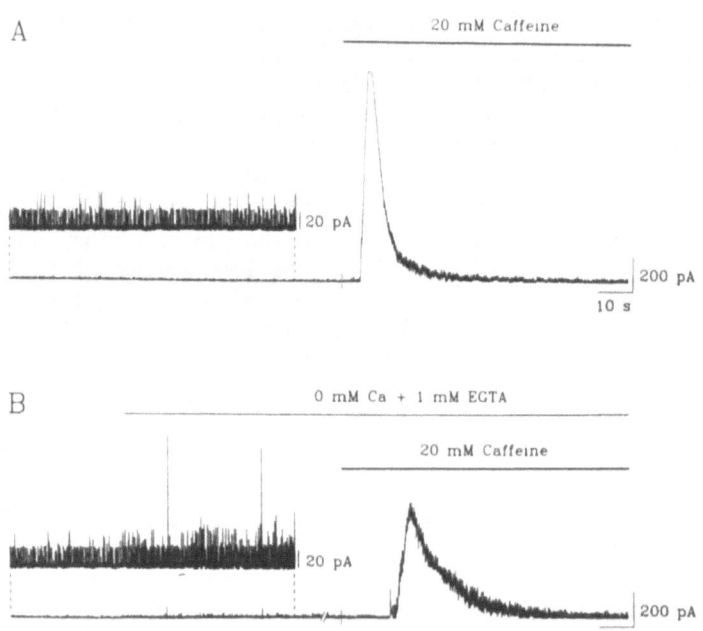

Figure 2. Caffeine-evoked outward currents in native endothelial cells obtained in the absence and presence of extracellular Ca^{2+}.
Outward currents recorded in normal (1.5 mM Ca^{2+}) PSS (A) and in Ca^{2+}-free (1 mM EGTA) external solution (B) before and during exposure to 20 mM caffeine. Holding potential, +20 mV. Horizontal bars indicate duration of exposure to caffeine and Ca^{2+}-free external solution.

In a Ca^{2+}-free, EGTA-containing solution, caffeine evoked a long-lasting transient outward current of smaller amplitude than that obtained in the presence of external Ca^{2+} (Rusko et al., 1992b; see **Figure 2B**). The ability of caffeine to evoke transient outward currents in the absence of external Ca^{2+} suggests that the intracellular Ca^{2+} store is an important source of Ca^{2+} for activation of K channels in native endothelial cells. The reduced amplitude of these currents, however, suggest that influx of extracellular Ca^{2+} is necessary for full development of transient currents evoked by agonists. Heparin, a relatively specific and potent competitive antagonist of the $InsP_3$ receptor, had no effect on caffeine-induced transient outward currents or STOCs in native endothelial cells suggesting the presence of another intracellular Ca^{2+} store (J. Rusko and D.J. Adams, unpublished observations).

CONCLUSIONS

The picture which has emerged from studies of Ca^{2+} entry pathways and Ca^{2+} release from intracellular stores in vascular endothelial cells is summarized in **Figure 3**. The intracellular ER Ca^{2+} stores play an important role in Ca^{2+} signalling and in creating intracellular Ca^{2+} gradients. Investigation of the proposed compartmentalization of the ER and the different types of Ca^{2+} release channels present in endothelial cells are required. The interpretation of studies to date on the relationship between the state of filling of the intracellular Ca^{2+} stores and Ca^{2+} influx, however, are limited by the specificity of agents (e.g., caffeine, thapsigargin, BHQ) used to perturb Ca^{2+} homeostasis in endothelial cells. The transfection of endothelial cells with the complimentary DNA for the Ca^{2+}-sensitive photoprotein, aequorin, fused in a frame encoding a non-mitochondrial organelle (ER) presequence as recently applied to monitor mitochondrial $[Ca^{2+}]$ (see Rizzuto et al., 1992), should provide insight into the the temporal response of changes in ER $[Ca^{2+}]$ upon receptor activation in endothelial cells. Similarly, improved temporal and spatial resolution of $[Ca^{2+}]_i$ in cells using fluorescent probes and digital imaging techniques (see Fay et al., 1989) may provide evidence for Ca^{2+} gradients in the endothelial cell as recently proposed for smooth muscle cells (van Breemen and Saida, 1989).

While considerable attention has focussed on Ca^{2+} entry pathways, little is known about the mechanisms of Ca^{2+} extrusion from endothelial cells. The removal of extracellular Na^+ has been reported to have little influence on $[Ca^{2+}]_i$ in cultured endothelial cells, indicating that Ca^{2+} extrusion is mediated primarily by the plasmalemmal Ca^{2+}-ATPase and not the Na^+-Ca^{2+} exchanger. However, further experiments are required to quantitatively describe the relative contributions of the Na^+-Ca^{2+} exchanger and Ca^{2+}-ATPase to Ca^{2+} extrusion during rest and excitation.

Much of our understanding of the ionic basis of cytoplasmic Ca^{2+} regulation in vascular endothelium has been obtained from studies on cultured endothelial cells. However, recent studies on freshly dissociated and intact endothelial cells have revealed striking differences between cultured and native cells with respect to the expression of the muscarinic ACh receptor, the inwardly rectifying K channel, and intracellular Ca^{2+} release mechanisms such as Ca^{2+}-induced Ca^{2+} release channels. These recently uncovered differences in the expression of cell surface receptors and ionic transport mechanisms imply that studies on cultured endothelial cells alone cannot provide a satisfactory understanding of Ca^{2+} signalling in the endothelium *in vivo*. Furthermore, many of the studies of endothelial cell physiology are based on cells derived from large conduit vessels, however, evidence suggesting specific properties of microvascular endothelial cells (e.g., Bossu et al., 1992) indicates a need for further studies to convincingly establish differences between endothelia derived from conductance and resistance arteries, veins and capillaries.

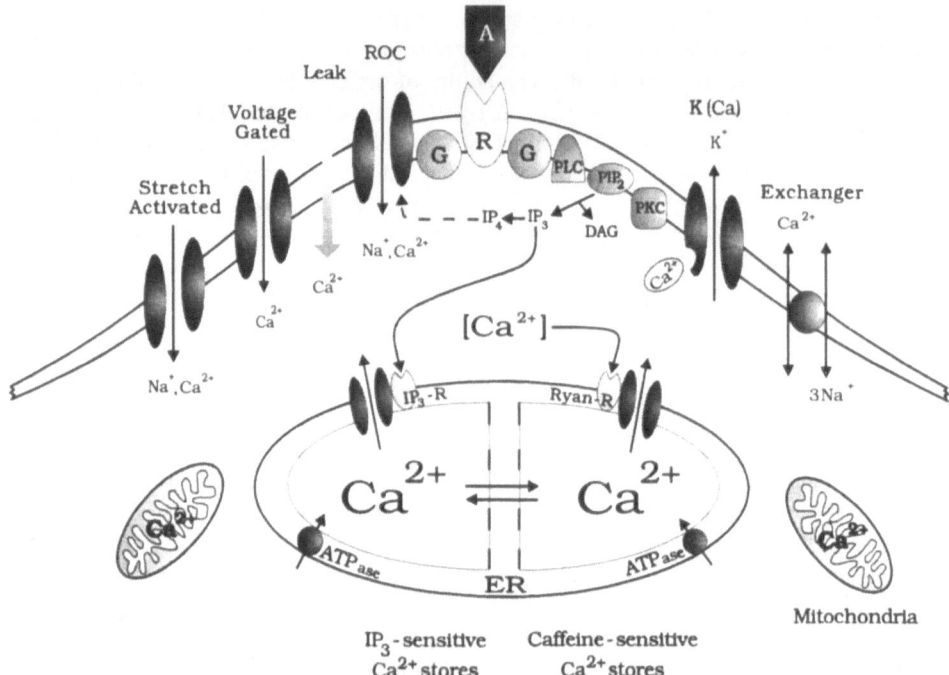

Figure 3. Schematic illustration of ionic transport pathways and intracellular Ca^{2+} stores mediating Ca^{2+} entry and release in vascular endothelial cells. Calcium entry may occur through a Ca^{2+} leak pathway, stretch-activated channels, and receptor-operated channels (ROC) according to the electrochemical gradient. The driving force for Ca^{2+} influx during activation is maintained primarily by the hyperpolarizing action of Ca^{2+}-activated K channels (K(Ca)). There is evidence to suggest Ca^{2+} influx may occur via the Na^{+}-Ca^{2+} exchanger operating in reverse mode and via voltage-gated Ca channels in capillary endothelium. Agonist (A) binding to a cell-surface receptor (R), which is coupled by a G-protein (G) to a specific phospholipase C (PLC), leads to the hydrolysis of phosphatidylinositol-4,5-bisphosphate(PIP_2) yielding diacylglycerol and inositol 1,4,5-trisphosphate ($InsP_3$). $InsP_3$ serves as a mediator of Ca^{2+} release from a compartment of the endoplasmic reticulum (ER). The caffeine-induced release of Ca^{2+} from the ER via activation of the ryanodine receptor (RyanR) suggests the presence of a Ca^{2+}-induced Ca^{2+} release mechanism in vascular endothelial cells. Elevated intracellular Ca^{2+} may be sequestered into the mitochondria (a low affinity, high capacity organelle) or into the ER compartments via an ATP-Ca^{2+} pump.

Acknowledgements

We gratefully acknowledge the collaboration of Dr. Casey van Breemen in our studies of endothelial calcium regulation over the past 6 years. Supported by National Institutes of Health grants RO1 HL39831 and RO1 HL35422.

REFERENCES

Adams, D.J., Barakeh, J., Laskey, R., and van Breemen, C., 1989, Ion channels and regulation of intracellular calcium in vascular endothelial cells. *FASEB J.* 3:2389-2400.

Ando, J., Komatsuda, T., and Kamiya, A., 1988, Cytoplasmic calcium response to fluid shear stress in cultured vascular endothelial cells. *In Vitro Cell. & Dev. Biol.* 24:871-877.

Benham, C.D. and Bolton, T.B., 1986, Spontaneous transient outward currents in single visceral and vascular smooth muscle cells of rabbit. *J. Physiol.* **381**:385-406.

Benham, C.D., Bolton, T.B., Lang, R.J. and Takewaki, T., 1986, Calcium-activated potassium channels in single smooth muscle cells of rabbit jejunum and guinea-pig mesenteric artery. *J. Physiol.* **371**:45-67.

Benham, C.D. and Tsien, R.W., 1986, Calcium-permeable channels in vascular smooth muscle: Voltage-activated, receptor-operated, and leak channels, *in:* "Cell Calcium and the Control of Membrane Transport", L.J. Mandel and D.C. Eaton, ed., pp 45-64, The Rockefeller University Press, New York.

Berridge, M.J. and Irvine, R.F., 1989, Inositol phosphates and cell signalling. *Nature* **341**:197-205.

Bezprozvanny, I., Watras, J., and Ehrlich, B.E., 1991, Bell-shaped calcium responses of inositol 1,4,5-trisphosphate-gated and calcium-gated channels from endoplasmic reticulum of cerebellum. *Nature* **351**:751-754.

Bolton, T.B. and Lim, S.P., 1989, Properties of calcium stores and transient outward currents in single smooth muscle cells of rabbit intestine. *J. Physiol.* **409**:385-401.

Bossu, J-L., Elhamdani, A., Feltz, A., Tanzi, F., Aunis, D. and Thierse, D., 1992, Voltage-gated Ca entry in isolated bovine capillary endothelial cells: evidence of a new type of Bay K 8644-sensitive channel. *Pflügers Arch* **420**:200- 207.

Bossu, J-L., Feltz, A., Rodeau, J-L. and Tanzi, F., 1989, Voltage-dependent calcium currents in freshly dissociated capillary endothelial cells. *FEBS Lett.* **255**:377-380.

Bregestovski, P., Bakhramov, A., Danilov, S., Moldobaeva, A., and Takeda, K., 1988, Histamine-induced inward currents in cultured endothelial cells from human umbilical vein. *Br. J. Pharmacol.* **95**:429-436.

Brunet, P.C. and Bény, J-L., 1989, Substance P and bradykinin hyperpolarize pig coronary artery endothelial cells in primary culture. *Blood Vessels* **26**:228-234.

Buchan, K.W. and Martin, W., 1991, Bradykinin induces elevations of cytosolic calcium through mobilization of intracellular and extracellular pools in bovine aortic endothelial cells. *Br. J. Pharmacol.* **102**:35-40.

Busse, R., Fichtner, H., Lückhoff, A., and Kohlhardt, M., 1988, Hyperpolarization and increased free calcium in acetylcholine-stimulated endothelial cells. *Am. J. Physiol.* **255**:H965-H969.

Cannell, M.B. and Sage, S.O., 1989, Bradykinin-evoked changes in cytosolic calcium and membrane currents in cultured bovine pulmonary artery endothelial cells. *J. Physiol.* **419**:555-568.

Carter, T.D. and Ogden, D., 1992, Kinetics of intracellular calcium release by inositol 1,4,5-trisphosphate and extracellular ATP in porcine cultured aortic endothelial cells. *Proc. R. Soc. Lond. B.* :235-241.

Changya, L., Gallacher, D.V., Irvine, R.F., Potter, B.V.L., and Petersen, O.H., 1989, Inositol 1,3,4,5-tetrakisphosphate is essential for sustained activation of the calcium-dependent K^+ current in single internally perfused mouse lacrimal acinar cells. *J. Membrane Biol.* **109**:85-93.

Chen, G. and Cheung, D.W., 1992, Characterization of acetylcholine-induced membrane hyperpolarization in endothelial cells. *Circ. Res.* **70**:257-263.

Colden-Stanfield, M., Schilling, W.P., Ritchie, A.K., Eskin, S.G., Navarro, L.T., and Kunze, D.L., 1987, Bradykinin-induced increases in cytosolic calcium and ionic currents in cultured bovine aortic endothelial cells. *Circ. Res.* **61**:632-640.

Colden-Stanfield, M., Schilling, W.P., Possani, L.D. and Kunze, D.L., 1990, Bradykinin-induced potassium current in cultured bovine aortic endothelial cells. *J. Membrane Biol.* **116**:227-238.

D'Amore, P. and Shepro, D., 1977, Stimulation of growth and calcium influx in cultured, bovine, aortic endothelial cells by platelets and vasoactive substances. *J. Cell Physiol.* **92**:177-184.

Danthuluri, N.R., Cybulsky, M.I. and Brock, T.A., 1988, ACh-induced calcium transients in primary cultures of rabbit aortic endothelial cells. *Am. J. Physiol.* **255**:H1549-H1553.

Derian, C.K. and Moskowitz, M.A., 1986, Polyphosphoinositide hydrolysis in endothelial cells and carotid artery segments (Bradykinin-2 receptor stimulation is calcium independent). *J. Biol. Chem.* **261**:3831-3837.

Dolor, R.J., Hurwitz, L.M., Mirza, Z., Strauss, H.C., and Whorton, A.R., 1992, Regulation of extracellular calcium entry in endothelial cells: role of intracellular calcium pool. *Am. J. Physiol.* **262**:C171-C181.

Endo, M., 1985, Calcium release from sarcoplasmic reticulum. *Curr. Topics Memb. Trans.* **25**:181-230.

Fay, F.S., Carrington, W., and Fogarty, K.E., 1989, Three-dimensional molecular distribution in single cells analysed using the digital imaging microscope. *J. Microscopy* **153**:133-149.

Ferris, C.D., Huganir, R.L., Supattapone, S., and Snyder, S.H., 1989, Purified inositol 1,4,5-trisphosphate receptor mediates calcium flux in reconstituted lipid vesicles. *Nature* **342**:87-89.

Fong, P., Turner, P.R., Denetclaw, W.F., and Steinhardt, R.A., 1990, Increased activity of calcium leak channels in myotubes of Duchenne human and *mdx* mouse origin. *Science* **250**:673-676.

Freay, A., Johns, A., Adams, D.J., Ryan, U.S., and van Breemen, C., 1989, Bradykinin and inositol-1,4,5-trisphosphate stimulated calcium release from intracellular stores in cultured bovine endothelial cells. *Pflügers Arch* **414**:377-384.

Furchgott, R.F. and Vanhoutte, P.M., 1989, Endothelium-derived relaxing and contracting factors. *FASEB J.* **3**:2007-2018.

Geiger, R.V., Berk, B.C., Alexander, R.W., and Nerem, R.M., 1992, Flow-induced calcium transients in single endothelial cells: spatial and temporal analysis. *Am. J. Physiol.* **262**:C1411-C1417.

Ghosh, T.K., Mullaney, J.M., Tarazi, F.I., and Gill, D.L., 1989, GTP-activated communication between distinct inositol 1,4,5-trisphosphate-sensitive and -insensitive calcium pools. *Nature* **340**:236-239.

Gordon, J.L. and Martin, W., 1983, Endothelium-dependent relaxation of the pig aorta: relationship to stimulation of ^{86}Rb efflux from isolated endothelial cells. *Br. J. Pharmacol.* **79**:531-541.

Griffith, T.M., Edwards, D.H., Newby, A.C., Lewis, M.J., and Henderson, A.H., 1986, Production of endothelium-derived relaxant factor is dependent on oxidative phosphorylation and extracellular calcium. *Cardiovasc. Res.* **20**:7-12.

Hallam, T.J., Jacob, R., and Merritt, J.E., 1988, Evidence that agonists stimulate bivalent-cation influx into human endothelial cells. *Biochem. J.* **255**:179-184.

Hallam, T.J., Jacob, R., and Merritt, J.E., 1989, Influx of bivalent cations can be independent of receptor stimulation in human endothelial cells. *Biochem. J.* **259**:125-129.

Hallam, T.J. and Pearson, J.D., 1986, Exogenous ATP raises cytoplasmic free calcium in fura-2 loaded piglet aortic endothelial cells. *FEBS Lett.* **207**:95-99.

Hoth, M. and Penner, R., 1992, Depletion of intracellular calcium stores activates a calcium current in mast cells. *Nature* **355**:353-355.

Iino, T., Kobayashi, T., and Endo, M., 1988, Use of ryanodine for the functional removal of the calcium store in smooth muscle cells of the guinea-pig. *Biochem. Biophys. Res. Comm.* **152**:417-422.

Irvine, R.F., Moor, R.M., Pollock, W.K., Smith, P.M., and Wreggett, K.A., 1988, Inositol phosphates: proliferation, metabolism and function. *Phil. Trans. R. Soc. B* **320**:281-298.

Irvine, R.F., 1992, Inositol phosphates and Ca^{2+} entry: toward a proliferation or simplification? *FASEB J.* **6**:3085-3091.

Jacob, R., 1990, Agonist-stimulated divalent cation entry into single cultured human umbilical vein endothelial cells. *J. Physiol.* **421**:55-77.

Jacob, R., Merritt, J.E., Hallam, T.J., and Rink, T.J., 1988, Repetitive spikes in cytoplasmic calcium evoked by histamine in human endothelial cells. *Nature* **335**:40-45.

Janigro, D., Gordon, E.L., and Winn, H.R., 1992, ATP-sensitive potassium channels in rat brain microvascular endothelial cells. *Soc. Neurosci. Abst.* **18**:1263.

Johns, A., Lategan, T.W., Lodge, N.J., Ryan, U.S., van Breemen, C., and Adams, D.J., 1987, Calcium entry through receptor-operated channels in bovine pulmonary artery endothelial cells. *Tissue & Cell* **19**:733-745.

Kass, G.E.N., Duddy, S.K., Moore, G.A., and Orrenius, S., 1989, 2',5'-Di(*tert*-butyl)-1,4-benzohydroquinone rapidly elevates cytosolic Ca^{2+} concentration by mobilizing the inositol 1,4,5-trisphosphate-sensitive Ca^{2+} pool. *J. Biol. Chem.* **264**:15192-15198.

Kuno, M. and Gardner, P., 1987, Ion channels activated by inositol 1,4,5-trisphosphate in plasma membrane of human T-lymphocytes. *Nature* **326**:301-304.

Lambert, T.L., Kent, R.S., and Whorton, A.R., 1986, Bradykinin stimulation of inositol polyphosphate production in porcine aortic endothelial cells. *J. Biol. Chem.* **261**:15288-15293.

Lansman, J.B., Hallam, T.J., and Rink, T.J., 1987, Single stretch-activated ion channels in vascular endothelial cells as mechano-transducers. *Nature* **325**:811-813.

Laskey, R.E., Adams, D.J., Johns, A., Rubanyi, G.M., and van Breemen, C., 1990, Membrane potential and Na^+-K^+ pump activity modulate resting and bradykinin-stimulated changes in cytosolic free calcium in cultured endothelial cells from bovine atria. *J. Biol. Chem.* **265**:2613-2619.

Lodge, N.J., Adams, D.J., Johns, A., Ryan, U.S., and van Breemen, C., 1988, Calcium activation of endothelial cells. *in:* "Resistance Arteries". W. Halpern, B.L. Pegram, J.E. Brayden, K. Mackey, M.K. McLaughlin, G. Osol eds. pp.152-161, Perinatology Press, N.Y.

Long, C.J. and Stone, T.W., 1985, The release of endothelium-derived relaxant factor is calcium dependent. *Blood Vessels* **22**:205-208.

Lopez-Jaramillo, P., Gonzalez, M.C., Palmer, R.M.J., and Moncada, S., 1990, The crucial role of physiological Ca^{2+} concentrations in the production of endothelial nitric oxide and the control of vascular tone. *Br. J. Pharmacol.* **101**:489-493.

Lückhoff, A. and Busse, R., 1990a, Refilling of endothelial calcium stores without bypassing the cytosol. *FEBS Lett.* **276**:108-110.

Lückhoff, A. and Busse, R., 1990b, Calcium influx into endothelial cells and formation of endothelium-derived relaxing factor is controlled by the membrane potential. *Pflügers Arch* **416**:305-311.

Lückhoff, A. and Busse, R., 1990c, Activators of potassium channels enhance calcium influx into endothelial cells as a consequence of potassium currents. *Naunyn-Schmiedeberg's Arch Pharmacol.* **342**:94-99.

Lückhoff, A. and Clapham, D.E., 1992, Inositol 1,3,4,5-tetrakisphosphate activates an endothelial Ca^{2+}-permeable channel. *Nature* **355**:356-358.

Lückhoff, A., Pohl, U., Mülsch, A., and Busse, R., 1988, Differential role of extra- and intracellular calcium in the release of EDRF and prostacyclin from cultured endothelial cells. *Br. J. Pharmacol.* **95**:189-196.

Mehrke, G. and Daut, J., 1990, The electrical response of cultured guinea-pig coronary endothelial cells to endothelium-dependent vasodilators. *J. Physiol.* **430**:251-272.

Mendelowitz, D., Bacal, K., and Kunze, D.L., 1992, Bradykinin-activated calcium influx pathway in endothelial cells. *Am. J. Physiol.* **262**:H942-H948.

Merritt, J.E. and Rink, T.J., 1987, Regulation of cytosolic free calcium in fura-2-loaded rat parotid acinar cells. *J. Biol. Chem.* **262**:17362-17369.

Mo., M., Eskin, S.G., and Schilling, W.P., 1991, Flow-induced changes in Ca^{2+} signaling of vascular endothelial cells: effect of shear stress and ATP. *Am. J. Physiol.* **260**:H1698-H1707.

Moncada, S., Palmer, R.M.J., and Higgs, E.A., 1991, Nitric oxide: Physiology, pathophysiology, and pharmacology. *Pharmacol. Rev.* **43**:109-142.

Morgan-Boyd, R., Stewart, J.M., Vavrek, R.J. and Hassid, A., 1987, Effects of bradykinin and angiotensin II on intracellular Ca^{2+} dynamics in endothelial cells. *Am. J. Physiol.* 253:C588-C598.

Morris, A.P., Gallacher, D.V., Irvine, R.F., and Petersen, O.H., 1987, Synergism of inositol trisphosphate and inositol tetrakisphosphate in activating Ca^{2+}-dependent K^+ channels. *Nature* 330:653-655.

Mullaney, J.M., Yu, M., Ghosh, T.K., and Gill, D.L., 1988, Calcium entry into the inositol 1,4,5-trisphosphate-releasable calcium pool is mediated by a GTP-regulatory mechanism. *Proc. Natl. Acad. Sci. USA* 85:2499-2503.

Nakache, M. and Gaub, H.E., 1988, Hydrodynamic hyperpolarization of endothelial cells. *Proc. Natl. Acad. Sci. USA* 85:1841-1843.

Nilius, B., 1990, Permeation properties of a non-selective cation channel in human vascular endothelial cells. *Pflügers Arch* 416:609-611.

Nilius, B., 1991, Regulation of transmembrane calcium fluxes in endothelium. *NIPS* 6:110-114.

Nilius, B. and Riemann, D., 1990, Ion channels in human endothelial cells. *Gen. Physiol. Biophys.* 9:89-112.

Nollert, M.U., Eskin, S.G., and McIntire, L.V., 1990, Shear stress increases inositol trisphosphate levels in human endothelial cells. *Biochem. Biophys. Res. Comm.* 170:281-287.

Olesen, S.-P., Clapham, D.E., and Davies, P.F., 1988, Haemodynamic shear stress activates a K^+ current in vascular endothelial cells. *Nature* 331:168-170.

Parker, I. and Ivorra, I., 1991, Caffeine inhibits inositol trisphosphate-mediated liberation of intracellular calcium in *Xenopus* oocytes. *J. Physiol.* 433:229-240.

Peach, M.J., Singer, H.A., Izzo, N.J., and Loeb, A.L., 1987, Role of calcium in endothelium-dependent relaxation of arterial smooth muscle. *Am. J. Cardiol.* 50:35A-43A.

Penner, R., Matthews, G., and Neher, E., 1988, Regulation of calcium influx by second messengers in rat mast cells. *Nature* 334:499-504.

Pirotton, S., Raspe, E., Demolle, D., Erneux, C. and Boeynaems, J.-M., 1987, Involvement of inositol 1,4,5-trisphosphate and calcium in the action of adenine nucleotides on aortic endothelial cells. *J. Biol. Chem.* 262:17461-17466.

Pollock, W.K., Wreggett, K.A., and Irvine, R.F., 1988, Inositol phosphate production and Ca^{2+} mobilization in human umbilical-vein endothelial cells stimulated by thrombin and histamine. *Biochem. J.* 256:371-376.

Popp, R. and Gögelein, H., 1992, A calcium and ATP sensitive nonselective cation channel in the antiluminal membrane of rat cerebral capillary endothelial cells. *Biochim. Biophys. Acta* 1108:59-66.

Popp, R., Hoyer, J., Meyer, J., Galla, H-J., and Gögelein, H., 1992, Stretch-activated non-selective cation channels in the antiluminal membrane of porcine cerebral capillaries. *J. Physiol.* 454:435-449.

Putney, J.W., 1986, A model for receptor-regulated calcium entry. *Cell Calcium* 7:1-12.

Putney, J.W., 1990, Capacitative calcium entry revisited. *Cell Calcium* 11:611-624.

Rizzuto, R., Simpson, A.W.M., Brini, M., and Pozzan, T., 1992, Rapid changes of mitochondrial Ca^{2+} revealed by specifically targeted recombinant aequorin. *Nature* 3358:325-327.

Ross, C.A., Meldolesi, J., Milner, T.A., Satoh, T., Supattapone, S., and Snyder, S.H., 1989, Inositol 1,4,5-trisphosphate receptor localized to endoplasmic reticulum in cerebellar Purkinje neurons. *Nature* 339:468-470.

Rubanyi, G.M., Freay, A.D., Kauser, K., Johns, A., and Harder, D.R., 1990, Mechanoreception by the endothelium: mediators and mechanisms of pressure- and flow-induced vascular responses. *Blood Vessels* 27:246-247.

Rusko, J., Tanzi, F., van Breemen, C., and Adams, D.J., 1992a, Calcium-activated potassium channels in native endothelial cells from rabbit aorta: Conductance, Ca^{2+} sensitivity and block. *J. Physiol.* 455:601-621.

Rusko, J., van Breemen, C., and Adams, D.J., 1992b, Caffeine-induced Ca^{2+} release from intracellular stores in freshly dissociated endothelial cells from rabbit aorta. *J. Physiol. (abst.)* (in press)

Sage, S.O., Adams, D.J., and van Breemen, C., 1989, Synchronized oscillations in cytoplasmic free calcium concentration in confluent bradykinin-stimulated bovine pulmonary artery endothelial cell monolayers. *J. Biol. Chem.* **264**:6-9.

Sage, S.O., van Breemen, C., and Cannell, M.B., 1991, Sodium-calcium exchange in cultured bovine pulmonary artery endothelial cells. *J. Physiol.* **440**:569-580.

Sakai, T., 1990, Acetylcholine induces Ca-dependent K currents in rabbit endothelial cells. *Jap. J. Pharmacol.* **53**:235-246.

Sauvé, R., Chahine, M., Tremblay, J., and Hamet, P., 1990, Single-channel analysis of the electrical response of bovine aortic endothelial cells to bradykinin stimulation: contribution of a Ca^{2+}-dependent K^+ channel. *J. Hypertens.* **8**:S193-S201.

Sauvé, R., Parent, L., Simoneau, C., and Roy, G., 1988, External ATP triggers a biphasic activation process of a calcium-dependent K^+ channel in cultured bovine aortic endothelial cells. *Pflügers Arch* **412**:469-481.

Schilling, W.P., Cabello, O.A., and Rajan, L., 1992, Depletion of the inositol 1,4,5-trisphosphate-sensitive intracellular Ca^{2+} store in vacsular endothelial cells activates the agonist-sensitive Ca^{2+}-influx pathway. *Biochem. J.* **284**:521-530.

Schilling, W.P., Rajan, L., and Strobl-Jager, E., 1989, Characterization of thee bradykinin-stimulated calcium influx pathway of cultured vascular endothelial cells: Saturability, selectivity and kinetics. *J. Biol. Chem.* **264**:12838-12848.

Schilling, W.P., Ritchie, A.K., Navarro, L.T., and Eskin, S.G., 1988, Bradykinin stimulated calcium influx in cultured bovine aortic endothelial cells. *Am. J. Physiol.* **255**:H219-H227.

Schwarz, G., Droogmans, G., and Nilius, B., 1992, Shear stress induced membrane currents and calcium transients in human vascular endothelial cells. *Pflügers Arch* **421**:394-396.

Shen, J., Luscinskas, F.W., Connolly, A., Dewey, C.F., and Gimbrone, M.A., 1992, Fluid shear stress modulates cytosolic free calcium in vascular endothelial cells. *Am. J. Physiol.* **262**:C384-C390.

Silver, M.R. and DeCoursey, T.E., 1990, Intrinsic gating of inward rectifier in bovine pulmonary artery endothelial cells in the presence or absence of internal Mg^{2+}. *J. Gen. Physiol.* **96**:109-133.

Singer, H.A. and Peach, M.J., 1982, Calcium- and endothelial-mediated vascular smooth muscle relaxation in rabbit aorta. *Hypertension* **4**(Suppl. II):II19-II25.

Streb, H., Bayerdorffer, E., Haase, W., Irvine, R.F., and Shulz, I., 1984, Effect of inositol 1,4,5-trisphosphate on isolated subcellular fractions of rat pancreas. *J. Membrane Biol.* **81**:241-253.

Sturek, M., Smith, P., and Stehno-Bittel, L., 1991, In vitro models of vascular endothelial cell calcium regulation. *in:* "Ion Channels of Vascular Smooth Muscle Cells and Endothelial Cells". N. Sperelakis and Kuriyama, eds. pp 349-364, Elsevier Science Publishing Co.

Takeda, K. and Klepper, M., 1990, Voltage-dependent and agonist-activated ionic currents in vascular endothelial cells: A review. *Blood Vessels* **27**:169-183.

Takeda, K., Schini, V., and Stoeckel, H., 1987, Voltage-activated potassium, but not calcium currents in cultured bovine aortic endothelial cells. *Pflügers Arch* **410**:385-393.

Takemura, H., Hughes, A.R., Thastrup, O., and Putney, J.W., 1989, Activation of calcium entry by the tumor promotor thapsigargin in rat parotid acinar cells. *J. Biol. Chem.* **264**:12266-12271.

Thuringer, D. and Sauvé, R., 1992, A patch clamp study of the Ca^{2+} mobilization from internal stores in bovine aortic endothelial cells. I. Effects of caffeine on intracellular Ca^{2+} stores. *J. Membrane Biol.* **130**:125-137.

Van Breemen, C. and Saida, K., 1989, Cellular mechanisms regulating $[Ca^{2+}]_i$ in smooth muscle. *Ann. Rev. Physiol.* **51**:315-329.

Volpe, P., Krause, K., Hashimoto, S., Zorzato, F., Pozzan, T., Meldolesi, J., and Lew, D.P., 1988, "Calciosome", a cytoplasmic organelle: the inositol 1,4,5-trisphosphate-sensitive Ca^{2+} store of nonmuscle cells? *Proc. Natl. Acad. Sci. USA* **85**:1091-1095.

Whorton, A.R., Willis, C.E., Kent, R.S., and Young, S.L., 1984, The role of calcium in the regulation of prostacyclin synthesis in porcine aortic endothelial cells. *Lipids* **19**:17-24.

Yamamoto, Y., Chen, G., Miwa, K., and Suzuki, H., 1992, Permeability and Mg^{2+} blockade of histamine-operated cation channel in endothelial cells of rat intrapulmonary artery. *J. Physiol.* **450**:395-408.

MECHANOSENSITIVE ION CHANNELS IN VASCULAR ENDOTHELIAL CELLS

Teryl R. Elam[1] and Jeffry B. Lansman[2]

[1]Department of Physiology
[2]Department of Pharmacology
School of Medicine
University of California
San Francisco, CA 94143-0450

INTRODUCTION

Endothelial cells form a monolayer lining the entire vascular system and are the target of a wide variety of blood-borne substances which regulate vascular tone (reviewed by Vanhoutte, 1987). The endothelium also mediates changes in vessel tone that occur in response to mechanical forces produced by blood flowing under pressure. Both the vasodilation that occurs in response to an increase in blood flow (Rubanyi et al., 1986; Cooke et al., 1990) and the vasoconstriction that occurs in response to an increase in vascular pressure (Katusic et al., 1987; Fischell et al., 1989) are abolished by removing the endothelium. The cellular mechanisms underlying endothelium-dependent responses to mechanical forces, however, are poorly understood.

Endothelial cells are ideally situated to sense and respond to the mechanical forces produced by blood flowing under pressure. It has been suggested that mechanosensitive ion channels in vascular endothelial cells transduce membrane deformation into a change in membrane permeability to specific ions and that this constitutes an early step in the the transduction process (reviewed by Lansman, 1988). We review here the basic functional properties of mechanosensitive ion channels in vascular endothelial cells. We suggest that these channels have properties consistent with the idea they provide a mechanism for transducing mechanical deformation of the the endothelial cell membrane to Ca^{2+} entry.

Ion Flux in Pulmonary Vascular Control, Edited by
E.K. Weir et al., Plenum Press, New York, 1993

SINGLE-CHANNEL RECORDINGS OF MECHANOSENSITIVE ION CHANNEL ACTIVITY IN AORTIC ENDOTHELIAL CELLS

Recordings of single-channel activity from cell-attached patches on aortic endothelial cells show discrete steps of inward current of ~2 pA at a holding potential of -70 mV when the patch electrode contained physiological saline. Resting channel activity varies from recording to recording, ranging from 0 to ~20%. Figure 1 shows that the channels producing short bursts of inward current at negative membrane potentials is enhanced by deforming the membrane by applying suction to the patch electrode. Applying pressure to the patch electrode increased channel activity and it returned promptly to its previous resting level after releasing the pressure. The response was rapid and followed the imposed pressure change within the delay expected from the equilibration time of the system. The time of onset of the application of pressure to the patch membrane can be seen in figure 1 as the small downward deflection of the holding current. Applying positive pressure to the patch electrode also increased channel activity (data not shown).

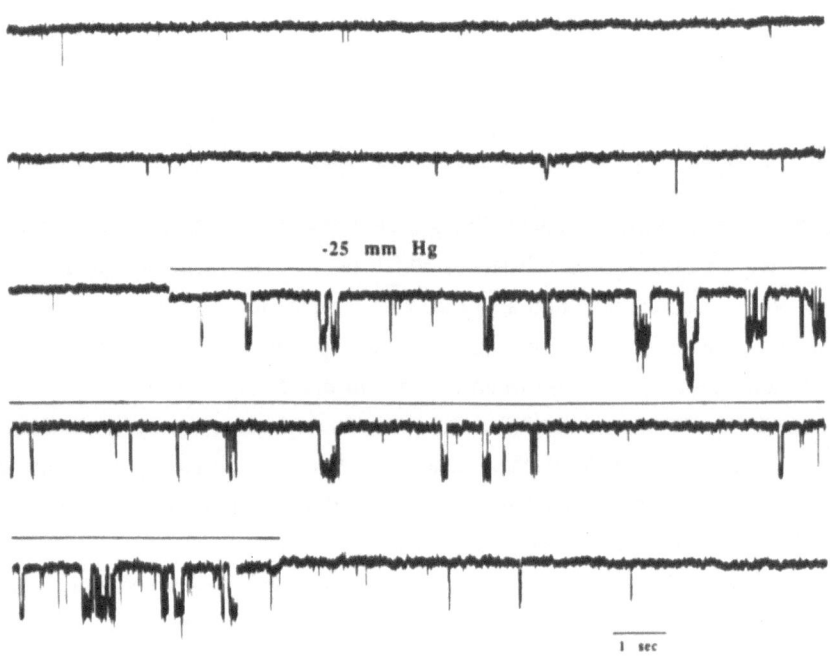

Figure 1. Record of single-channel activity at a constant holding potential of -50 mV. The channel open probability measured shortly after seal formation was 0.04. Applying -25 mm Hg of suction pressure to the patch electrode increased channel open probability to ~ 0.1 (time during which pressure was applied indicated by bar above current record). Upon release of the pressure, activity returned to a low level (open probability = 0.05). Currents were filtered at 500 Hz and sampled at 1 ms.

Ion selectivity

To compare the mechanosensitive ion channels in endothelial cells with those in other types of cells, we investigated the ion selectivity of the single-channel activity . With isotonic KCl in the patch electrode, single-channel currents appeared as discrete steps of inward current ~-2.0 pA at -70 mV. Figure 2a shows the single-channel current-voltage (i-V) relation. The slope conductance for the inward current was 25.1 ± 3.9 pS (n=7). The slope conductance was smaller for outward current through the channel (13.8 ± 1.2 pS). The single-channel i-V relation measured with isotonic KCl in the patch electrode was identical whether recordings were made from either cell-attached or excised membrane patches. Figure 2b shows the single-channel i-V relation measured with 150 NaCl in the patch electrode. The slope conductance was 21 ± 2 pS (n=4), similar to that measured with KCl in the electrode. In the presence of either electrode filling solutions, the single-channel currents reversed near 0 mV, indicating that the channel is not strongly selective for either of these monovalent cations.

We also investigated the permeability of the channel to divalent cations to determine whether the channels could provide a pathway for Ca^{2+} entry into endothelial cells. Divalent cations carried significant current through the channel. With 110 mM $CaCl_2$ in the patch electrode, the single-channel activity recorded at -70 mV appeared as inward current steps of ~1 pA. Figure 2c shows the single-channel i-V relation measured with 110 mM $CaCl_2$ in the patch electrode. The single-channel conductance was 13.1 ± 1.8 pS and current reversed at ~$+22 \pm 7$ mV (n=4). Figure 2d shows the i-V relation measured with 110 mM $BaCl_2$ in the electrode. The slope conductance was 14.0 ± 1.0 pS and current reversed at $+19 \pm 5$ mV (n=6). To determine how much Ca^{2+} would enter the cell through the channel, we calculated the relative permeability of Ca^{2+} to K^+ from the constant field equation modified for the condition in which divalent cations carry charge (see Lee and Tsien, 1985). The relative permeability $P_{Ca2+}/P_{K+} = $ ~4, indicating that the channel is somewhat more permeable to divalent than to monovalent cations. Thus, Ca^{2+} would be expected to carry significant charge through the channel, particularly at negative membrane potentials where the driving force for ion entry is large.

Mechanosensitive gating

Figure 3 shows the effects of different amounts of suction applied to the patch electrode on the activity of mechanosensitive channels. Channel open probability increased with an increase in the applied pressure. The relation between pressure and channel open probability was sigmoidal. The curve represents a fit to a Boltzmann relation with a half maximal channel open probability of 34 mm Hg and a slope that changed ~e-fold/8 mm Hg.

Both the steepness and the half-activation level of the relation varied in different recordings (data not shown), perhaps representing variability in the actual tension at the membrane surface.

Voltage-sensitive gating

The mechanosensitive ion channels in endothelial cells also show a voltage dependent gating mechanism in which channel opening increases with membrane depolarization. Figure 4 shows a recording of the activity of a single channel at different steady holding potentials. Channel opening increased as the holding potential was made more positive approaching unity at +80 mV. The single-channel current was inward at 0 mV, but reversed and became clearly outward at +20 mV. The steepest increase in channel opening occurred between +40 and +60 mV. The inward and outward current is likely to be carried by the same channel because the presence of inward current was always associated with the presence of outward current and there was a gradual change in open channel probability as the current changed

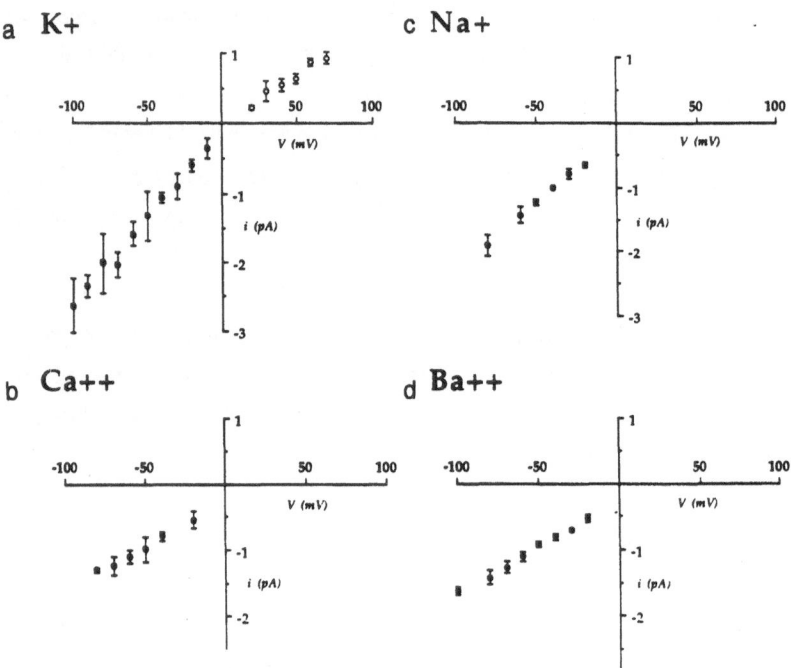

Figure 2. Current-voltage (i-V) relation in the presence of monovalent and divalent cations. a) I-V relation in the presence of either 150 mM NaCl (filled symbols) or KCl open symbols). In the presence of KCl, the slope conductance for inward current was 25.1 ± 3.9 pS (n=7); the slope conductance for outward current 13.8 ± 1.2 pS. In the presence of NaCl, the slope conductance for inward current is 20.9 ± 2.0 pS (n=4). b) I-V relation measured with either 110 mM $CaCl_2$ or $BaCl_2$ in the electrode. The slope conductances are 13.1 ± 1.8 pS (n=3) and 13.8 ± 1.3 pS (n=4), respectively.

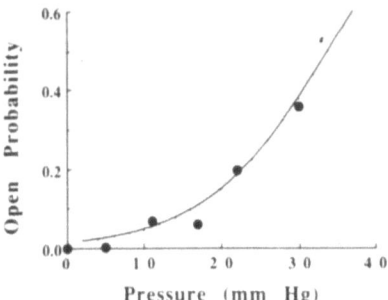

Figure 3. Relation between channel open probability and the pressure applied to the patch electrode. Mean channel open probability was half maximal at a pressure of -34 mm Hg with a slope of 8 mm Hg/e-fold change in activity. Mean channel open probability was calculated by dividing the total time a channel was open by the integral of open channel current over the same time period.

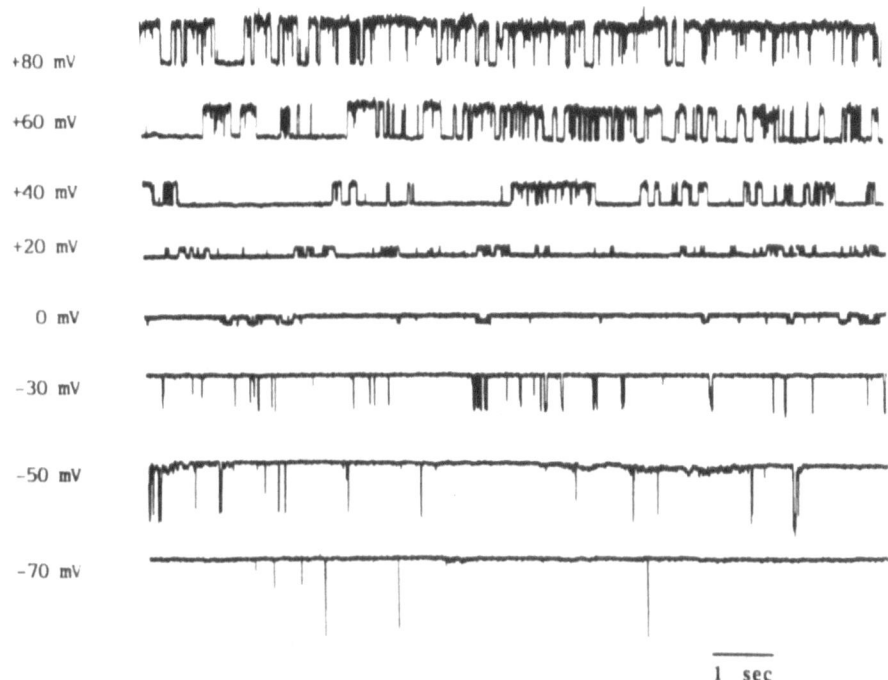

Figure 4. Activity of a single mechanosensitive channel recorded at the indicated steady holding potentials.

from inward to outward which is reflected in the monotonic change in the durations of open and closed times with membrane potential.

The results indicate that the channels have both a mechanosensitive and voltage sensitive gating mechanism. Whether these represent independent energetic contributions to the transition from resting to open channels was investigated by examining the effects of the holding membrane potential on the relationship between the pressure applied to the patch electrode. The effect of membrane potential on the pressure dependence of channel opening was to shift the threshold and half-activation pressure to lower values at more positive membrane potentials, consistent with a joint dependence of channel gating on membrane potential and pressure.

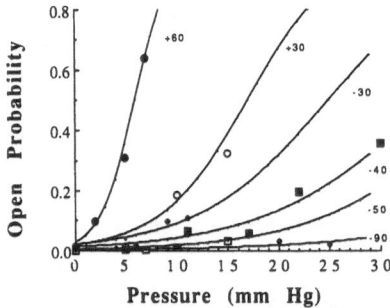

Figure 5. The effects of membrane potential on the sensitivity to pressure of mechanosensitive channel activity.

CONCLUSIONS

Recordings of single-channel activity from bovine aortic endothelial cells show the activity of mechanosensitive ion channels that carry inward current at negative holding potentials. The channels are similar to mechanosensitive ion channels studied in other cell types in being permeable to both monovalent and divalent cations (skeletal muscle: Guharay and Sachs, 1984; Franco and Lansman; 1990; frog lens epithelial cells: Cooper et al., 1986; gastric smooth muscle: Kirber et al.,1988; mammalian fibroblasts: Stockbridge and French, 1988; ventricular myocytes: Craelis, 1988; frog eggs: Taglietti and Toselli, 1988). The channels in endothelial cells from bovine aorta differ from those in muscle and oocytes, however, in that they show little selectivity for K^+ over Na^+ (Guharay and Sachs, 1984, 1985; Taglietti and Toselli, 1988). In general, however, the channel in bovine endothelial cells can be considered to belong to the general class of cation-selective, mechanosensitive ion channels.

The existence of endothelium-dependent responses to mechanical stimuli has suggested that endothelial cells possess a mechanism for sensing and responding to the mechanical forces produced by blood flowing under pressure (Lansman, 1988). The energy required to open channels can be compared with the forces generated by the wall shear stress produced by blood flow. If the patch of membrane within the electrode forms is a hemisphere with a diameter of 5-15 μm^2, the pressures required to open the channels described here would require an input of ~25 dynes/cm^2, assuming simple pressure-tension relations. This value falls within the range of shear forces that are sufficient to cause endothelium-dependent vasodilation (5 - 30 dynes/cm^2 for laminar flow conditions; Dewey, 1979). Although the distribution of forces applied to the membrane by the patch electrode is not well understood, the force exerted by the shearing action of flowing blood may be sufficient to open a fraction of channels. Distension of the aorta during diastolic filling, moreover, would also be expected to deform the endothelial surface as the vessel expands and the extent of deformation would depend on vessel compliance. Thus, both shear stress and distension of compliant vessels would be expected to open channels allowing and influx of Na^+ and Ca^{2+}. Mechanosensitive channel gating in endothelial cells may, therefore, provide a signaling pathway for the extracellular Ca^{2+}-dependent release of endothelium-derived relaxing factor (Singer and Peach, 1982; Long and Stone, 1985).

ACKNOWLEDGEMENTS

Supported by the NIH and the Office of Army Research.

REFERENCES

1. Bregestovski, P., A. Barkhramov, S. Danilov, A. Moldovaeva, and K. Takeda, K. Histamine-induced inward currents in cultured endothelial cells from human umbilical vein. *British Journal of Pharmacology* 95(2): 429-436, 1988.

2. Cooke, J.P., J. Stamler, N. Andon, P.F. Davies, G. McKinley, and J. Loscalzo. Flow stimulates endothelial cells to release a nitrovasodilator that is potentiated by reduced thiol. *American Journal of Physiology* 259(3:2): H804-812, 1990.

3. Cooper, K.E., J.M. Tang, J.L. Rae, and R.S. Eisenberg. A cation channel in frog lens epithelia responsive to pressure and calcium. *Journal of Membrane Biology* 93: 259-269, 1986.

4. Dewey, C.F. In, Advances in Experimental Medicine and Biology Ed., S. Wolf and N.T. Werthessen. New York: Plenum Press, 1977, pp 55-103.

5. Fichtner, H., U. Frobe, R. Busse, and H. Kohlhardt, H. Single nonselective cation channels and Ca^{2+} activated K$^+$ channels in aortic endothelial cells. *Journal of Membrane Biology* 98(2): 125-133, 1987.

6. Fischell, T.A., U. Nellessen, D.E. Johnson, and R. Ginsburg. Endothelium-dependent arterial vasoconstriction after balloon angioplasty. *Circulation* 79(4): 899-910, 1989.

7. Franco, A. and J.B. Lansman. Stretch-sensitive channels in developing muscle cells in a mouse cell line. *Journal of Physiology* 427: 361-380, 1990.

8. Griffith, T.M., D.H. Edwards, M.J. Lewis, A.C. Newby, A.H. Henderson. Production of endothelium-derived relaxing factor is dependent on oxidative phosphorylation and extracellular calcium. *Cardiovascular Research* 20: 7-12, 1986

9. Guharay, F., and F. Sachs. Stretch-activated single ion channel currents in tissue-cultured embryonic chick skeletal muscle. *Journal of Physiology* 352: 685-701, 1984

10. Guharay, F. and F. Sachs. Mechanotransducer ion channels in chick skeletal muscle: the effects of extracellular pH. *Journal of Physiology* 363: 119-134, 1985

11. Hamill, O.P., A. Marty, E. Neher, B. Sakmann and F.J. Sigworth. Improved patch clamp techniques for high resolution current recordings from cells and cell-free membrane patches. *Pflügers Archiv* 391: 85-100, 1981.

12. Johns, A., T.W. Lategan, N.J. Lodge, U.S. Ryan, C. Van Breemen, and D.J. Adams. Calcium entry through receptor-operated channels in bovine pulmonary artery endothelial cells. *Tissue and Cell* 19(6): 733-745, 1987.

13. Katusic, Z.S., J.T. Shepherd, and P.M. Vanhoutte. Endothelium-dependent contraction to stretch in canine basilar arteries. *American Journal of Physiology* 252(3:2): H671-673, 1987.

14. Kirber, M.T., J.V. Walsh Jr., and J.J. Singer. Stretch-activated ion channels in smooth muscle: a mechanism for the initiation of stretch-induced contraction. *Pflugers Archiv.* 412(4): 339-345, 1988.

15. Langille, B.L., and F. O'Donnell. Reductions in arterial diameter produced by chronic decreased in blood flow are endothelium-dependent. *Science* 231: 405-407, 1986.

16. Lansman, J.B., T.J. Hallam, and T.J. Rink. Single stretch-activated ion channels in vascular endothelial cells as mechanotransducers? *Nature* 325: 811-813, 1987.

17. Lansman, J.B. Endothelial mechanosensors: Going with the flow. *Nature* 331: 481-482, 1988.

18. Lee, K.S. and R.W. Tsien. High selectivity of calcium channels in single dialysed heart cells of the guinea-pig. *Journal of Physiology* 354: 253-272, 1984.

19. Long, C.J. and T.W. Stone. The release of endothelium-derived relaxant factor is calcium dependent. *Blood Vessels* 22: 205-208, 1985.

20. Nilius, B. and D. Riemann. Ion channels in human endothelial cells. *General Physiology and Biophysics* 9(2): 89-111, 1990.

21. Olesen, S.P., P.F. Davies and D.E. Clapham. Muscarinic-activated K$^+$ current in bovine aortic endothelial cells. *Circulation Research* 62(6): 1059-1064, 1988.

22. Pohl, U., J. Holtz, R. Busse and E. Bassenge. Crucial role of endothelium in the vasodilator response to increased flow in vivo. *Hypertension* 8(1): 37-44, 1986.

23. Rubanyi, G.M., J.C. Romero, and P.M. Vanhoutte. Flow-induced release of endothelium-derived relaxing factor. *American Journal of Physiology* 250(6:2): H1145-H1149, 1986.

24. Singer, H.A. and M.J. Peach. Calcium and endothelial-mediated vascular smooth muscle relaxation in rabbit aorta. *Hypertension* 4 (Supp II): 19-25, 1982.

25. Stockbridge, L.L. and A.S. French. Stretch-activated cation channels in human fibroblasts. *Biophysical Journal* 54(1): 187-190, 1988.

26. Taglietti, V. and M. Toselli. A study of stretch-activated channels in the membrane of frog oocytes: Interaction with Ca^{2+} ions. *Journal of Physiology* 407: 311-328, 1988.

27. Takeda, K., V. Schini and H. Stoeckel. Voltage-activated potassium, but not calcium currents in cultured bovine aortic endothelial cells. *Pflügers Archiv.* 410(4-5): 385-393, 1987.

28. Vanhoutte, P.M. Endothelium and control of vascular function. State of the Art lecture. *Hypertension* 13(6:2): 658-667, 1989.

MEMBRANE HYPERPOLARIZATION AS A MECHANISM
OF ENDOTHELIUM-DEPENDENT VASODILATION

Joseph E. Brayden

Department of Pharmacology
University of Vermont Medical Research Facility
55A South Park Drive
Colchester, VT 05446

INTRODUCTION

Vascular endothelial cells can generate and release a variety of vasoactive substances in response to physiological stimuli such as receptor activation, shear stress, hypoxia and vascular wall distension (31). Endothelium-dependent vasodilation may be mediated by prostacyclin, endothelium-derived relaxing factor (nitric oxide, NO$\cdot$) or by other substances. One or more of the factors released from the endothelium can hyperpolarize the underlying vascular smooth muscle cells.

Hyperpolarization of vascular smooth muscle cells may be involved in one of the mechanisms of endothelium-dependent vasodilation. Hyperpolarization, induced by whatever means, would be expected to close voltage-dependent calcium channels. Calcium influx through voltage-dependent calcium channels is centrally involved in the process of vascular force generation, particularly in the cerebral and coronary circulations and in resistance arteries throughout the body (24). Therefore, reduction in calcium influx by this route should lead to attenuation of vascular tone. The objective of this chapter is to describe some of the characteristics of endothelium-dependent hyperpolarization and to review the evidence in support of the proposal that membrane hyperpolarization is an important mechanism of endothelium-dependent vasodilation.

EVIDENCE FOR THE EXISTENCE OF EDHF

Prior to the discovery of EDRF by Furchgott (15), several investigators had noted the hyperpolarizing effects of acetylcholine in isolated strips of vascular smooth muscle (19,21). In 1984, Bolton et al. (4) reported that carbachol-induced hyperpolarizations of vascular smooth muscle cells in the guinea-pig mesenteric artery were endothelium-dependent. Since then, numerous other investigators have noted a similar effect of endothelium-dependent vasodilators such as acetylcholine (5,6,10,11,12,18), ADP (5), histamine (10), the calcium ionophore A23187 and substance P (3). The factor responsible for the hyperpolarizing response has been termed endothelium-derived hyperpolarizing factor (EDHF) (10).

Subsequent studies supported the proposal that EDHF, like EDRF is a diffusible factor. For instance, Feletou and Vanhoutte (14) found that exposure of a canine coronary artery segment to ACh resulted in hyperpolarization of a de-endothelialized coronary artery present in the same bath; when both donor and recipient were denuded, ACh had no effect on membrane potential. This demonstration that EDHF is a diffusible factor was later confirmed using arteries with and without endothelium that were perfused in series (18) or placed together in a sandwich type preparation (12). An example of this type of "donor/recipient" experiment using a rabbit aorta as the donor and a cerebral artery as the recipient is shown in figure 1. In this experiment perfusion of the donor aorta with ACh resulted in a large hyperpolarization of an endothelium-denuded segment of middle cerebral artery.

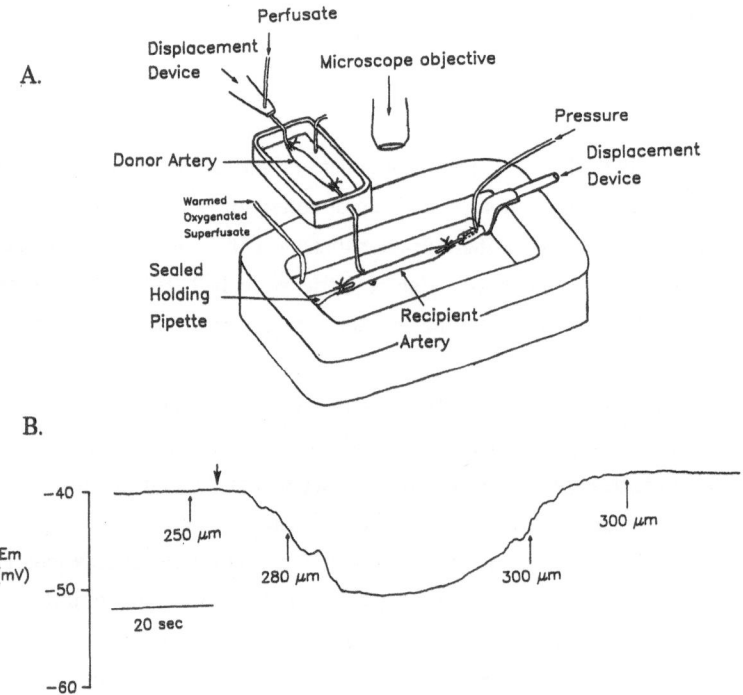

Figure 1. EDHF is a diffusable factor. Abdominal aorta from a cat was used as a donor artery [arranged as indicated in (A)]. A pressurized feline middle cerebral artery without endothelium was exposed to aortic perfusate. Diameter and membrane potential of the cerebral artery were monitored (5). Addition of ACh to the perfusate released EDHF from the aorta which hyperpolarized the cerebral artery (B). Cerebral artery diameter at given times is indicated by the values at the arrows. Note transient hyperpolarization, but prolonged dilation of the cerebral artery.

288

PHARMACOLOGICAL AND CHEMICAL FEATURES OF EDHF

Since the initial description of EDRF, many investigators have confirmed the hypothesis that EDRF is distinct from prostacyclin. Inhibitors of cyclooxygenase have no effect on endothelium-dependent vasodilation in the vast majority of cases (31). Similarly, although prostacyclin does hyperpolarize vascular smooth muscle (27), indomethacin has no effect on the hyperpolarizing action of endothelium-dependent vasodilators (5,12). Thus, EDHF must be a non-prostanoid endothelial factor. An obvious question that has been addressed by many investigators is the possibility that EDRF and EDHF are the same substance, namely NO' or an NO'-containing compound. Several studies indicate that EDHF is not NO'. Inhibitors of the formation of NO' such as nitroarginine or L-NMMA do not affect the action of EDHF in most tissues (12,23). Likewise, agents such as hemoglobin or methylene blue which inhibit the actions of NO' seldom inhibit endothelium-dependent hyperpolarization (5,10). In many studies, NO' itself has been found to have little if any direct effect on membrane potential of vascular smooth muscle (2,6,17). There are of course exceptions to these observations. For instance, in the guinea-pig uterine artery (30) and in mesenteric resistance arteries of the rat (16), high concentrations of NO' hyperpolarize the vascular smooth muscle cells. However, because the concentrations of NO' required for hyperpolarization usually are far in excess of those required for maximal vasodilation to this agent, on balance the evidence would argue that EDHF and NO' are distinct endothelial factors in most instances. To date there are no published studies that provide convincing evidence (e.g., half-life or mimicry by other substances) as to the characteristics or identity of EDHF. One report has suggested that EDHF may be an arachidonic acid metabolite derived from a cytochrome P-450-dependent pathway (26), but direct proof of this hypothesis is not available.

CHARACTERISTICS OF THE HYPERPOLARIZING RESPONSES

Despite uncertainties about the identity of EDHF, this factor hyperpolarizes vascular smooth muscle primarily by increasing membrane potassium conductance. This conclusion is based on studies of the effects of altered extracellular potassium ions (1,4,23) and of K^+-channel blockers (6,12,13,18) on endothelium-dependent hyperpolarization and of the ability of endothelium-dependent hyperpolarizing agents to increase the efflux of labeled rubidium (11), a marker of K^+-efflux. EDHF appears to activate K_{ATP}-channels in the vascular smooth muscle cells of middle cerebral arteries isolated from the rabbit (6). Glibenclamide (3 μM), a sulphonyl urea shown to block K_{ATP}-channels in many tissues including vascular smooth muscle (28), abolished the endothelium-dependent hyperpolarizing action of ACh in this artery. This was also the case for ADP-induced responses in mesenteric and skeletal muscle resistance arteries isolated from the rabbit (8) (figure 2).

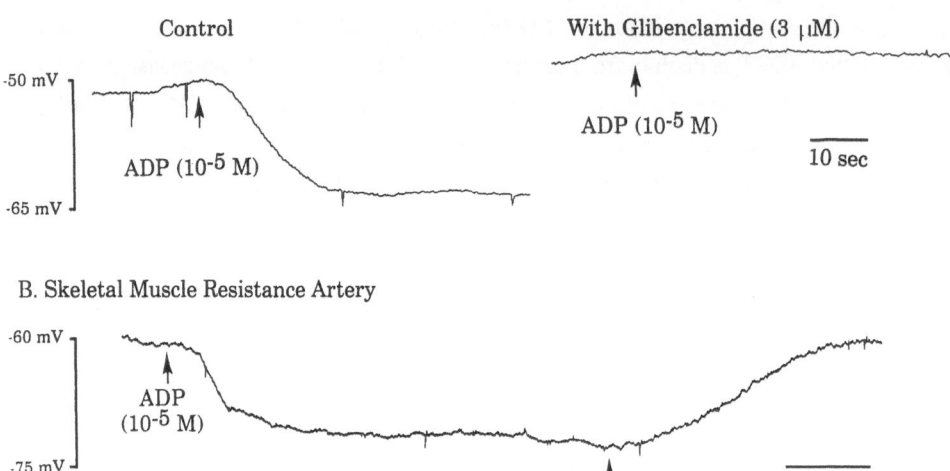

Figure 2. Glibenclamide blocks ADP-induced hyperpolarizations of rabbit resistance arteries. Pre-treatment with glibenclamide prevented the response to ADP in a small mesenteric artery (A). ADP-induced hyperpolarizations were sustained for many minutes in a resistance artery from skeletal muscle and addition of glibenclamide fully reversed the response (B).

In blood vessels from other species glibenclamide does not inhibit endothelium-dependent hyperpolarization, suggesting a role for other K^+-channels in this response (12,13,16). Some evidence indicates a role for large-conductance calcium-activated potassium channels in the hyperpolarizing response (12,13,18). Also, it has been suggested that activation of Na/K ATPase may in some arteries account for the hyperpolarization induced by endothelium-dependent vasodilators (1,5,14), but this effect may be species or tissue dependent as other have seen no effect of ouabain on these responses (12,29).

Substantial differences in the time course of endothelium-dependent hyperpolarization have been reported. For instance, the response in large arteries tends to be transient in most instances, whereas the response in cerebral arteries and resistance arteries is more sustained (5,6). Chen and Suzuki (10) have suggested that the transient nature of the response is related to desensitization of receptors on the endothelial cell for ACh or histamine for instance, rather than desensitization of receptors for EDHF. It has been proposed that a transient hyperpolarization may help initiate endothelium-dependent vasodilation, which is then sustained by other mechanisms (20). In tissues where the hyperpolarizing response is sustained (5,6) such changes in membrane potential could of course help determine steady-state levels of vascular tone.

RELATIONSHIP BETWEEN THE HYPERPOLARIZATION AND VASODILATION

Whereas hyperpolarization of vascular smooth muscle, through its effect on voltage-dependent calcium channels, is a logical and proven mechanism of vasodilation for several substances, this can be difficult to demonstrate directly in some instances. This is particularly so when more than one pathway of dilation is activated, as is the case for endothelium-dependent as well as for other endogenous, endothelium-independent vasodilators. In order to demonstrate the importance of membrane hyperpolarization in these cases, one must find a way to selectively inhibit the hyperpolarization or otherwise dissect out the component of dilation that may be directly mediated by the changes in membrane potential. Several investigators have taken such an approach and have demonstrated a role for hyperpolarization as one of the mechanisms of endothelium-dependent vasodilation.

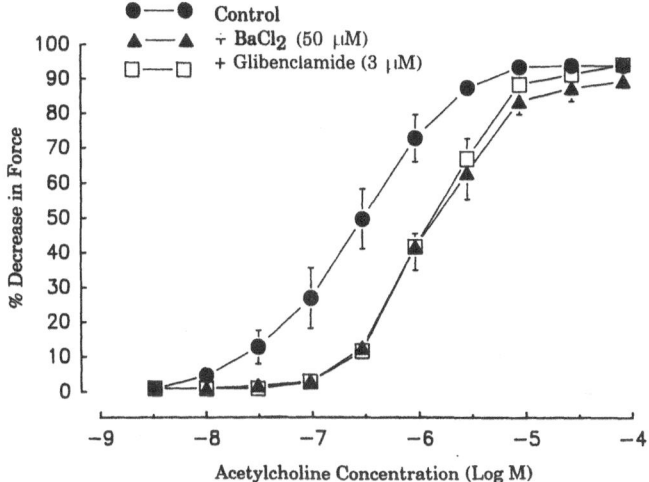

Figure 3. Dose/response curves for ACh-induced vasodilation of middle cerebral arteries isolated from the rabbit. Inhibitors of K_{ATP}-channels barium chloride (50 μM) and glibenclamide (3 μM) caused substantial shifts in the relationship indicating a significant role for membrane hyperpolarization in the vasodilator response (from reference 6).

In the rabbit middle cerebral artery glibenclamide abolishes ACh-induced endothelium dependent hyperpolarization and significantly inhibits the vasodilator response (6) (figure 3).

A combination of glibenclamide and methylene blue nearly abolishes the dilator response (6). These data indicate that multiple pathways are involved in the endothelium-dependent dilation in this artery, and clearly demonstrate an important role for hyperpolarization of the smooth muscle in one of the mechanisms of dilation. Similar observation were made using mesenteric and skeletal muscle resistance arteries from the rabbit (8) in experiments using ADP as the endothelium-dependent vasodilator (Figure 4.)

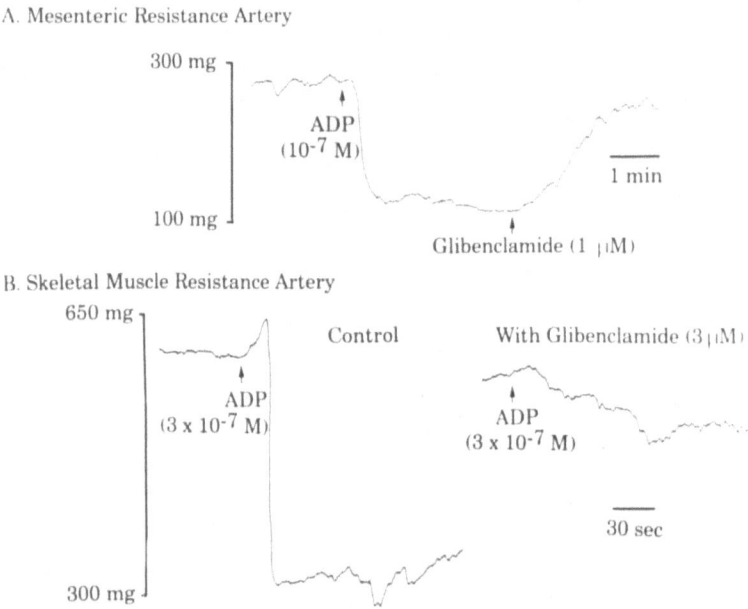

Figure 4. ADP-induced relaxation of a mesenteric resistance artery is reversed upon exposure to glibenclamide (A). Pre-treatment of a resistance artery with glibenclamide significantly inhibited the relaxant effect of a slightly higher concentration of ADP (B). Arteries with resting diameters of 150 to 175 μm were mounted in a resistance artery myograph for recording of force and membrane potential as previously described (8).

Endothelium-dependent responses in some arteries are not inhibited by glibenclamide (12,13,16). In these cases blockers of other potassium channels, in particular substances such as TEA^+ and charybdotoxin which block large-conductance calcium-activated potassium channels, inhibit the hyperpolarizations and dilator responses (12,13). Other K-channel blockers, for instance apamin (1), may also reduce endothelium-dependent dilations, although this has not been a consistent observation (13). Caution should be exercised in interpreting these experiments, however, because in all cases the site (endothelial cell vs. smooth muscle cell) of inhibition by the K^+-channel blockers could not be determined with certainty. If activation of calcium-activated K^+-channels is involved in the release of EDHF or other endothelial factors, as has been suggested (9,22), then blockers of this channel might inhibit the dilator response independently of their effects at the level of the smooth muscle cell. Further experiments using donor/recipient tissue configurations could shed light on this question.

In several arteries the endothelium-dependent hyperpolarization is transient but the relaxation is sustained. Chen and Suzuki (10) found, however, that when the action of EDRF was blocked the relaxation response became transient, matching the pattern of the hyperpolarization and suggesting a close correlation between the membrane potential change and the tissue response under these conditions. Others have found that when isolated arteries are contracted by elevating extracellular potassium, which will clamp the membrane potential at depolarized potentials, no hyperpolarization can be elicited and the vasodilator response under this condition is abolished by inhibitors of EDRF synthesis (23). In the same vessels contracted by agonists such as NE or PGF_{2a}, inhibitors of nitric oxide synthesis reduce the vasodilator response, but do not abolish it. In the canine coronary artery, endothelium-dependent vasodilator responses to bradykinin are largely unaffected by inhibitors of EDRF but are abolished by agents that block the bradykinin-induced hyperpolarization (23) suggesting that nearly all of the response to bradykinin in this artery is mediated by membrane hyperpolarization. Thus, although it is apparent that the contribution of EDHF-induced hyperpolarization to the vasodilation induced by various endothelium-dependent vasodilators can vary significantly among blood vessels and species, membrane hyperpolarization appears to make some contribution to the overall dilation in most instances.

VOLTAGE-DEPENDENCE OF FORCE

Very recent data from the author's laboratory have provided detailed information regarding the quantitative relationship between membrane potential and force in isolated cerebral arteries (7). These observations reinforce the suggestion that very small changes in membrane potential in response to EDHF can be expected to be associated with large changes in force. Cerebral arteries from the rabbit depolarized and contracted in response to exogenous serotonin. In the presence of nitroarginine to block formation of nitric oxide, arteries were exposed to various concentrations of ACh. Membrane potential and force were measured simultaneously and force was plotted as a function of membrane potential. The resulting data were well-described by an exponential function of the form: $F = Fmax/(1 + exp[(Vm-V_{0.5})/k])$, where F is the measured force, Fmax is the maximum force, V is the measured membrane potential, and $V_{0.5}$ is the membrane potential where force is half maximal. The steepness factor (k) of this relationship indicated that force changed by e-fold (2.7 fold) for a 4 mV change in membrane potential. This sensitivity of force to changes in membrane potential in cerebral arteries was somewhat larger than that observed for systemic arteries (25) suggesting additional factors which combine to determine the importance of membrane potential changes in determining vascular tone in different parts of the circulation.

SUMMARY

A factor (or factors) released from the vascular endothelium in response to physiological stimuli hyperpolarizes the underlying vascular smooth muscle cells. Because the voltage-dependence of calcium channels in vascular smooth muscle is steep, even a small hyperpolarization will close some of these calcium channels and force will be reduced. Multiple pathways of endothelium-dependent dilation have been identified (figure 5) and several may be activated by a single dilator agonist. Depending on the particular artery and species from which the arteries are obtained, hyperpolarization may represent a major mechanism of endothelium-dependent vasodilation.

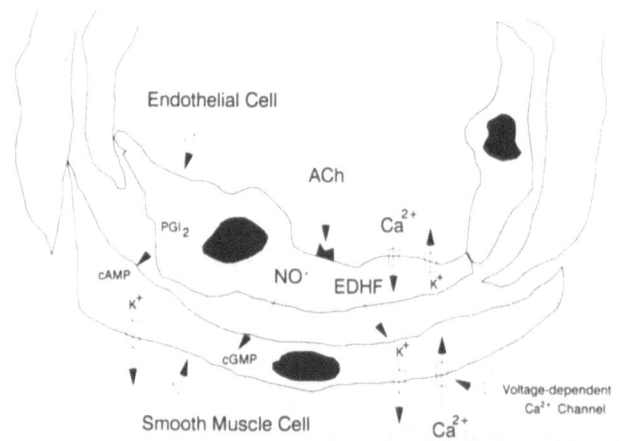

Figure 5. Endothelium-dependent vasodilators such as ACh cause release of multiple factors from the endothelium. Release of these substances is probably due to hyperpolarization of endothelial cells (increased K^+ conductance) and resultant Ca^{2+} influx down a steep electrochemical gradient. EDHF increases K^+ conductances in smooth muscle cells and the resulting hyperpolarization closes voltage-dependent calcium channels.

ACKNOWLEDGEMENTS. This work was supported by NIH grant HL35911. J.E.B. is an Established Investigator of the American Heart Association.

REFERENCES

1. Adeagbo, A.S.O., and K.U. Malik. Endothelium-dependent and BRL 34915-induced vasodilatation in rat isolated perfused mesenteric arteries: role of G-proteins, K^+ and calcium channels. *Br. J. Pharmacol.* 100:427-434, 1990.

2. Beny, J.-L., and P.C. Brunet. Neither nitric oxide nor nitroglycerin accounts for all the characteristics of endothelially mediated vasodilation of pig coronary arteries. *Blood Vessels* 25:308-311, 1988.

3. Bolton, T.B., and L.H. Clapp. Endothelial-dependent relaxant actions of carbachol and substance P in arterial smooth muscle. *Br.J. Pharmacol.* 87:713-723, 1986.

4. Bolton, T.B., R.J.Lang, and T.Takewaki. Mechanism of action of noradrenaline and carbachol on smooth muscle of guinea-pig anterior mesenteric artery. *J. Physiol. Lond.* 351:549-572, 1984.

5. Brayden, J.E., and G. C. Wellman. Endothelium-dependent dilation of feline cerebral arteries: role of membrane potential and cyclic nucleotides. *J. Cereb. Flood Flow Metab.* 9:956-963, 1989.

6. Brayden, J.E. Membrane hyperpolarization is a mechanism of endothelium-dependent cerebral vasodilation. *Am. J. Physiol.* 259:H668-H673, 1990.

7. Brayden, J.E. Quantitative analysis of the role of membrane hyperpolarization as a mechanism of cerebral vasodilation. *Circulation* 84 (Suppl. II):II-406, 1991 (abstract).

8. Brayden, J.E. Hyperpolarization and relaxation of resistance arteries in response to adenosine diphosphate. Distribution and mechanism of action. *Circ. Res.* 69:1415-1420, 1991.

9. Chen, G., and D.W.Cheung. Characterization of acetylcholine-induced membrane hyperpolarization in endothelial cells. *Circ. Res.* 70:257-263, 1992.

10. Chen, G., and H. Suzuki. Some electrical properties of the endothelium-dependent hyperpolarization recorded from rat arterial smooth muscle cells. *J. Physiol. Lond.* 410:91-106, 1989.

11. Chen, G., H. Suzuki, and A.H. Weston. Acetylcholine releases endothelium derived hyperpolarizing factor and EDRF from rat blood vessels. *Br. J. Pharmacol.* 95:1165-1174, 1988.

12. Chen, G., Y. Yamamoto, K. Miwa, and H. Suzuki. Hyperpolarization of arterial smooth muscle induced by endothelial humoral substances. *Am. J. Physiol.* 260:H1888-H1892, 1991.

13. Eckman, D.M., J.D. Frankovich, and K.D. Keef. Comparison of the actions of acetylcholine and BRL 38227 in the guinea-pig coronary artery. *Br. J. Pharmacol.* 106:9-16, 1992.

14. Feletou, M., and P.M. Vanhoutte. Endothelium-derived factor(s) and membrane potential of vascular smooth muscle. *Br. J. Pharmacol.* 93:515-524, 1988.

15. Furchgott, R.F., and J.V. Zawadski. The obligatory role of endothelial cells in the relaxation of arterial smooth muscle by acetylcholine. *Nature Lond.* 288:373-376, 1980.

16. Garland, C.J., and G.L. McPherson. Evidence that nitric oxide does not mediate the hyperpolarization and relaxation to acetylcholine in the rat small mesenteric artery. *Br. J. Pharmacol.* 105:429-435, 1992.

17. Huang, A.H., R. Busse, and E. Bassenge. Endothelium-dependent hyperpolarization of smooth muscle cells in rabbit femoral arteries is not mediated by EDRF (nitric oxide). *Naunyn-Schmiedeberg's Arch. Pharmacol.* 338:438-442, 1988.

18. Kauser, K., W.J. Stekiel, G. Rubanyi, and D. R. Harder. Mechanism of action of EDRF on pressurized arteries: effect on K^+ conductance. *Circ. Res.* 65:199-204, 1989.

19. Kitamura, K, and H. Kuriyama. Effects of acetylcholine on the smooth muscle cell of isolated main coronary artery of the guinea-pig. *J. Physiol. Lond.* 293: 119-133, 1979.

20. Komori, K., and P.M. Vanhoutte. Endothelium-derived hyperpolarizing factor. *Blood Vessels* 27:238-245, 1990.

21. Kuriyama H., and H. Suzuki. The effects of acetylcholine on the membrane and contractile properties of smooth muscle cells of the rabbit superior mesenteric artery. *Br. J. Pharmacol.* 64:493-501, 1978.

22. Mehrke, G., and J. Daut. The electrical responses of cultured coronary endothelial cells to endothelium-dependent vasodilators. *J. Physiol. Lond.* 430:251-272, 1990.

23. Nagao T., and P.M. Vanhoutte. Hyperpolarization as a mechanism for endothelium-dependent relaxations in the porcine coronary artery. *J. Physiol. Lond.* 445:355-367, 1992.

24. Nelson, M.T., J.B. Patlak, J.F. Worley, and N.B. Standen. Calcium channels, potassium channels and the voltage-dependence of force. *Am. J. Physiol.* 259:C3-C18, 1990.

25. Nelson, M.T., N.B. Standen, J.E.Brayden, and J.F. Worley. Noradrenaline contracts arteries by activating voltage-dependent calcium channels. *Nature Lond.* 336:382-385, 1988.

26. Rubanyi, G., and P.M. Vanhoutte. Nature of endothelium-derived relaxing factor: are there two relaxing mediators? *Circ. Res.* 61 (Suppl. 2):61-67, 1987.

27. Siegel, G.F., F. Schnalke, J. Grote, and G. Stock. Membrane physiological mechanisms of vasorelaxation. In: *Resistance Arteries*, Edited by W. Halpern, B.L. Pegram, J.E. Brayden, K. Mackey, M.K. McLaughlin, and G. Osol. Ithaca:Perinatology Press, 1988, p.170-178.

28. Standen, N.B., J.M. Quayle, N.W. Davies, J.E. Brayden, Y. Huang, and M.T. Nelson. Hyperpolarizing vasodilators activate ATP-sensitive potassium channels in arterial smooth muscle. *Science* 245:177-180, 1989.

29. Suzuki, H. The electrogenic Na-K pump does not contribute to endothelium-dependent hyperpolarization in the rabbit ear artery. *Eur. J. Pharmacol.* 156:295-297, 1988.

30. Tare, M., H.C. Parkington, H.A. Coleman, T.O. Neild, and G.J. Dusting. Hyperpolarization and relaxation of arterial smooth muscle caused by nitric oxide derived from the endothelium. *Nature Lond.* 346:69-71, 1990.

31. Vanhoutte, P.M., G.M. Rubanyi, V.M. Miller, and D.S. Houston. Modulation of vascular smooth muscle contraction by the endothelium. *Annu. Rev. Physiol.* 48:307-320, 1986.

THE INTERACTION BETWEEN FLOW-INDUCED VASOCONSTRICTION AND VASODILATION: A MECHANISM FOR SETTING THE LEVEL OF VASCULAR TONE

John A. Bevan

Vermont Center for Vascular Research
Department of Pharmacology
University of Vermont
College of Medicine
Burlington, VT 05405-0068

THE TWO COMPONENTS OF THE FLOW RESPONSE - CONSTRICTION AND DILATION

Although the possibility that changes in blood flow might influence vascular tone was first described more than 40 years ago (36), there are still differences of opinion regarding the nature of the changes and their role in the regulation of the diameter of small blood vessels. In this chapter we will develop the hypothesis first put forward in 1990 (7) that flow initiates two responses, constriction and dilation, which depend on separate cellular mechanisms and that the final manifestation of the response to flow is the consequence of their interaction. Under isometric conditions flow causes contraction, constriction - an increase in wall tone when the smooth muscle cells are in a low state of tone, i.e. when the blood vessel is relatively relaxed. On the otherhand, when wall tone is raised and the blood vessel is more constricted, flow commonly causes relaxation, dilation - the result of an active inhibition of wall tone. The exact definition of "relatively relaxed" and "more constricted" varies depending on factors that are poorly understood. There are some observations that suggest that when wall tone is "excessively" raised, the predominant response to intraluminal flow is contraction. What is excessive, remains at the moment, undefined. It may be that the flow-induced constriction seen at high levels of wall tone simply represents an extension of the contraction component seen at low tone levels but without concomitant masking flow-induced dilation. The level of wall tone when flow-contraction and flow-dilation responses are in balance has been termed - the flow balance or set point.

Ion Flux in Pulmonary Vascular Control, Edited by
E.K. Weir et al., Plenum Press, New York, 1993

Our working hypothesis is that intraluminal flow tends to buffer the consequences of constrictor and dilator influences on the blood vessel wall, modifying their effect, buffering their influence, shifting the level of wall force to the "balance" point for that vessel. Many of the influences that contribute to the tone of small blood vessels are general to at least the major part of the circulation, for example, the myogenic events that follow a rise in arterial pressure, activation of the sympathetic outflow following stress, the presence in the circulation of a vasoactive peptide, etc. A locally effective flow-dependent mechanism provides a means whereby these effects can be locally modified to conform to a varying and often spatially heterogeneous tissue demand.

Most blood vessel segments tested in vitro respond to changes in intramural flow, but the relative and absolute size of the two components seem quite variable even in an anatomically identical artery from different animals. Most of our initial observations were made on resistance branches of the central ear and the middle cerebral arteries of the rabbit. Both these segments exhibit flow-contraction and flow-dilation of comparable size in a reasonably consistent manner. Pulmonary artery segments of similar diameter show some dilation, but only an occasional and small constriction to flow (Table 1) (unpublished data, Bevan and Joyce). Assuming that this is confirmed in a more comprehensive study, this pattern seems to be etiologically correct. It would be functionally disadvantageous if flow elicited a constrictor effect in the only vascular connection between the right and left sides of the heart.

The evidence from functional in vitro studies of vascular segments supporting a flow null or balance point will be briefly presented. The observation that the flow response can change direction with change in wall tone is considered supportive of the null or balance theory. During the initial responses elicited in many of our experiments, a reversal of the flow response related to change in the level of wall force is commonly

Table 1. Summary of Effects of Intraluminal Flow on Various Rabbit Vascular Segments.

Artery	Flow-Contraction	Flow-Dilation
Coronary	+	+
Femoral (muscle branch)	+++	+
Pulmonary	+/0	+*
Pial	+++	+++
Renal	+	+*
Mesenteric	++	+++
Ear	+++	+++
Ear Vein	+++	++

*Sometimes Absent

The effects of intraluminal flow of physiological saline solution through segments of blood vessels from various vascular beds mounted under isometric conditions in a myograph. Flow-contraction represents the response to infusion without tone and flow dilation that when wall force is raised by agonists to 50-70% of maximum obtainable force.

seen. For this reason, most experiments are conducted with the pre-flow tone level distant from the null point.

It should be pointed out that a significant number of arterial segments do not show flow response reversal but only flow-contraction or flow-dilation. The explanation of this is uncertain. It could be that the cellular processes that balance out at the null point interact either above or below the cellular level of activity that is manifest in a change in tone. There is some support for this point of view. Later in this chapter the electrophysiological changes associated with flow-constriction and dilation will be presented. It will be argued that if the membrane potential of the cell is too negative or not sufficiently negative, flow-induced changes in membrane potential will not be translated into changes in tone.

SUMMARY OF EVIDENCE IN SUPPORT OF THE BALANCE POINT

(a) Different levels of wall force were obtained in rabbit ear resistance arteries ($200\mu m$ unstretched diameter) by varying the concentration of NE. At each level the flow response to physiological saline solution (PSS) was obtained. The null point of flow was established by trial and error - by simply manipulating wall tone with NE. Levels of NE-induced tone that before flow deviated from the null by a mean of 35% were reduced to 14%. In this series of five segments, the mean balance point for NE initiated tone was 69% of the maximum achievable tone level; it varied from 32-84% of maximum (7).

(b) Many segments of human pial arteries obtained during neurosurgery - ranging in diameter from 250 to $680\mu m$ from patients in the age range 15 to 73 years, exhibited flow reversal. In some segments, changes in tone were effected by agonists such as $PGF_{2\alpha}$ but others exhibited variation in spontaneous, presumably myogenic activity. Flow effects were elicited at different levels of wall force. An approximation of the flow null point was obtained by determining the regression of the level of the equilibrium flow response - both constrictor and dilator against wall tone. The null was considered the level of wall tone when the flow response as indicated by the line was zero. In a series of 9 human pial arteries, this was $13.3 \pm 13.2\%$ (n=8) tissue maximum response (11).

(c) It is possible that in an experiment in which the null point is determined primarily by varying flow contraction, the level of wall tone when flow no longer causes contraction, represents a "ceiling" response effect. That this possibility is unlikely is indicated by the contraction to flow that occurs when wall tone is raised by passive stretch to the level when flow has no effect. This demonstrates that the vessel (Figure 1) is capable of responding to flow above the wall force level indicated by the original null. The artery is also able to dilate beyond the null response level obtained by flow-dilation experiments when wall force is altered by passive adjustment.

(d) There is further evidence that flow initiates two opposite responses, i.e. that the overall response is bidirectional and that these two components can be independently manipulated. Although not as persuasive as those experiments in which the null is established experimentally, the existence of two opposite responses that can be influenced independently, is considered strong evidence.

(i) Typically, the rate of tone recovery from flow-contraction is much faster than for flow-dilation. Consequently, on many occasions upon cessation of flow made when wall tone is close to the balance point, there is an immediate, rapid relaxation when tone falls below the preflow level - representing presumably recovery from flow-constriction, which is then followed by a phase of slower contraction towards the preflow level. This latter phase is interpreted as recovery from flow-dilation (7).

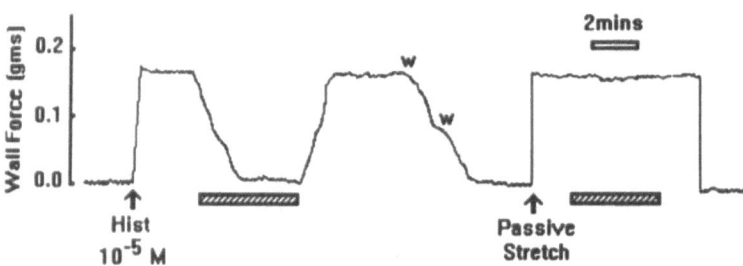

Figure 1: Part of a tracing of wall force of a resistance branch of the rabbit central ear artery. After tone was raised by histamine (10^{-5}M) intraluminal infusion of physiological salt solution (hatched bar) also containing histamine caused dilation. After washing, when wall force was increased by passive stretch, flow caused no change in wall force.

(ii) Although both flow-contraction and dilation are both inhibited by the strictly zero calcium conditions achieved by the addition of EGTA to the zero calcium PSS, only flow-dilation is inhibited by "nominal" zero calcium PSS conditions - exclusion of calcium from the PSS. Exposure of small arteries to nominal zero PSS calcium conditions results in the reversal of flow-dilation to flow-contraction (1).

(iii) In the myogenically active rabbit facial vein - a vessel where flow-dilation only occurs when the endothelium is intact, a dilator response to flow is reversed to constriction by nitric oxide synthase inhibitors such as LNNA (25).

(iv) Flow dilation is diminished or inhibited by methylene blue, reflecting presumably the dependency of flow-dilation on guanylate cyclase activity. Exposure of isolated vascular segments to methylene blue under certain circumstances can reverse flow-dilation to constriction (25).

(v) Contractions and dilations resulting from intraluminal infusion made when wall tone is close to but not identical with the balance point are often associated with periods of vascular smooth muscle depolarization followed by hyperpolarization or hyperpolarization followed by depolarization respectively (see below). We interpret this as reflecting the interaction between the membrane potential changes associated with the two opposing but simultaneously elicited flow-induced responses. The electrophysiological data clearly indicate that the depolarizing and hyperpolarizing events are independent and separate (40).

(vi) The effect of chronic sympathetic and also sensory denervation on flow-induced changes in smooth muscle tone has been examined under in vitro conditions in a branch of the rabbit ear artery (200-400μm I.D.). In the denervated compared with the contralateral innervated segments, flow-induced contraction was increased. This is an effect that is reminiscent of non-specific, non-deviational hypersensitivity of denervated vascular smooth muscle to constrictor influences. By contrast there was a significantly diminished flow-related dilation in the denervated arterial segment. In the segments examined, an intraluminal flow rate could be found that resulted in dilation in the innervated, but was without effect in the matched, denervated tissue. Endothelial mediated dilation to acetylcholine was also impaired by chronic sympathectomy.

THE SODIUM SENSITIVITY OF FLOW-INDUCED TONE CHANGE

Upon reduction in sodium from the level found normally in PSS (145 down to 119mM) by substitution of NaCl by either NMDG or sucrose (6,10), there occurs a selective and comparable diminution in flow contraction and relaxation (Figures 2a,b). The changes in flow are most likely due to a reduction in sodium, as one substitute is the chloride salt of NMDG. Because the consequences of both substitutions - that of NMDG and sucrose cannot be distinguished from each other, it is unlikely that the alterations are associated with the addition of the substitute molecule. The solutions used had the same osmolality as control. A sodium reduction of 10 and 20% did not influence the contraction of the artery segment to NE, nor its relaxation to acetylcholine or papaverine. These are endothelial-dependent and endothelium-independent vasodilator agents, respectively. NMDG and sucrose are commonly used substitutes for Na in a variety of functional and biochemical studies of vascular tissue (see 5). The finding that both flow contraction and relaxation are reduced to the same extent by Na^+ reduction, leading to relaxation and contraction respectively, suggests that the sodium sensitive mechanism may be found at a site common to both flow effects. One possibility is a flow-sensor.

It is of interest to speculate regarding the physiological relevance of these findings, specifically the consequence of a reduction in extracellular Na on flow-induced tone in vivo. We have proposed that the response to flow represents the interaction of both contractile and relaxant processes (7). These together tend to modify wall force bringing it toward an intermediate set level. If other factors remain constant, it seems reasonable to conclude that the relative size of the two components of the flow response would not change. Rather the effectiveness of the flow mechanisms modifying wall force would be lessened.

Sodium-related ionic mechanisms that are considered to influence vascular smooth muscle tone are relatively insensitive to changes in extracellular sodium. A few examples will be briefly mentioned.

(i) **Sodium-Calcium Exchange**: This mechanism has been assessed in a variety of tissue preparations. Blaustein et al (12) (Fig. 3a) have documented changes in rates of contraction and relaxation of the rat aorta with alteration in extracellular Na^+. Contraction rate decreases and relaxation increases as sodium concentration is increased experimentally up to physiological levels. For both processes the maximum rate of change occurs at approximately 25mM. There was little alteration in these responses with changes in Na^+ concentrations around the normal level found in PSS. Similar changes in extracellular sodium had little effect on cytoplasmic free calcium in primary and passaged cultures of rat aortic muscle cells, whether or not they had been treated with ouabain, nor on their calcium content (30) (Fig. 3b). The latter observations are consistent with the reported absence of change in the size of norepinephrine-induced contraction of small artery segments in vitro with change in extracellular Na^+ (5).

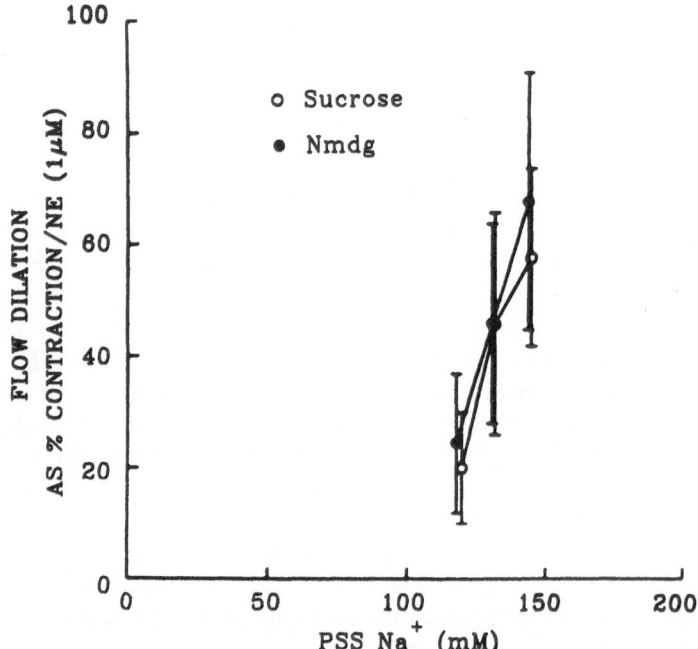

Figure 2a: Relationship between flow-induced contraction of resistance artery branches of rabbit central ear artery and concentration of Na in physiological saline solution (PSS). Responses are expressed as percentage of tone increase to norepinephrine (NE, $10^{-6}M$). Relaxation was elicited in arteries precontracted to this concentration of NE. Na in PSS was changed by substitution of NaCl either by sucrose or NMDG. (See reference 5).

(ii) **Sodium-Hydrogen Exchange:** In a number of studies that employed a variety of measurements considered to reflect sodium-hydrogen exchange activity, the curves relating these to external sodium concentration show little or no alteration over the range of 100-150 mM (31). Data from a study of the rate of pH recovery of human smooth muscle cells after acid load is shown (Fig. 3c).

(iii) **Sodium, Potassium and Chloride-Cotransport:** Potassium influx into cultured human fibroblast cells changes relatively little with alterations in external sodium greater than 100mM. The concentration of sodium at which the rate of change was half-maximum was 40-60mM (32) (Fig. 3d).

A number of other studies of the changes that occur in various parameters with alteration in external sodium concentrations have been summarized (2). These include - agonist relaxation rates, reduction in tone with increases in calcium concentration above those found in PSS and with decreased potassium concentration, calcium uptake into sarcoplasmic vesicles, calcium efflux initiated by angiotensin, cytoplasmic free intracellular calcium concentration, total calcium content, intracellular sodium concentration, angiotensin-induced calcium efflux, potassium induced relaxation after a prior potassium free state and potassium stimulated sodium efflux (2).

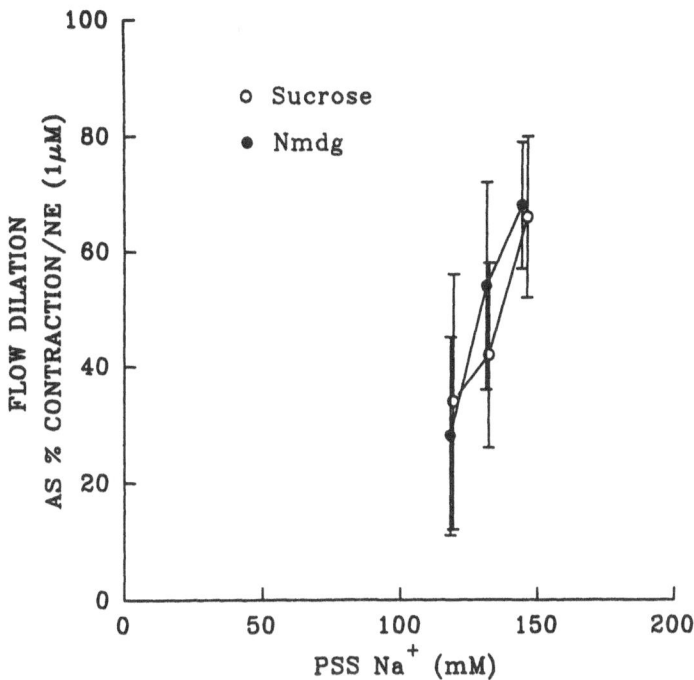

Figure 2b: Relationship between flow-induced dilation of resistance artery branches of rabbit central ear artery and concentration of Na in physiological saline solution (PSS). Responses are expressed as percentage of tone increase to norepinephrine (NE, 10^{-6}M). Relaxation was elicited in arteries precontracted to this concentration of NE. Na in PSS was changed by substitution of NaCl either by sucrose or NMDG. (See reference 5).

SITE OF THE SODIUM SENSITIVE FLOW-DEPENDENT MECHANISM

The data cited above suggest that the small changes in extracellular sodium of the order that influences flow, do not significantly effect intravesicular and transplasmalemmal processes considered to influence vascular wall force. If this is true, then the location of the sodium sensitive effect could be located extracellularly - outside the cell, at least external to lipid cell membrane.

The majority of sodium in the blood vessel wall - variously estimated to be 70-90% is extracellular (37). Friedman (17) adapted ion exchange methods to quantitatively estimate sodium distribution and found that 65-70% of the wall sodium is extracellular and probably bound to glycosaminoglycans (GAG). Siegel described four sites of location of sodium - in the extracellular fluid, two distinct extracellular binding sites and a rapid exchanging fraction that may represent sodium in pinocytic vesicles. Sodium, calcium and other cations are mostly bound to extracellular polyionic cations - the glycosaminoglycans in the vessel wall that form an enveloping sheath around each cell. This sheath covering has connections through the cell membrane to the cytoskeleton and with the extracellular supporting vascular matrix. Sodium competes with other monovalent and divalent ions for sites on the sulphated and carboxylated polysaccharides that are intrinsic components of the polyanionic macromolecules.

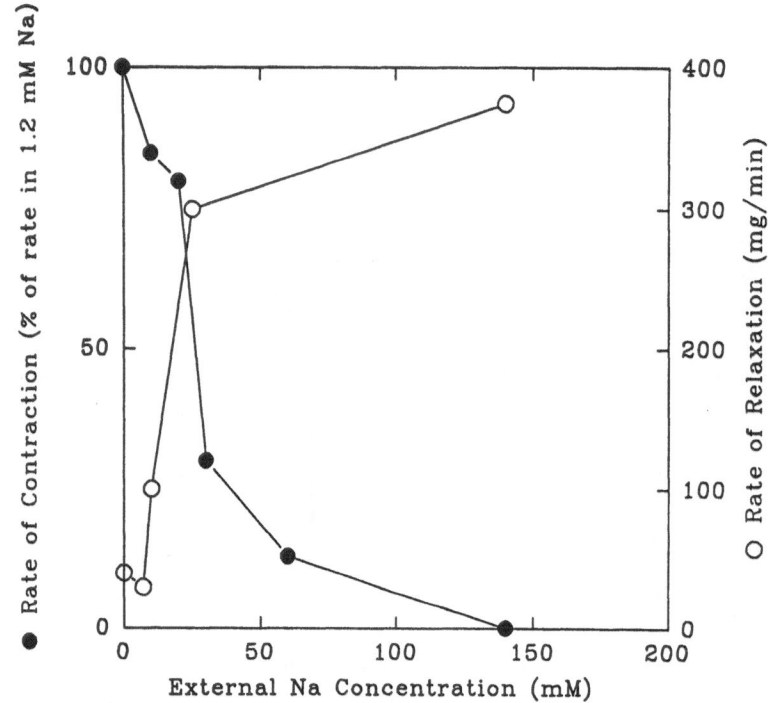

Figure 3a: Relationship between $[Na^+]_o$ and the relative rate of contraction (●) and relaxation (o) in rat aorta. (See reference 12).

Studies of tissue water conductivity show that these glycosaminoglycan molecular systems, though they form only a relatively small percentage by weight of the extracellular matrix contribute to the low value found in this type of tissue (27) presumably as a result of their association with the collagen framework. Friedman graphically describes the blood vessel wall as containing "smooth muscle cells... laced together by a network of collagen fibers floating in a gel-like polysaccharide sac - presumably with a higher sodium content."

Changes in vascular wall sodium have commonly been discussed in the context of hypertension. An increase in sodium would be associated with increased cellular water content. Tobian and Binion (39) described their excessive presence in hypertension as "water logging." Such an effect would most likely result in changes in arterial cellular and mechanical properties and so influence the cellular response to physical stimuli and to agonists. Changes in free sodium in the paracellular matrix could alter the transmembrane sodium gradient and thus alter reactivity to tone-producing agents and to pressure and flow (shear stress). In the context of this discussion, we think it likely that changes of sodium in the glycocalyx could change responsiveness of the vascular wall to flow, and indeed to pressure. Such effects might lead to secondary changes in response to vasoactive substances.

Manning (29) has elaborated an explanation of the electrostatic interaction of cations with anionic polysaccharides. Four phases of condensed counterions on the micrion unit are envisaged. The first represents cation binding to specific ion sites. The next outer zone, of microion-microion interactions, when condensed counterions have little or no translational mobility relative to the polyelectrolyte; a highly residual field due to the net charge of the polyelectrolyte modifying the behavior of "free ions" and the

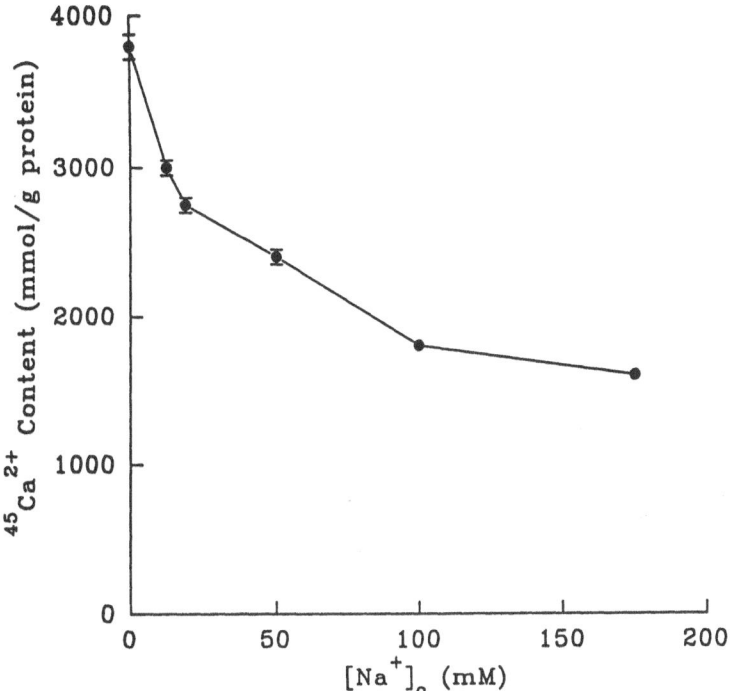

Figure 3b: Effect of varying $[Na^+]_o$ on $^{45}Ca^{2+}$ uptake and constant into monolayers of cultured rat aortic vascular smooth muscle cells exposed to varying $[Na^+]_o$. Each point represents the mean ± SEM of seven determinations. (See reference 30).

outermost zone - some modification of interaction between mobile ions. Thus, with changes in sodium ion concentration not only are the concentrations or binding of counterions - sodium (and calcium) changed, but the consequences of this extends for varying distances beyond the polyionic macromolecule.

Gustavsson et al (19) have investigated the interaction of sodium and calcium and other cations with glycosaminoglycans. Using NMR techniques to study specific polyanionic proteoglycans, they found that calcium can cause conformational changes in these molecules resulting in changes in the mobility of the polysaccharide chains, possibly due to intra- or intermolecular cross-linkage. This change in turn has an effect on the binding of sodium. Thus, calcium is involved in more than a competitive

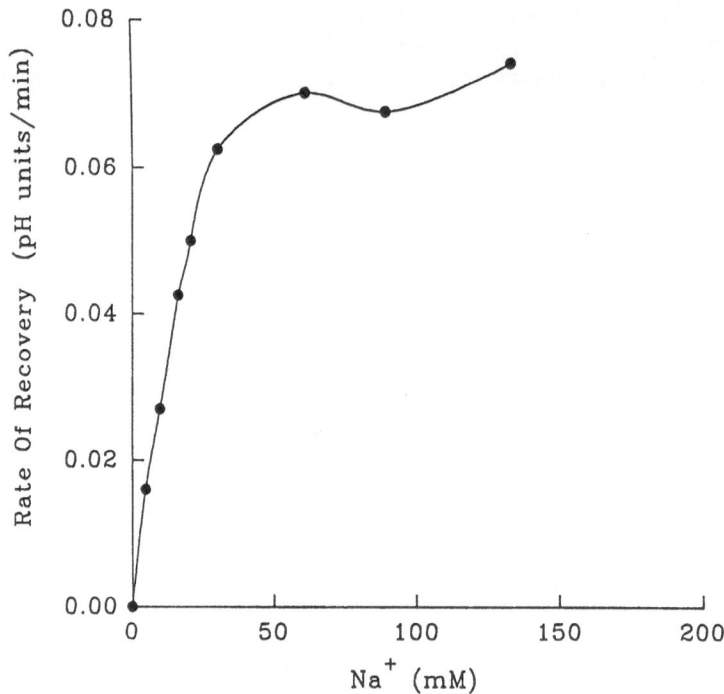

Figure 3c: Dependence of the recovery of human smooth muscle cells from 15mM NH_4Cl-induced intracellular acidosis in CO_2/HCO^-_3-buffered physiological salt solution (PSS) (pH 7.4) on extracellular Na^+ concentration. (See reference 31).

interaction with sodium at the primary binding site, but may influence the conformation of the polysaccharide complex, with secondary effects on ionic binding.

THE FLOW-SENSOR

In 1991 Bevan and Siegel (9) proposed that glycosaminoglycan molecules may represent the essential element of the flow-sensor in the blood vessel wall. This proposal was based on a number of parallel observations by Siegel and colleagues (37 and prior references), on the characteristics of sodium and calcium binding to multi-chain peptidoglycans using NMR, and our own observations on the ionic dependency of flow-induced changes in tone in resistance arteries. These considerations are briefly summarized below:

(i) Flow-induced constriction and dilation are both uniquely dependent on extracellular sodium, suggesting that the sodium-dependent site is associated with a pathway common to the two flow effects. The majority of blood vessel sodium is extracellular, bound to polyanionic mucopolysaccharides and associated cell supporting systems. It is the main counterion. Other sodium-dependent systems in vascular smooth muscle do not share this sensitivity.

(ii) Flow-induced changes can be elicited not only during intraluminal flow in the intact artery, but also after endothelium-removal and by flow over the external artery surface. Thus, "flow-sensing" is not limited to any one cellular layer. This conclusion is consistent with a distribution of the sensing system to the entire extracellular vascular wall matrix.

(iii) Enzymes that disrupt components of the extracellular matrix influence the flow effect. Pohl et al (33) have demonstrated this in the intact coronary artery bed of the dog using neuraminidase.

(iv) Flow-induced changes in tone appear to depend on the sodium/calcium ratio. The response to flow does not reflect the antagonism expected between these ions for binding sites. Over comparable concentration ranges, there is cooperativity in the binding between sodium and calcium on proteoglycan systems.

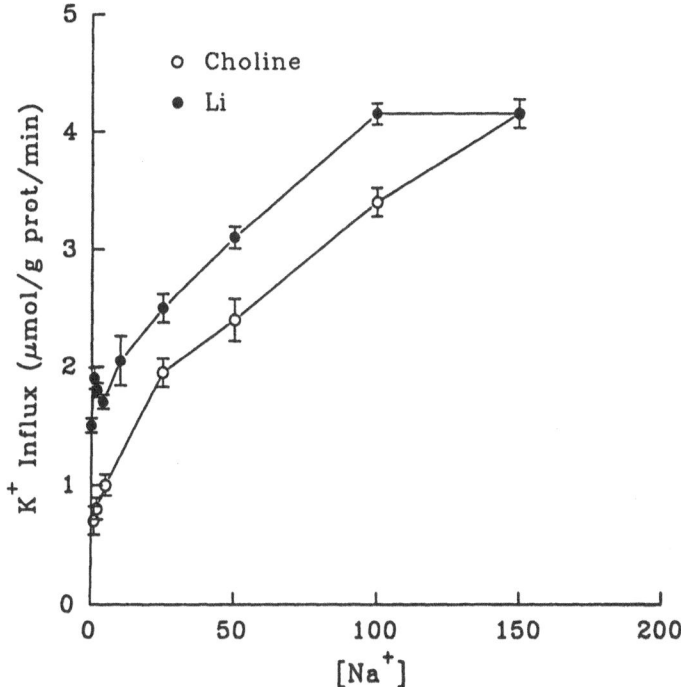

Figure 3d: K$^+$ titration curve. Following serum deprivation, HSWP cells were preincubated in a Hepes-buffered medium containing 20 μM digitoxin and the designated K$^+$ concentration for 5 min. Cells were then assayed in identical medium containing 1 Ci/ml ^{86}Rb$^+$ for 5 min. Values represent mean ± S.E. of quadruplicate determinations from four separate experiments. (See reference 32).

(v) During NMR studies of ion binding to these polyanionic mucopolysaccharides carried out at constant calcium concentrations, shaking of the solution - resulting in fluid movement and shear stress of the suspended molecules, results in a decrease in sodium binding.

Our hypothesis is that flow through hydraulic drag or shear, deforms the extracellular elements in the vascular wall. When the endothelium is intact this results in the release of EDRF (15) and possibly other substances, including prostanoids (26) from the endothelium. In addition, the shear effect is extended in part, at any rate, mechanically to the underlying cellular layers in the vascular wall. The same mechanism is operative when the stimulus is applied to the external surface of the artery wall. Shear or mechanical stress causes conformational changes in the glycosaminoglycans by stretching or extending them from a randomly coiled aggregated state to a more elongated condition along the line of flow. This elongation results in exposure of an increased number of cationic binding sites on the GAG molecule leading to increased sodium binding. The extent of the conformation change is influenced by the concentration of calcium, an ion that not only competes with sodium at specific binding sites, but cross links the polysaccharide chains of the protein saccharide complex. These complex interactions might account for the uncharacteristic cooperative, interaction of sodium and calcium over the physiological concentration range. The amount of sodium binding is influenced by changes in external sodium concentration which is reflected in the sodium sensitivity of the flow response. By mechanisms not yet identified, these changes in sodium binding signal the level of shear stress to the cell. This possibility includes changes in the ionic field around the polyions, in the local concentration of electrolytes in the intracellular clefts and conformation changes in the intracellular components of integral GAGs. We have proposed that in part, at any rate, this may be related to the trans-cell membrane sodium concentration gradient (6).

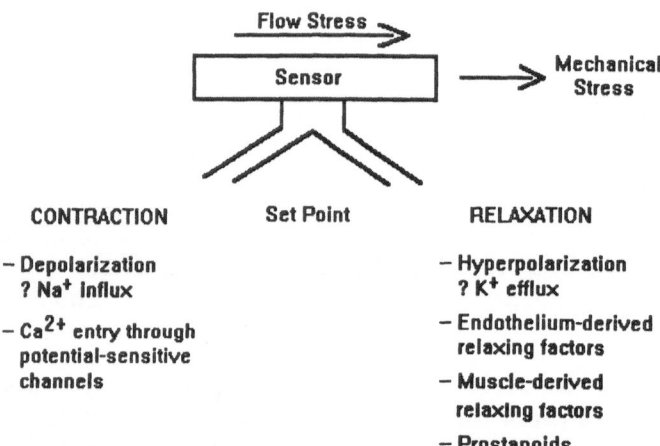

Figure 4: Summary of hypothesis of flow-induced changes in vascular tone (for details see text).

THE COUPLING MECHANISM

Our hypothesis - embodied in the model (Figure 4) is that flow causes deformation of specific surface molecules in the glycocalyx and/or the extracellular matrix either through shear stress, and/or by mechanical stretch or deformation mediated through the intracellular cytoskeleton and its connections with the extracellular matrix.

As a result there is an increased entry and possibly cycling of cations into vascular smooth muscle cells. The most likely candidates are Na^+ and Ca^{2+}. Flow causes its effects through conformational changes in anionic biopolyelectrolytes, exposing previously masked ionic sites allowing an increased sodium binding. The extent of conformational change would be sensitive to the extracellular concentration of calcium, that not only competes with sodium at specific binding sites on the polyanionic molecule, but which also stabilizes the macromolecule through cross-linkage (29). When the glycosaminoglycan is an integral protein these changes might result in alteration in the conformation of its intracellular component (24,35). Changes in sodium binding might also be expected to be associated with alteration in local electrical field (14,35) and in cationic concentrations in the fine interstitium in and around the supporting matrix elements close to the smooth muscle cell plasmalemma (38). It is possible that these mechanisms might alter the properties of the vascular smooth muscle cell cytoskeleton leading to alteration in cellular function. Whatever the local mechanism, the resultant change in tone presumably follows a change in ionic mechanisms, probably the cycling of cations through cells.

The hypothesis that flow effects are dependent at least to some extent on sodium entry derives from observations that both flow-contraction and dilation are inhibited by monensin and by long term exposure (> 2 hour) to ouabain - both conditions that elevate intracellular sodium, leading presumably to a decrease in Na influx. This conclusion is consistent with the inhibitory effect of amiloride (ID_{50} 3 x 10^{-5}M) on both flow effects (5,6,10). Increased $^{22}Na^+$ entry into the blood vessel wall during flow has been recently observed (Henrion et al, unpublished observations). Rosati and Garey (34) measured an increase in sodium content of A10 vascular smooth muscle cells subjected to turbulent flow in a culture chamber. This uptake was prevented by the calcium channel blocker, nitrendipine.

The role and mechanism of calcium in coupling, if any, is not known. One possibility is that sodium entry causes smooth muscle cell membrane depolarization and that as a consequence Ca^{2+} enters through voltage gated Ca^{2+} channels. Whether calcium enters the cell through an additional path such as a non-specific cationic channel as part of the flow activation process contributing to cellular depolarization is not known.

FLOW-CONTRACTION

The cellular basis of flow-contraction has not been established. What is known is that vascular smooth muscle cells with a membrane potential more negative than -58mV, usually depolarize during flow contraction (40). During contraction there is increased calcium entry and retention by the vessel wall that is proportional to the magnitude of the contraction (22). The magnitude of the calcium uptake per unit increase in force is similar to that associated with agonist-induced contraction. As flow and high potassium contractions are effected quantitatively the same by diltiazem, nifedipine and verapamil and have a pharmacological profile different from myogenic (stretch) and agonist-induced tone (41), calcium entry is probably through voltage-gated Ca^{2+} channels. Although a variety of endothelium-derived contracting factors have been proposed (28), as flow-contraction is unchanged by endothelium removal, these are not responsible for flow-initiated tone.

FLOW-DILATION

Flow-dilation remains somewhat diminished after endothelium removal. The endothelium-dependent contribution in most blood vessels appears to be due to EDRF (see 1,3 for references). In some small arteries, flow effects are mediated by prostanoids. The endothelium-independent component, a sizeable part of flow-dilation, at least in some vascular beds - such as the brain, observed in arteries in vitro probably results from the elaboration of an unidentified endogenous muscle-derived relaxing factor (MDRF) that activates cyclic GMP (18).

FLOW-INDUCED MODULATION OF OTHER SOURCES OF TONE

In a myograph mounted arterial preparation, the preflow level of the membrane potential appears to determine the direction of the flow response in animal pial arteries. When more negative than -58mV, contraction ensues upon flow and when less negative, dilation (40). The reported vascular smooth muscle cell membrane potential range in vitro is -60 to -75mV (23); although recently, somewhat less negative values have been reported. In vivo data, although scarce, suggest that membrane potentials are 10-25mV less negative than in vitro. On the basis of these figures, it would be predicted that in vivo, flow-dilation would be the commonly observed response - as almost invariably seems to be the case (see 3 for summary).

The origins or bases of the basal tone of the resistance vasculature are multiple. Myogenic (stretch-induced) tone contributes significantly and, it is speculated, is associated with calcium entry through stretch-dependent potential-sensitive channels (8). Harder et al (20) subjected the middle cerebral artery of the cat to increasing pressure over 40-120mmHg. E_m decreased from -53.06 ± 2.7 to -22.6 ± 1.4mV. At the equivalent of physiological pressures (80mmHg) the mean membrane potential in this artery was of the order of -35 to -40mV. Brayden and Wellman (13) found a resting membrane potential of -63mV in feline cerebral arteries of 400μm O.D. At 50 mmHg, the membrane potential averaged -48mV. In a rabbit pial artery flow-contraction is associated with depolarization (40). Thus, the level of the membrane potential in vivo will be influenced by the intramural pressure and the rate of the intraluminal flow. The membrane potential will also be modified by endothelial-derived factors, EDHF is a case in point, by neurotransmitters, and by locally manufactured and systemic circulating vasoactive substances.

It has been frequently stated that an important drawback to the idea that myogenic tone is the main determinant of basal tone is that it represents an open positive feedback loop and therefore cannot be solely responsible for regulation. This criticism does not apply to flow-regulation, because it induces opposing responses that interact with each other and tend to move wall tone to an intermediate tone level.

Upstream changes in diameter will alter flow and pressure in the distal branching system which then will cause changes in diameter. These changes will in turn will modify the effect of other constrictor or dilator influences acting on the distal vessels. Thus, it might be argued that pressure and flow are the final common vascular effectors. Such an arrangement permits the consequences of changes in flow (and pressure) to be distributed and prioritized according to local tissue need through a vascular distribution system that is not homogeneous. Such a concept has been elaborated for the cerebral circulation (4).

There has been some speculation regarding the mechanism of integration and coordination of the responses of successive branches of the vascular bed to changes in flow and in the spatial distribution of tissue demand. Likely candidates to integrate responses are flow, complemented by pressure. The circulation must ideally provide sufficient blood flow to match local tissue need. It is reasonable that flow itself is an important, if not, the primary regulating influence.

COMPARABLE SODIUM SENSITIVITY OF FLOW-INDUCED TONE AND ARTERIAL PRESSURE

Friedman has long considered that the cellular transmembrane distribution of sodium is a major determinant of blood pressure. In a recent study (16) he both increased and decreased plasma sodium in mice by intraperitoneal dialysis of various solutions. When corrected for changes in osmotic pressure, increases in plasma sodium resulted in increases in blood pressure, both systolic and diastolic (Fig. 5a).

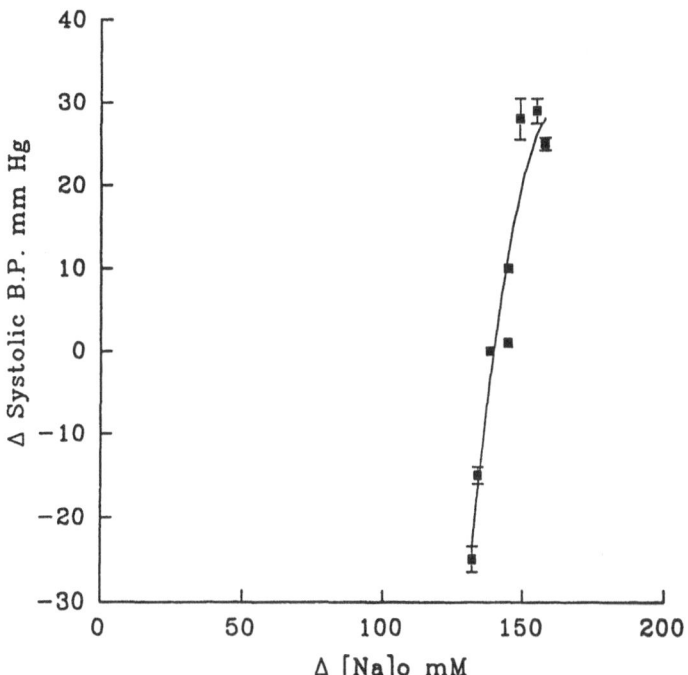

Figure 5a: Change in systolic blood pressure (B.P.) at 2h of intraperitoneal dialysis, plotted as a function of extracellular sodium concentration ($[Na]_o$). Each point represents the average of eight rats, and s.e.m. is indicated by verticals. Redrawn assuming normal values $[Na]_o$ of 142mM. (See reference 16).

Lowering sodium had an opposite effect. The change of pressure paralleled the change in plasma sodium. It is too simplistic to attempt to relate these findings to the sodium-sensitivity of flow-induced changes of tone. Possibly, the most arresting feature of Friedman's results is the <u>sensitivity</u> of the vascular response to changes in sodium. Significant changes in pressure are associated with relatively modest changes in plasma sodium. The sodium concentration difference across the cell membranes increased as external sodium was raised and decreased as it was lowered (Fig. 5b).

Our data shows that at least under the conditions of our experiments the constrictor and dilator components of flow are equally sensitive to sodium and that flow tends to nudge the level of wall tone towards an intermediate level or set point. A change in sodium might be expected to alter the effectiveness with which flow does this. Thus, it could be that under some circumstances, there would be no change in flow as plasma sodium is altered; in other situations a change in sodium would cause both

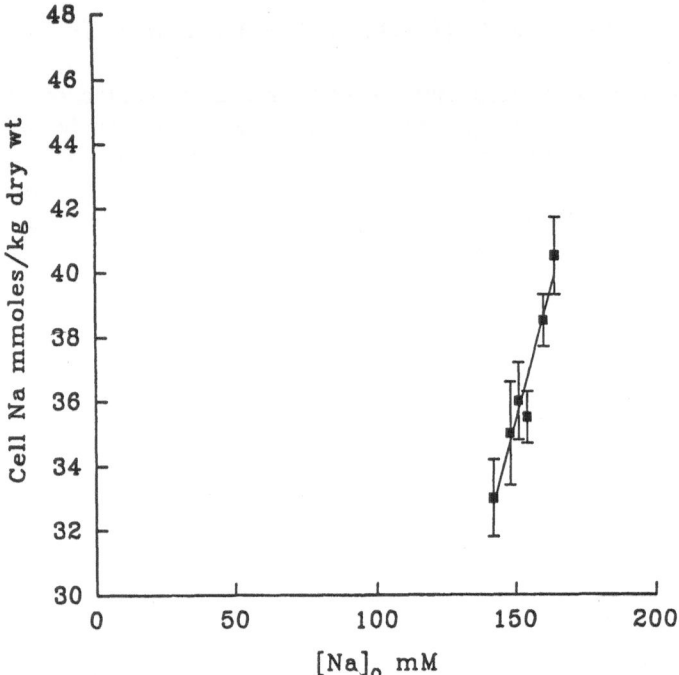

Figure 5b: The relation of stepwise increases in extracellular Na concentration ($[Na]_0$) to cell Na, measured in tail arteries rapidly excised 2h after the start of dialysis with high-Na solutions. Each point represents the average of 10 arteries, and s.e.m. is indicated by verticals. (See reference 16).

vasoconstriction or vasodilation. Friedman argues that *in vivo* there are a complex group of homeostatic mechanisms serving to maintain extracellular sodium relatively constant.

Thus, if the sodium gradient is to change it must be the result of alterations in intracellular sodium, which might follow from changes in sodium permeability and/or sodium pump activity. There is the possibility that changes in extracellular sodium concentration possibly mediated through changes in surface charge can affect vascular smooth muscle tone. Preliminary experiments show that myogenic activity is also exquisitely sensitive to extracellular sodium (21). A decrease in sodium increases myogenic tone and vice versa. Thus, there are several mechanisms that significantly contribute to the regulation of arterial pressure which can be modulated by sodium concentration change.

REFERENCES

1. Bevan, J.A. Flow-induced contraction and relaxation: their role in the regulation of vascular tone. Results of an *in vitro* study. In: Flow Dependent Regulation of Vascular Function in Health and Disease, Editors J. Bevan, G. Kaley and G. Rubanyi, 1993 In Press.

2. Bevan, J.A., and E.H. Joyce. Calcium dependence of flow-induced dilation: cooperative interaction with sodium. Hypertension (1993, In press).

3. Bevan, J.A. Flow-dependent vascular tone. In: The Resistance Vasculature, Editors J.A. Bevan, W. Halpern, and M.J. Mulvany, Clifton, NJ, Humana Press, 1991, pp. 169-191.

4. Bevan, J.A., and R.D. Bevan. Is the innervation of the cerebral circulation a primary regulator of cerebral blood flow. News In Physiological Sciences (1993, In Press).

5. Bevan, J.A., and E.H. Joyce. Comparable sensitivity of flow contraction and relaxation to Na reduction may reflect flow-sensor characteristics. Am. J. Physiol. 263(Heart Circ. Physiol. 32): H182-H187, 1992.

6. Bevan, J.A., and E.H. Joyce. Flow-induced relaxation in a resistance artery is associated with sodium dependent mechanism in vascular smooth muscle. J. Vasc. Med. Biol. 2: 281-288, 1990.

7. Bevan, J.A., and E.H. Joyce. Flow-induced resistance artery tone: balance between constrictor and dilator mechanisms. Am. J. Physiol. 258(Heart Circ. Physiol. 27): H663-H668, 1990.

8. Bevan, J.A., and I. Laher. Pressure and flow-dependent vascular tone. FASEB J. 5: 2267-2273, 1991.

9. Bevan, J.A., and G. Siegel. Blood vessel wall matrix flow-sensor; support and speculation. Blood Vessels 28: 552-556, 1991.

10. Bevan, J.A., G.C. Wellman, and E.H. Joyce. Flow-induced constriction of rabbit resistance artery is sodium-dependent. Blood Vessels 27: 369-372, 1990.

11. Bevan, J.A.: Myogenic (stretch-induced) and flow-regulated tone of human pial arteries. In: The Human Brain Circulation, Editors J.A. Bevan and R.D. Bevan, Humana Press, NJ, 1993, In Press.

12. Blaustein, M.P., T. Ashida, W.F. Goldman, W.G. Wier, and J.J. Hamlyn. Sodium/calcium exchange in vascular smooth muscle: a link between sodium metabolism and hypertension. Ann. N.Y. Acad. Sci. 488: 199-216, 1987.

13. Brayden, J.E., and G.C. Wellman. Endothelium dependent dilation of feline cerebral arteries: role of membrane potential and cyclic nucleotides. J. Cereb. Blood Flow Metabol. 9: 256-263, 1989.

14. Comper, W.D., and T.C. Laurent. Physiological function of connective tissue polysaccharides. Physiol. Rev. 58(1): 255-303, 1978.

15. Cooke, J.P., J.S. Stamler, N. Andon, P.R. Davies, and J. Loscalzo. Flow stimulates endothelial cells to release a nitrovasodilator that is potentiated by reduced thiol. Am. J. Physiol. 28(Heart Circ. Physiol.): H804-H812, 1990.

16. Friedman, S.M., McIndoe, R.A., Tanaka, M. The relation of blood sodium concentration to blood pressure in the rat. J. Hypertension 8: 61-66, 1990.

17. Friedman, S.M. Sodium ions and regulation of vascular tone. Adv. Microcirc. 11; 20-42, 1982.

18. Gaw, A.J., and J.A. Bevan. Mechanisms of the flow-induced relaxation of the rabbit middle cerebral artery. In: Resistance Arteries: Structure and Function, Editors M.J. Mulvany, M.J. C. Aalkjaer,

A.M. Heagerty, N.C.B. Nyborg, and S. Strandgaard, The Netherlands, Elsevier Sci. Publ. B.V., 1992, pp. 20-23.

19. Gustavsson, H., G. Siegel, B. Lindman, and L.-A. Fransson. $^{23}Na^+$-NMR studies of cation binding to multi-chain glycosaminoglycan peptides. Biochim. et Biophys. Acta 677: 23-31, 1981.

20. Harder, D.R., C. Sanchez-Ferrer, K. Kauser, W.J. Stekiel, and G.M. Rubanyi. Pressure releases a transferable endothelial contractile factor in cat cerebral arteries. Circ. Res. 65: 193-198, 1989.

21. Henrion, D., A. Klaasen, and J.A. Bevan. Extracellular sodium modulates flow and stretch-induced constriction in opposite directions. FASEB J. 6(4, Part 1): A1545, 1992 (Abstract).

22. Henrion, D.H., I. Laher, and J.A. Bevan. Intraluminal flow increases vascular tone and $^{45}Ca^{2+}$ influx in the rabbit facial vein. Circ. Res. 71: 339-345, 1992.

23. Hirst, G.D.S., and F.R. Edwards. Sympathetic neuroeffector transmission in arteries and arterioles. Physiol. Rev. 69(2): 546-595, 1989.

24. Jackson, R.L. Busch, S.J., and A.D. Cardin. Glycosaminoglycans: molecular properties, protein interactions, and role in physiological processes. Physiol. Rev. 71(2): 481-522, 1991.

25. Joyce, E.H., Xiao, X.-H., Bevan, R.D., and J.A. Bevan. Flow-induced change in tone of resistance vessels is the result of concurrent dilation and contraction. FASEB J. In Press, 1993.

26. Koller, A., and G. Kaley. Prostaglandins mediate arteriolar dilation to increased blood flow velocity in skeletal muscle microcirculation. Circ. Res. 67: 529-534, 1990.

27. Levick, J.R. Flow through interstitium and other fibrous matrices. Quart. J. Exp. Physiol. 72: 409-438, 1987.

28. Luscher, T.F., C.M. Boulanger, Y. Dohi, and Z. Yang. Endothelium-derived contracting factors. Hypertension 19: 117-130, 1992.

29. Manning, G.S. Counterion binding in polyelectrolyte theory. Acc. Chem. Res. 12: 443-449, 1979.

30. Nabel, E.G., B.C. Berk, T.A. Brock, and T.W. Smith. Na^+-Ca^{2+} exchange in cultured vascular smooth muscle cells. Circ. Res. 62: 486-493, 1988.

31. Neylon, C.B., P.J. Little, E.J. Cragoe, Jr., and A. Bobik. Intracellular pH in human arterial smooth muscle. Regulation by Na^+/H^+ exchange and a novel 5-(N-ethyl-N-isopropyl)amiloride-sensitive Na^+ - and HCO_3^--dependent mechanism. Circ. Res. 67: 814-825, 1990.

32. Owen, N.E., and M.L. Prastein. Na/K/Cl cotransport in cultured human fibroblasts. J. Biol. Chem. 260(3): 1445-1451, 1985.

33. Pohl, U., K. Herlan, A. Huang, and E. Bassenge. EDRF-mediated shear-induced dilation opposes myogenic vasoconstriction in small rabbit arteries. Am. J. Physiol. 261(Heart Circ. Physiol. 30): H2016-H2023, 1991.

34. Rosati, C., and R. Garay. Flow-dependent stimulation of sodium and cholesterol uptake and cell growth in cultured vascular smooth muscle. J. Hypertens. 9: 1029-1033, 1991.

35. Schmidt, A., Buddecke, E. Cell-associated proteoheparan sulfate from bovine arterial smooth muscle cells. Exptl. Cell Res. 178: 242-253, 1988.

36. Schretzenmayr, A. Uber kreislaufregulatorische Vorgange an den grossen Arterien bei der Muskerlarbeit. Pfluegers Arch. Ges. Physiol. 232: 743-748, 1933.

37. Siegel, G., A. Walter, W. Rettig, Ch. Kampe, B.J. Ebeling, and O. Bertsche. Sodium compartments in the arterial wall. In Intravascular Electrolytes and Arterial Hypertension, Editors H. Zumkley, H. Losse, N.Y., Thieme-Stratton Inc., 1980, pp. 30-50.

38. Siegel, G., A. Walter, M. Thiel, and B.J. Ebeling. Local regulation of blood flow. Adv. Exp. Med. Biol. 169: 515-540, 1984.

39. Tobian, L., and J.T. Binion. Tissue cations and water in arterial hypertension. Circ. V: 754-758, 1952.

40. Wellman, G.C., and J.A. Bevan. Flow-induced contraction and relaxation; electrophysiology of the balance/set point. (1993, Submitted).

41. Xiao-Hui, X.-H., and J.A. Bevan. Flow and potassium-induced contraction of the rabbit facial vein are similarly affected by calcium channel agonists and antagonists. (1993, Submitted).

CORRELATION OF ACUTE PROSTACYCLIN RESPONSE IN PRIMARY (UNEXPLAINED) PULMONARY HYPERTENSION WITH EFFICACY OF TREATMENT WITH CALCIUM CHANNEL BLOCKERS AND SURVIVAL

Bertron M. Groves , David B. Badesch, Darya Turkevich,
Roy V. Ditchey, Kathleen Donnellan, Kristine Wynne,
Alastair D. Robertson, Walker A. Long, Norbert Voelkel,
and John T. Reeves

Department of Medicine, Divisions of Cardiology and Pulmonary Sciences
and the Cardiovascular Pulmonary Research Laboratory, University of
Colorado Health Sciences Center, Denver, CO 80262; Burroughs Wellcome
Co., Research Triangle Park, NC 27709

INTRODUCTION

Unexplained, or primary, pulmonary hypertension, is usually a rapidly fatal disease, in which the use of vasodilator therapy is controversial (14, 22, 25). While some authors have proposed that there may be a significant vasoconstrictive component to the increased pulmonary arterial pressure and resistance (2, 7, 11, 19), others have indicated that such a component is usually small and/or unimportant (13). This controversy has likely arisen because the disease is relatively uncommon, its pathogenesis is not understood, it is usually diagnosed late in the natural history, and the rate of progression is variable between patients. Yet investigation into the problem is important because persons afflicted are often young and otherwise healthy. Needed was a prospective, longitudinal study in a relatively large group of patients, where the magnitude of the vasoconstrictive component could be measured acutely and compared to results of long term treatment.

Beginning in 1981, when prostacyclin became available, our approach was to serially study patients with unexplained pulmonary hypertension. In each, the presence and magnitude of pulmonary vasoconstriction was assessed acutely using intravenous prostacyclin, an effective and safe pulmonary vasodilator (6, 20). For chronic relief of vasoconstriction, we initially selected diltiazem, a calcium channel blocking agent which was available and seemed to be both effective in some patients with PPH (10). This decision was supported by experimental evidence that calcium antagonists were effective in blocking hypoxic pulmonary hypertension (12). Near the end of the study, after nifedipine had been reported to be effective in treating PPH (16, 17), patients were arbitrarily given nifedipine rather than diltiazem.

We assumed that in patients with primary pulmonary hypertension, prostacyclin would abolish the vasoconstrictive component of the elevated resistance, and the relatively fixed component (1, 15, 24) would remain, thereby displaying the spectrum of the two components in our population. If the magnitude of the vasoconstrictive component related directly to the efficacy of chronic therapy using a calcium antagonist, then prostacyclin administered acutely might predict the results of chronic therapy.

Ion Flux in Pulmonary Vascular Control, Edited by
E.K. Weir *et al.*, Plenum Press, New York, 1993

317

METHODS

Forty-four of 50 consecutive patients with unexplained pulmonary hypertension who were studied at the University of Colorado Health Sciences Center between September 1981 and March 1988 gave consent and were enrolled in the study. Six patients were excluded because of hemodynamic instability in 1, mental retardation in 1, coexistent coronary artery disease in 2, diffuse left ventricular hypocontractility in 1, and requested exclusion of 1 patient. The 44 patients ranged in age from 12 to 65 years of age and 29 (66%) were female, Table 1. The diagnosis of unexplained pulmonary hypertension was made after pulmonary function tests, ventilation perfusion lung scan and/or pulmonary angiography, echocardiography, and cardiac catheterization revealed no identifiable etiology for pulmonary hypertension. Examination of lung tissue was available in 22 patients. Although autopsy findings in patients #16 and 17 were consistent with pulmonary veno-occlusive disease and patient #51 was found at transplant to have necrotizing sarcoid granulomatosis, they were included because the clinical and laboratory findings were indistinguishable from those of the other patients.

Right heart catheterization was performed with a "guidewire" Swan-Ganz balloon tipped catheter (Baxter Healthcare, model 93A-821H) which we have found useful in patients with severe pulmonary hypertension associated with marked right ventricular dilation and tricuspid insufficiency (4). A 5-Fr, 23 cm catheter (North American Medical Instrument Company, model 91100900) was inserted percutaneously into a femoral artery for withdrawing blood and continuously monitoring systemic arterial pressure. Heart rate was monitored continuously. Cardiac output was measured in triplicate by thermodilution, a method we have previously studied in patients with severe pulmonary hypertension and reduced cardiac output (23). In patients who had atrial right to left shunts via a patent foramen ovale as determined by an early appearance of indocyanine green in the aorta following right atrial injection, thermodilution pulmonary blood flow measurements were accepted only when confirmed by the Fick method utilizing measured oxygen uptake and lung arterio-venous oxygen content difference.

"Acute" Prostacyclin

After baseline measurements were recorded, intravenous prostacyclin was begun at 1.0 ng/kg/min with increments of 1.0 to 2.0 ng/kg/min every 5-15 minutes to a maximum dose of 12.0 ng/kg/min. After stopping prostacyclin, measurements were repeated to confirm that the hemodynamic effects of prostacyclin had resolved.

"Short-Term" (Five Day) Calcium Channel Blocker Treatment

Patients then had electrocardiographic monitoring in the Clinical Research Center for up to 5 days during the initiation of oral diltiazem therapy. In the first 18 patients (Table 2A) we attempted to give 360 mg/day diltiazem in 4 divided doses. Based upon the report by Rich in 1987 which suggested better efficacy of "high dose" calcium channel blocker treatment in PPH (16), we attempted to increase the dosage to 720 mg/day of diltiazem in 12 patients. Toward the end of the study, 5 patients were arbitrarily treated with "high dose" oral nifedipine during hemodynamic monitoring in the intensive care unit by giving 20 mg/hr for a maximum of 12 hours to achieve a total dosage of 240 mg (17). After 5 days of diltiazem or 12 hours or less of nifedipine administration, catheterization was repeated in 35 patients to assess the initial hemodynamic effects of "short term" calcium channel blocker treatment, Table 2A.

"Chronic" (Eight Week) Calcium Channel Blocker Treatment

After 8 weeks of calcium blocker treatment (diltiazem in 19 and nifedipine in 1), the remaining 20 patients, had repeat cardiac catheterization to assess the efficacy of "chronic" treatment, Table 2B.

TABLE 1. Primary pulmonary hypertension: Baseline hemodynamics and change with prostacyclin

Pt #	Age Sex	BSA M2	RL	HR beats/min	Diff	RAM mmHg	Diff	PAM mmHg	Diff	SAM mmHg	Diff	QP L/min	Diff	TPR Wood units	% Diff	TSR Wood units	% Diff
1. 06	34M	2.09		100	-3	10	0	110	0	80	-12	3.6	1.3	30.6	-27	22.2	-38
2. 08	28F	1.90		86	-4	4	1	61	-24	105	-20	2.8	2.8	21.8	-70	37.5	-60
3. 09	22F	1.66		86	16	9	-6	42	-5	95	-22	3.0	3.1	14.0	-57	31.7	-62
4. 10	29F	1.63		88	24	1	0	40	-23	98	-3	4.4	2.8	9.1	-74	22.3	-41
5. 12	42M	1.91		84	2	5	0	50	-7	90	-20	3.1	1.6	16.1	-43	29.0	-49
6. 13	25F	1.68		109	9	13	-1	40	3	82	-16	2.2	0.2	18.2	-2	37.3	-26
7. 14	35F	1.30	+	93	2	1	0	50	0	76	-8	2.0		25.0		24.5	
8. 15	34F	1.58		82	13	2	0	76	-11	83	-10	2.6	1.6	29.2	-47	31.9	-46
9. 16	33M	2.30		73	18	2	0	47	-3	85	-11	4.1	1.9	11.5	-36	20.7	-41
10. 17	41M	1.90		87	13	14	3	59	-10	94	-21	3.1	1.9	19.0	-49	30.3	-52
11. 18	45F	2.03		90	2	15	-5	78	-3	88	-8	4.2	1.3	18.6	-27	21.0	-31
12. 19	65M	1.86		88	5	18	0	58	-5	82	-24	3.6	0.9	16.1	-27	22.8	-43
13. 20	29F	1.48		71	-5	8	-2	55	-10	75	-28	3.0	2.0	18.3	-51	25.0	-62
14. 22	33M	2.23		74	14	11	4	76	-12	74	-9	4.0	3.5	19.0	-55	18.5	-53
15. 23	50F	1.66	+	88	-2	4	-2	53	-5	120	-46	1.9	0.6	27.9	-31	63.2	-53
16. 24	40F	1.57		90	0	10	1	112	-2	125	-28	1.9	0.7	58.9	-28	65.8	-43
17. 25	46M	1.83		83	10	5	2	53	7	83	-16	4.1	0.8	12.9	-5	20.2	-33
18. 26	63F	1.52		66	18	5	3	55	7	80	-13	2.7	1.0	20.4	-18	29.6	-39
19. 27	36F	1.68	+	72	-14	17	1	51	-3	96	-17	3.1		16.5		31.0	
20. 28	37M	1.84		66	30	18	-2	56	-1	95	-11	2.5	2.3	22.4	-49	38.0	-54
21. 29	12F	1.09		117	5	7	-1	75	3	90	-10	1.8	0.4	41.7	-15	50.0	-27
22. 30	52M	1.99	+	66	16	11	1	63	5	82	-14	2.8	0.3	22.5	-3	22.2	-41
23. 33	41F	1.64		84	9	13	-2	65	3	100	-8	3.3	0.6	19.7	-12	30.3	-22
24. 34	32F	1.39	+	82	2	7	-4	58	-25	103	-26	2.2	2.5	26.4	-73	46.8	-65
25. 36	14F	1.59		103	27	6	3	73	8	106	-27	2.7	0.1	27.0	7	39.3	-28
26. 37	43F	1.66		68	12	16	-1	64	1	85	-18	2.4	0.0	26.7	2	35.4	-21
27. 38	45F	1.97		81	21	3	0	42	-6	98	-5	3.3	3.7	12.7	-60	29.7	-55
28. 39	39M	1.76	+	80	12	5	3	84	6	110	-20	2.0	1.4	42.0	-37	23.9	-15
29. 40	42F	1.54	+	87	7	12	-4	76	-7	103	-38	2.3		33.0		22.4	
30. 41	48F	1.96		93	8	15	-2	54	-4	135	-67	3.2	0.8	16.9	-26	42.2	-60
31. 42	28F	1.91	+	65	33	10	3	62	8	108	-23	2.2	0.2	28.2	4	27.0	-50
32. 43	30F	1.55		75	4	6	-1	50	-13	103	-31	2.5	1.8	20.0	-57	41.2	-59
33. 44	38F	1.84		106	11	20	0	77	4	125	-28	3.3	1.5	23.3	-28	37.9	-47
34. 45	27F	2.12		75	31	4	0	43	-20	87	-24	4.5	0.1	9.6	-48	19.3	-29
35. 46	30M	1.95		62	28	6	-3	32	-2	110	-10	7.1	3.3	4.5	-36	15.5	-38
36. 47	33M	1.99		65	7	9	1	77	11	97	-4	2.6	1.3	29.6	-24	37.3	-36
37. 48	58F	1.66		88	18	3	3	25	3	128	-3	4.7	1.9	5.3	-20	27.2	-31
38. 49	38F	1.95	+	103	13	21	-1	72	0	120	-25	2.4	1.4	30.0	-37	50.0	-50
39. 50	32F	1.65		89	27	5	1	74	10	108	-13	3.2	1.1	23.1	-16	33.8	-35
40. 51	34M	1.83		72	24	10	-1	73	2	105	-30	3.5	2.1	20.9	-36	30.0	-55
41. 52	62F	1.93		102	8	23	3	103	-2	94	-19	3.1	0.7	33.2	-20	30.3	-35
42. 53	27M	1.96		88	23	10	0	23	-7	105	-40	6.6	0.9	3.5	-39	15.9	-46
43. 54	44F	1.65		102	4	20	-5	67	-5	120	-48	3.7	1.0	18.1	-27	32.4	-53
44. 55	38M	2.11		72	18	10	2	48	10	87	-27	3.9	2.4	12.3	-25	22.3	-57
Mean				84	12	10	0	61	-3	98	-20	3.2	1.5	21.7	-32	31.5	-43
SEM				2	2	1	0.3	3	1	2	2	0.2	0.2	1.6	3	1.7	2

Pt # = patient number, RL + = right to left shunt, HR = heart rate, RAM = right atrial mean
pressure, PAM = pulmonary arterial mean pressure, SAM = systemic arterial mean pressure
QP = pulmonary flow, TPR = total pulmonary resistance, TSR = total systemic resistance
Diff = absolute difference from baseline, % Diff = percent difference from baseline, SEM = 1
standard error of the mean

"Long-Term" (> One Year) Calcium Channel Blocker Treatment

Five patients with improved symptomatology had one or more repeat catheterizations after "long-term" diltiazem treatment for a year or more (#10 after 1, 2, 4, and 6 yrs; #12 2 and 6 yrs; #20 1 and 5 yrs; #45 1.3 yrs; and #46 1.1 yrs). Three patients who developed late clinical deterioration despite continued diltiazem treatment for one or more years had repeat catheterization prior to referral for transplantation (#34 after 3 yrs, #43 1 yr, and #46 4 yrs), Table 2C.

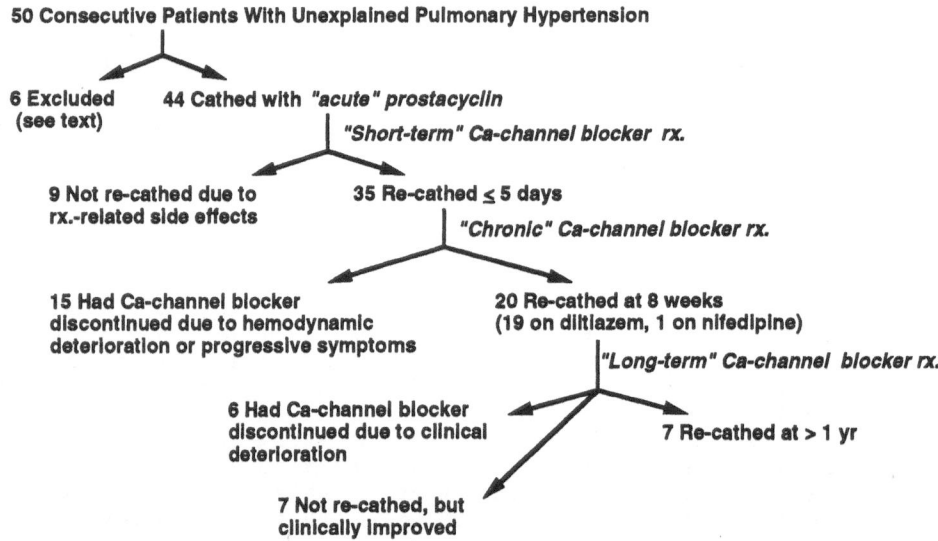

Figure 1. Summary of cardiac catheterization procedures performed in UCHSC PPH cohort during calcium channel blocker treatment.

Statistics

Group data are presented as mean values and one standard error of the mean (SEM) in the text and tables. Changes in individuals within a group were considered significant when p<0.05 for the paired t test. Relationship of two variables by linear regression was considered to exist when the correlation coefficient indicated p<0.05. The survival status as of 10/1/92 for each patient was established by personal or telephone contact, Table 3. The predicted survival at 1(P1), 2 (P2), and 3 (P3) years for each patient was calculated using the NIH Registry formula derived from the 194 patient cohort using the hemodynamic variables of mean right atrial and pulmonary arterial pressures and pulmonary flow (3). In addition, the predicted 50% survival (T50) was interpolated, when possible, from the 1 to 3 year predicted survival values, Table 3.

RESULTS

Baseline mean right atrial pressures were 8 mmHg or above in 25 patients, but there was no clear relation of right atrial pressure to mean pulmonary arterial pressure (r=.22, p=NS), Table 1. Baseline pulmonary capillary wedge pressures in all subjects were below 15 mmHg, (mean 7.3 ± 0.4 mmHg). Higher heart rates occurred in patients with higher mean pulmonary arterial pressures (r=0.30, p<0.02). The cardiac index appeared to be sharply reduced with elevation of mean pulmonary arterial pressure above 40 mmHg, but for pressures above 60 mmHg, there appeared to be little further reduction in cardiac index.

"Acute" Prostacyclin

Intravenous prostacyclin acutely administered to 44 patients with PPH increased heart rate, lowered pulmonary and systemic arterial pressure, and increased pulmonary blood flow, Table 1. Right atrial and wedge pressures were unchanged by prostacyclin. The mean maximum tolerated dose of 8.0 ± 0.4 ng/kg/min was continued for 17.3 ± 1.5 min to achieve a steady state. Larger decreases in pulmonary arterial pressure with prostacyclin were associated with larger increases in flow. For the group, total pulmonary resistance decreased. The decrease in pulmonary resistance was greater the higher the initial pulmonary arterial pressure ($r=-0.36$, $p<.02$). Because wedge pressure was unchanged during prostacyclin infusion and could not always be remeasured at the maximum prostacyclin doseage, total pulmonary resistances (mean pulmonary arterial pressure / pulmonary blood flow) are reported. In 3 of the 9 patients with right to left shunts via a patent foramen ovale pulmonary blood flow could not be measured during the maximum prostacyclin infusion.

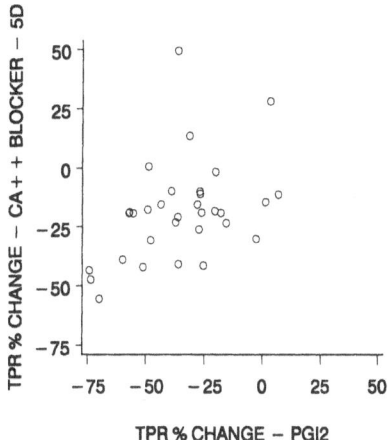

Figure 2. Prostacyclin (PGI2) decreased total pulmonary resistance (TPR) more than 5 day (D) calcium channel blocker treatment in PPH.

"Short-Term" Calcium Channel Blocker Treatment

After 5 days or less calcium channel blocker treatment, repeat catheterization was performed in 35 of the 44 patients who received prostacyclin. Nine patients were not recatheterized because of intolerance to diltiazem related side effects including nausea, marked sinus bradycardia (with or without atrioventricular dissociation or impaired AV nodal conduction), orthostatic hypotension, or progressive worsening of right ventricular decompensation. The mean total daily diltiazem dosage of 350 ± 7 mg (range 240 to 360 mg) in the first 18 patients was increased to a mean total daily dosage of 620 mg (range 480 to 720 mg) in the last 12 patients. In the 5 patients who received oral nifedipine, the mean total tolerated doseage was 130 mg given during a 5.4 hour interval (total dose ranged from 20 mg/1 hour to 240 mg/12 hours). In response to 5 days treatment with diltiazem, pulmonary flow increased and pulmonary arterial pressure and total pulmonary resistance decreased, Table 2A. The fall in pulmonary resistance was related to the fall which occurred acutely with prostacyclin, Figure 2.

However, individual measurements with diltiazem or nifedipine tended to lie above the line of identity, consistent with a smaller percent decrease in total pulmonary resistance with calcium blocker treatment ($-18 \pm 4\%$) than with prostacyclin ($-32 \pm 2\%$) ($p<0.01$). Heart rate decreased in response to diltiazem reflecting its known effect of decreasing sinus node activity possibly combined with its pulmonary vasodilator effect. The fall in heart rate, coupled with an increase in cardiac output, resulted in an increased stroke volume. Both the stroke volume ($r=-0.58$, $p<0.01$) and the change in stroke volume ($r=-0.40$, $p<0.05$) were inversely related to the pulmonary arterial pressure after 5 days of diltiazem therapy. There was no change in wedge pressure. As expected, in the 5 patients treated with nifedipine the mean heart rate increased slightly from 91 to 97 beats/min.

The slowing of the heart rate from diltiazem and the known adverse effects of calcium antagonists on cardiac contractility increased mean right atrial pressure only 1 mmHg, Table 2A, although there was a large individual variation ranging from +9 to -4 mmHg. The patients showing large decrements in total pulmonary resistance, decreased their right atrial pressures. Therefore, the effects of diltiazem acting to increase right atrial pressure were opposed by lowered pulmonary resistance acting to decrease right atrial pressure.

"Chronic" Calcium Channel Blocker Treatment

Twenty patients were recatheterized after 8 weeks of treatment with diltiazem. The group decrease in total pulmonary resistance with diltiazem (-5.2 ± 0.8 Wood units) was less ($p<0.01$) than with prostacyclin given acutely (-7.4 ± 1.2 Wood units). For the group, diltiazem treatment for 8 weeks caused no further hemodynamic improvement from that observed at the 5 day catheterization, Table 2A&B. Fifteen of the 35 patients were not continued on chronic calcium blocker treatment after the 5 day catheterization because of documented hemodynamic deterioration or progressive symptoms of right ventricular dysfunction after discharge. They had higher pulmonary resistances after 5 days of treatment (21.8 ± 2.5 Wood units) than did the 20 patients who tolerated 8 weeks of treatment (18.1 ± 1.7 Wood units). Also during the baseline catheterization, these 15 patients had a smaller decrease in pulmonary resistance in response to prostacyclin ($-24 \pm 4\%$) than did the 20 patients ($-43 \pm 5\%$). The change from baseline in total pulmonary resistance after 8 weeks treatment with diltiazem was related to the change seen with prostacyclin acutely. However, in the treatment interval from 5 days to 8 weeks, not all patients maintained their reductions in total pulmonary resistance. Patients #8, 9, 14, 30, 36, and 46 showed hemodynamic deterioration as measured by an increase in pulmonary arterial and/or right atrial pressures, and an increase in total pulmonary resistance, Table 2B.

"Long-Term" Calcium Channel Blocker Treatment

Six of the 7 patients who were catheterized after a year or more treatment with diltiazem had sustained clinical and hemodynamic improvement, Table 2C. Patient #34 initially had a dramatic hemodynamic response to prostacyclin and diltiazem and remained clinically improved for over 2 years. Because of progressive symptoms, she was recatheterized after 2.8 years of diltiazem treatment and had a persistent 28% decrease in mean pulmonary artery pressure and 53% decrease in total pulmonary resistance. Nonetheless, within 6 months her condition continued to deteriorate and she was referred for single lung transplantation which was successfully performed 14 months later. Patient #43 who had significant clinical and hemodynamic benefit after 8 weeks of treatment had noted the onset of progressive clinical deterioration after 10 months of treatment with hemodynamic confirmation by catheterization after 1.3 years of treatment. She was immediately referred for transplantation but died 6 weeks later before a donor became available. Despite being improved by catheterization after treatment with diltiazem for 1.1 years, patient #46 had the onset of life threatening hemoptysis associated with severe progression of familial primary pulmonary hypertension which led to recatheterization after 4 years of diltiazem treatment. On 10/01/92, four patients (#10, 12, 20, and 45) continued to have sustained clinical benefit on diltiazem treatment after a mean of 8.6 ± 1.1 years (range 5.5 to 10.2 yrs) and were enjoying productive lifestyles including near-normal physical activities!

TABLE 2A. Hemodynamics after 5 day treatment with calcium blocker - versus baseline

Pt #	Drug Rx mg/day	HR beats /min	Diff	RAM mmHg	Diff	PAM mmHg	Diff	SAM mmHg	Diff	QP L/min	Diff	TPR Wood units	% Diff	TSR Wood units	% Diff
1. 06	Dil 360	81	-19	10	0	93	-17	85	5	3.4	-0.2	27.4	-11	25.0	13
2. 08	Dil 360	87	1	4	0	44	-17	90	-15	4.5	1.7	9.8	-55	20.0	-47
3. 09	Dil 360	78	-8	7	-2	40	-2	93	-2	3.5	0.5	11.4	-18	26.6	-16
4. 10	Dil 360	81	-7	2	1	29	-11	94	-4	5.6	1.2	5.2	-43	16.8	-25
5. 12	Dil 360	78	-6	10	5	48	-2	90	0	3.5	0.4	13.7	-15	25.7	-11
6. 14	Dil 360	81	-12	2	1	48	-2	69	-7	2.6	0.6	18.5	-26	18.2	-26
7. 16	Dil 360	85	12	8	6	57	10	80	-5	3.3	-0.8	17.3	51	24.2	17
8. 17	Dil 360	68	-19	14	0	50	-9	80	-14	2.6	-0.5	19.2	1	30.8	2
9. 18	Dil 360	86	-4	12	-3	79	1	95	7	4.7	0.5	16.8	-10	20.2	-4
10. 20	Dil 360	64	-7	6	-2	46	-9	80	5	4.3	1.3	10.7	-42	18.6	-26
11. 22	Dil 360	70	-4	15	4	71	-5	74	0	4.6	0.6	15.4	-19	16.1	-13
12. 23	Dil 360	80	-8	0	-4	53	1	91	-29	1.7	-0.2	31.2	14	53.5	-15
13. 26	Dil 360	64	-2	9	4	53	-2	71	-9	3.2	0.5	16.6	-19	22.2	-25
14. 27	Dil 360	83	11	26	9	60	9	93	-3						
15. 28	Dil 240	63	-3	14	-4	50	-6	87	-8	2.7	0.2	18.5	-17	32.2	-15
16. 30	Dil 360	58	-8	10	-1	57	-6	78	-4	3.6	0.8	15.8	-30	18.1	-18
17. 34	Dil 300	75	-7	3	-4	42	-16	88	-15	3.0	0.8	14.0	-47	29.3	-37
18. 36	Dil 360	94	-9	6	0	70	-3	94	-12	2.9	0.2	24.1	-11	32.4	-17
19. 37	Dil 720	68	-22	16	0	64	2	85	-7	2.4	0.4	26.7	-14	35.4	-23
20. 38	Dil 480	78	-3	7	4	43	1	98	0	5.5	2.2	7.8	-39	17.8	-40
21. 39	Dil 600	60	-20	10	5	78	-6	90	-20	2.4	0.4	32.5	-23	37.5	57
22. 40	Dil 720	80	-9	18	6	71	-6	86	-14						
23. 41	Dil 720	83	-10	22	7	55	1	99	-36	4.0	0.8	13.8	-19	24.8	-41
24. 42	Dil 720	70	5	15	5	58	-4	80	-28	1.6	-0.6	36.3	29	50.0	85
25. 43	Dil 600	74	-1	9	3	52	2	102	-1	3.2	0.7	16.3	-19	31.9	-23
26. 44	Nif 80	94	-14	19	0	68	-17	78	-22	3.2	-0.2	21.3	-15	24.4	-17
27. 45	Dil 600	70	-5	7	3	34	-9	93	6	5.1	0.6	6.7	-30	18.2	-6
28. 46	Dil 480	64	2	8	2	28	-4	85	-25	7.8	0.7	3.6	-20	10.9	-30
29. 48	Dil 480	76	-4	3	1	22	1	100	-28	6.0	1.3	3.7	-18	16.7	-39
30. 50	Dil 720	88	-1	7	2	73	-1	90	-18	4.1	0.9	17.8	-23	22.0	-35
31. 51	Nif 240	90	3	8	-2	59	-7	83	-22	3.9	1.3	15.1	-40	21.3	-28
32. 52	Nif 20	107	5	27	6	83	-7	70	-24	2.8	-0.2	29.6	-1	25.0	-11
33. 53	Dil 600	64	-24	6	-4	18	-5	95	-10	5.7	-0.9	3.2	-9	16.7	5
34. 54	Nif 160	98	3	13	-1	58	-5	75	-29	2.6	0.5	22.3	-26	28.8	-42
35. 55	Nif 150	97	8	9	-2	51	-5	84	-10	5.1	1.8	10.0	-41	16.5	-42
Mean		78	-5	10	1	54	-4	86	-11	3.8	0.5	16.7	-18	25.1	-15
SEM		2.0	1.5	1.0	0.5	2.9	1.0	1.5	1.9	0.2	0.1	1.4	3.5	1.6	4.6

Drug Rx = vasodilator treatment, Dil = diltiazem, Nif = nifidepine, Diff = absolute difference between 5 day calcium blocker and baseline, % Diff = percent difference between 5 day calcium blocker and baseline

TABLE 2B. Hemodynamics after 8 weeks treatment with calcium blocker - versus baseline

Pt #	Drug Rx mg/day	HR beats/min	Diff	RAM mmHg	Diff	PAM mmHg	Diff	SAM mmHg	Diff	QP L/min	Diff	TPR Wood units	% Diff	TSR Wood units	% Diff
1. 06	Dil 360	70	-30	10	0	97	-13	88	8	4.0	0.4	24.3	-21	22.0	-1
2. 08	Dil 360	88	2	5	1	63	2	95	-10	4.3	1.5	14.7	-33	22.1	-41
3. 09	Dil 360	84	-2	7	-2	48	6	83	-12	3.4	0.4	14.1	1	24.4	-23
4. 10	Dil 360	80	-8	1	0	25	-15	95	-3	6.1	1.7	4.1	-55	15.6	-30
5. 12	Dil 360	70	-14	3	-2	42	-8	77	-13	3.8	0.7	11.1	-32	20.3	-30
6. 14	Dil 360	82	-11	4	3	57	7	81	5	2.2	0.2	25.9	4	27.9	14
7. 18	Dil 360	88	-2	16	1	70	-8	87	-1	5.7	1.5	12.3	-34	15.3	-27
8. 20	Dil 360	60	-11	6	-2	40	-15	77	2	4.5	1.5	8.9	-52	17.1	-32
9. 22	Dil 360	70	-4	3	-8	65	-11	78	4	5.2	1.2	12.5	-34	15.0	-19
10. 23	Dil 360	80	-8	0	-4	53	1	91	-29	2.7	0.8	19.6	-28	33.7	-47
11. 28	Dil 240	66	0	10	-8	53	-3	84	-11	3.2	0.7	16.6	-26	26.3	-31
12. 30	Dil 360	60	-6	16	5	70	7	90	8	3.5	0.7	20.0	-11	20.0	-10
13. 34	Dil 300	81	-1	2	-5	44	-14	98	-5	3.2	1.0	13.8	-48	30.6	-35
14. 36	Dil 360	76	-27	15	9	80	7	97	-9	3.0	0.3	26.7	-1	32.3	-18
15. 38	Dil 480	81	0	3	0	40	-2	98	0	5.0	1.7	8.0	-37	19.6	-34
16. 43	Dil 600	72	-3	5	-1	45	-5	105	2	4.0	1.5	11.3	-44	26.3	-36
17. 45	Dil 600	68	-7	5	1	31	-12	99	12	4.6	0.1	6.7	-30	21.5	11
18. 46	Dil 480	62	0	8	2	31	-1	105	-5	7.3	0.2	4.2	-6	14.4	-7
19. 48	Dil 480	80	0	4	2	23	2	98	-30	6.6	1.9	3.5	-22	14.8	-46
20. 51	Nif 240	84	12	17	7	77	4	63	-42	3.5	0.0	22.0	-5	18.0	-40
	Mean	75	-6	7	0	53	-4	89	-6	4.3	0.9	14.0	-26	21.9	-24
	SEM	2	2	1	1	4	2	2	3	0.3	0.1	1.6	4	1.3	4

Table 2C. Hemodynamics after 1 year or more treatment with calcium blocker - versus baseline

Pt #	Drug Rx mg/day	Yrs Drug Rx	HR beats/min	Diff	RAM mmHg	Diff	PAM mmHg	Diff	SAM mmHg	Diff	QP L/min	Diff	TPR Wood units	% Diff	TSR Wood units	% Diff
1. 10	Dil 360	1.0	68	-20	2	1	27	-13	98	0	5.9	1.5	4.6	-49	16.6	-26
10	Dil 360	6.0	78	-10	5	4	35	-5	98	0	6.2	1.8	5.6	-38	15.8	-29
2. 12	Dil 360	2.5	58	-26	6	1	45	-5	88	-2	4.1	1.0	11.0	-32	21.5	-26
12	Dil 360	5.9	60	-24	8	3	45	-5	88	-2	5.3	2.2	8.5	-47	16.6	-43
3. 20	Dil 360	1.0	51	-20	8	0	38	-17	65	-10	3.9	0.9	9.7	-47	16.7	-33
20	Dil 360	6.0	58	-13	5	-3	43	12	88	13	4.1	1.1	10.5	-43	21.5	-14
4. 34	Dil 300	3.0	63	-19	2	-5	42	-16	92	-11	3.4	1.2	12.4	-53	27.1	-42
5. 43	Dil 600	1.3	80	5	15	9	67	17	85	-18	2.7	0.2	24.8	24	31.5	-24
6. 45	Dil 600	1.2	60	-15	6	2	30	-13	90	3	5.5	1.0	5.5	-43	16.4	-15
7. 46	Dil 360	1.2	61	1	7	1	31	-1	95	-15	7.1	0.0	4.4	-2	13.4	-14
46	Dil 360	3.9	60	-2	9	3	63	31	86	-24	4.0	-3.1	15.8	251	21.5	39

Yrs Drug Rx = years of treatment with calcium channel blocker

Survival

Table 3 summarizes the survival status for the 44 patient cohort on 10/01/92. Of the 28 patients who were dead (64% mortality), 4 had undergone heart/lung and 2 single lung transplantation. Seven of the 15 patients who were alive had survived heart/lung transplantation a mean of 4.2 years (range 2.6 to 6.4 yrs). One patient (#26) was lost to follow up. Eight of the 31 non-transplanted patients were alive with a mean survival of 7.0 years (range 4.8 to 10.2 yrs). Seven of the 8 non-transplanted survivors remained on diltiazem treatment because of documented hemodynamic benefit. The remaining survivor (#42), who was refractory to prostacyclin and diltiazem and had declined referral for transplantation, was still alive 5.9 years later in a debilitated condition on no vasodilator treatment!

DISCUSSION

In patients with unexplained pulmonary hypertension, pulmonary resistance during acute prostacyclin administration related to pulmonary resistance observed subsequently during diltiazem therapy. Prostacyclin, because of its relative safety based upon a biologic half-life of 3 to 5 minutes, has a well defined end-point. The drug was given to tolerance, and the magnitude of the decrease in total pulmonary resistance was similar to that previously observed (20). Normal wedge pressures before and after the administration of prostacyclin indicated that the changes were not the result of an effect on the left ventricle. The finding in some patients of a large increase in pulmonary flow associated with a significant decrease in pulmonary pressure, indicated prostacyclin could produce significant pulmonary vasodilation. Thus we considered that an adequate prostacyclin effect had been achieved in a population appropriate for study.

Diltiazem therapy was more complicated to interpret. For our first 18 patients, the target dose was 360 mg per day. After greater potential effect with "high-dose" calcium blocker treatment (up to 720 mg diltiazem and 240 mg nifedipine total daily dose) was reported (16, 17), we attempted to achieve these maximal daily doses of diltiazem (n=12) or nifedipine (n=5). In this sequentially enrolled group of PPH patients we were unable to achieve a greater lowering of pulmonary arterial pressure and resistance using the higher doses of diltiazem or nifedipine. Another potential problem in the therapeutic trial with diltiazem was selection bias toward less severely ill patients. For example, prostacyclin caused a greater vasodilation in patients who could be, compared to those who could not be, treated for 8 weeks with diltiazem or nifedipine. It is likely that those who received the longer course of diltiazem treatment had a relatively large pretreatment vasoconstrictive component and a relatively small fixed component contributing to their high pulmonary resistance. Thus the comparison of prostacyclin and diltiazem is limited by the fact that the most effective dose of diltiazem is not clear and further, the analysis is weighted toward those patients with PPH who were in a potentially more reversible stage of the disease.

Even so, the comparison indicated that the decrease in resistance with diltiazem was related to that achieved with prostacyclin. The relationship was present both after 5 days and 8 weeks of treatment which suggested that the finding was reliable. Diltiazem given chronically, appeared to be a less potent pulmonary vasodilator when compared to prostacyclin given acutely, which could reflect insufficient doses of diltiazem, and/or greater effectiveness of prostacyclin, even when given acutely. Because of its short half-life, intravenous prostacyclin could also be safely increased to "maximal tolerated" doses which may not have been achieved with oral diltiazem. The findings suggest that prostacyclin can be used to indicate those patients most likely to respond to calcium blocker vasodilator treatment. While the present study showed that prostacyclin predicted the effects of diltiazem given for 5 days and 8 weeks in most patients, future studies must determine the accuracy with which it will predict the chronic response to the same or other agents given in different doses and for longer durations.

Patients with a favorable acute response to prostacyclin usually benefited from calcium blocker treatment. Some have sustained the clinical and hemodynamic improvement from calcium blocker treatment for several years (#10, 12, 20, 45, and 53). However, our group also included patients (#8, 9, 15, 17, 22, 34, 38, and 43) who had a

TABLE 3. Primary pulmonary hypertension: Predicted versus observed survival

Pt #	Age Sex	Date 1st UCHSC Cath	Predicted Survival				Date Xplnt	Date Death	Died Post Cath	Died Post Xplnt	Curr Rx	Alive Post Cath	Alive Post Xplnt
			P1	P2	P3	T50							
1. 06	34M	09/30/81	0.54	0.38	0.28	1.2		08/17/82	0.9				
2. 08	28F	01/19/82	0.73	0.61	0.51	3.2		10/08/82	0.7				
3. 09	22F	06/22/82	0.71	0.58	0.49	2.8	HL 04/16/85	06/08/86	4.0	1.1			
4. 10	29F	07/20/82	0.85	0.77	0.71						Dil	10.2	
5. 12	42M	12/10/82	0.73	0.61	0.52	3.2					Dil	9.8	
6. 13	25F	01/11/83	0.61	0.46	0.36	1.7		01/14/84	1.0				
7. 14	35F	01/14/83	0.82	0.73	0.66			08/06/91	8.6				
8. 15	34F	04/26/83	0.74	0.63	0.54	3.5	HL 01/18/87					9.4	5.7
9. 16	33M	05/10/83	0.78	0.68	0.59	4.5		07/06/83	0.2				
10. 17	41M	07/12/83	0.61	0.46	0.36	1.7		08/29/85	2.1				
7 18	45F	08/23/83	0.57	0.41	0.31	1.4		10/15/84	1.1				
12. 19	65M	09/23/83	0.55	0.39	0.29	1.3		12/13/84	1.2				
13. 20	29F	01/05/84	0.71	0.59	0.50	3.0					Dil	8.7	
14. 22	33M	03/13/84	0.57	0.42	0.31	1.4		01/16/88	3.8				
15. 23	50F	06/05/84	0.69	0.57	0.47	2.6		08/11/86	2.2				
16. 24	40F	09/25/84	0.48	0.32	0.21	0.9		04/10/85	0.5				
17. 25	46M	10/23/84	0.79	0.69	0.61	4.8	L 03/26/90	12/23/90	6.2	0.7			
18. 26	63F	12/06/84	0.73	0.61	0.52	3.2		LTF					
19. 27	36F	02/06/85						03/02/86	1.1				
20. 28	37M	04/09/85	0.49	0.33	0.23	0.9	HL 05/14/86					7.5	6.4
21. 29	12F	05/21/85	0.66	0.52	0.42	2.2		03/01/86	0.8				
22. 30	52M	08/16/85	0.64	0.50	0.40	2.0		11/10/88	3.2				
23. 33	41F	03/18/86	0.62	0.48	0.37	1.8		11/22/89	3.7				
24. 34	32F	04/01/86	0.68	0.56	0.46	2.5	L 10/23/90	06/19/92	6.2	1.7			
25. 36	14F	08/05/86	0.68	0.55	0.45	2.5	HL 02/18/90					6.2	2.6
26. 37	43F	09/30/86	0.50	0.34	0.24	1.0	HL 01/03/88					6.0	4.7
27. 38	45F	09/30/86	0.77	0.66	0.58	4.1		03/04/92	5.4				
28. 39	39M	10/03/86	0.74	0.63	0.54	3.6		11/05/90	4.1				
29. 40	42F	10/14/86	0.70	0.57	0.48	2.7		12/16/88	2.2				
30. 41	48F	10/14/86	0.58	0.43	0.32	1.5	HL 11/11/89					6.0	2.9
31. 42	28F	11/11/86	0.68	0.55	0.45	2.5						5.9	
32. 43	30F	01/09/87	0.72	0.59	0.50	3.0		06/08/88	1.4				
33. 44	38F	03/03/87	0.45	0.29	0.19	0.7	HL 08/11/89	09/11/89	2.5	0.1			
34. 45	27F	03/31/87	0.78	0.69	0.60	4.7					Dil	5.5	
35. 46	30M	04/03/87	0.86	0.79	0.73						Dil	5.5	
36. 47	33M	06/16/87	0.59	0.44	0.33	1.5	HL 05/31/89	07/02/89	2.0	0.1			
37. 48	58F	07/07/87	0.86	0.79	0.73						Dil	5.2	
38. 49	38F	07/14/87	0.35	0.20	0.11	0.4		10/12/87	0.2				
39. 50	32F	09/15/87	0.71	0.59	0.49	2.9	HL 07/18/88					5.0	4.2
40. 51	34M	09/29/87	0.64	0.50	0.39	2.0	HL 07/18/89					5.0	3.2
41. 52	62F	10/06/87	0.30	0.15	0.08	0.2		04/01/88	0.5				
42. 53	27M	12/04/87	0.80	0.71	0.64	6.1					Dil	4.8	
43. 54	44F	02/09/88	0.52	0.37	0.26	1.1	HL 01/18/89	03/27/90	2.1	1.2			
44. 55	38M	03/01/88	0.69	0.56	0.46	2.5		01/27/92	3.9				
		Mean	0.66	0.53	0.43	2.4			2.6	0.8		6.7	4.3
		SEM	0.02	0.02	0.02	0.2			0.4	0.2		0.5	0.5

P1,P2,P3 = probability of surviving 1 to 3 years, T50 = predicted 50% survival in years
Xplnt = transplant, HL = heart/lung, L = single lung, Curr Rx = current treatment
Dil = diltiazem, all time intervals reported in years

favorable acute response to prostacyclin and short-term calcium blocker treatment but subsequently failed calcium blocker therapy, suggesting that the capacity for acute vasodilation did not guarantee a good long term result. While the acute administration of prostacyclin provided a measure of vasoconstriction present at one time in the natural history of the disease, it did not establish the nature or activity of the disease process. In terms of patient management, the utilization of the 5 day and 8 week catheterizations provided us with early evidence of decreasing effectiveness of therapy in some patients. Such early discovery could enhance success of an altered treatment strategy in rapidly progressive disease. In particular, if heart/lung or single lung transplantation is being considered, and given the time lag before transplantation can be carried out because of inadequate donor supply, time gained in the assessment of medical management could be of vital importance for the patient.

A correlation of acute prostacyclin with diltiazem effects occurred despite both the potential differences in the actions of each drug and the different durations of administration for each. In addition, if diltiazem caused continued relief of vasoconstriction for days or weeks, one might expect further fall in pressure and resistance if medial hypertrophy of the pulmonary vessels regressed. Diltiazem has also been shown to inhibit hypoxic pulmonary vasoconstriction which occurs in some patients with PPH (5).

Recent reports indicate that continuous intravenous prostacyclin infusion via an indwelling subclavian vein catheter may offer a needed "bridge to transplant" for patients who are refractory to conventional vasodilator therapy (8, 9, 21). Chronic administration of intravenous prostacyclin in primary pulmonary hypertension may achieve greater reduction in pulmonary resistance than acute administration (8).

When projected and observed survival curves for our 44 patients were compared with the NIH PPH Registry survival curve derived from a database of 194 patients, the UCHSC cohort appeared to have improved survival whether or not transplanted patients were considered as "lost to follow up" or dead at the time of transplantation, Figure 3.

Using the NIH PPH Registry formula which calculates projected survival from the baseline right atrial and pulmonary artery pressures and the cardiac index, the one (P1), two (P2), and three year (P3) mean projected survivals for the UCHSC total cohort were 0.66, 0.53,

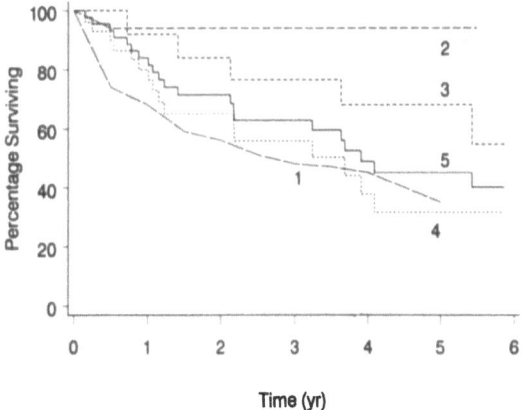

Figure 3. Comparative survival of PPH patients. The historical controls of the NIH PPH Registry (curve 1, n=194) are plotted against the University of Illinois responders (curve 2, n=17), UCHSC responders (curve 3, n=13), UCHSC non-responders (curve 4, n=31), and the UCHSC total cohort (curve 5, n=44). The rate of survival of the UCHSC responders was greater than both the UCHSC non-responders and the NIH Registry patients, but was less than the University of Illinois responders.

and 0.43 years (Table 3). The projected mean 50% survival (T50) for the UCHSC cohort was 2.4 years. However, when compared to the University of Illinois PPH "high dose" calcium blocker treatment survival curve for 17 "responders" out of their total group of 64 PPH patients (18), the "responders" in our cohort appeared to have a position of intermediate survival despite similar treatment, Figure 3. The difference between these two groups remains unexplained. The subset of 7 patients in our cohort who remain alive on longterm diltiazem treatment have demonstrated sustained clinical improvement and apparent prolonged survival compared to their predicted survival derived from the NIH PPH Registry data. Thus it appears that calcium blocker treatment may prolong life in some patients with PPH who respond to long-term vasodilator treatment with a sustained increase in cardiac output associated with a decrease in pulmonary artery pressure. However, other PPH patients who demonstrate a favorable initial response to vasodilator treatment have progression of their disease after one or more years of treatment and become candidates for lung transplantation, Figure 4.

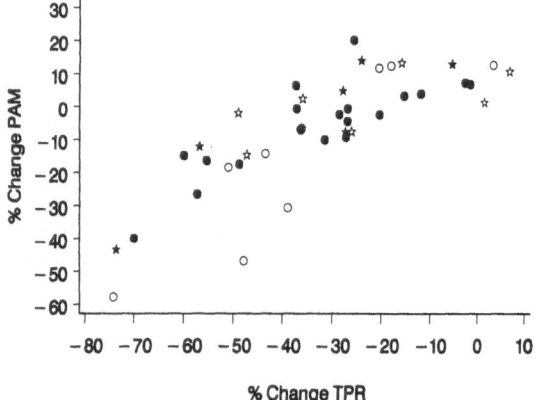

Figure 4. Acute hemodynamic effects of prostacyclin on total pulmonary resistance (TPR) and mean pulmonary artery pressure (PAM) versus survival and/or need for transplantation.   Patient symbols: white circle = non-transplanted/living; black circle = non-transplanted/dead; white star = transplanted/living; black star = transplanted/dead. Five of 13 responders (>10% decrease PAM and >30% decrease TPR) are living without transplant. Only 4 of 28 non-responders are living without transplant and 2 of them are debilitated in need of transplant, 1 was lost to follow up, and 1 is stable.

It is our current approach to acutely assess pulmonary vasoreactiviy in patients with PPH using intravenous prostacyclin or iloprost (an investigational prostacyclin analog) in the cardiac catheterization laboratory. If the patient has not previously failed calcium channel blocker treatment, hemodynamic monitoring in the intensive care unit for 8 to 12 hours during a calcium channel blocker challenge is performed using maximum tolerated oral dosage of diltiazem (720 mg/day) or nifedipine (240 mg/day) to select appropriate patients for chronic oral vasodilator treatment. We arbitrarily define "responders" as PPH patients who have a >10% decrease in mean pulmonary artery pressure and a >30% decrease in pulmonary resistance in response to prostacyclin and/or calcium channel blocker. It seems likely that those who have marked pulmonary vasodilation acutely but fail chronic therapy have a relatively virulent form of the disease. While calcium channel blocker therapy may ameliorate the course of PPH in some patients, it is clearly not effective in those with advanced disease. Therefore, symptomatic PPH patients in New York Heart Association functional classes 3 and 4 who are refractory to calcium channel blocker treatment are offered continuous intravenous prostacyclin protocol treatment as a "bridge" to single lung transplantation.

Acknowledgements. Supported in part by grants from the National Institutes of Health (HL 14985, HL 07171, and the Clinical Research Center PHS RG5MO1RR00051), Burroughs Wellcome Co., and Marion Laboratories Inc. We wish to thank Dr. Giuseppe G. Pietra for reviewing the histopathology of the lung in patient #51 who had necrotizing sarcoid granulomatosis with angiitis.

REFERENCES

1. Anderson, E. G., G. Simon, and L. Reid. Primary and thromboembolic pulmonary hypertension: a quantitative pathological study. *J. Pathol.* 110: 273-293, 1973.

2. Daoud, F.S., D.B. Kelly, and J.T. Reeves. Isoproteronol as a potential pulmonary vasodilator in primary pulmonary hypertension. *Am. J. Cardiol.* 42: 817-823, 1978.

3. D'Alonzo, G.E., R.J. Barst, S.M. Ayers, E.H. Bergofsky, B.H. Brundage, K.M. Detre, A.P. Fishman, R.M. Goldring, B.M. Groves, J.T. Kernis, P.S. Levy, G.G. Pietra, L.M. Reid, J.T. Reeves, S. Rich, C.E. Vreim, G.W. Williams, and M. Wu. Survival in patients with primary pulmonary hypertension. Results from a national prospective registry. *Ann. Intern. Med.* 115: 343-349, 1991.

4. Groves, B.M., R.V. Ditchey, J.T. Reeves, B.H. Brundage, C.R. McKay, E. Powers, R. Barst, L.J. Rubin, L.J. Aker, R. Siegel, and H. Zadaca. Multicenter trial of a new guidewire thermodilution catheter. *J. Am. Coll. Cardiol.* 3: 599, 1984.

5. Groves, B.M., K. Donnellan, A.D. Robertson, and J.T. Reeves. Diltiazem inhibits hypoxic pulmonary vasoconstriction in primary pulmonary hypertension. In: Sutton, J.R., Coates, G., and Remmers, J.E. eds., Hypoxia: The adaptations, pages 163-169, Toronto, B.C. Decker, Inc., 1990.

6. Groves, B.M., Rubin, L.J., Frosolono, M.F., and J.T. Reeves. A comparison of the hemodynamic effects of prostacyclin and hydralazine in primary pulmonary hypertension. *Am. Heart J.* 110: 1200-1204, 1985.

7. Groves, B.M., Turkevich, D., Donnellan, K., Voelkel, N., Robertson, A.D., and Reeves, J.T. Current approach to treatment of primary pulmonary hypertension. *Chest* 93: 175S-178S, 1988.

8. Higenbottam, T. The place of prostacyclin in the clinical management of primary pulmonary hypertension. *Am. Rev. Resp. Dis.* 136: 782-785, 1987.

9. Jones, D.K., T.W. Higenbottam, and J. Wallwork. Treatment of primary pulmonary hypertension with intravenous epoprostenol (prostacyclin). *Br. Heart J.* 57: 270-278, 1987.

10. Kambara, H., K. Fujimoto, A. Wakabayashi, and C. Kawai. Primary pulmonary hypertension: beneficial therapy with diltiazem. *Am. Heart J.* 101: 21-23, 1981.

11. Lupi-Herrera, E., Sandoval, J. Seoane, M., and D. Bialostozky. The role of hydralazaine therapy for pulmonary arterial hypertension of unknown cause. *Circulation* 65: 645-650, 1982.

12. Mc Murtry, I.F., Davidson, A.B., Reeves, J.T., and R.F. Grover. Inhibition of hypoxic pulmonary vasoconstriction by calcium antagonists in isolated rat lungs. *Circ. Res.* 38: 99-104, 1976.

13. Packer, M. Greenberg, B., Massie, B., and H. Dash. Deleterious effects of hydralazine in patients with pulmonary hypertension. *N. Engl. J. Med.* 306: 1326-1331, 1982.

14. Reeves, J. T., B.M. Groves, and D. Turkevich. The case for treatment of selected patients with primary pulmonary hypertension. *Am. Rev. Resp. Dis.* 134: 342-346, 1986.

15. Reeves, J.T., and J.A. Noonan. Microarteriographic studies of primary pulmonary hypertension. *Arch. Pathol.* 95: 50-55, 1973.

16. Rich, S., and B.H. Brundage. High-dose calcium channel-blocking therapy for primary pulmonary hypertension: evidence for long-term reduction in pulmonary arterial pressure and regression of right ventricular hypertrophy. *Circulation* 76: 135-141, 1987.

17. Rich, S., and E. Kaufmann. High dose titration of calcium channel blocking agents for primary pulmonary hypertension: guidelines for short-term drug testing. *J. Am. Coll. Cardiol.* 18: 1323-1327, 1991.

18. Rich, S., E. Kaufmann, and P.S. Levy. The effect of high doses of calcium-channel blockers on survival in primary pulmonary hypertension. *N. Engl. J. Med.* 327: 76-81,1992.

19. Rubin, L.J. and R.H. Peter. Oral hydralazine therapy for primary pulmonary hypertension. *N. Engl. J. Med.* 302: 69-73, 1980.

20. Rubin, L.J., B.M. Groves, J.T. Reeves, M. Frosolono, F. Handel, and A.E. Cato. Prostacyclin induced acute pulmonary vasodilation in pulmonary hypertension. *Circulation* 66: 334-338, 1982.

21. Rubin, L.J., J. Mendoza, M. Hood, M. McGoon, R. Barst, W.B. Williams, J.H. Diehl, J. Crow, and W. Long. Treatment of primary pulmonary hypertension with continuous intravenous prostacyclin (epoprostenol). Results of a randomized trial. *Ann. Intern. Med.* 112: 485-491, 1990.

22. Voelkel, N.F., and J.T. Reeves. Primary pulmonary hypertension. In: K.M. Moser, ed. Pulmonary Vascular Disease, pp 573-628. New York: Marcel Dekker, 1979.

23. Von Grondelle, A., R.V. Ditchey, B.M. Groves, W.W. Wagner, and J.T. Reeves. Thermodilution method overestimates low cardiac output in humans. *Am. J. Physiol.* 14: H690-692, 1983.

24. Wagenvoort, C.A., and N. Wagenvoort. Primary pulmonary hypertension. *Circulation* 42: 1163-1184, 1970.

25. Weir, E.K. Diagnosis and management of primary pulmonary hypertension. In: Pulmonary Hypertension. E.K. Weir and J.T. Reeves, eds. pp 115-168. Mt. Kisco, NY: Futura, 1984.

Steven H. Abman[1], David N. Cornfield[1], and John P. Kinsella[2]

[1] Section of Pediatric Pulmonary Medicine,
[2] Section of Neonatology,
Department of Pediatrics,
University of Colorado School of Medicine and
The Children's Hospital,
Denver, CO 80218-1088

INTRODUCTION

Recent advances in vascular biology have led to greater insight into the role of the endothelial
cell in regulating vascular function and structure. Several potent vasoactive products, including
endothelium-derived relaxing factor or nitric oxide (endothelium-derived nitric oxide; EDNO),
prostacyclin, endothelin, and others, have been shown to contribute substantially to vascular tone
and growth in many experimental and clinical settings.[18,26,28,32] Although several studies have
demonstrated the importance of endothelial products in the regulation of vascular tone in various
adult circulations, relatively little is known about their functional role in the perinatal lung or
maturation-related changes. Over the past several years, our laboratory and others have
examined the potential role of endothelial-derived products, especially EDNO, in modulating
vascular tone in the late-gestation fetal and transitional pulmonary circulations. These findings
have in part contributed to recent clinical studies of inhaled NO in the treatment of newborns with
severe hypoxemic respiratory failure and pulmonary hypertension, who have failed to achieve the
normal decline in pulmonary vascular resistance (PVR) at birth (Persistent Pulmonary
Hypertension of the Newborn, PPHN). In the following chapter, we present a brief overview of
data from our laboratory regarding the potential role of EDNO activity in the normal fetal and
transitional pulmonary circulations, as well as data from our clinical experience in using inhaled
NO in treating newborns with severe PPHN.

PHYSIOLOGIC CHANGES IN THE NORMAL PERINATAL PULMONARY
CIRCULATION AT BIRTH

In the late-gestation fetus, pulmonary blood flow is low, pressure is high and the lung only
receives 8-10% of combined ventricular output.[35] High fetal PVR allows most of the right

ventricular output to cross the ductus arteriosus to the descending aorta, thereby enhancing umbilical-placental flow. As gas exchange occurs at the placenta, pulmonary blood flow remains sufficient for providing substrates and nutrients to promote lung growth and development *in utero*.[38] In addition to high basal PVR, the fetal pulmonary circulation is characterized by its capacity for autoregulation (Figure 1). Whereas many simuli, including increased PO_2, several pharmacologic agents, and shear stress, increase fetal pulmonary blood flow, vasodilation is often transient, as flow decreases toward baseline during prolonged exposure to many stimuli.[1-3,7,8] These findings led to the hypothesis that mechanisms exist in the fetus which counteract pulmonary vasodilation despite continuous exposure to many dilator stimuli.

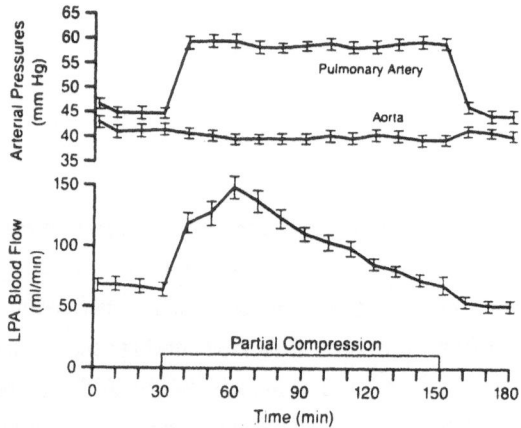

Figure 1: Myogenic responses in the fetal pulmonary circulation during partial compression of the ductus arteriosus *in utero*. Although initial ductus compression increases mean pulmonary artery pressure and flow, pulmonary blood flow decreases toward basal values while maintaining pulmonary artery pressure constant. Similar responses are found with several vasodilator stimuli but not with some endothelium-independent agonists (see text; from ref 1, with permission).

Mechanisms which oppose vasodilation and maintain high PVR *in utero* are unknown, but may include low oxygen tension (roughly 20 torr), mechanical factors (such as the absence of ventilation with gas in the fluid-filled fetal lung), increased myogenic tone, decreased ability to release or sustain production of vasodilator substances, or the presence of vasoconstrictors. Interestingly, some pharmacologic agents which act by directly stimulating smooth muscle cyclic

GMP (8-bromo-GMP, atrial natriuretic peptide, and inhaled nitric oxide) and the K+-ATP channel agonist, lemakalim, cause sustained pulmonary vasodilation during prolonged treatment.[2,13] In contrast, endothelium-dependent agonists (acetylcholine, bradykinin, histamine, tolazoline, oxygen and others) are unable to sustain vasodilation.[1-3,7,8] It appears that autoregulation may in part be due to the inability of some stimuli to sustain EDNO release. Fetal pulmonary arterial smooth muscle seems responsive to direct-acting vasodilator stimuli during late gestation,[2,13,20] and that the inability of the endothelial cell to release or sustain release of dilator substances or enhanced release of a vasoconstrictor (such as endothelin) may play important roles in autoregulation in the fetal pulmonary circulation.

At birth, the pulmonary circulation undergoes a dramatic fall in PVR, as blood flow increases 8 -10 fold and pulmonary artery pressure steadily declines.[9,10,15,35] There is an immediate fall in PVR within minutes after delivery, most likely due to the combined effects of vasodilation, recruitment and vascular distension, which are associated with rapid structural reorganization of small pulmonary arteries.[19] Right-to-left shunting of blood flow across the ductus arteriosus *in utero* switches to predominantly left-to-right flow with the fall in PVR, further enhancing pulmonary blood flow in the first hours of life. PaO2 increases from fetal levels of 20 torr to over 50 torr. Over time, pulmonary artery pressure slowly decreases during infancy until the adult level of PVR is attained. Mechanisms which contribute to postnatal adaptation in the normal lung include physical factors, such as drainage of fetal lung liquid, establishment of an air-liquid interface, rhythmic distension, increased PO2, and shear stress.[35] Physical birth-related stimuli act in part by altering pulmonary vascular production of vasoactive mediators, increasing release of vasodilators such as prostacyclin and endothelium-derived relaxing factor/nitric oxide (EDNO), or perhaps by decreasing production of vasoconstrictors, such as leukotrienes or endothelin-1.[11] Multiple homeostatic mechanisms appear to have overlapping contributions in the normal decline in PVR at birth, as no single simulus appears to be the sole "mediator" of this process. For example, physical stimuli such as removal of fetal lung liquid or rhythmic distension of the fetal lung can lower fetal PVR without changing arterial blood gas tensions.[10,15,35] Similarly, increased oxygenation to levels achieved at birth in the absence of ventilation or removal of fetal lung liquid can also increase pulmonary blood flow to levels in the newborn.[9,27] These physical stimuli appear to operate directly and indirectly, by stimulating the release of vasoactive products. For example, cyclooxygenase blockade attenuates the decrease in PVR during delivery of perinatal lambs and goats, which was later shown to due to ventilation-, but not oxygen-, induced release of prostacyclin.[36] Although changes in oxygen tension had no apparent affect on prostacyclin release, an inhibitor of EDNO activity blocked the decline in PVR by increased oxygen during the transition (see below).[4] Thus, physical stimuli can act on the endothelium to selectively release vasoactive substances in the perinatal lung. The roles of EDNO activity in modulating basal fetal pulmonary vascular tone and the normal transition of the fetal lung at birth are described below.

MATURATION-RELATED CHANGES IN ENDOTHELIUM-DERIVED NITRIC OXIDE ACTIVITY IN THE DEVELOPING LUNG

To examine mechanisms underlying fetal pulmonary vascular vasodilator responses and maturational changes in endothelial function, we studied the effects of endothelium -dependent and -independent vasodilators on tone of intralobar conduit pulmonary artery rings isolated from late-gestation fetal, newborn and adult sheep *in vitro* (Figure 2).[5] Rings from fetal pulmonary arteries had little relaxation to endothelium-dependent agonists, including acetylcholine, adenosine diphosphate, and A23187 (Figure 2). In contrast, the endothelium-independent agent, sodium nitroprusside (SNP), caused complete relaxation. In comparison with fetal pulmonary artery rings, rings from newborn and adult animals had significantly greater relaxation in response to the endothelium-dependent agonists. SNP-induced relaxations were not different between fetal, newborn and adult rings. These findings suggest that fetal conduit pulmonary arteries have diminished EDNO activity as assessed *in vitro*, and that maturational changes in endothelial cell function may contribute to ontogenetic differences in pulmonary vasoreactivity. Teleologically, diminished EDNO activity *in utero* would potentially favor maintenance of high PVR in the normal fetus.[1] Normal intrauterine conditions potentially attenuate or inhibit EDNO activity in the fetal lung. In comparison with pulmonary arteries from postnatal animals, fetal pulmonary arteries exist in an environment of low oxygen tension, high pressure, and low flow. Increasing oxygen tension enhances EDNO activity in fetal pulmonary artery rings *in vitro*,[33] and EDNO antagonism blocks O_2-induced fetal pulmonary vasodilation in chronically-prepared fetal lambs[24] and at birth (see below).[12] In addition, experimental studies suggest that high pressure impairs endothelium- dependent dilation in adult cerebral arteries,[37] and chronic elevations of blood flow may augment EDNO activity.[25] We speculate that EDNO activity is relatively suppressed in the normal fetal lung, and that adaptation to postnatal conditions (high O_2, low pressure, and high flow) may induce greater EDNO activity after birth. Whether these stimuli play important roles in the maturation of EDNO activity during postnatal life remains speculative.

Previous whole animal studies have demonstrated that several endothelium-dependent agonists, such as acetylcholine, bradykinin, and histamine, cause fetal pulmonary vasodilation during late-gestation,[23,27] suggesting the capacity for stimulated EDNO activity *in vivo*. Recent studies reported that intrapulmonary infusion of the EDNO antagonist, L-NA, increased fetal PVR, suggesting that basal EDNO modulates pulmonary vascular tone in the late-gestation fetus.[4] L-NA blocked acetylcholine-, but not atrial natriuretic peptide-, induced fetal pulmonary vasodilation, demonstrating its selective effects on inhibiting EDNO-mediated vasodilation in the fetal lung.[4] The timing of onset and changes in EDNO activity in the fetal pulmonary circulation during lung development and mechanisms regulating its activity are poorly understood. Past physiologic studies of pulmonary vascular responses in the ovine fetus to the endothelium-dependent agonists, acetylcholine and increased PO_2, demonstrate progressive vasodilator potency between 115 - 140 days (or, 0.77 to 0.93) gestation (full term = 150 days).[23,27] These

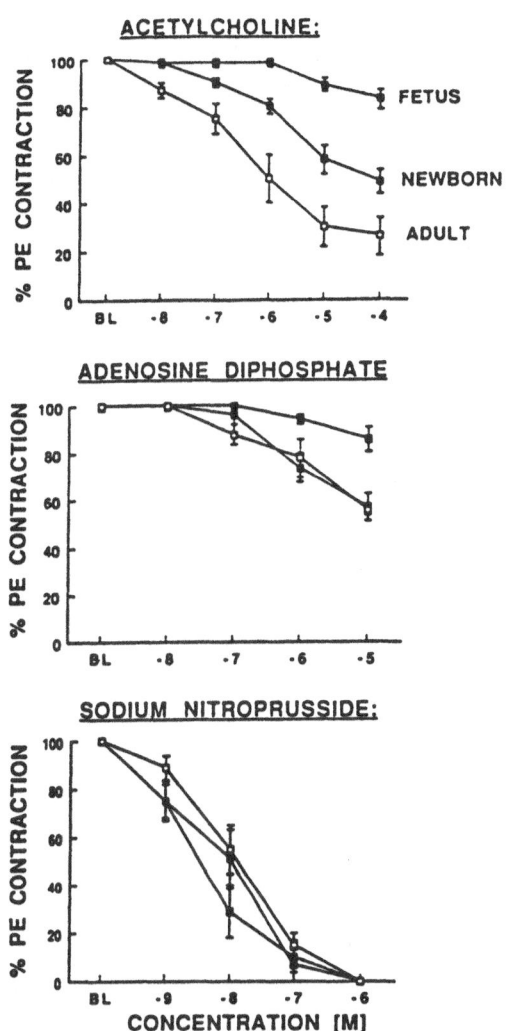

Figure 2. Maturation-related changes in endothelium-dependent and -independent activity in ovine conduit (third generation) pulmonary arteries *in vitro*. The effects of endothelium-dependent vasodilators, acetylcholine and adenosine diphosphate, and the endothelium- independent agonist, sodium nitroprusside, are shown in the upper, middle and lower panels, respectively. Relaxation is expressed as percent of phenylephrine (PE)-induced contraction. In comparison with postnatal arteries, relaxation to acetylcholine and adenosine diphosphate, but not sodium nitroprusside, are diminished in fetal pulmonary artery rings (from ref 5, with permission).

physiologic findings indirectly suggest that capacity for stimulated EDNO activity increases during this period of lung growth and development. We have recently observed that L-NA causes fetal hypertension as early as 0.77 term, and blocks mechanical ventilation -induced fetal pulmonary vasodilation at this age (Kinsella JP, Ivy DD, Abman SH; unpublished observations). These physiologic findings correlate with preliminary data from our laboratory demonstrating the presence of type III (endothelial) NO synthase protein in the fetal lung by western blot and immunostaining at 0.77 term but not at 0.5 term. Interestingly, fetal pulmonary vascular responsiveness to exogenous NO is intact relatively early in gestation. Unlike several vasodilator stimuli, such as acetylcholine and oxygen, inhaled NO (20 ppm) is a potent fetal pulmonary vasodilator as early as 0.75 gestation. Thus, as partly supported by *in vitro* and *in vivo* studies, we hypothesize that the ability of developing pulmonary vascular smooth muscle to respond to exogenous NO precedes the immature endothelium's ability to release or sustain release of EDNO.

Mechanisms regulating EDNO activity in the late gestation fetal lung are unclear. As stated, increased PO_2 stimulates EDNO activity at least transiently in the chronically-prepared fetus.[24] Similarly, intrapulmonary infusions of the precursor of EDNO, L-arginine, cause modest fetal pulmonary vasodilation at relatively high doses.[24] However, L-arginine infusions do not enhance or prolong vasodilator responses to acetylcholine or oxygen, suggesting that substrate limitation is unlikely to explain the inability to sustain pulmonary vasodilation during prolonged exposure to these endothelium-dependent stimuli in the ovine fetus.

In summary, in the developing pulmonary circulation, endogenous EDNO activity influences basal fetal pulmonary vascular tone, can be stimulated by physiologic stimuli (shear stress, ventilation, and increased oxygen tension) and pharmacologic agonists, contributes to the pulmonary vasoreactivity as early as 0.75 term, and may be enhanced during early postnatal life.[4,14,16]

ENDOTHELIUM-DERIVED NITRIC OXIDE ACTIVITY DURING THE TRANSITION OF THE PULMONARY CIRCULATION AT BIRTH

As described above, mechanisms contributing to the normal decline in PVR at birth include ventilation, increased oxygen, shear stress, and altered release of vasoactive products.[35] To determine whether EDNO activity contributes to the normal decline in PVR, we studied the effects of the EDNO antagonist, L-NA, on changes in PVR during cesarean-section delivery of near-term fetal lambs.[4] L-NA pretreatment attenuated the decline in PVR by nearly 50% despite ventilation with 100% O_2 (Figure 3). To more specifically determine distinct effects of specific birth-related stimuli on EDNO activity, we examined the effects of L-NA on sequential changes in ventilation (or, "rhythmic distension") without changing fetal arterial oxygen tension, followed by increased PO_2.[12] Vasodilation during ventilation with fetal PO_2 was attenuated by L-NA, as was the fall in pulmonary artery pressure during administration of 100% O_2 (Figure 4, A). To determine whether flow- or shear stress- induced vasodilation is EDNO-mediated, we studied the effects of L-NA treatment during acute compression of the ductus arteriosus.[12] As shown, the progressive rise in pulmonary blood flow is inhibited by L-NA (Figure 4, B). These findings suggest that ventilation, increased oxygen, and shear stress are independently capable of stimulating EDNO activity.

To further examine responsiveness of the fetal pulmonary circulation to exogenous NO, we studied the effects of inhaled NO in the near-term fetal pulmonary circulation.[20] Without

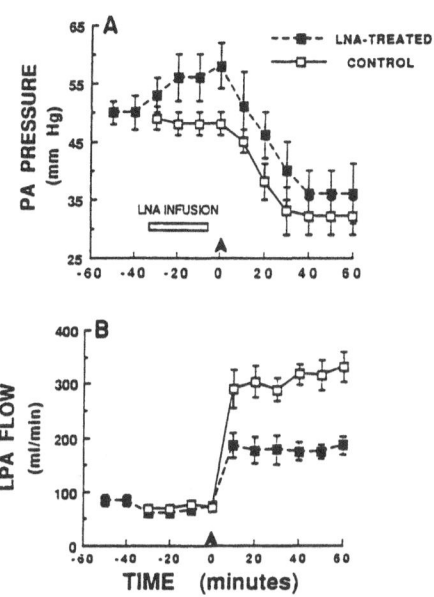

Figure 3. Role of EDNO activity during the transition of the pulmonary circulation in late- gestation fetal lambs. Pretreatment with an EDNO antagonist, nitro-L-arginine, attenuated the rise in left pulmonary artery (LPA) blood flow during ventilation with high FiO2 (from ref 4, with permission).

changing fetal PO2, inhaled NO caused potent, selective and sustained pulmonary vasodilation (Figure 5). Vasodilation achieved during exposure to inhaled NO without increasing fetal PO2 was similar to that achieved by the combined effects of treatment with ventilation and high FiO2 with or without NO.[20]

Although EDNO activity contributes to the normal decline in PVR at birth, it remains unknown whether diminished EDNO activity or the inability to release or sustain release of EDNO at birth contributes to the failure of postnatal adaptation.[34] We have previously demonstrated that partial compression of the ductus arteriosus for 8-12 days *in utero* alters perinatal pulmonary vascular

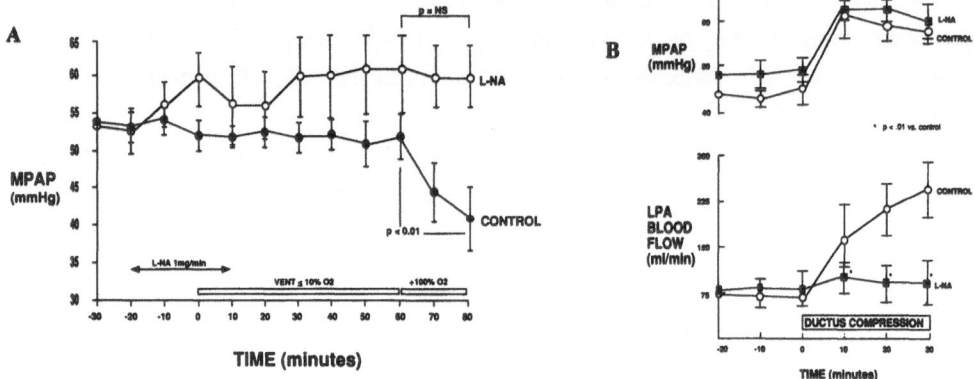

Figure 4: Effects of birth-related stimuli (oxygen and shear stress) on pulmonary vascular EDNO activity in the late- gestation fetal lamb. In the left panel (A), fetal lambs were ventilated initially with < 10% O2 to maintain PaO2 aat 20 torr. With the addition of 100% O2 (at 60 min), mean pulmonary artery pressure (MPAP) decreased in control but not in animals pre-treated with the EDNO antagonist, nitro-L-arginine (L-NA). In the right panel (B), L-NA blocked the progressive 3-fold rise in pulmonary blood flow during partial compression of the ductus arteriosus *in utero* (from ref 12, with permission).

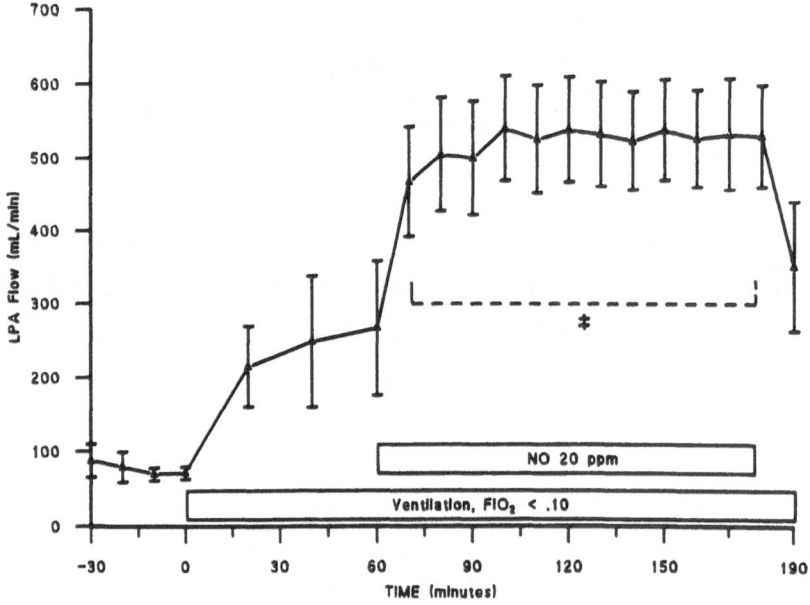

Figure 5: Pulmonary vascular effects of inhaled nitric oxide (NO) in the late-gestation ovine fetus. After ventilating the lung with low oxygen to maintain arerial blood gas tensions at fetal levels (20 torr), inhaled NO (20 ppm) increased pulmonary blood flow to similar levels as that achieved with 100% O2 with NO. As shown, NO-induced fetal pulmonary vasodilation was sustained for the entire 2 hour exposure period (from ref 20, with permission).

structure and function, providing an experimental model for studying the pathophysiology of PPHN.[6] In this model, chronic intrauterine pulmonary hypertension causes right ventricular hypertrophy, pulmonary vascular remodeling, altered vasoreactivity, and sustained elevation of PVR after cesarean-section delivery.[6] Whether EDNO activity is diminished in this animal model is currently under investigation; however, chronic intrauterine pulmonary hypertension blunts oxygen-induced fetal pulmonary vasodilation,[6] which is at least partly due to EDNO activity.[12,24] In addition, preliminary studies suggest that endothelium-dependent vasodilation with acetylcholine is preferentially impaired during chronic hypertension, but responsiveness to the endothelium-independent agonists, atrial natriuretic peptide and inhaled NO remain relatively intact. Whether decreased EDNO activity in this model is due to downregulation of type III NOS gene expression or protein synthesis, inactivation of EDNO, insufficient substrate, altered smooth muscle cell responsiveness, or other mechanisms is unknown.

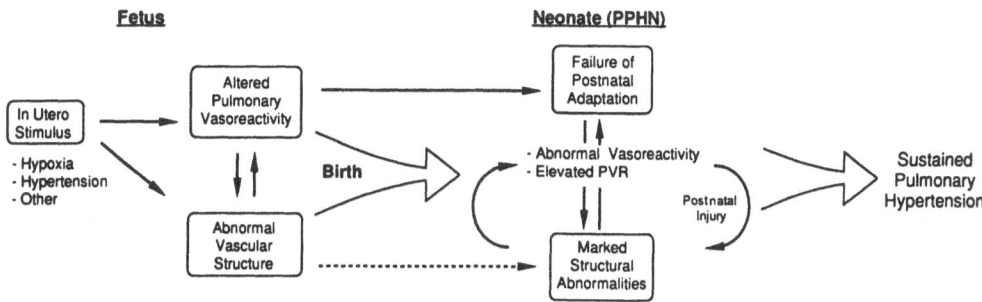

Figure 6. Schematic illustrating potential pathogenetic and pathophysiologic mechanisms underlying PPHN. Intrauterine stimuli, especially chronic hypertension, can alter fetal pulmonary vascular reactivity and structure, leading to the failure of the normal decline in PVR at birth. Whether chronic hypoxia alters pulmonary vascular reactivity or structure remains controversial. Postnatal lung injury may further exacerbate pulmonary vasoreactivity and structure (from ref. 34, with permission).

ROLE OF INHALED NITRIC OXIDE IN THE TREATMENT OF PERSISTENT PULMONARY HYPERTENSION OF THE NEWBORN (PPHN)

PPHN represents the failure of the pulmonary circulation to achieve or sustain the normal decline in PVR during the early perinatal period.[22] PPHN is a clinical *syndrome* associated with a wide variety of neonatal cardiopulmonary disorders, including asphyxia, sepsis, meconium aspiration syndrome, congenital diaphragmatic hernia, respiratory distress syndrome, lung hypoplasia, and others, or can be "idiopathic" (the so-called "persistent fetal circulation). Despite the diversity of clinical disorders, PPHN is characterized by common pathophysiologic features, including severe pulmonary hypertension and abnormal pulmonary vasoreactivity (Figure 6).[34]

Since PVR exceeds systemic vascular resistance, right-to-left shunting occurs across the ductus arteriosus and/or foramen ovale, causing critical hypoxemia. Current therapy has been limited in part by the lack of understanding of basic mechanisms which regulate high PVR in the normal fetus and contribute to the normal decline in PVR at birth. Whether PPHN reflects the failure of mechanisms which contribute to the normal decline in PVR (such as decreased EDNO activity[4]), excessive production of vasoconstrictors (such as endothelin-1[31]), or altered vascular responsiveness to dilator stimuli are unknown.

Despite aggressive therapy with hyperoxia, hyperventilation, alkalosis, and cardiotonic drugs, morbidity and mortality remain high. Therapeutic interventions often include attempts to lower PVR with vasodilator agonists, including tolazoline, sodium nitroprusside, prostaglandin E1, and others. Vasodilator drug therapy is often ineffective, however, due to concomitant systemic hypotension (due to the lack of selectivity for the pulmonary circulation), an inability to acutely lower PVR, the inability to sustain vasodilation (tachyphylaxis), or other side effects. Over time, the clinical course is often characterized by progressive hypoxemia which becomes refractory to treatment, and is associated with the development of superimposed lung injury due to barotrauma, hyperoxia, and lung inflammation. Patients who fail conventional therapy require treatment with extracorporeal membrane oxygenator (ECMO; or, cardiac bypass) therapy. Although ECMO has improved survival in refractory PPHN, it is labor intensive, costly, has multiple side effects, requires ligation of the right common carotid artery and internal jugular vein, and may be associated with long term neurologic sequelae.

Based on studies demonstrating potent and selective pulmonary vasodilation during treatment with inhaled NO in juvenile[17] and fetal[20] sheep, and a clinical report demonstrating the selective decrease of PVR in adults with primary pulmonary hypertension during brief inhalation of NO,[29] we studied the acute effects of inhaled nitric oxide therapy in newborns with severe PPHN. Patients were eligible for NO therapy if they had critical hypoxemia refractory to conventional therapy, echocardiographic evidence of pulmonary hypertension, and fulfilled criteria for ECMO therapy. NO was administered directly into the afferent limb of the ventilator circuit; NO and NO2 were measured directly on line by chemiluminescence. We found that low doses of inhaled NO (10-20 ppm) acutely improved oxygenation and pulmonary hypertension, as assessed by serial blood gas tensions and echocardiographic studies.[21] As reported in our first 6 patients who were treated for $\geq$ 24 hours, low dose NO (at 6 ppm) caused sustained improvement in oxygenation without causing systemic hypotension or increased methemoglobin levels (Figure 7). Since these initial studies, we have successfully treated 15/17 consecutive patients with severe PPHN without needing ECMO therapy. Two patients improved oxygenation during NO therapy, but subsequently required ECMO therapy because of concomitant cardiac insufficiency and multiorgan failure associated with "sepsis syndrome."

Thus, we found that inhaled NO effectively and selectively lowers PVR and improves oxygenation in newborns with severe PPHN, decreasing the need for ECMO therapy.[21] Clinical improvement during inhaled NO therapy occurs with many disease-specific entities associated with PPHN, including congenital diaphragmatic hernia, meconium aspiration syndrome, severe sepsis, idiopathic, and others.[21,30] In many cases, however, greater clinical improvement is achieved with inhaled NO *in combination with* high frequency oscillatory ventilation (HFOV) than with inhaled NO or HFOV alone.[21] Multicenter studies are currently underway to determine the overall efficacy, potential toxicity and relative role of inhalational NO therapy in the clinical managment of severe PPHN.

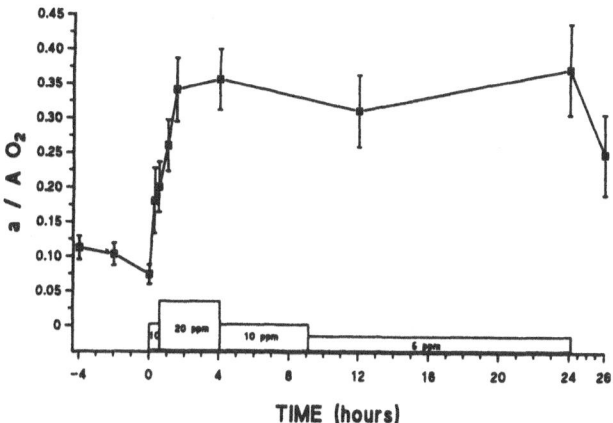

Figure 7: Improved oxygenation during inhaled NO therapy in severe PPHN. As shown, treatment with low concentrations of inhaled NO (6 ppm) caused sustained clinical improvement as reflected by serial changes in arterial/alveolar oxygen tension ratio (a/A O2) in 6 patients (from ref 21, with permission).

CONCLUSIONS

Although its exact roles in the perinatal pulmonary physiology and regulation of EDNO are incompletely understood, recent studies suggest that endogenous EDNO: 1) modulates basal pulmonary vascular tone in the normal late-gestation fetus; 2) is released by pharmacologic (acetylcholine) and physiologic (shear stress, increased PO2, ventilation) stimuli; 3) contributes to the normal decline in PVR at birth; 4) and maintains low resistance and modulates vasoreactivity in the postnatal lung. More recent studies further suggest that type III NO synthase protein content and activity increase during lung development and may increase with postnatal maturation. Mechanisms determining type III NO synthase gene expression or protein synthesis and regulating EDNO activity in the normal and hypertensive perinatal lung remain unclear. Despite the crucial physiologic role of EDNO in the developing pulmonary circulation, it remains unknown whether decreased EDNO activity contributes to the pathophysiology of clinical PPHN. Further studies are needed to determine whether responsiveness of newborns with severe hypoxemia and pulmonary hypertension to very low doses of exogenous (inhaled) NO suggests diminished endogenous EDNO production as a causal factor in this syndrome, or simply reflects the potent vasodilator effects of inhaled NO without a specific relationship to the primary pathogenesis of PPHN.

ACKNOWLEDGMENTS

The authors are grateful for the hard work, support and contributions of Stacia Koppenhafer, Marilee Horan and Drs. Barbara A. Chatfield, John A. McQueston, David M. Rodman, D. Dunbar Ivy, Ann Halbower, Frank J. Accurso and Ivan F. McMurtry.

REFERENCES

1. Abman SH and FJ Accurso. Acute effects of partial compression of the ductus arteriosus on the fetal pulmonary circulation. Am J Physiol 257: H626-634, 1989.

2. Abman SH and FJ Accurso. Sustained fetal pulmonary vasodilation during prolonged infusion of atrial natriuretic factor and 8-bromo-guanosine monophosphate. Am J Physiol. 260: H183-192, 1991.

3. Abman SH, FJ Accurso, RM Ward and RB Wilkening. Adaptation of fetal pulmonary blood flow to local infusion of tolazoline. Pediatr Res. 20:1131-1135, 1986.

4. Abman SH, BA Chatfield, SL Hall, and IF McMurtry. Role of endothelium-derived relaxing factor during transition of pulmonary circulation at birth. Am J Physiol. 259:H1921-1927, 1990.

5. Abman SH, BA Chatfield, DM Rodman, SL Hall, and IF McMurtry. Maturation-related changes in endothelium-dependent relaxation of ovine pulmonary arteries. Am J Physiol. 260:L280-285, 1991.

6. Abman SH, PF Shanley, and FJ Accurso. Failure of postnatal adaptation of the pulmonary circulation after chronic intrauterine pulmonary hypertension in fetal lambs. J Clin Invest. 83:1849-1858, 1989.

7. Accurso FJ, B Alpert, RB Wilkening, RG Petersen and G Meschia. Time-dependent response of fetal pulmonary blood flow to an increase in fetal oxygen tension. Respir Physiol. 63:43-52, 1986.

8. Accurso FJ and RB Wilkening. Temporal response of the fetal pulmonary circulation to pharmacologic vasodilators. Proc Soc Exp Biol Med 187:89-98, 1988.

9. Assali NS, TH Kirschbaum and PV Dilts. Effects of hyperbaric oxygen on uteroplacental and fetal circulation. Circ Res. 22:573-588, 1968.

10. Cassin S, GS Dawes, JC Mott, BB Ross and LB Strang. Vascular resistance of the foetal and newly ventilated lung of the lamb. J Physiol. 171:61-79, 1964.

11. Chatfield BA, IF McMurtry, SL Hall and SH Abman. Hemodynamic effects of endothelin-1 on the ovine fetal pulmonary circulation. Am J Physiol. 261:R182-187, 1991.

12. Cornfield DN, BA Chatfield, JA McQueston, IF McMurtry, and SH Abman. Effects of birth-related stimuli on L-arginine -dependent pulmonary vasodilation in the ovine fetus. Am J Physiol. 262:H1474-1481, 1992.

13. Cornfield DN, JA McQueston, IF McMurtry, DM Rodman and SH Abman. Role of ATP-sensitive K+ channels in ovine fetal pulmonary vascular tone. Am J Physiol. 263:H1363-1368, 1992.

14. Davidson D and A Eldemerdash. Endothelium-derived relaxing factor: evidence that it regulates pulmonary vascular resistance in the isolated newborn guinea pig lung. Pediatr Res. 29:538-542, 1991.

15. Dawes GS and JC Mott. The vascular tone of the foetal lung. J Physiol 164:465-477, 1962.

16. Fineman JR, MA Heymann, SJ Soifer. Nitro-L-arginine attenuates endothelium-dependent pulmonary vasodilation in lambs. Am J Physiol. 260:1299-1306, 1991.

17. Frostell C, M-D Fratacci, JC Wain, R Jones and WM Zapol. Inhaled nitric oxide: a selective pulmonary vasodilator reversing hypoxic pulmonary vasoconstriction. Circulation. 83:2038-2047, 1991.

18. Furchgott RF and JV Zawadski. The obligatory role of endothelial cells in the relaxation of arterial smooth muscle by acetylcholine. Nature. 288:373-376, 1980.

19. Haworth SG. Pulmonary vascular remodeling in neonatal pulmonary hypertension: state of the art. Chest. 93:133S-138S, 1988.

20. Kinsella JP, JA McQueston, AA Rosenberg, and SH Abman. Hemodynamic effects of exogenous nitric oxide in ovine transitional pulmonary circulation. Am J Physiol. 263:H875-880, 1992.

21. Kinsella JP, S Neish, E Shaffer, and SH Abman. Low dose inhalational nitric oxide in persistent pulmonary hypertension of the newborn. Lancet. 340:819-820, 1992.

22. Levin DL, MA Heymann, JA Kitterman, GA Gregory, RH Phibbs, and AM Rudolph. Persistent pulmonary hypertension of the newborn. J Pediatr 89:626-633, 1976.

23. Lewis AB, MA Heymann, and AM Rudolph. Gestational changes in pulmonary vascular responses in fetal lambs in utero. Circ Res. 39:536-541, 1976.

24. McQueston JA, DN Cornfield, IF McMurtry, and SH Abman. Effects of oxygen and exogenous L-arginine on endothelium-derived relaxing factor activity in the fetal pulmonary circulation. Am J Physiol. In Press, 1993.

25. Miller VM, LL Aarhus and PM Vanhoutte. Modulation of endothelium-dependent responses by chronic alterations of blood flow. Am J Physiol. 251:H520-527, 1986.

26. Moncada S, RMJ Palmer, and EA Higgs. Nitric Oxide: physiology, pathophysiology, and pharmacology. Pharmacol Reviews 43:109-142, 1991.

27. Morin FC, EA Egan, W Ferguson and CEG Lundgren. Development of pulmonary vascular response to oxygen. Am J Physiol. 254:H542-546, 1988.

28. Nathan C. Nitric oxide as a secretory product of mammalian cells. FASEB J. 6:3051-3064 1992.

29. Pepke-Zaba J, TW Higenbottam, AT Dinh Xuan, D Stone and J Wallwork. Inhaled nitric oxide as a cause of selective pulmonary vasodilation in pulmonary hypertension. Lancet. 338:1173-1174, 1991.

30. Roberts JD, DM Polaner, P Lang and WM Zapol. Inhaled nitric oxide in persistent pulmonary hypertension of the newborn. Lancet. 340:818-819, 1992.

31. Rosenberg AA, J Kennaugh, SL Koppenhafer, M Loomis, and SH Abman. Increased immunoreactive endothelin-1 levels in persistent pulmonary hypertension of the newborn. J Pediatr. In Press, 1993.

32. Ryan US and G Rubanyi. Endothelial regulation of vascular tone. NY: Marcel Dekker, 1992.

33. Shaul PW, MA Farrer and TM Zellers. Oxygen modulates endothelium-derived relaxing factor production in fetal pulmonary arteries. Am J Physiol. 262:H355-364, 1992.

34. Stenmark KR, SH Abman, and FJ Accurso. Etiologic mechanisms of persistent pulmonary hypertension of the newborn, in, Pulmonary vascular physiology and pathophysiology Weir EK and JT Reeves (eds). NY: Marcel-Dekker, 1989, p. 335.

35. Tod ML and S Cassin. Fetal and neonatal pulmonary circulation, in: The lung: scientific foundations. Crystal R and JB West (ed). NY: Raven Press, 1991. pp1687-1698.

36. Velvis H, P Moore, and MA Heymann. Prostaglandin inhibition prevents the fall in pulmonary vascular resistance as the result of rhythmic distension of the lungs in fetal lambs. Pediatr Res 30:62-67, 1991.

37. Yang S-T, WG Mayhan, FM Faraci and DD Heistad. Endothelium-dependent responses of cerebral blood vessels during chronic hypertension. Hypertension. 17:612-618, 1991.

38. Wallen LD, SF Perry, JT Alston, and JE Maloney. Morphometric study of the role of pulmonary arterial flow in fetal lung growth in sheep. Pediatr Res. 27:122-127, 1990.

INDEX

Histamine, 252, 261, 288, 290, 333, 334
Hydrogen peroxide, 196
Hydroxyeicosatetranoic acid (HETE), 106-109
Hypoxia, 5, 105, 138, 149, 150, 177-181, 189-201, 206-220, 240-242, 339
Hypoxic pulmonary vasoconstriction, 5, 189-201, 205-220, 240, 317

Iloprost, 160-174, 328
Indo-1, 50
Inositol-tetrakisphosphate (InsP4), 252, 255, 261
Inositol triphophate, 51, 106, 116, 120, 146, 224, 230, 231-233, 260, 266-269
Inside-out patch, 24
Ionomycin, 238
Isoproterenol, 83, 99-101

Leiotonin, 111
Levcromakalim (Lemakalim), 6, 129-131, 134-138, 333
Lipopolysaccharide, 231
Lung transplantation, 318-320, 322-329

Manganese, 238, 252-254, 262
Manoalide, 8
Markov process, 34
Mechanosensitive channels, 277-283
Membrane capacitance, 29
Mesenteric smooth muscle cells, 211
Methemoglobin, 340
Microfluorescence spectroscopy, 19
Minoxidil sulphate, 159
Mitochondria, 269
Myosin light chain, 111, 112
Myosin light chain kinase (MLCK), 111

NADPH oxidase, 4, 185
Neomycin, 121
Neuraminidase, 307
Nicorandil, 129, 159
Nifedipine, 83, 87, 88, 101, 144, 212, 322-328
Nisoldipine, 192
Nitric oxide, 3, 8, 146, 223-233, 239, 250, 287, 289, 331, 337-342
    measurement, 229, 230, 259, 340
Nitric oxide synthase, 223-233, 259, 336, 342
    inhibitors, 225, 240, 289, 293, 300, 336-338
Norepinephrine, 232, 293, 299, 303

Ouabain, 68, 302
Oxygen, 4, 231
Oxygen radicals, 190
Oxytocin, 83

Papaverine, 301
Patch clamp headstage, 20
Patch mode
    inside-out, 24
    outside-our, 27
Perforated patch, 26
Persistent fetal circulation, 339

Persistent pulmonary hypertension of the newborn (PPHN), 331-342
Pertussis toxin, 159, 168
Phencyclidine, 131
Phenylephrine, 227, 335
Phorbol ester, 84, 91, 122
Phospholamban, 50, 67
Phospholipase-C (PL-C), 91, 108, 224, 266
Pial arteries, 299
Pinacidil, 129, 159, 250
Platelet activating factor, 3
Potassium (K$^+$) channels
    ATP-sensitive, 5, 24, 113-115, 129-138, 218, 264, 289
    Ca^{2+}-activated, 24, 32, 49, 112-115, 23, 131, 149, 169, 191, 233, 263-265, 292
    delayed rectifier, 36, 133, 169
    inward-rectifying, 247-250, 264, 269
    outward rectifying, 26, 113
    oxygen-sensitive, 177
Prostacyclin, 3, 146, 159-175, 225, 259, 289, 317-329, 331
Prostaglandin E$_1$, 340
Protein kinase (PK)
    PK-A, 92
    PK-C, 84, 85, 91, 106
    PK-G, 92, 96
Pulmonary artery smooth muscle cells, 2-4, 18, 189, 206
Pulmonary hypertension, 1
    chronic hypoxic 3, 229
    primary 2, 317, 329

Quinidine, 9
Quinine, 9

Renal artery smooth muscle cells, 106
Renal circulation, 225-230
Resting membrane potential (RMP), 16, 25, 31, 33
Ryanodine, 51, 116, 120, 265, 268
    receptors (CICR), 51, 270

Saponin, 116
Sarcoplasmic reticulum, 50-53, 57-64, 98, 113, 116-123, 205, 216, 224
Saxitoxin, 81
Seal, 23
Shear stress, 4, 5, 252, 263, 283, 305, 307, 336-338, 341
Sodium (Na$^+$) channels, 34, 36, 77-89, 101
Sodium-calcium exchange, 57, 58, 61-65, 231, 263, 269, 301
Sodium-hydrogen exchange, 302
Sodium-potassium ATPase, 68, 290
Sodium dithionite, 199, 208
Sodium nitroprusside (SNP), 334, 340
Spontaneous transient outward currents (STOCs), 113, 265, 268, 269
Staurosporine, 84
Stretch-activated channels, 260, 262, 269, 277-283
Substance P, 114, 228, 264, 288